Springer Series in Statistics

John W. Pratt
Jean D. Gibbons

Concepts of
Nonparametic Theory

With 23 Figures

Springer-Verlag
New York   Heidelberg   Berlin

John W. Pratt
Graduate School of Business
Administration
Harvard University
Boston, Massachusetts 02163
USA

Jean D. Gibbons
Graduate School of Business
Administration
University of Alabama
University, Alabama 35486
USA

AMS Classification (1981): 62Gxx

Library of Congress Cataloging in Publication Data
Pratt, John W. (John Winsor), 1931–
    Concepts of nonparametric theory.
    (Springer series in statistics)
    Includes index.
    1. Nonparametric statistics. I. Gibbons, Jean
Dickinson, 1938–   . II. Title. III. Series.
QA278.8.P73   519.5   81-8933

Printed in the United States of America.

9 8 7 6 5 4 3 2 1

ISBN 0-387-90582-0 Springer-Verlag New York Heidelberg Berlin
ISBN 3-540-90582-0 Springer-Verlag Berlin Heidelberg New York

To Joy and Jean

To Jack and John

Preface

This book explores both nonparametric and general statistical ideas by developing nonparametric procedures in simple situations. The major goal is to give the reader a thorough intuitive understanding of the concepts underlying nonparametric procedures and a full appreciation of the properties and operating characteristics of those procedures covered. This book differs from most statistics books by including considerable philosophical and methodological discussion. Special attention is given to discussion of the strengths and weaknesses of various statistical methods and approaches. Difficulties that often arise in applying statistical theory to real data also receive substantial attention.

The approach throughout is more conceptual than mathematical. The "Theorem-Proof" format is avoided; generally, properties are "shown," rather than "proved." In most cases the ideas behind the proof of an important result are discussed intuitively in the text and formal details are left as an exercise for the reader. We feel that the reader will learn more from working such things out than from checking step-by-step a complete presentation of all details.

Those who are interested in applications of nonparametric procedures and not primarily in the mathematical side of things, but who would like to have a general understanding of the theoretical bases and properties of these techniques, will find this book useful as both a reference and a text. In order to follow most of the main ideas and concepts presented, the reader should have a good knowledge of the basic concepts of probability theory and statistical inference at the level of introductory books with a prerequisite of one or two years of calculus. More advanced topics require more mathematical and statistical sophistication. The particularly advanced sections are indicated by an asterisk and may be omitted. The many exercises at the end of each chapter also vary in level, from a straightforward data analysis to a complicated proof. They are designed to supplement, complement, and illustrate the materials covered in the text. The extensive references provide ample sources for further study. The nonparametric area is still a fertile field for research, and the interested reader will find no dearth of topics for further study; this book might provide an impetus for additional research in nonparametric inference.

The instructor who adopts this book for classroom use can proceed in various directions and at various levels, as appropriate to the level and interests of the students. If this course is the student's first exposure to nonparametric methods, we recommend coverage of selected (unstarred) portions of Chap. 1–7. If the student has already had an elementary survey course in nonparametric methods, this book can be used for a second course to provide more advanced material and deeper coverage of the properties of the procedures already known to the student. In assigning problems, the instructor should indicate how much rigor is expected in the solution. Appropriate references could be assigned for reports on selected topics. The book could be supplemented by outside readings from some of the references given.

The book does not attempt to provide a complete compendium of all the nonparametric methods presently available; only the most important procedures for testing and estimation that are applicable to the one-sample and two-sample situations are included. However, those procedures covered are treated in considerable detail.

This book originated from notes which provided the basis for a course in nonparametric statistics first given in 1959 at Harvard University. Over the years, many readers have made valuable comments and suggestions. As there are too many to name individually, we can only acknowledge a large collective debt to all readers in the past.

The authors are particularly grateful to the Office of Naval Research, the National Science Foundation, the Guggenheim Foundation, The Associates of the Harvard Business School, the Kyoto Institute of Economic Research, Kyoto University, and the Board of Visitors of the Graduate School of Business at The University of Alabama, for support; to Robert Schlaifer for computation of some entries in Tables 8.1 and 11.1 of Chapter 8; to Arthur Schleifer, Jr. for computation of the entries in Table C; and to the Literary Executor of the late Sir Ronald A. Fisher, F.R.S., to Dr. Frank Yates, F.R.S., and to Longman Group Ltd., London, for permission to reprint Table IIi from their book *Statistical Tables for Biological, Agricultural and Medical Research* (6th edition, 1974).

A Note to the Reader

A two digit reference system is used throughout this book (with the exception of problems). The first digit denotes a section number within a chapter. For subsections, equations, theorems, figures and tables within each section, a second digit is added. If a reference is made to a different chapter, the chapter number is included, but within the same chapter, it is omitted. Numerical references, except to equations, are preceded by an appropriate designation, like Section or Table. Equation numbers always appear in parentheses and are referred to in this way, e.g., (3.4) of Chap. 2 means Eq. (3.4), which is the fourth numbered equation in Sect. 3 of Chap. 2. Problems are given at the end of each chapter; they are numbered sequentially.

Justification of a result, even when entirely heuristic, is sometimes labeled proof and separated from the rest of the text so that the reader who is not interested can skip that portion. The end of a proof is indicated by a □ when it seems helpful. References in the text are given by surname of author and date. The full citations for these and other pertinent references are given in the Bibliography.

Throughout the book, more difficult material is indicated by an asterisk * at the beginning and end. These portions may be omitted without detriment to understanding other parts of the book.

Contents

CHAPTER 2
One-Sample and Paired-Sample Inferences Based on the Binomial Distribution   82

CHAPTER 5

Two-Sample Rank Procedures for Location 231

Tables   425

Bibliography   445

Index   455

CHAPTER 1
Concepts of Statistical Inference
and the Binomial Distribution

1 Introduction

Most readers of this book will already be well acquainted with the binomial probability distribution, since it arises in a wide variety of statistical problems, is simple to understand and use, and is extensively tabled. Our study of nonparametric statistics will begin with a rather thorough discussion of the basic concepts of statistical inference, developed and explained in the context of the binomial model. This approach has been chosen for two reasons. First, some important nonparametric procedures lead to the binomial model, and the properties of these nonparametric procedures therefore depend on properties of binomial procedures. Second, the binomial model provides a familiar and easy context for the illustration of many of the concepts, terms and notations which are necessary for an understanding of the nonparametric procedures developed later in this book. Some of these ideas will be familiar to the reader, but many belong especially to the area of nonparametric statistics and will require more careful study. The reader may also find that even the "simple" binomial situation is less simple than it may have seemed on previous acquaintance.

In this first chapter, after a brief introduction to probability distributions, we will discuss the basic concepts and principles of point estimation, hypothesis testing and interval estimation. The various inference techniques will be described, with an emphasis on problems arising in their interpretation. In the process of illustrating the procedures, we will study many properties of the binomial probability distribution, including approximations using other distributions.

2 Probability Distributions

Suppose that the possible outcomes of an experiment are distinguished only as belonging to one of two possible categories which we call success and failure. The two categories must be mutually exclusive, but the terms success and failure are completely arbitrary and are used solely for convenience. (For example, if the experiment involves administering a drug to a patient, we might assign the label "success" to the event that the patient dies. This choice might be convenient, not merely macabre, because tables are sometimes limited to the situation where the probability of success does not exceed 0.50.) We denote the probability of success by p and the probability of failure by q, where $q = 1 - p$ for any $p, 0 \leq p \leq 1$. The set of all possible outcomes of this simple experiment could be written as {Success, Failure}, but it will be more convenient to write $\{1, 0\}$ where 1 denotes a success and 0 denotes a failure.

When an experiment of this type is repeated, the trials are called *Bernoulli trials* if they are independent and the probability p of success is identical on every trial. Consider a sequence of n Bernoulli trials where n is fixed. Then the possible outcomes of this compound experiment can be written as n-tuples

$$(X_1, \ldots, X_n) \quad \text{where } X_i = 0, 1 \text{ for } i = 1, \ldots, n.$$

The ith component of the n-tuple denotes the outcome of the ith trial. The set of all possible outcomes contains 2^n points. The probability of any particular point (outcome) whose n-tuple has exactly r 1's and $n - r$ 0's is

$$p^r(1 - p)^{n-r}$$

for any $r, r = 0, 1, \ldots, n$. This probability is the same for every arrangement of exactly r 1's and $n - r$ 0's. There are

$$\binom{n}{r} = \frac{n!}{r!(n - r)!}$$

different such arrangements, all mutually exclusive.

Let S denote the number of 1's which occur in the experiment. Then the probability that S is equal to r, written as $P(S = r)$, is the sum of the probabilities for all those points with exactly r 1's. Since each of these points has the same probability, the sum is the number of points multiplied by that probability, or

$$P(S = r) = \binom{n}{r} p^r(1 - p)^{n-r}.$$

This holds for $r = 0, 1, \ldots, n$. The probability is zero for all other values of r. This result is useful whenever we want to distinguish the possible outcomes of the compound experiment only according to the value of S, i.e., the number of 1's irrespective of the order in which they appear.

More formally, what we have done is to map the set of n-tuples into a set of nonnegative integers which represent the number of 1's. The function can be denoted by $S(x_1, \ldots, x_n)$, with a range of $\{0, 1, \ldots, n\}$. The function S is then called a *random variable*. This means that it is a function whose domain is the set of all possible outcomes of an experiment, each outcome of which has a probability, known or unknown.

This illustrates the usual mathematical definition of a random variable as a "function on a sample space." Intuitively, a random variable is any uncertain quantity to which one is willing to attach probability statements. The "sample space" can be refined if necessary so that all such quantities are functions thereon. A measurement, or the outcome of an experiment, is a random variable, provided the probabilities of its possible values are subject to discussion. A random variable may be multidimensional—when several one-dimensional uncertain quantities are considered simultaneously. Thus a large set of measurements may be considered *a* random variable, but as a vector or an n-tuple.

Any function of a random variable—for instance the sum of a set of measurements—is also ipso facto a random variable. Any function of observable random variables is called a *statistic*. For instance, in Bernoulli trials, a random variable describing the outcome of the ith trial could be defined as

$$X_i = \begin{cases} 0 & \text{if the } i\text{th outcome is failure,} \\ 1 & \text{if the } i\text{th outcome is success.} \end{cases}$$

A statistic of interest is the number of successes in n Bernoulli trials, which is the sum of these random variables, $\sum_1^n X_i = S$, since this is the number of successes, discussed above.

It is often convenient to speak of "observations" rather than "random variables." The term "observation" is more suggestive of the real world, and "random variable" of the mathematical definition. In common usage, *observation* may be either the observed value of a random variable, or a random variable which is going to be observed.

Confusion sometimes arises if a random variable is not distinguished from the actual value observed. This is the distinction between a function and its value. For example, if the observation S is the number of successes in 10 independent tosses of a fair coin, we may say that

$$P(S = 0) = \frac{1}{1024}. \tag{2.1}$$

If we observe $S = 0$, however, we cannot substitute 0 for S in (2.1), since in common sense $P(0 = 0) \neq 1/1024$. Similarly, if we observe $S = 5$, we cannot substitute 5 for S in (2.1), since $P(5 = 0) \neq 1/1024$. This distinction is not always as trivial as it seems here, as we shall see later. Nevertheless, we will sometimes use terminology, such as "observation," which does not make the distinction, if it does not lead to ambiguity.

It is conventional to denote a random variable by a capital letter, such as S above, and an arbitrary value assumed by that random variable as the corresponding letter or some other letter in lower case. Thus if the random variable S denotes the number of successes in n Bernoulli trials, we may write

$$P(S = s) = \binom{n}{s}p^s(1 - p)^{n-s} \quad \text{for } s = 0, 1, \ldots, n. \qquad (2.2)$$

If S satisfies (2.2) it is said to *follow the binomial distribution*, or to be *binomially distributed*, with parameters n and p. The term *parameter* here means any characteristic of the population or theoretical distribution. The binomial is then a family of distributions, with a particular member of the family specified according to the values assigned to the parameters n and p. The letter q is commonly used to denote $1 - p$, in (2.2) for instance. While q is also a parameter, it is known whenever p is known.

The term *distribution*, or more specifically *probability distribution*, will be used in this book for any function which determines all probabilities relating to a random variable. It could refer to a frequency function (discrete case), a density function (continuous case), a cumulative distribution function, a graph of any of these, or something else. The term will be used by itself only when it does not matter which interpretation is given.

A random variable or its probability distribution is called *discrete* if it is confined to a finite or countably infinite number of values whose probabilities sum to one. Equivalently, the values with positive probability account for all of the probability (Problem 7). The probability distribution can then be given by its *frequency function* (sometimes called a mass function) defined by $f(x) = P(X = x)$. Thus the binomial distribution is discrete, with frequency function given by $f(x) = \binom{n}{x}p^x(1 - p)^{n-x}$ for $x = 0, 1, \ldots, n$ and $f(x) = 0$ otherwise, as developed in (2.2).

In practice, all observable random variables are discrete, because of limitations on precision in measurement. However, it is frequently convenient to use distributions which are not discrete as approximations to the distribution of, for instance, a very fine measurement, or the sum or average of several measurements, or an "ideal" measurement.

The simplest kind of nondiscrete random variable is one whose probability distribution is completely specified by a *probability density function*. The values of this density function are not probabilities because the probability assigned to any particular single value of such a random variable is zero. Rather, the density function is integrated to find probabilities for sets of values of the random variable. Hence a real random variable X is said to have a density f if

$$P(a \leq X \leq b) = \int_a^b f(x)dx \quad \text{for all } a, b, a \leq b.$$

In particular we must have both[1]

$$f(x) \geq 0 \quad \text{and} \quad \int_{-\infty}^{\infty} f(x)dx = 1.$$

The value of $P(a \leq X \leq b)$ is then the area under the density function $f(x)$ and above the x axis, between a and b. The area $P(X \leq z)$ for arbitrary z is shown in Fig. 2.1 as the hatched region. Generalization to vector random variables is straightforward.

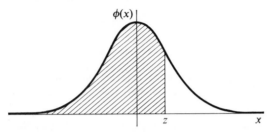

$\phi(x)$

z

x

Figure 2.1

The particular density function graphed in Figure 2.1 is called the *standard*, or *unit*, *normal* density and is given by the formula

$$\phi(x) = \frac{1}{\sqrt{2\pi}} e^{-x^2/2}. \tag{2.3}$$

The area under this curve from z to infinity, or $P(X \geq z)$, is given by Table A for $z \geq 0$. Because the density is symmetric about 0, we have $P(X \leq -z) = P(X \geq z)$. Thus the probability to the left of a negative number, that is, the area from minus infinity to $-z$ for $z \geq 0$, can also be read directly from this table. If X has a normal distribution with mean μ and standard deviation σ, then $Z = (X - \mu)/\sigma$ has the standard normal density ϕ above.

The *cumulative distribution function*, or *c.d.f.*, of any random variable X, is defined as $F(x) = P(X \leq x)$, so that

$$F(x) = P(X \leq x) = \begin{cases} \sum_{t \leq x} f(t) & \text{if } X \text{ has frequency function } f, \\ \int_{-\infty}^{x} f(t)dt & \text{if } X \text{ has density function } f. \end{cases}$$

Note that $F(-\infty) = 0$ and $F(\infty) = 1$, while $F(a) \leq F(b)$ for all $a \leq b$. It is customary to denote the c.d.f. by the capital of that letter which in lower case denotes the frequency or density function. The c.d.f. of a discrete random variable jumps upward by an amount equal to the value of the frequency

---

[1] This book omits measurability and "almost everywhere" qualifications. Anyone who ought to care about them should have no difficulty deciding where they are appropriate.

function at each point where the latter is positive; elsewhere it is horizontal. A random variable is called *continuous* if its c.d.f. is continuous. This holds if it has a density. Although there are also continuous random variables without densities, no explicit examples will arise in this book.

The c.d.f. F of a binomial random variable S satisfies

$$F(s) = P(S \leq s) = \sum_{i=0}^{s} \binom{n}{i} p^i (1 - p)^{n-i} \qquad (2.4)$$

for $s = 0, 1, \ldots, n$, and in general for any real number x, we have

$$F(x) = \begin{cases} 0 & \text{if } x < 0 \\ F([x]) & \text{if } 0 \leq x \leq n \\ 1 & \text{if } x > n \end{cases}$$

where $[x]$ is the largest integer not exceeding x, or the integral part of x. Numerical values of the cumulative distribution in (2.4) are given in Table B, for $2 \leq n \leq 20$ and selected values of p, and for $21 \leq n \leq 30$ when $p = 0.50$. Several more extensive tables are available, for instance, National Bureau of Standards [1949] or Harvard University [1955]. For $n > 20$ the normal probabilities in Table A may be used to approximate the binomial distribution in various ways. One is explained at the end of Table B; the theoretical relevance of the normal distribution will be given in Sect. 9.

3 Estimators and Their Properties

Suppose we have a sequence of n Bernoulli trials in which the probability p of success is unknown. Then the distribution of the number of successes belongs to the binomial family, but it is not completely specified even though we assume n is known. What can we say about p on the basis of the number of successes observed? To start with, we could estimate the unknown probability of success by the proportion of successes observed in the n trials, S/n. This is a natural estimate, since one intuitive meaning of "probability" is "long-run value of the observed proportion." (Some people think no other meaning is appropriate.)

Sometimes it is useful to distinguish between the function determining an estimate and the actual value of the estimate in a particular situation. Then the function is called an *estimator*, and an *estimate* is an observed value of the function. Strictly speaking then, S/n should be called an estimator of p; if $S = 3$ is observed, then the estimate is $3/n$, and in general if $S = s$ is observed, the estimate is s/n. Thus an estimator is a random variable, or statistic, used to estimate a parameter, while an estimate is an observed value of an estimator.

What are the properties of S/n as an estimator of p? (We did not say "estimate," since we cannot say how good an actual estimate is, but we can say something about an estimator or a method of estimation.) Several properties will now be discussed.

3.1 Unbiasedness and Variance

One property of the estimator S/n of the parameter p in the binomial distribution is that its expected value, or mean, is exactly p. Denoting this expectation by E, we write this statement as

$$E\left(\frac{S}{n}\right) = p. \tag{3.1}$$

If we want to emphasize the fact that the distribution of S depends on the parameter p, and hence that the expected value depends on p, we may add a subscript to E, and write

$$E_p\left(\frac{S}{n}\right) = p. \tag{3.2}$$

In general, an estimator T of a parameter θ is called unbiased for θ if its expectation is exactly the parameter being estimated, that is, $E(T) = \theta$, for every possible distribution under consideration. Thus, for all binomial distributions, S/n is an unbiased estimator of p. The unbiasedness of S/n for p is of little significance by itself. For example, the observation X_i on any individual trial is also unbiased for p since $E(X_i) = p$. A property which differentiates these two unbiased estimators is the spread of their values about p. One measure of spread is the *variance*, defined in general as

$$\text{var}(T) = E\{[T - E(T)]^2\} = E(T^2) - [E(T)]^2. \tag{3.3}$$

The variance of X_i is $p(1 - p)$, and that of S/n is

$$\text{var}_p\left(\frac{S}{n}\right) = \frac{p(1 - p)}{n}. \tag{3.4}$$

Hence, for any $n \geq 2$, the variance of S/n is smaller than the variance of the single observation X_i for all values of p except 0 and 1 (where both variances are 0).

In fact, the same comparison holds between S/n and any other unbiased estimator, so that S/n is the unique *minimum variance unbiased estimator* of p. This property will be further discussed and proved in Sect. 3.4.

When reporting the value of an estimator T of a parameter θ, one should report also some measure of its spread or an estimate thereof. For this purpose, it is better to use the square root of the variance, the *standard deviation*, because it has the same units as θ and T. For many theoretical purposes, however, like that just mentioned, the variance is slightly more convenient.

Of course, the variance or standard deviation is a measure of the spread of the estimator around the parameter only if the estimator is unbiased. For an arbitrary estimator T it is also useful to define the *bias* as the difference between the expected value of the estimator and the parameter θ being estimated, or $E(T) - \theta$, and the *mean squared error* as $E[(T - \theta)^2]$. The latter can be written as

$$E[(T - \theta)^2] = E\{[T - E(T)]^2\} + [E(T) - \theta]^2 = \text{var}(T) + [\text{bias}(T)]^2.$$

If the bias contributes a negligible proportion of the mean squared error, then the lack of unbiasedness is of little consequence. Of course the bias, variance, and expectations above depend on the distribution of T, which may or may not be completely determined by θ under any given assumptions.

3.2 Consistency

In the binomial estimation problem the variance of S/n approaches zero as n approaches infinity, as is obvious from (3.4) since p is a constant. The combination of this property with the fact that S/n is unbiased implies that S/n is a *consistent* estimator of p. Intuitively, an estimator is consistent if the error in the estimate probably becomes small as the sample size increases. A more formal general definition of consistency is that for a sequence of estimators T_n depending on n, the probability that the estimator differs from θ, the parameter estimated, by more than any arbitrary number approaches zero as n approaches infinity. In symbols, the definition is that for any $\varepsilon > 0$, we have

$$\lim_{n \to \infty} P(|T_n - \theta| > \varepsilon) = 0. \tag{3.5}$$

It is easily shown by application of Chebyshev's inequality (Problem 1) that if the bias and variance of a sequence of estimators both approach zero as n approaches infinity, then the sequence is consistent. Unless otherwise stated, these properties are to hold for each member of whatever family of distributions is under discussion.

3.3 Sufficiency

It seems natural to assume that the estimation procedure in the binomial situation should depend only on the number S (or proportion S/n) of successes. But why not also take into account which trials are successes? Specifically, let $X_j = 1$ or 0 according as the jth outcome is a success or failure. The number of successes can then be written as $S = \sum_1^n X_j$. Should we consider as estimates only functions of S alone, such as S/n? What happens if we consider procedures which are functions of $X_1, \ldots, X_n$, but not necessarily functions of S? For example, what about the unbiased estimator $(X_1 + X_n)/2$? The following discussion will answer this and related questions.

If the value of S is given, say $S = 3$, then it is intuitively obvious and easily proved (Problem 2) that the three successes are equally likely to have occurred on any particular three of the n trials. This is true regardless of the value of p. More generally, given $S = s$, for any p, the s success are equally likely to have occurred on each set of s of the n trials. Consequently, once the number of successes is known, it appears intuitively that no further information is gained about p by knowing which trials produced these successes. The meaning and implications of this intuitive idea can be made more explicit as follows.

Suppose that we (the authors) know the outcomes of the individual trials $X_1, \ldots, X_n$ while you (the reader) know only S. It might seem that we have an advantage by access to more complete information about the experiment. Suppose, upon observing $S = s$, however, that you choose s out of the n trials at random and arbitrarily call those trials successes and the rest failures, getting what might be called simulated trials $Y_1, \ldots, Y_n$. Then, whatever the value of p, your simulated trials $Y_1, \ldots, Y_n$ have the same distribution as the trials $X_1, \ldots, X_n$ which actually took place and whose outcomes we know. (Proof: Whatever the value of p, the X's and the Y's have the same conditional joint distribution given S; and of course S is common to both sets of trials. Consequently $Y_1, \ldots, Y_n$ have the same unconditional joint distribution as $X_1, \ldots, X_n$ for every p.)

It is now evident that any inference about p which we can make knowing $X_1, \ldots, X_n$, you can mimic knowing only S. More explicitly, suppose we use a certain procedure depending on $X_1, \ldots, X_n$, such as an estimator of p, or an inference statement about p, or a forecasting statement about future observations $X_{n+1}, \ldots$, or a decision rule whose payoff depends on p and/or $X_{n+1}, \ldots$. Suppose you use the same procedure, but applied to the simulated trials $Y_1, \ldots, Y_n$ in place of $X_1, \ldots, X_n$. Then, although we may not get the same result in a particular instance, your procedure will have exactly the same probabilistic behavior as ours regardless of the value of p. The probability of any event defined in terms of $X_1, \ldots, X_n$ depends only on the value of p. The same event defined in terms of the simulated trials $Y_1, \ldots, Y_n$ will have the same probability, whatever the value of p, because the Y's have the same distribution as the X's for all p. If, for instance, we estimate p by a function of $X_1, \ldots, X_n$ and you estimate p by the same function of $Y_1, \ldots, Y_n$, then your estimator will have the same bias, variance, and distribution as ours for all p. In short, our procedure and yours have the same operating characteristics, where the term *operating characteristic* means any specific aspect of the probabilistic behavior of a procedure. Introducing the vector notations $X = (X_1, \ldots, X_n)$ and $Y = (Y_1, \ldots, Y_n)$ for convenience, we state the following properties.

(1) X has various possible distributions, one for each value of p.
(2) S is a function of X, namely $S = \sum_1^n X_j$.
(3) The conditional distribution of X for a given value of S does not depend on p.
(4) The conditional distribution of Y for a given value of S is exactly the same as that of X.

The property of S that made it possible to define Y (Property 3) is called "sufficiency." In general, a function S of a random variable X is called *sufficient for a family of possible distributions of X*, or *sufficient for X*, if the conditional distribution of X given S is the same regardless of which member of the family is actually the distribution of X.

It is possible to define Y so as to duplicate the distribution of X without knowing which of its possible distributions X has if and only if S is sufficient. This can be seen by examining the definition of Y. (See the proofs below.) More explicitly, a function S of a random variable X is sufficient for X if and only if there exist random variables $Y(s)$ with completely specified distributions (which therefore do not depend on the distribution of X), one such random variable for each value s of S, such that $Y(S)$ has the same distribution as X whichever of its possible distributions X has.

We see in general, therefore, that if S is sufficient for X, then given any procedure based on X there exists one based on S having the same probabilistic behavior. We simply define Y so as to duplicate the distribution of X and then apply the procedure to Y instead of X. There is one catch, however — the definition of Y involves additional randomness, because for a given value s of S, $Y(s)$ is a random variable. Thus we must allow "randomized procedures," which are considered in Sect. 5.

Is there an easy method by which a statistic can be checked for sufficiency? The following *factorization criterion* is an almost immediate consequence of the definition of sufficiency. Suppose that X has the family of density or frequency functions $f(x; \theta)$, where, for convenience, θ indexes the possible distributions of X. A function S of X is sufficient for X if and only if $f(x; \theta)$ can be written as the product of a function of x alone and a function of S and θ, that is, if and only if $f(x; \theta)$ can be factored in the form

$$f(x; \theta) = g[S(x); \theta]h(x) \quad \text{for all } x. \tag{3.6}$$

Here $S(x)$ is the value of S when X has value x. It is important to note that this form of the product must hold for all real vectors x. This means in particular that the domains of the functions may not depend on θ, nor may the region where $h(x) \neq 0$.

The factors may be, but need not be, the density or frequency function of S, say g_0, and the density or frequency function of the conditional distribution of X given S, say h_0. To see the idea, note that always $f = g_0 h_0$, and g_0 is a function of s and θ. Thus $f(x; \theta) = g_0(s; \theta)h_0(x|s, \theta)$ where $s = S(x)$. If S is sufficient for X, then by definition h_0 does not depend on θ in any way, so it is a function of x and s and hence a function of x alone, and $f = g_0 h_0$ is a factorization of the form (3.6). Conversely, if a factorization of the form (3.6) exists, then h_0 does not depend on θ, as can be shown by computing it as in (3.7)–(3.10) below, so the conditional distribution of X given S does not depend on θ.

Thus we have developed the following three equivalent conditions for sufficiency of S.

(1) $f(x; \theta)$ factors into a function of S and θ times a function of x for all real vectors x.
(2) The conditional distribution of X given S does not depend on θ.
(3) There is a (random) function Y of S such that, for all θ, $Y(S)$ has the same distribution as X.

Any of these might be considered a justification for the intuitive explanation that S is sufficient for X if the distribution of X depends on θ only through S, or S contains all the information about θ, so it is fortunate that they agree. We will now prove that (1) implies (2), and (2) implies (3). The converse proofs, and the special case of the binomial, are left for the reader in Problem 3. Some of these proofs have been given in part already.

PROOFS. To avoid technicalities in the general definition of conditional distributions, we will assume that X has a discrete distribution with frequency function $f(x; \theta)$.

Suppose that f factors as in (3.6). To show that S is sufficient, we compute the conditional distribution of X given $S = s$ as follows.

$$P_\theta(X = x \mid S = s) = \frac{P_\theta(X = x, S = s)}{P_\theta(S = s)}$$

$$= \begin{cases} \dfrac{P_\theta(X = x)}{P_\theta(S = s)} & \text{if } S(x) = s, \\ 0 & \text{otherwise.} \end{cases} \tag{3.7}$$

Using (3.6), we have

$$P_\theta(X = x) = f(x; \theta) = g[S(x); \theta]h(x) \tag{3.8}$$

$$P_\theta(S = s) = \sum_{S(x')=s} P_\theta(X = x') = g(s; \theta) \sum_{S(x')=s} h(x'), \tag{3.9}$$

where the sums are over those x' for which $S(x') = s$. Substituting (3.8) and (3.9) into (3.7), we find

$$P_\theta(X = x \mid S = s) = \begin{cases} \dfrac{h(x)}{\sum_{S(x')=s} h(x')} & \text{if } S(x) = s \\ 0 & \text{otherwise.} \end{cases} \tag{3.10}$$

Since the right-hand side does not depend on θ, the conditional distribution of X given S does not depend on θ, and S is sufficient.

Now suppose that the conditional distribution of X given S does not depend on θ. We will define $Y(s)$ so that it has the same distribution as X for every θ. Let

$$f_s(x) = P_\theta(X = x \mid S = s). \tag{3.11}$$

This is the conditional frequency function of X given $S = s$, and does not depend on θ, by assumption.

For each s let $Y(s)$ be a random variable with frequency function $f_s(x)$. We must show that $Y(S)$ has the same distribution as X for all θ. This follows from the fact that it has the same conditional distribution given S as X has, for all θ. Explicitly,

$$
\begin{aligned}
P_\theta[Y(S) = x] &= \sum_s P_\theta(S = s)P_\theta[Y(S) = x|S = s] \\
&= \sum_s P_\theta(S = s)f_s(x) \\
&= \sum_s P_\theta(S = s)P_\theta(X = x|S = s) \\
&= P_\theta(X = x),
\end{aligned}
\tag{3.12}
$$

so $Y(S)$ has indeed the same distribution as X for all θ. $\square$

3.4 Minimum Variance

It was stated in Sect. 3.1 that S/n is the unique, minimum variance unbiased estimator of the parameter p of the binomial distribution. In this section we will first discuss this important property briefly and then prove the statement.

In general, an unbiased estimator T of a parameter θ is called a *minimum variance unbiased estimator* if no other unbiased estimator has smaller variance for any distribution under discussion, so that T minimizes the variance, among unbiased estimators, simultaneously for every distribution of whatever family has been assumed. This sounds like a splendid property for an estimator to have, and a minimum variance unbiased estimator is ordinarily a good one to choose. Note, however, that no such estimator need exist. Furthermore, even if one does, nothing in the definition precludes the possibility that some other estimator, though biased, has much smaller mean squared error. Also, a minimum variance unbiased estimate is sometimes smaller than the smallest possible value of the parameter, or larger than the largest (Problem 4). When this happens, the estimate seems clearly unreasonable, and replacing it by the smallest or largest possible value of the parameter as appropriate obviously reduces estimation error, though it makes the estimator biased.

Thus as a concept of optimality in estimation, minimum variance unbiasedness is not completely satisfactory. But neither is any other concept. Mean squared error cannot ordinarily be minimized for more than one distribution at a time (Problem 5). However, seeking a truly satisfactory concept is taking the point estimation problem too seriously. Formal versions of it do not correspond at all closely to any real problem of inference. There is, after all, no need to give just a single estimate, "optimal" or not. (In making actual decisions, treating a decision as an estimate or vice versa is more confusing than clarifying.) A full-fledged inference must somehow reflect the uncertainty in the situation. An estimate is just an intuitive first step.

When formalization leads to difficulty, it is best to give it up and and turn to other methods permitting richer sorts of inference statements (as we do in Sects. 4 and 6 below).

Returning to the choice of an estimator in the binomial case, observe that, since S is a sufficient statistic, we can restrict consideration to estimators which are functions of S. This means that, to mimic a given estimator, when we observe $S = s$, we may need to choose a distribution depending on s and draw a random variable U_s from this distribution to use as our estimate. To show we can do better without randomization, let $T(s) = E(U_s)$ for each s. Then no matter what p is, the nonrandomized estimator $T(S)$ has the same mean as the randomized estimator Y_S, and smaller variance (equal variance if $U_s = T(S)$ with probability one) (Problem 6). Therefore we cannot reduce the variance by allowing randomized estimators, as one would expect intuitively.

The following proof shows that S/n is the only function of S which is an unbiased estimator of p. From this and the previous paragraph, it follows that S/n is the unique, minimum variance unbiased estimator of p.

PROOF. Suppose that two functions of S are both unbiased estimators of p. Then their difference $\Delta(S)$ has expected value $p - p = 0$ no matter what the true value of p. Therefore

$$E_p[\Delta(S)] = \sum_{s=0}^{n} \Delta(s)P_p(S = s)$$

$$= \sum_{s=0}^{n} \Delta(s)\binom{n}{s}p^s(1 - p)^{n-s} = 0 \qquad (3.13)$$

for all p. Dividing by $(1 - p)^n$ for $p \neq 1$ and replacing $p(1 - p)^{-1}$ by y gives the polynomial equation in y

$$\sum_{s=0}^{n} \Delta(s)\binom{n}{s}y^s = 0 \qquad (3.14)$$

for all $y \geq 0$. But a polynomial vanishes identically (in fact, at more points than its degree) if and only if all coefficients vanish. Hence

$$\Delta(s)\binom{n}{s} = 0 \qquad (3.15)$$

for all s. Since $\binom{n}{s} > 0$, it follows that $\Delta(s) = 0$ for all s. This says that the difference of two functions of S which are both unbiased estimators of the binomial parameter p is always 0; accordingly, there is only one such function. $\square$

*The main part of this proof showed that a function of S having expected value 0 for all p must be identically 0, that is, the binomial family of distributions with n fixed is "complete." A family of distributions is called *complete* if it admits no nontrivial unbiased estimator of 0. (A trivial estimator of 0 is 0 itself or any other estimator which equals 0 with probability 1 under every distribution of the family.) Once one knows that the family of possible distributions of any random variable X is complete, then any funcion of X is an essentially unique unbiased estimator of its expected value, for if there were a nontrivially different one, their difference would be a nontrivial unbiased estimator of 0.*

4 Hypothesis Testing

4.1 Tests and Their Interpretation

As we have already seen, in the binomial situation, a statement about the parameter p can be made on the basis of the observed S by simply using S/n as an estimator of p. Another way to make a statement about p is to perform a statistical *test*, usually called a *significance test* or *test of an hypothesis*. A statistical test consists of a *null hypothesis* and a *rejection rule*.

Null Hypothesis

The null hypothesis (called "null" to distinguish the hypothesis under test from alternative possible hypotheses) is a statement about the distribution of the observations. Here it might be that "the number of successes is binomial with $p \le 0.10$," for example. As long as the binomial family is clearly understood in the context of the problem, the statement might be given as simply "$p \le 0.10$." It is customary to denote the null hypothesis by H_0.

A distribution of the observations is called a *null distribution* if it satisfies the null hypothesis, and an *alternative distribution* otherwise. Although we defined a null hypothesis as a statement about the distribution of the observations, we can also define it as the set of all distributions satisfying that statement, that is, the set of all null distributions. It will be convenient to allow both usages. An *alternative hypothesis* may be defined similarly as a set of alternative distributions. If the null hypothesis completely specifies the null distribution including all parameters, that is, the set contains only one particular distribution, then the null hypothesis is called *simple*. Otherwise, it is called *composite*. An alternative hypothesis may also be either simple or composite.

Rejection Rule

The rejection rule is a criterion saying when the null hypothesis should be rejected; it is "accepted" (see below) otherwise. The rule may depend on the observations in any way, but must not depend on any unknown parameters.

It determines the *rejection region*, often called the *critical region*, given usually as a range or region of values of a *test statistic*. A test statistic may be any function of the observations. For example, here the rule might be "reject if $S \leq 3$," or "reject if $|S/n - 0.3| > 0.10$," when stated in terms of the test statistics S and S/n. A test is said to be a *one-tailed* or a *two-tailed* test based on a statistic S if it rejects in one or both tails of S, that is, it rejects for S outside some interval but not for S inside the interval. Each end of the interval may be closed or open, finite or infinite. (More complicated regions are occasionally useful, but in this book the term "test statistic" will always imply a one- or two-tailed test.) The least extreme value of a test statistic in the rejection region is called its *critical value*. For instance, if the rejection region is $S \leq 3$, then 3 is the critical value of S. A two-tailed test has a critical value in each tail, called the *lower* and *upper critical values*.

It is sometimes convenient to represent a test based on a random variable X with observed value x by the *critical function* $\phi(x)$, $0 \leq \phi(x) \leq 1$ for all x. The rejection region corresponds to $\phi(x) = 1$, that is, those values x for which $\phi(x) = 1$, while the "acceptance" region corresponds to $\phi(x) = 0$. When randomization is considered (Sect. 5), the value of $\phi(x)$ is the probability that, given the observed x, the test will choose to reject the null hypothesis, while $1 - \phi(x)$ is the probability that it will not. Regions where the test may either reject or "accept" thus correspond to values x such that $0 < \phi(x) < 1$. In any case, if the distribution of X is F, the probability of rejection by a test with critical function $\phi(x)$ is $E_F[\phi(x)]$.

Interpretation of Test Conclusions

In an actual application, when the observations are such that the value of the test statistic lies in the critical region, it is customary to announce that H_0 is rejected by the test, or that the set of observations is *statistically significant* by the test, or simply that the *result is significant*. In the contrary case, one may say that the null hypothesis is not rejected by the test, or that the set of observations is not statistically significant. Of course (Problem 8), a result which is not statistically significant may still appear significant for practical purposes, especially if the test is weak, and vice versa, especially if the test is strong (technically, powerful—see Sect. 4.2).

If a null hypothesis is rejected by a "reasonable" statistical test, then one is presumably justified in concluding, at least tentatively in the absence of other evidence, that the null hypothesis is false. (Unfortunately, this statement is either very vague or merely a definition of "reasonable.") If the null hypothesis is not rejected, this does not ordinarily justify a conclusion that the null hypothesis is true. We will find it convenient to say that the null hypothesis is "accepted" whenever it is not rejected, but will use quotation marks to emphasize that "accepting" the null hypothesis does not justify concluding it is true in the same sense that rejecting it justifies concluding it is false.

"Accepting" the null hypothesis is not tantamount to rejection of all other possibilities; in fact, it might be considered tantamount to drawing no conclusion whatever. Tests are thus intended to make it rare that strong conclusions will be drawn prematurely.

For example, suppose that S is the number of successes in 10 Bernoulli trials, and the null hypothesis $H_0: p \geq 0.6$ is to be tested by the rule "reject if $S \leq 3$." If $S = 2$ is observed, then the test calls for rejection of the null hypothesis, and we would probably feel justified in concluding that the null hypothesis is false. On the other hand, if $S = 5$ is observed, then the null hypothesis is not rejected. By our definition, H_0 is "accepted." However, we would not feel justified in concluding that the null hypothesis is true; in fact, $S = 5$ seems more evidence that the null hypothesis is false than true. Would $S = 10$ justify concluding that the null hypothesis is true? Maybe, but then we must treat $S = 10$ differently from $S = 5$ and use a rule with at least three possibilities—concluding that the null hypothesis is false, concluding that it is true, and concluding neither. Such an interpretation will be discussed in Sect. 4.6. For the present, we are considering rules with only two possibilities, rejection and "acceptance." Some observations in the "acceptance" region justify no conclusion, and this must therefore be the interpretation of "acceptance."

In the authors' opinion, the foregoing interpretation of hypothesis tests is appropriate as they are typically used in practice, especially in situations calling for nonparametric methods. Some people favor the following, somewhat stronger, interpretation of "accept." Suppose an action has to be taken, and only two actions are available, one of which is better if the null hypothesis is true and the other otherwise. Then one could set up a test and act as if the null hypothesis were true or false according as the test "accepts" or rejects the null hypothesis. The interpretation of the test would then be that *if one had to treat the null hypothesis as either true or false and a verdict of "no conclusion" were not allowed*, then one would treat the null hypothesis as true if the test "accepts" it and false if the test rejects it. Sometimes this interpretation is suggested in terms of conclusions even when no action is required. In practice, however, tests are seldom chosen appropriately for any such interpretation. Furthermore, actions are seldom taken as a direct result of tests. When they are, as in acceptance sampling and quality control procedures, for example, special considerations (the relative seriousness of the two types of error, if nothing else) usually have an important bearing on the choice of the rejection rule.

In any case, according to all usual rationales, the problem should be formulated so that rejecting the null hypothesis when it is true is a serious error, at least more serious than the reverse. The test then controls the probability of such an error. (See Sect. 4.2.) "Preliminary" tests made to check a model or assumption on which a later primary analysis will be based seldom satisfy the previous condition. They should be judged on the basis of their effect on the total analysis and employed, if at all, not in a routine fashion.

There are problems, such as discrimination between two hypotheses or classification into one of two categories, where the two types of error are really alike. It is possible but unnatural to adapt the essentially unsymmetric framework of testing hypotheses to these situations. If some items may be left unclassified, the three-decision interpretation of two-tailed tests (Sect. 4.6) may be relevant.

It is not entirely obvious that tests usually are appropriately interpreted in any of the foregoing ways. This will be discussed later in Sect. 4.4 in connection with *P*-values, an alternative method of presenting test results. For further discussion, both comforting and alarmist, see also, for instance, Cox [1958a], Kruskal [1968], Neyman [1950, 1957], Pearson [1962], Pratt [1965], and Savage [1954, especially Sect. 16.3]. See also Kadane [1976], Kempthorne [1976], Neyman [1976], Pratt [1976], and Roberts [1976].

4.2 Errors

Types of Errors and Their Probabilities

When a statistical test is performed, two kinds of error are possible. We may reject the null hypothesis when it is true, making a *Type I error* (or *error of the first kind*). On the other hand, we may "accept" (fail to reject) the null hypothesis when it is false, making a *Type II error* (or *error of the second kind*). The types of errors and correct decisions, which cover all four possibilities, are shown in the diagram below.

	Conclusion	
True Situation	"Accept" H_0	Reject H_0
H_0 true	No error	Type I error
H_0 false	Type II error	No error

If we commit a Type I error, we are concluding that the null hypothesis is false when it is actually true. With a Type II error, we are drawing essentially no conclusion when in fact the null hypothesis is false. The two types of error thus differ in kind, but both are undesirable. Statistics is concerned with situations where, unfortunately, we cannot be certain to avoid both types of error.

The probability of a Type I error is evaluated using a null distribution and hence is frequently called a *null probability*, while the probability of a Type II error is evaluated from an alternative distribution. The calculation is easily illustrated for the binomial distribution. Suppose we use the rule "reject if

$S \leq 3$," where S is the number of successes in 10 Bernoulli trials. Then the probability of rejection $\alpha(p)$ depends on p and is given by

$$\alpha(p) = P_p(S \leq 3) = \sum_{i=0}^{3} \binom{10}{i} p^i (1 - p)^{n-i}, \qquad (4.1)$$

which may be looked up in Table B. The upper curve in Fig. 4.1, labeled $S = 3$, shows a graph of this probability for all values of p. If the null hypothesis under test is $H_0: p = 0.5$, then the probability of a Type I error, rejecting H_0 when it is true, is simply $\alpha(0.5)$ on the curve, which is 0.172. On the other hand, if we test the null hypothesis $H_0': p \geq 0.6$ with this same rejection region, then the probability of a Type I error is given by that part of the curve $\alpha(p)$ in Fig. 4.1 where $p \geq 0.6$. Since the curve never rises above 0.055 for $p \geq 0.6$, this probability is never more than 0.055.

The probability of a Type II error is calculated in a similar manner. Since the null hypothesis is "accepted" whenever it is not rejected, the probability of "acceptance" is one minus the probability of rejection. Hence the probability of "acceptance" in the example of the previous paragraph is given by

$$1 - \alpha(p) = 1 - P_p(S \leq 3)$$

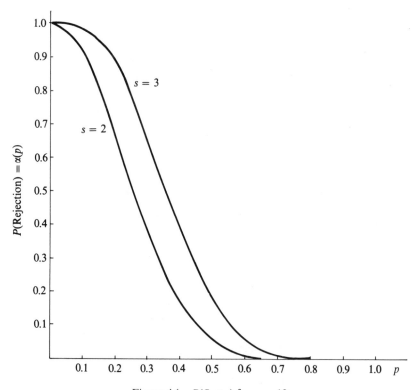

Figure 4.1 $P(S \leq s)$ for $n = 10$.

where $\alpha(p)$ is defined as before. This quantity is the probability of a Type II error for all values of p not in the null hypothesis. Thus, if the null hypothesis is $H_0: p = 0.5$, the probabilities of a Type II error can also be read off Fig. 4.1, as the complements of the ordinates of the curve for all $p \neq 0.5$. Similarly, if the null hypothesis is $H_0': p \geq 0.6$, the probabilities of a Type II error are the complements of the ordinates where $p < 0.6$. The maximum (strictly, supremum) probability of a Type II error in this case is 0.945, and it decreases to zero as p decreases.

Level

Ordinarily, for any statistical test, the probability of rejecting the null hypothesis depends on the distribution of the observations. It can be calculated as long as this distribution is known. If the null hypothesis is true, this probability may be a fixed number or may still depend on the distribution of the observations. If the null hypothesis is simple, like $H_0: p = 0.5$ above, then the probability of rejecting the null hypothesis when it is true necessarily has only one value. If the null hypothesis is composite, then the probability of rejecting the null hypothesis may or may not be the same for all distributions allowed by the null hypothesis. For instance, in the previous example, the probability of rejecting the null hypothesis $H_0': p \geq 0.6$ when true depends on p.

    If for a test of a particular null hypothesis, simple or composite, the probability of a Type I error is less than or equal to some selected value α for all null distributions, then the test is said to have *level* α or to be at level α (0.05 and 0.01 are popular values for α). The level may be described as *conservative* to emphasize that this kind of level is meant rather than nominal or exact level (defined below). Thus the test above, which rejects the null hypothesis $H_0': p \geq 0.6$ when $S \leq 3$, has level 0.10. It also has level 0.08 and level 0.06. It does not quite have level 0.05, however, because there is a distribution satisfying the null hypothesis and giving a probability of rejection greater than 0.05; for example, $p = 0.6$ gives probability 0.055 of rejection. If a test has level α, it is natural to say also that *the level of the test* is α. The word "the" here is somewhat misleading, though not seriously so in practice, since, as the foregoing example illustrates, the level of a test is not unique.

    If a test of the null hypothesis $H_0': p \geq 0.6$ is desired at level 0.10, the test rejecting when $S \leq 3$ might be selected. The number 0.10 would then be called the "nominal level" of the test; 0.055 is called the "exact level" because it is the maximum probability of rejection for $p \geq 0.6$. In general, the *nominal level* of a test is the level which one set out to achieve, while the *exact level* is the actual maximum probability of rejection under the null hypothesis. The exact level is the smallest conservative level the test actually has. It would perhaps be simpler mathematically to define only exact levels, but conservative and nominal levels are needed in practice and must be discussed, and it is convenient to have names for them.

The term *size* is sometimes used instead of level, but it will not be used in this book. Connotatively, "size" seems to place more emphasis on the rejection region, and less on the null hypothesis, than the word "level." A test is sometimes called *valid* at level α if it has level α; this terminology is especially useful when a null hypothesis has been changed or broadened. Sometimes *significance level* is used instead of simply level, to distinguish it from confidence level which will be defined in Sect. 6.

Interpretation of Level

The level α of a test can be interpreted as bounding the probability of drawing an erroneous conclusion, if failing to reject the null hypothesis is regarded as drawing no conclusion and hence as not drawing an erroneous conclusion. This does not mean, however, that a conclusion to reject the null hypothesis for one particular set of observations has probability at least $1 - \alpha$ of being correct. A particular conclusion is not a random variable, but rather is the value of a random variable. If only an "objective" or "frequentist" concept of probability is used, the probability that a conclusion to reject is correct once the observations have been made must be either 0 (if the conclusion is incorrect) or 1 (if the conclusion is correct), and this probability is unknown (except in the trivial cases where the conclusion is either impossible or surely true).

The probability that any particular conclusion is correct can be computed if the Bayesian philosophy of statistics is adopted and the additional input it requires is provided. This philosophy favors using probability to measure the uncertainty about any quantity which cannot be determined exactly from available evidence, including the true p in the binomial situation. Probability is then interpreted as subjective (more rarely objective) but rational degree of belief. A "prior" probability distribution must be assigned to the parameters before observation. It is then straightforward to compute the "posterior" probability, given the observations, that any possible conclusion is correct. The results may be more or less influenced by the "prior" distribution, depending on how definitive the data are. Strictly, the prior distribution should represent prior judgment, but in practice may be chosen conventionally as a reference point for inferences. There are strong arguments in favor of the Bayesian philosophy which will not be discussed here (e.g., Savage [1954]; Lindley [1971]; Pratt et al. [1964]). It is a disquieting warning about tests of hypotheses to find that they rarely seem appropriate from a Bayesian point of view, and that in most situations, the level of a test is not closely related to the posterior probability of a correct conclusion (Pratt [1964]; see also the discussion of the interpretation of P-values in Section 4.4).

Power

The probability of rejection when the null hypothesis is false also depends on the distribution of the observations, and is called the *power* of the test. Power is evaluated using an alternative distribution. If the alternative is simple, the power is a single number; otherwise it is a function.

Consider again, for instance, the test above for $n = 10$, which rejects H'_0: $p \geq 0.6$ when $S \leq 3$. Its power against the alternative $p < 0.6$ is given by the function (4.1) for values of $p < 0.6$. The power curve is then represented by that portion of the curve in Fig. 4.1 for which $p < 0.6$. Specifically, the power of the test is 0.172 when $p = 0.5$; 0.650 when $p = 0.3$; 0.987 when $p = 0.1$; etc. Recall that the remaining portion of the curve, where $p \geq 0.6$, represents the probability of a Type I error, and the complements of the ordinates for that portion where $p < 0.6$ represent the probability of a Type II error. Clearly, the power is always one minus the probability of a Type II error.

Use of Power in Choosing a Test Statistic

The power should be considered, at least informally, in choosing a test statistic and the significance level. Even after a test statistic has been chosen, the power can be increased by changing the critical value, but the probability of a Type I error will also increase. Thus the power and significance level, or the probabilities of the two types or errors, must be traded off against each other. In principle, both of these must also be traded off against the size and kind of experiment if they are subject to choice (Pearson [1962]). All these tradeoffs are very difficult to make, especially in a "frequentist" framework, and are consequently more honored in theory than in practice. It can be argued that varying the significance level according to circumstances impedes communication and that conventional levels provide an appropriate tradeoff between "noise" in the knowledge system and suppression of valid, useful conclusions (Bross [1971]).

Even when a God-given nominal level is to be used, if the test statistic is discrete, power considerations should enter the decision of whether to use a conservative test (unless controlling the Type I error is considered paramount, in which case why not make it infinitesimal?). For example, if the nominal level of a test is 0.05 and the two choices of exact levels nearest 0.05 are 0.001 and 0.06, the power of the conservative test will generally be much smaller than the power of the test at exact level 0.06. The latter test may then be the more desirable one. The practice of reporting the P-value, as explained in Sect. 4.4, partly sidesteps this kind of decision.

Power is also the basis for choosing among different types of test or test statistics. Different tests may have relatively large or small power against different alternatives. Hence in choosing among them, one might favor a test which provides large power against those alternatives which are of particular interest because of their practical importance or ("subjective!") probability of occurrence.

Power comparisons of different tests should ordinarily be made at the same exact levels. Otherwise they may be seriously misleading, because the power of a test can be increased by increasing the probability of a Type I error. Confusion generally attends comparisons of tests which have the same nominal or conservative levels but different exact levels.

Two tests, say A and B, are called *equivalent* if test A rejects if and only if Test B rejects. Equivalent tests necessarily have the same exact level and the same power against all alternatives, but the converse is not true (Problem 11). Correspondingly, two test statistics are called equivalent if any test based on either statistic is equivalent to a test based on the other. This holds whenever the test statistics are strictly monotonically related (Problem 12).

4.3 One-Tailed Binomial Tests

The null hypothesis $p \geq 0.6$ and the alternative $p < 0.6$ are each one-sided, in an obvious sense. The test which rejects when $S \leq 3$ is also called one-sided, or one-tailed. Explicitly, it is called lower-tailed or left-tailed since the rejection region is at the lower end of the range of the test statistic S.

More generally, suppose S is the number of successes in n Bernoulli trials and we wish to test the null hypothesis $p = p_0$ or $p \geq p_0$ against the alternative $p < p_0$. One rule for testing either of these null hypotheses is to reject when $S \leq s_c$, where the critical value s_c is chosen so that the level of the test is some preselected number α. Let s_c be the largest integer possible, subject to the restriction that the left-tail probability $P(S \leq s_c)$ is less than or equal to α when $p = p_0$. This critical value is easily found from Table B. Algebraically it is the largest integer s_c for which

$$P_{p_0}(S \leq s_c) = \sum_{i=0}^{s_c} \binom{n}{i} p_0^i (1 - p_0)^{n-i} \tag{4.2}$$

does not exceed the nominal level. (The subscript on P indicates that the probability is to be computed for $p = p_0$.) Given s_c, a simple comparison of the observed value of S with s_c determines whether the observations are significant or not.

For the null hypothesis $p = p_0$, the exact level of this test is, of course, the exact value of the probability in (4.2). Furthermore, it has the same exact level for the null hypothesis $p \geq p_0$ (with of course the same power against alternatives $p < p_0$). Intuitively, this is because testing the null hypothesis $p \geq p_0$ against alternatives $p < p_0$ is the same as testing the "least favorable case" $p = p_0$ against alternatives $p < p_0$. For the binomial distribution, this intuition is correct (see Problem 13).

The power of this test, namely $P_p(S \leq s_c)$ where p is any value less than p_0, is also easily found from Table B, and can be expressed algebraically as

$$P_p(S \leq s_c) = \sum_{i=0}^{s_c} \binom{n}{i} p^i (1 - p)^{n-i} \quad \text{for any } p < p_0. \tag{4.3}$$

What about a one-sided hypothesis testing situation in which the alternative lies on the other side of the null hypothesis? The null hypotheses $p = p_0$ and $p \leq p_0$ may be tested against the alternative $p > p_0$ at level α by rejecting when $S \geq s_c$, where s_c is chosen to be the smallest integer possible subject to the restriction that $P(S \geq s_c)$ does not exceed α when $p = p_0$. Thus s_c is the smallest integer such that

$$P_{p_0}(S \geq s_c) = \sum_{i=s_c}^{n} \binom{n}{i} p_0^i (1 - p_0)^{n-i} \leq \alpha \tag{4.4}$$

and again Table B can be used. For example, when $n = 10$ and the alternative is $p > 0.6$, rejecting the null hypothesis $p = 0.6$ or $p \leq 0.6$ when $S \geq 9$ gives an exact level of 0.046. This is specifically called an upper-tailed or right-tailed test procedure. In fact, these testing problems and procedures correspond to the one-sided binomial testing problems and procedures discussed earlier by the simple exchange of the definitions of "failure" and "success." For instance, previously we tested the null hypothesis that the probability of success is equal to 0.6 against the alternative that it is less than 0.6. Using the rule "reject when the number of successes is 3 or fewer in 10 trials," the exact level was 0.055. This is precisely the same as testing the null hypothesis that the probability of *failure* is equal to 0.4 against the alternative that it is greater than 0.4, by rejecting if there are 7 or more failures in the 10 trials. Since "success" and "failure" are completely arbitrary designations anyway, we can rename the failures "successes." This is therefore a right-tailed test, and properties of the two types correspond, with the exact level again 0.055.

4.4 P-values

Definition of P-values

We now discuss an important variation in the method of carrying out a test and reporting the result. The procedure we have described for a conservative test is to select a critical region in such a way that the probability of a Type I error is not more than some nominal level chosen in advance. We then report whether or not the observations are significant at the particular level chosen. We could instead, however, find and report the smallest level at which the observations are significant, the level of just significance or the critical level, also called the *P-value*. If this value is smaller than the nominal level, the observations are significant, and otherwise not significant, according to the procedure just described. However, if no decision is actually required, it is not necessary to choose a nominal level or form a specific rejection rule; significance is a secondary question and reporting the P-value is more informative and avoids arbitrary choices. Ordinarily the P-value can be found as a tail probability computed under the null hypothesis using the observed value of the test statistic. Hence the name P-value. These ideas will now be explained in more detail.

As an example, in the binomial case with 10 trials and $p_0 = 0.6$, $S = 3$ is significant by all lower-tailed tests with critical value $s_c \geq 3$ but not by those with $s_c \leq 2$. Since from Table B we have

$$P_{0.6}(S \leq 2) = 0.0123$$
$$P_{0.6}(S \leq 3) = 0.0548,$$

we see that $S = 3$ is significant by tests at all levels $\alpha \geq 0.0548$, but not at levels $\alpha < 0.0548$. Thus 0.0548 is the smallest level at which $S = 3$ is significant. Similarly, $S = 2$ is just significant at the 0.0123 level, $S = 4$ at the 0.1662 level, etc.

More generally, by a lower-tailed test for $H_0: p = p_0$ or $H_0': p \geq p_0$ in a binomial problem, a value $S = s$ is just significant at the level

$$\alpha_s = P_{p_0}(S \leq s). \tag{4.5}$$

In other words, the borderline level of significance is the tail probability of s or less under p_0. This is the P-value.

To generalize to an arbitrary situation, suppose that the possible outcomes x can be ordered according to how "extreme" they are, as by a test statistic, and that critical regions at all possible significance levels are the tails of this ordering. If a value x is observed, the smallest critical region to which x belongs is the tail which just includes x. Hence the smallest significance level at which x is significant equals the maximum probability, under null distributions, of the tail which just includes x. This maximum probability is called the P-value (or exact P-value or P-value at x). It is also frequently called the tail probability including x, or the associated probability at x, since it represents the probability associated with an outcome equal to or more extreme than that actually observed. Of course, it is also the critical level, and if a decision about significance were required, the outcome x would be judged significant at all levels greater than or equal to the P-value at x, but not significant at smaller levels.

When the possible outcomes are ordered as above, significance at one level implies significance at all larger levels. Conversely, suppose one has a test for each level α, $0 < \alpha < 1$, and significance at one level implies significance at all larger levels. Then the outcomes can be ordered according to the levels at which they are just significant and the critical regions are the tails of this ordering. Furthermore, the P-value itself may be viewed as a test statistic on which the tests are based.

For some well-behaved problems, the P-value as a tail probability and the critical level as the level of just significance can be sensibly defined and are equal. This value can therefore be reported and provides more information than a statement that the observations are or are not significant at a pre-selected level. It is possible in some problems, even by some kinds of "optimum" tests, for a set of observations to be significant at one level but not at a larger level, for instance at the 0.01 level but not at the 0.05 level. Then rejection regions at different levels would not be nested, and P-values and critical levels would be difficult to interpret, even if they were defined. No such situations arise in this book; Chernoff [1951] gives an example which illustrates the pathology.

Interpretation of P-values

The P-value may perhaps be interpreted as a kind of measure of the extent to which the observations contradict or support the null hypothesis. One must be cautious about such an interpretation, however. It certainly is true, but rather trivial, that in a single experiment, the more extreme the P-value the more extreme the outcome, and hence the more the null hypothesis is contradicted, as long as "extreme" is properly defined. But this would also be true for any strictly monotonic function of the P-value, and in particular, for the test statistic itself.

The real question is, can one compare P-values across sample sizes or experiments? The answer, unfortunately, has to be that such comparisons usually have little meaning. We mention three justifications for this assertion. First, the *more* powerful a test is, the more extreme a P-value is to be expected if the null hypothesis is false, and hence the *less* a given P-value contradicts the null hypothesis. Second, convincing arguments both within and outside the "objective" or "frequentist" theory of probability show that the extent of contradiction of the null hypothesis is *not* determined by the P-value, but rather by the likelihood function, a very different animal. (See Birnbaum [1962].) Third, P-values can be very discordant with the direct answers to questions of interest when such answers are available, as they are in the Bayesian framework once the relevant prior distributions have been provided (see *Interpretation of Level* in Sect. 4.2).

Often a null hypothesis such as $p = p_0$ is almost certainly not literally true. Then the P-value is obviously not a useful measure of the plausibility of this literal null hypothesis. Usually the question of real interest relates to whether such a null hypothesis is nearly true. In the Bayesian framework, however, the "posterior" probability of this, given the observations, may vary widely, depending on the sample size and the problem, for a fixed P-value and a fixed "prior" probability before observation. Even in practical problems, if a null hypothesis analogous to $p = p_0$ has prior probability close to 0.5 of being nearly true, the posterior probability after observation may well be as small as 6 times or as large as 12 times the P-value for P-values between 0.001 and 0.05, although it is seldom less than 3 times or more than 30 times the P-value (according to Good [1958]). These figures are rough, but based on considerable though unpublished evidence. See also Jeffreys [1961] and Lindley [1957]. For interesting examples with discussion, see Good [1969], Efron [1971], and Pratt [1973]. In this framework then, if the value of a test statistic is just significant at the 0.05 level, there is still a substantial chance (at least 0.15) that the null hypothesis is nearly true. This suggests that bare significance at the 0.05 level, a P-value just below 0.05, is at best not a very strong justification for concluding that the null hypothesis is appreciably false. Of course, significance substantially beyond the 0.05 level is another matter. We note that this illustrates again the disadvantage of a mere statement of significance or nonsignificance.

In the special case of a one-tailed test of a truly one-sided null hypothesis, such as $p \leq p_0$ (not $p = p_0$) or a multiparameter analogue, the P-value may often be expected to be close to the posterior probability of the null hypothesis (see, e.g., Pratt [1965]). It can be argued that in all other cases, both P-values and tests should be interpreted with great caution.

Discreteness and P-values

We now discuss some difficulties connected with discrete distributions. In binomial problems, and many others, it is customary to use conservative tests, that is, to choose the critical value so that the exact level is as near the nominal level as possible subject to the condition that it be no greater. (Most available tables and computer programs giving critical values are so constructed.) With this custom, an outcome is considered significant at the nominal level α if and only if the exact P-value is less than or equal to α.

An alternative method of selecting critical values in discrete problems is to choose them so that the exact level is as near the nominal level as possible on either side (greater or smaller). Some tables and perhaps computer programs use this criterion. It should be remembered that whenever the exact level is larger than the nominal level, this procedure has greater power than a conservative test.

Consider, for instance, the previous binomial example where $n = 10$ and the null hypothesis is $p \geq 0.6$. If the nominal level is 0.05, the conservative critical value is $s_c = 2$. The exact level nearest the nominal level is given by $s_c = 3$, since

$$P_{0.6}(S \leq 2) = 0.0123 \quad \text{and} \quad P_{0.6}(S \leq 3) = 0.0548$$

and the latter is nearer 0.05. The probability of rejection for these two tests is graphed in Fig. 4.1 and shows how much greater is the power of the test using $s_c = 3$.

When the exact level nearest the nominal level is used, what is the borderline level of significance? Suppose, for example, that $S = 3$ is observed. Then the exact P-value is 0.0548. The probability $P_{0.6}(S \le 2) = 0.0123$ will be called the next P-value. The average of these two numbers is $(0.0548 + 0.0123)/2 = 0.0336$, called the *mid-P-value*; this is the borderline level of significance since a nominal level greater than 0.0336 is nearer to 0.0548 than to 0.0123, while one less is nearer to 0.0123.

For a test of the null hypothesis $p \ge p_0$ in a binomial problem, by the same rule, $S = s$ is significant at nominal levels greater than, and not significant at nominal levels smaller than, the mid-P-value

$$\frac{P_{p_0}(S \le s) + P_{p_0}(S < s)}{2}. \tag{4.6}$$

In general, as long as the possible outcomes x can be ordered according to how extreme they are, the mid-P-value is defined as the arithmetic average of the exact P-value and the next P-value, and is the borderline level of significance[2] or critical level according to the rule of "exact level nearest the nominal level." Here the next P-value, also called the tail probability beyond x, is the maximum probability under null distributions of an outcome more extreme than x (see Lancaster [1952] for further discussion).

Summary Recommendations

As mentioned earlier, reporting the P-value for the outcome observed gives more information than a report of simply significant or not significant, and in effect permits everyone to choose his own significance level.

For test statistics with discrete null distributions, however, there is still the question of whether to report the exact P-value or the mid-P-value when they differ appreciably. This seems a matter of taste, as long as it is made clear which is being done. If there is a custom for the particular type of problem, this should be followed. For some audiences, it may be desirable to report both the exact and next P-values (the tail probabilities including and beyond x). Approximations based on continuous distributions generally approximate the mid-P-value rather than the exact P-value unless a "correction for continuity" is made. This sometimes makes the mid-P-value a little more convenient to compute. Some people, especially if they believe that the P-value can be given a precise interpretation, may feel that some one number should be chosen for the purpose and that there are fundamental grounds for choice between the exact P-value and the mid-P-value (or something else). See the discussion of randomized P-values in Sect. 5 and, for instance, Lancaster [1961].

---

[2] Whether the outcome would be significant at precisely this nominal level depends on whether one chooses the larger or smaller when the nominal level is halfway between the two nearest exact levels.

Even though it is not appropriate to interpret a P-value as more than a measure of the extent to which the observations contradict or support the null hypothesis in a single experiment, the method is well justified and advised on the grounds that it contains information about the experimental results which is not reflected in a simple statement of significance at a preselected level. P-values are discussed further in Gibbons and Pratt [1975].

4.5 Two-Tailed Test Procedures and P-values

Tests Against Two-Sided Alternatives

In the binomial problem, how should we test the null hypothesis $p = 0.6$ against the alternative $p \neq 0.6$? This alternative is a combination of the two alternatives $p < 0.6$ and $p > 0.6$, and is *two-sided* in an obvious sense. A test might be performed by combining the left-tail and right-tail tests discussed previously. Thus, for $n = 10$, one might reject when $S \leq 3$ and also when $S \geq 9$. Since the two tails are mutually exclusive, the exact level of this test can be computed as the sum of the two tail probabilities under the null hypothesis $p = 0.6$. These two tail probabilities, the exact levels of the two one-tailed tests, are respectively 0.055 and 0.046. The exact level of this two-tailed test is thus

$$P_{0.6}(S \leq 3 \text{ or } S \geq 9) = P_{0.6}(S \leq 3) + P_{0.6}(S \geq 9)$$
$$= 0.055 + 0.046 = 0.101.$$

Similarly, given that the nominal levels of the two individual tests were both 0.10, the nominal level of this two-tailed test is 0.20.

In general, it is always true that combining individual tests at levels α_1, $\alpha_2, \ldots$ for the same null hypothesis H_0 gives a test at level $\alpha_1 + \alpha_2 + \cdots$ for H_0. The exact level of the combined test is the sum of the exact levels of the individual tests if the null hypothesis is simple (or the same distribution is "least favorable" for all tests) *and* the tests are mutually exclusive (that is, no possible set of observations is rejected by more than one of the tests). Otherwise the exact level may be less than (but cannot be more than) the sum (Problem 19).

For a binomial null hypothesis $H_0: p = p_0$, the standard two-tailed test at level α rejects when either one-tailed test at level $\alpha/2$ rejects. Thus, specifically it rejects if $S \leq s_\ell$ or if $S \geq s_u$, where s_ℓ is the largest integer which satisfies

$$P_{p_0}(S \leq s_\ell) \leq \frac{\alpha}{2} \tag{4.7}$$

and s_u is the smallest integer which satisfies

$$P_{p_0}(S \leq s_u) \leq \frac{\alpha}{2}. \tag{4.8}$$

While this test has nominal level α, its exact level is the sum of the actual values of the left-hand sides of Eqs. (4.7) and (4.8). Similarly, its power function is the sum

$$P_p(S \leq s_\ell) + P_p(S \geq s_u),$$

calculated under any alternative $p \neq p_0$.

It is not automatic that a test against a combined alternative should be constructed by combining the tests which would be chosen for the individual alternatives, nor even that a test against a two-sided alternative should be two-tailed in form just because a one-tailed test would have been used with each of the one-sided alternatives. It can be shown, however, that two-tailed tests are (in a sense to be made precise later) the only ones that need be considered against two-sided alternatives in binomial problems and indeed in most practical problems. (See Sect. 8.3.)

Even if attention is restricted to two-tailed tests, the critical values s_ℓ and s_u can be chosen in ways other than that above. They may be critical values for any one-tailed levels α_1 and α_2 such that $\alpha_1 + \alpha_2 \leq \alpha$. In other words, they need only satisfy

$$P_{p0}(S \leq s_\ell) + P_{p0}(S \geq s_u) \leq \alpha. \tag{4.9}$$

Various possibilities, including an "optimality criterion" which is really a convention, will be discussed later, when the properties of two-tailed tests are investigated in Sect. 8. It is difficult, however, to give a convincing justification in the frequentist framework for choosing a particular two-tailed test among those at level α.

For two-tailed tests, then, we have the problem of selecting not only a significance level α, but also the upper and lower critical values, or α_1 and α_2, for a given significance level. For the latter, like the former, except by adoption of some convention, the usual methodology of hypothesis testing tells us only to "look at the power functions and make a choice," but sheds no light on how to do so. It is easier said than done.

P-values for Two-Tailed Tests

For two-tailed, as for one-tailed, tests, we can avoid the problem of choosing a significance level by reporting a P-value. Unfortunately, however, the very definition of the P-value for two-tailed tests presents a problem equivalent to that of choosing α_1 and α_2 for a given significance level. (This problem has no counterpart for one-tailed tests.)

One possibility is to report the one-tailed P-value even for a two-tailed test, and remark that the two-tailed P-value, while depending on what kind of two-tailed critical region would have been formed, is presumably about twice as large as the one-tailed P-value reported. Some people go further and claim that P-values are not appropriate in two-sided situations, but that seems an inappropriate dismissal of a problem which is not trivial and should be examined.

To obtain a precise two-tailed P-value, one would have to add the probability of a value equal to or more extreme than that observed in the same tail and *some* probability from the opposite tail. However, a single observed result can give various P-values depending on what probability is added to represent the other tail. The most common procedure is to report a two-tailed P-value as twice the one-tailed P-value, but there are other possibilities.

To simplify the discussion, we assume that the test statistic has only one relevant null distribution, either because only one is possible or because its other possible null distributions have tail probabilities at least as small. When this null distribution is symmetric, like the binomial for $p = 0.5$, all procedures lead to a two-tailed P-value which is twice the one-tailed P-value.

When the null distribution is not symmetric, doubling the one-tailed P-value corresponds to the test with conservative level $\alpha/2$ in each tail, as at (4.7) and (4.8). If the null distribution is discrete, the exact level of this procedure will ordinarily be strictly less than α (Problem 20), clouding the interpretation of the P-value. A modification which avoids this problem is to define the two-tailed P-value as the sum of the one-tailed P-value and the largest attainable probability in the other tail which does not exceed the one-tailed P-value. This is the exact level if the foregoing test is carried out at a nominal level equal to twice the one-tailed P-value. It is also the exact P-value of a two-tailed test in which the values of the test statistic are ordered according to their one-tailed P-values (Problem 21). If instead we add the nearest attainable probability in the other tail, whether larger or smaller than the one-tailed P-value, two points are often added to the rejection region simultaneously, unnecessarily reducing the number of exact levels available, and the resulting P-values need not correspond to any test procedure (Problem 22).

Another approach might be called the "principle of minimum likelihood." If the value $S = s$ is observed, the P-value at s is found by summing the null probabilities of all values of S which do not exceed the probability $P(S = s)$. If S has a density f under the null hypothesis, the P-value is the probability of the region where the density is $f(s)$ or less. This gives a two-tailed probability as long as the null distribution is unimodal. It is equivalent to ordering the values of S according to its mass or density function under the null hypothesis. It corresponds to the most powerful test against the alternative that S is uniformly distributed.

Still another possibility is to locate the cutoff of the two tails at an equal distance from some specified location parameter μ of the null distribution, such as the mean, median, or mode. This corresponds to ordering values of the test statistic S according to $|S - \mu|$. Then the P-value is $P(S \geq s) + P[S \leq \mu - (s - \mu)]$ when s is in the upper tail. This procedure seems appealing if one interprets the P-value as a measure of agreement or disagreement between the observed value of the test statistic and some central value under the null hypothesis. Unfortunately, some sort of skewness correction is needed for severely asymmetric null distributions, just those where the choice of procedure matters most.

We illustrate these procedures in the binomial case with $n = 10$ and H_0: $p = 0.6$. For convenience, the point probabilities for S under H_0 are given in Table 4.1.

Suppose that $S = 3$ is observed. The appropriate one-tailed P-value is lower-tailed, and $P_{0.6}(S \leq 3) = 0.055$. One procedure is simply to report this, with the comment that the two-tailed P-value is presumably about 0.110. This is the borderline level of significance if the standard two-tailed test with level $\alpha/2$ in each tail is used. Since no upper tail probability equals 0.055, the first procedure suggested above would add the next smaller upper-tail probability, which is $P_{0.6}(S \geq 9) = 0.046$, and report a two-tailed P-value of $0.055 + 0.046 = 0.101$. This is the exact level of the standard two-tailed test at the borderline nominal level 0.110. It is also the borderline level of significance of a test based on the one-tailed P-value. The nearest attainable upper-tail probability is the next smallest in this case and hence gives the same result.

For the minimum likelihood procedure, the values of S with probability smaller than $S = 3$ under the null hypothesis are 0, 1, 2, 9, and 10. Hence the P-value is $P_{0.6}(S \leq 3) + P_{0.6}(S \geq 9) = 0.101$ again.

Since the mean, median, and mode of the null distribution each equal 6, locating the two tails at equal distances from any of these also gives the same result of 0.101. Using the midrange of 5, however, gives a P-value of $P_{0.6}(S \leq 3) + P_{0.6}(S \geq 7) = 0.055 + 0.382 = 0.437$. If the observed value of S had been in the upper tail, this procedure with the midrange would have given a two-tailed P-value smaller than twice the one-tailed P-value in this example since the null distribution is skewed left here.

We mention one other procedure that could be used for statistics with a finite range. It places an equal number of possible values in each tail if the distribution is discrete. If the possible values are equally spaced, this also makes the tails equal in length, and this is the procedure used for continuous distributions also. Except for discrete distributions with unequally spaced values, which are unusual in practice, this procedure is equivalent to locating the tails at equal distances from the midrange, as in the example of the previous paragraph. Unfortunately, this procedure not only is restricted to statistics with a finite range, but also, even when defined, can give absurd results if the null distribution is highly skewed. For example, in the binomial case with H_0: $p = 0.1$, suppose that $S = 7$ is observed with $n = 10$. The one-tailed P-value is then $P_{0.1}(S \geq 7) = 0.000$. When an equal number of extreme

Table 4.1 Binomial Probabilities for $p = 0.6$, $n = 10$

s	0	1	2	3	4	5	6	7	8	9	10
$P_{0.6}(S = s)$	0.000	0.002	0.010	0.043	0.111	0.201	0.251	0.215	0.121	0.040	0.006

values are placed in the lower tail, the two-tailed P-value becomes $P_{0.1}(S \leq 3)$ $+ P_{0.1}(S \geq 7) = 0.987$. Even though $S = 7$ strongly contradicts H_0, a P-value of 0.987 would lead to the conclusion that H_0 is highly "acceptable."

Intuitively, the extent to which the data contradict a null hypothesis should not change sharply if the hypothesis is changed slightly. However, for all of these procedures except doubling the one-tailed P-value, a slight change in the null hypothesis can lead to a sharp change in the P-value because the P-value is a discontinuous function of the null hypothesis (Problem 23). The authors consider this counterintuitive property less a flaw in the methods of forming two-tailed P-values than a symptom of the fundamental difficulty of interpreting P-values as measuring the support or contradiction of the null hypothesis. The extent of contradiction depends in part on the congruence of the data with plausible alternatives, whereas the P-value depends only on the null distribution.

In summary, a reasonable two-tailed P-value can be obtained in most situations by either doubling the one-tailed P-value, or adding to it the largest attainable probability not exceeding it in the other tail. The minimum likelihood method may also be satisfactory. The practice of doubling the one-tailed P-value is perhaps the most popular, but that may be more the result of habit than a thoughtful consideration of the merits. When all is said and done, however, we find the game of defining a precise two-tailed P-value not worth the candle. If a single procedure is to be recommended as appropriate for two-tailed tests based on any distribution and any outcome, we prefer reporting the one-tailed P-value and the direction of the observed departure from the null hypothesis. The primary basis for this recommendation is that the P-value then retains its clearest interpretation. Further, when the one-tailed P-value is small, the sample outcome is extreme in a particular direction and a one-sided conclusion will usually be desired. On the other hand, if the one-tailed P-value is moderate or not small, the null hypothesis will be "accepted" whether it is doubled or not. In borderline cases, the appropriate conclusions require careful thought, not blind adherence to some rule. Careful thought is perhaps best encouraged by reporting a one-tailed P-value with suitable commentary attached. For further discussion, see Gibbons and Pratt [1975]. The recommendation for reporting the one-tailed P-value even with a two-tailed test is further reinforced when we consider the test procedures which allow a greater variety of conclusions to be reached, as our next topic.

4.6 Other Conclusions in Two-Tailed Tests

If it is necessary to reach a conclusion on the basis of a two-tailed test, the two-tailed P-value, however computed, can be used as a critical level to define the test just as in the one-tailed case. Whether a two-tailed test is defined in this way or by setting up a rejection region in terms of upper and lower critical values of a test statistic, the general interpretation of tests given in Sect. 4.1 still applies. Specifically, one may conclude, at least tentatively in the absence of other evidence, that the null hypothesis is false if it is rejected by a "reasonable" two-tailed test, while one may draw essentially no conclusion if it is not rejected. This two-conclusion interpretation is indeed appropriate in some situations. For example, rejection may amount to deciding from a preliminary experiment that further study is worthwhile. Alternatively, it may mean that a simple model is inadequate, in circumstances where it is not necessary to conclude how the model might be made adequate.

In many situations, however, more definite conclusions are desirable. For instance, when the null hypothesis $p = p_0$ is rejected by a two-sided binomial test, we might want to conclude that $p < p_0$ if $S \leq s_\ell$ and that $p > p_0$ if $S \leq s_u$. Then there would be three possible conclusions, namely $p < p_0$, $p > p_0$, and "no conclusion" (which corresponds to "accepting" the null hypothesis). Table 4.2 gives the probability of drawing each conclusion in each kind of situation where it is erroneous. These probabilities are bounded by the one-tailed levels α_1 and α_2. For instance, if $p > p_0$, the probability of concluding that $p < p_0$ depends on p but is less than α_1. There are no entries in the third column because "accepting" the null hypothesis is regarded as drawing no conclusion and hence cannot be erroneous. No matter what the true situation, a two-tailed test, with this three-conclusion interpretation, leads to an erroneous conclusion with probability at most $\alpha = \alpha_1 + \alpha_2$, the ordinary two-tailed significance level.

Now suppose we modify the test procedure so that instead of concluding that $p < p_0$ when $S \leq s_\ell$, we conclude that $p \leq p_0$. This leads to the probabilities of erroneous conclusions given in Table 4.3. No matter what the true situation, the probability of an erroneous conclusion is now at most the larger of α_1 and α_2.

For example, if the two one-tailed tests each have level 0.05, Table 4.3 shows that this two-tailed test procedure will lead to an erroneous conclusion with probability at most 0.05 (the one-tailed level). The procedure of Table 4.2 permits a more refined conclusion in one case, but at the cost of increasing the bound on the probability of an erroneous conclusion to 0.10 (the two-tailed level).

If we would be just as happy to conclude that $p \leq p_0$ as $p < p_0$, the second procedure would be much better than the first because of its much lower error

Table 4.2 Probabilities of Erroneous Conclusions

α_1 = exact level of lower-tailed test
α_2 = exact level of upper-tailed test

| Observed: | $S \leq s_\ell$ | $S \geq s_u$ | $s_\ell < S < s_u$ | |
Conclusion:	$p < p_0$	$p > p_0$	no conclusion	
True Situation				Total
$p < p_0$		$< \alpha_2$		$< \alpha_2$
$p = p_0$	α_1	α_2		$\alpha_1 + \alpha_2$
$p > p_0$	$< \alpha_1$			$< \alpha_1$

Table 4.3 Probabilities of Erroneous Conclusions

α_1 = exact level of lower-tailed test
α_2 = exact level of upper-tailed test

Observed:	$S \leq s_\ell$	$S \geq s_u$	$s_\ell < S < s_u$	
Conclusion:	$p \leq p_0$	$p > p_0$	no conclusion	
True situation				Total
$p < p_0$		$< \alpha_2$		$< \alpha_2$
$p = p_0$		α_2		α_2
$p > p_0$	$< \alpha_1$			$> \alpha_1$

rate. One might add p_0 to the upper-sided conclusion instead of, or in addition to, the lower-sided, making the first two conclusions $p < p_0$ and $p \geq p_0$, or (symmetrically) $p \leq p_0$ and $p \geq p_0$. The probability that the procedure will lead to an erroneous conclusion is still at most the larger of α_1 and α_2 in each case, although the appropriate tables will differ somewhat from Table 4.3 (Problem 24). There is also a procedure with a similar property allowing all five conclusions mentioned above (Problem 25).

The validity of Tables 4.2 and 4.3, and thus of the alternative interpretations of two-tailed tests considered here, follows from the fact that the probability that $S \geq s_u$ is larger when $p = p_0$ than when $p < p_0$, and similarly in the other tail. For all two-tailed tests which are used in practice, this fact and consequently Tables 4.2 and 4.3 remain valid when S is replaced by the test statistic and p by a suitable parameter θ. Thus the corresponding alternative interpretations of two-tailed tests are always valid in practice.

In summary, when reporting a conclusion from a two-tailed test, a conclusion at the appropriate one-tailed level may be more descriptive of the true probability of erroneous rejection than the two-tailed level, unless it is clear that rejection requires the conclusion $\theta \neq \theta_0$ or one of the conclusions $\theta < \theta_0$ and $\theta > \theta_0$, not $\theta \leq \theta_0$ or $\theta \geq \theta_0$. From this point of view, a one-tailed P-value is also more descriptive even in the two-tailed test situation. This further suggests the desirability of reporting a one-tailed P-value so that when a definite conclusion rather than a P-value is required, the choice of the two-tailed procedure which best fits the purposes and problem at hand is left to the ultimate decision-maker.

5 Randomized Test Procedures

5.1 Introduction: Motivation and Examples

We have seen that discrete distributions present difficulty for carrying out a test of hypothesis at a desired level α. The "conservative" procedure is to determine the critical value of the test statistic so that the exact level does not exceed the nominal level α. In reporting whether the observations are significant, the exact level might be stated instead of, or in addition to, the nominal level.

Consider, for example, a lower-tailed binomial test with $n = 10$, $p_0 = 0.6$ and $\alpha = 0.10$. The "conservative" procedure has critical value $s_c = 3$ with an exact level of 0.055. If the rejection region could be enlarged without increasing the exact level above 0.10, the power would increase. However, the next possibility is $s_c = 4$ and $P_{0.6}(S \leq 4) = 0.166$. If we reject when $S \leq 4$, the exact level increases to 0.166, which considerably exceeds the nominal level of 0.10. However, the conservative test is too conservative. Even if the critical value is chosen to give the exact level nearest the nominal level, the test remains the same. The exact level 0.055 is far smaller than we would like, while 0.166 is far larger. What then shall we do?

There is a definite theoretical answer to this question, but unfortunately it is not a satisfactory practical alternative because it introduces an irrelevant random variable. This is a procedure called a randomized test. Consider again the binomial problem, but now suppose that $n = 6$ and we wish to test H_0: $p \geq 0.5$ versus H_1: $p < 0.5$, at the level $\alpha = \frac{6}{64}$. (This example is used because it leads to simpler arithmetic than the previous one. The ideas are the same for both.) For $p = 0.5$, $n = 6$, the binomial probabilities are given in Table 5.1.

We naturally plan to reject when $S = 0$. If we reject when $S = 1$ as well, the exact level is $\frac{7}{64}$, larger than the value $\frac{6}{64}$ selected for α. However, if we reject only when $S = 0$, the exact level is only $\frac{1}{64}$. How can we enlarge the rejection region in order to increase the power? A good solution might appear to be to reject the null hypothesis when $S = 1$, but not when $S = 0$. This test has greater power for most values of $p < 0.5$ than the test rejecting only when $S = 0$. This is not an appropriate solution, however, since another procedure is clearly superior, as will be shown shortly.

The respective power functions for any p are

$$6p(1 - p)^5 \text{ for the test rejecting when } S = 1 \text{ only,} \qquad (5.1)$$

$$(1 - p)^6 \text{ for the test rejecting when } S = 0 \text{ only.} \qquad (5.2)$$

These two functions are graphed in Fig. 5.1. For p very small, a case where rejection is especially desirable, the power of the test rejecting when $S = 1$ only is smaller than for $S = 0$ only, and in fact decreases to zero, while the power of the test rejecting only when $S = 0$ increases to one. From (5.1) and (5.2), the power of the $S = 1$ only test is greater than the power of the $S = 0$ only test for all $p > \frac{1}{7}$ (Problem 26). A test which is a combination of these two tests and

Table 5.1 Binomial Probabilities for $p = 0.5$, $n = 6$

s	0	1	2	3	4	5	6
$P_{0.5}(S = s)$	$\frac{1}{64}$	$\frac{6}{64}$	$\frac{15}{64}$	$\frac{20}{64}$	$\frac{15}{64}$	$\frac{6}{64}$	$\frac{1}{64}$

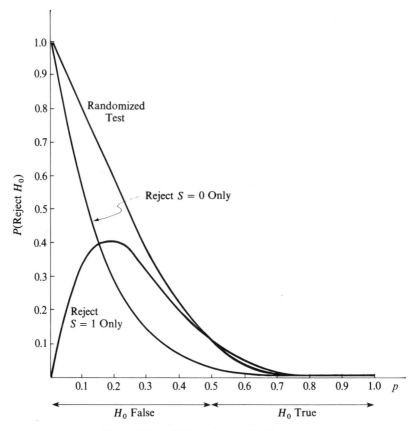

Figure 5.1 $P(\text{Reject } H_0: p \geq 0.5)$ for $n = 6$.

has power everywhere greater than either of them would be desirable; this can be accomplished using a *randomized test procedure*.

Specifically, consider a test which rejects when $S = 0$ is observed, rejects with probability $\frac{5}{6}$ when $S = 1$ is observed, and "accepts" otherwise. For instance, when $S = 1$ we might roll a fair die, reject if the number of spots is less than 6, and "accept" if it is equal to 6. This procedure makes the probability of a Type I error when $p = 0.5$ equal to

$$(1)P_{0.5}(S = 0) + (\tfrac{5}{6})P_{0.5}(S = 1) = (\tfrac{1}{64}) + (\tfrac{5}{6})(\tfrac{6}{64}) = \tfrac{6}{64},$$

which is exactly the desired value.

For any p, the probability of rejection by this randomized test is

$$(1)(1 - p)^6 + (\tfrac{5}{6})p(1 - p)^5.$$

This function is also plotted in Fig. 5.1 for all p. The figure shows that this probability is smaller than $\frac{6}{64}$ for all $p > 0.5$. It also shows that the randomized test has power everywhere greater than either of the other two tests, and has smaller probability of a Type I error than the test which rejects when $S = 1$ only, except at $p = 0.5$ where the probabilities are the same. Later we will show that similar properties hold more generally.

Whether or not one would ever use a randomized test in practice, the fact that there is a randomized test everywhere better than the test rejecting when $S = 1$ only shows that the latter test should not be used. In the next subsection, we will explain what is meant by randomized procedures generally and why it is useful to talk about them even though no claim is made that people do or should carry out irrelevant randomizations.

5.2 Randomized Tests: Definitions

The basic idea behind any *randomized procedure* is that we decide what action to take, or what inference to make, or how to report the results, not only on the basis of an observed random variable X as previously, but also at least in part on the basis of some irrelevant random experiment. When we observe $X = x$, we may reject the null hypothesis, or we may not. We may even decide at random what to do. That is, we may reject the null hypothesis with probability $\phi(x)$, say, and "accept" it otherwise, where $\phi(x)$ may be any value, $0 \leq \phi(x) \leq 1$. This kind of procedure is called a randomized test, and $\phi(x)$ is its critical function, as already defined in Sect. 4.1. If such a test were carried out repeatedly, in the long run the null hypothesis would be rejected in a proportion $\phi(x)$ of those cases in which the observation is x. The randomized test discussed in Sect. 5.1 which rejected always when $S = 0$ and with probability $\frac{5}{6}$ when $S = 1$, and "excepted" otherwise, is given by $\phi(0) = 1$, $\phi(1) = \frac{5}{6}$, and $\phi(s) = 0$ for $s \geq 2$.

Ordinary (nonrandomized) tests are equivalent to randomized tests for which $\phi(x)$ takes on only the values 0 and 1. Specifically, the nonrandomized test with rejection region R is given by $\phi(x) = 1$ for x in a region R and $\phi(x) = 0$ otherwise. Thus we reject the null hypothesis with probability 1 for all $x \in R$ and we "accept" it with probability 1 for all $x \notin R$.

The randomization necessary to perform a randomized test could be carried out by drawing a random variable U (independent of X) from the uniform distribution between 0 and 1 and rejecting the null hypothesis if $U \leq \phi(x)$ but not otherwise. Such a U may be obtained, to any desired accuracy, from a table of random digits or generated by a computer. Since the event $U \leq \phi(x)$ has probability $\phi(x)$, this procedure rejects the null hypothesis with probability $\phi(x)$ when x is observed. We note incidentally that this makes any randomized test based on X equivalent to a nonrandomized test based on (X, U). Instead of carrying out the randomization, one might report the value of $\phi(x)$ for the x observed.

In the general binomial problem, suppose that S is the number of successes in n independent Bernoulli trials with probability p of success on each trial.

A *lower-tailed randomized test* is to reject always if $S < s_\ell$ and reject with probability ϕ_ℓ if $S = s_\ell$. The exact level of such a test for $H_0: p = p_0$ is

$$\alpha_1 = P_{p_0}(S < s_\ell) + \phi_\ell P_{p_0}(S = s_\ell). \tag{5.3}$$

For any $\alpha_1 (0 \leq \alpha_1 \leq 1)$ there is (Problem 28) exactly one lower-tailed randomized test with exact level α_1. (All tests with the same critical function are considered the same, since they can differ only in the method of carrying out the randomization.) Similarly, an *upper-tailed randomized test* is to reject always if $S > s_u$ and to reject with probability ϕ_u if $S = s_u$; its exact level is

$$\alpha_2 = P_{p_0}(S > s_u) + \phi_u P_{p_0}(S = s_u), \tag{5.4}$$

and α_2 determines the test (critical function) uniquely.

A *two-tailed randomized test* is the combination of an upper-tailed and a lower-tailed randomized test and rejects if either of the one-tailed tests rejects but not otherwise. That is, a two-tailed randomized test is to reject always if $S < s_\ell$ or $S > s_u$, reject with probability ϕ_ℓ if $S = s_\ell$ and with probability ϕ_u if $S = s_u$, and "accept" otherwise.[3] The exact level α of such a test for H_0: $p = p_0$ is the sum of the exact levels α_1 and α_2 above, or

$$\alpha = P_{p_0}(S < s_\ell) + \phi_\ell P_{p_0}(S = s_\ell) + P_{p_0}(S > s_u) + \phi_u P_{p_0}(S = s_u). \tag{5.5}$$

There is an infinite number of two-tailed tests at a given exact level α. For each $\alpha_1, 0 \leq \alpha_1 \leq \alpha$, there is one given by the lower-tailed test at exact level α_1 and the upper-tailed test at exact level $\alpha_2 = \alpha - \alpha_1$. The difficulty of choosing among them was pointed out in Section 4.5.

5.3 Nonrandomized Tests Equivalent to Randomized Tests

Some nonrandomized tests amount in a sense to randomized tests. In our numerical example where $n = 6$, $p_0 = 0.5$, $\alpha = \frac{6}{64}$, suppose now that the original Bernoulli trials are distinguishable (say by the order in which they were made) and that the individual outcomes are available. Consider the test that rejects the null hypothesis if there is no success or if there is just one success and it occurs after the first trial. This test has level exactly $\frac{6}{64}$, and in fact it has the same conditional probability of rejection given S (and hence the same unconditional probability of rejection) as the randomized test above, for every p (Problem 29). This relates to the fact (see below) that S is a sufficient statistic. Even though this test is nonrandomized with respect to the original observations, its behavior for all p can be duplicated by a randomized test based on S, and hence it is just as reasonable or unreasonable. Rejecting if there is a single success and it is not on the first trial amounts to rejecting with probability $\frac{5}{6}$ whenever there is a single success, where the trial number of the success is used to carry out the randomization.

---

[3] We assume that $s_\ell \leq s_u$ and $\phi_\ell + \phi_u \leq 1$ if $s_\ell = s_u$ so that the upper and lower tails are mutually exclusive.

This type of duplication was discussed further in connection with sufficiency in Sect. 3.3. We showed there how any procedure based on the original observations can be duplicated by a procedure based on the sufficient statistic S. Randomized procedures may be required, however. No nonrandomized procedure based on S duplicates the foregoing nonrandomized procedure based on the original observations and having level exactly $\frac{6}{64}$. If randomized procedures were not permitted even theoretically, it might be necessary to look at the original observation rather than the sufficient statistic, not because the original observations provide more information, but simply because their distribution is finer and allows randomization to be smuggled in.

5.4 Usefulness of Randomized Tests in Theory and Practice

Randomization removes all theoretical problems stemming from discreteness. It permits tests to be made at any arbitrary level α. Suppose we want a lower-tailed test at exact level 0.01 in the binomial problem with $n = 6$ and $p_0 = 0.5$. Then we should reject with probability 0.64 when $S = 0$. The level of a nonrandomized test that rejects for $S = 0$ is $\frac{1}{64}$. The only nonrandomized procedure at a level less than $\frac{1}{64}$ is that which never rejects the null hypothesis. This is true even if the original observations are available.

"Optimum" tests in discrete problems are usually randomized. For example, the randomized test in Sect. 5.3 at level $\frac{6}{64}$ is the best test at its level, in the sense that for every $p < 0.5$, its power is at least as great as that of any other test, randomized or not, based on the same observations. (This will be shown in Sect. 7). Accordingly, if we do not want to use randomized tests, the test used might be regarded as an approximation to this best randomized test. Then we will presumably choose between rejecting only for $S = 0$, giving exact level $\frac{1}{64}$, and rejecting for $S = 0$ or 1, giving much greater power but exact level $\frac{7}{64}$, which is larger than we wanted. In any case, only one-tailed tests need be considered, and not, for example, the test rejecting only when $S = 1$ but not when $S = 0$. Without randomized tests, this is much more difficult to see.

One can argue that there is certainly no harm in introducing randomized procedures. If they are inferior, the theory should show that. They serve many useful purposes, as we have already seen. They show clearly that some procedures should be excluded from consideration. They justify restricting consideration to procedures which are based on a sufficient statistic, and they clarify the nature of procedures which are not (such as the one that rejects for no successes or one success not on the first trial). They permit a simple and reasonable formulation of theoretical problems (such as "What is the best test at level $\frac{6}{64}$ of the null hypothesis $p \geq 0.5$?"). The solutions of these theoretical problems provide insight about the choice of procedures, and about the limitations of this kind of theory for choosing procedures. Further, randomized procedures are useful for power comparisons of tests which cannot be performed at the same level using nonrandomized tests. As mentioned in Sect. 4.2, power comparisons of tests may be uninterpretable, or at least misleading, unless the exact levels are the same.

The problem of what to do in practice still remains. It does not seem that one should be required to use a randomized procedure in a practical statistical problem, although if two procedures are equally good one might not object to letting an irrelevant randomization choose between them. (Reasons can be given for requiring randomized procedures, for example, in a game against an opponent or in writing quality control contracts, where a specific α may be desirable. However, these reasons do not apply to ordinary statistical problems.) Even when only a randomized test will provide the desired level, the authors do not recommend that a randomized procedure be actually performed. To do so without explanation would be a deception, and once it is explained, any reader can easily carry it out if he thinks it useful. The process of performing a randomization does not add to the information provided by the observations.

In the case of one-tailed tests, where the P-value is uniquely defined, following the procedure of reporting the P-value essentially eliminates the problem, although it remains under the rug in that a discrete P-value is implicitly conservative. For two-tailed tests, the problems discussed in Sect. 4.5 are relevant. Reporting the one-tailed P-value, but clearly labeled as such, is probably the simplest solution in practice. In either case, one might also report the next P-value, the tail probability beyond the observed value.

In the binomial example above with $n = 6$ and $H_0: p \geq 0.5$ (Table 5.1), suppose that $S = 1$. The one-tailed test results could be reported by giving the exact P-value $\frac{7}{64} = 0.109$ and perhaps also the next P-value $\frac{1}{64} = 0.016$. That is, one could report significance at the exact level 0.109 and perhaps also that the next more extreme result would have been significant at the exact level 0.016. If the test were two-tailed for the null hypothesis $p = 0.5$ and the alternative $p \neq 0.5$, the same P-values could be reported but labeled as one-tailed.

*5.5 Randomized P-values

Extending the earlier definition of the P-value to randomized tests gives what we shall call the *randomized P-value*, which is uniformly distributed between the exact P-value and the next P-value (Problem 32). If the P-value is to measure the extent to which the data support the null hypothesis (see Sect. 4.4 for difficulties with this interpretation), a single number is presumably desired for each possible outcome. The mid-P-value is suggested by the fact that the distribution of the randomized P-value is symmetric about it (and in particular has mean and median equal to it). The observations are significant at nominal levels above this value and not significant at nominal levels below it if the test chosen at nominal level α is the nonrandomized test having greatest probability of agreeing with the randomized test at exact level α, or, as mentioned in Sect. 4.4, that with exact level nearest to the nominal level.

This does not alter the recommendation made above to report the exact P-value as well as the next P-value. Anyone who thinks that the mid-P-value is of special interest can then compute it.*

6 Confidence Regions

We now introduce the concept of confidence regions. This form of inference, like estimation, refers to any parameter value, not just a preselected one, yet also, like a significance test, provides an exact statement of error probability. We shall lead up to confidence regions by way of tests to remove the mystery of their construction and to emphasize the intimate relationship between the two concepts.

6.1 Definition and Construction in the Binomial Case

Consider again the situation of n Bernoulli trials with unknown probability p of success on each trial. We have seen how to perform a one- or two-sided test at level α for the simple null hypothesis $H_0: p = p_0$. Suppose now that we have no reason to choose a particular value of p_0 to be tested in H_0. Then we might prefer to determine for each value p_0 whether or not the null hypothesis $p = p_0$ would be rejected. This procedure will give the set of values of p_0 which would be "accepted" if a (nonrandomized) one- or two-sided test were performed at fixed level α. For standard binomial tests this set turns out to be an interval of values for p_0 (which may extend to 0 or 1). In terminology to be introduced formally later, this interval is called a confidence region or interval with confidence level $1 - \alpha$.

Suppose, for example, that $n = 10$ and $\alpha = 0.05$. If $S = 3$ successes are observed, a lower-tailed test of $H_0: p = p_0$ would call for rejection of H_0 if $P_{p_0}(S \leq 3) \leq 0.05$ and not otherwise. We find (for instance, by interpolation in extensive tables of the binomial distribution such as those published by the National Bureau of Standards [1949] or Harvard University [1955]) that

$$P_{0.607}(S \leq 3) = 0.05$$

$$P_{p_0}(S \leq 3) < 0.05 \quad \text{for} \quad p_0 > 0.607$$

$$P_{p_0}(S \leq 3) > 0.05 \quad \text{for} \quad p_0 < 0.607.$$

The decision would therefore be to reject $H_0: p = p_0$ for all $p_0 \geq 0.607$ and to "accept" it for all $p_0 < 0.607$.

More generally, if we observe s successes in n trials, the null hypothesis $p = p_0$ would be rejected by a lower-tailed test at the level α if the lower tail probability satisfies $P_{p_0}(S \leq s) \leq \alpha$, and would be "accepted" otherwise.

Since, for given s, this probability is a strictly decreasing function of p_0 (Problem 13) there exists a value p_u, which depends on s, such that

$$P_{p_0}(S \leq s) = \alpha \quad \text{for } p_0 = p_u \tag{6.1}$$

$$P_{p_0}(S \leq s) < \alpha \quad \text{for } p_0 > p_u \tag{6.2}$$

$$P_{p_0}(S \leq s) > \alpha \quad \text{for } p_0 < p_u. \tag{6.3}$$

The lower-tailed test at level α would then reject $H_0: p = p_0$ for all $p_0 \geq p_u$, and would "accept" it for all $p_0 < p_u$.

We now change the emphasis slightly. Suppose that $p = p_0$. Notice that p_u is a random variable; it is a function of the random variable S, and is sometimes written $p_u(S)$ to emphasize this fact. Since $p_u \leq p_0$ if and only if $H_0: p = p_0$ is rejected at level α, the probability that $p_u \leq p_0$ is the probability of rejection, which is at most α. Similarly, the probability of the complementary event $p_u > p_0$ is at least $1 - \alpha$. These statements hold for any value of p_0. In short,

$$P_{p_0}[p_u(S) \geq p_0] \geq 1 - \alpha \quad \text{for all } p_0. \tag{6.4}$$

When this property holds, we call the random variable $p_u(S)$ an *upper confidence limit* (or *upper confidence bound*) for p at confidence level $1 - \alpha$. This limit was obtained from a *lower*-tailed test. The confidence property says that, whatever the true value p_0 may be, before S is observed, the probability is at least $1 - \alpha$ that the upper confidence limit $p_u(S)$ will be at least p_0.

Similarly, a *lower confidence limit* (or *lower confidence bound*) for p at confidence level $1 - \alpha$ is defined as a random variable $p_\ell(S)$ such that

$$P_{p_0}[p_\ell(S) \leq p_0] \geq 1 - \alpha \quad \text{for all } p_0. \tag{6.5}$$

This lower confidence limit corresponds to an *upper*-tailed test at level α.

For any S, a lower confidence limit and an upper confidence limit for p, each at level $1 - \alpha$, when taken together, are said to form a two-sided *confidence interval* for p at confidence level $1 - 2\alpha$ because if $p_\ell(S)$ and $p_u(S)$ satisfy (6.5) and (6.4) respectively, then

$$P_{p_0}[p_\ell(S) \leq p_0 \leq p_u(S)] \geq 1 - 2\alpha \quad \text{for all } p_0. \tag{6.6}$$

As an example, Table 6.1 shows the lower and upper 90% confidence limits for p, corresponding to upper-tailed and lower-tailed tests respectively, each at

Table 6.1 Lower and Upper 90% Binomial Confidence Limits for $n = 5$

s	0	1	2	3	4	5
$p_\ell(s)$	0.000	0.021	0.112	0.247	0.416	0.631
$p_u(s)$	0.369	0.584	0.753	0.888	0.979	1.000

level 0.10, for $n = 5$ (Problem 35). The upper confidence limits are found following a procedure analogous to (6.1)–(6.3), for each possible value of S. The lower limits are found similarly. Notice that half of the values in Table 6.1 can be obtained by subtraction, since $p_\ell(s) = 1 - p_u(n - s)$ for any s (Problem 37). This example will be discussed further in Section 6.4.

The results in Table 6.1 are plotted as points on two curves in Fig. 6.1. The principle of construction for this graph can be extended to any sample size; this has been done by Clopper and Pearson [1934] to produce the well-known Clopper–Pearson charts. The chart for confidence level 0.95 is reproduced as Fig. 6.2 in order to illustrate the format. These charts provide a convenient method for finding upper, lower, or two-sided confidence limits in binomial problems. They also provide a graphic version of the foregoing derivation. Consider the region between the curves for a given n. The horizontal sections are the values of s/n for which each given p would be "accepted." The vertical sections are the values of p which would be "accepted" for a given s/n. The vertical section corresponding to the observed s/n covers the true p if and only if the horizontal section corresponding to the true p covers the observed s/n. The relations between tests and confidence regions and between their error probabilities follow.

Though slightly less convenient than graphs, Table C is compact and allows greater accuracy in finding binomial confidence limits. It includes five common levels α in the range $0.005 \le \alpha \le 0.100$. For each s, for $s/n < 0.50$, the tabulated values are n times the confidence limits, and hence are simply divided by n to obtain the confidence limits. For $s/n > 0.50$, Table C is entered with $1 - (s/n)$ and lower and upper are interchanged; the corresponding entries are then divided by n and subtracted from 1 to find the confidence limits. The values $s = 0$ and $s = n$ are special cases, as explained in the table. To illustrate the use of Table C, consider $n = 5$ and $s = 2$. Then $s/n = 0.40$, and the table entries for $\alpha = 0.10$ are 0.561 and 3.77. These numbers are

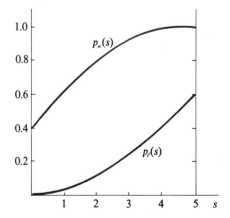

Figure 6.1   Lower and upper 90% confidence limits for p when $n = 5$.

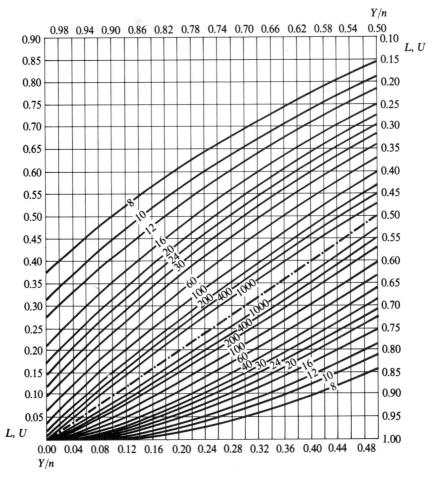

Figure 6.2   Chart providing confidence limits for p in binomial sampling, given a sample fraction Y/n, confidence coefficient, $1 - 2\alpha = 0.95$. The numbers printed along the curves indicate the sample size n. If for a given value of the abscissa Y/n, L, and U are the ordinates read from (or interpolated between) the appropriate lower and upper curves, then $P(L \leq p \leq U) \geq 1 - 2\alpha$. (Adapted from Table 41, pp. 204–205, of E. S. Pearson and H. O. Hartley, Eds. (1962), *Biometrika Tables for Statisticians*, Vol. 1, Cambridge University Press, Cambridge, with permission of the Biometrika Trustees.)

divided by 5 to get 0.112 and 0.754 as lower and upper limits, which agree (except for rounding) with Table 6.1.

Binomial confidence limits are quantiles of the beta distribution. They are also ratios of linear functions of quantiles of the F distribution. The approximation at the end of Table C results from a transformation of the F distribution derived from the cube root transformation of the chi-square distribution (see Wilson and Hilferty [1931], Paulson [1942], Camp [1951] and Pratt [1968]).

6.2 Definition of Confidence Regions and Relationship to Tests in the General Case

The concept and derivation of confidence limits explained above for binomial problems generalize directly to confidence regions for any parameter θ in any situation. Suppose that we have a test based on any random variable S of the null hypothesis $\theta = \theta_0$, for each value of θ_0 and level α. For any fixed θ_0 and α, the test will reject for certain values of S and not for others. The set of values of S for which the null hypothesis $\theta = \theta_0$ would be "accepted" will be denoted by $A(\theta_0)$. Once $S = s$ is observed, we are interested in the converse question: for which values of θ_0 would the null hypothesis $\theta = \theta_0$ be "accepted?" The set of all such values of θ_0 is a region $C(s)$, defined by

$$\theta_0 \in C(s) \quad \text{if and only if} \quad s \in A(\theta_0). \tag{6.7}$$

If the test of the null hypothesis $\theta = \theta_0$ has level α, then the probability of "accepting" $\theta = \theta_0$ when it is true is at least $1 - \alpha$, that is

$$P_{\theta_0}[S \in A(\theta_0)] \geq 1 - \alpha. \tag{6.8}$$

(If $\theta = \theta_0$ allows more than one distribution of S, this holds for all of them, and similarly hereafter.) But the event $S \in A(\theta_0)$ is, by the definition (6.7) of C, equivalent to the event $\theta_0 \in C(S)$. Substituting this equivalence in (6.8), we have

$$P_{\theta_0}[\theta_0 \in C(S)] \geq 1 - \alpha. \tag{6.9}$$

If, for each θ_0, the test of the null hypothesis $\theta = \theta_0$ is a test at level α, then (6.9) holds for every θ_0. This is the defining condition for a confidence region $C(S)$ at confidence level $1 - \alpha$. We see that a test at level α for each θ_0 leads immediately to a corresponding *confidence region* at *confidence level* $1 - \alpha$. The left-hand side of (6.9) is called the *true confidence level* and is discussed in Sect. 6.4. The *exact confidence level* is the maximum value of $1 - \alpha$ such that (6.9) holds for all possible distributions of S. This is the minimum (or infimum) of the left-hand side, the probability that $C(S)$ includes the true parameter value, over all possible distributions. *Nominal* and *conservative* levels are defined in the obvious way.

Conversely, if one has a confidence region $C(S)$ at confidence level $1 - \alpha$, then for each θ_0 a test of the null hypothesis $\theta = \theta_0$ at level α may be performed by "accepting" if the confidence region $C(S)$ includes θ_0 and rejecting otherwise. This is equivalent to defining the "acceptance" region $A(\theta_0)$ by (6.7), "accepting" the null hypothesis $\theta = \theta_0$ if $S \in A(\theta_0)$ and rejecting it otherwise.

Thus there is an exact correspondence between a confidence region for θ at confidence level $1 - \alpha$ and a family of tests of null hypotheses $\theta = \theta_0$, each test at significance level α. It is conventional, if not particularly convenient, to measure significance levels as (Type I) error probabilities and confidence levels as 1 minus the error probability. Typical significance levels are 0.10, 0.05, 0.01, and the corresponding typical confidence levels are 0.90, 0.95, 0.99. We generally adhere to this convention, although context would determine the meaning anyway. Notice that the definition of a confidence region, that (6.9) holds for all θ_0, makes no reference to hypothesis testing. Nevertheless, the relationship is so intimate that it should never be lost sight of.

For the parameter p of the binomial distribution, the confidence regions defined above agree with the confidence limits derived in Sect. 6.1. Specifically, the confidence region corresponding to the family of lower-tailed tests at level α is the interval $p < p_u$, where p_u is the upper confidence limit for p at level $1 - \alpha$, and similarly for upper-tailed and two-tailed tests. Verification beyond what has already been given is left to the reader (Problem 40). Some confidence procedures for the binomial parameter which correspond to other two-tailed tests are discussed later, in Sect. 8.2.

6.3 Interpretation of Confidence Regions

The random region $C(S)$ must be distinguished from the particular region $C(s)$ obtained for the particular value s of S. $C(S)$ is a random variable, because it is a function of a random variable. The distinction between $C(S)$ and $C(s)$ is another case of the distinction between a random variable and its value, like that between estimator and estimate in Sect. 3. Unfortunately, the term confidence region is standard for both the random variable and its value, and it must be determined from the context which is meant. Intuitively, a confidence region is random before the observations are made, but not afterwards.

If $C(S)$ is a confidence region at confidence level $1 - \alpha$, then, by definition, the event $\theta_0 \in C(s)$ has probability at least $1 - \alpha$ whatever the true value θ_0. However, once a value $S = s$ is observed, the probability that the true value θ_0 lies in $C(s)$ is not necessarily $1 - \alpha$. In the "frequentist" framework of probability, one can say only that the probability that the true value lies in $C(s)$ is unknown, but must be either 0 or 1. (A similar point was made in Sect. 4.2 in connection with hypothesis tests.) The confidence level is a property of the confidence procedure, not of the confidence region for a particular outcome.

One interpretation of a confidence region is as those values which cannot be rejected, with a testing interpretation of rejection. Some people are careful to limit themselves to this interpretation. Most, however, probably go somewhat further in allowing the connotations of the word "confidence" to enter their thinking, without necessarily claiming any very strong justification for doing so,

The Bayesian framework sometimes justifies confidence in observed confidence regions. When probabilities are used in a Bayesian framework to represent "degrees of belief" (see Sect. 4.2), prior belief may have much less influence than the observations. In this case, if an "appropriate" confidence procedure is used, then the probability, after the observations are made, that the true value is included in the confidence region is approximately equal to the confidence level $1 - \alpha$. (Here the true value is the random variable, while the confidence region is fixed by the observations.) A rather similar statement is that, in the absence of any information other than that provided by the data, after seeing the observations and computing a confidence region at level $1 - \alpha$, approximately fair odds for a bet that the region includes the true value would be $1 - \alpha$ to α. This statement can be made precise enough to judge its truth only by using probabilities for degrees of belief. "Absence of information" is especially difficult to represent mathematically.

The defining property of a confidence procedure does not in itself justify any such interpretation after the observations have been made, and indeed there are confidence procedures for which such interpretations are demonstrably unreasonable. However, when little information is available a priori, we can say that such interpretations "generally" hold "approximately" for "appropriate" procedures in most problems. Regardless of statistical philosophy, one would not ordinarily make a confidence statement if one had reason to doubt the appropriateness of the confidence level as an approximate measure of the confidence to be placed in it after seeing the observations.

The foregoing discussion provides a very natural "approximately Bayesian" way to interpret confidence regions. No such simple interpretation of tests or P-values is possible in general. Further, a confidence region conveys greater information than a test, since it is equivalent to a family of tests. These two facts make the reporting of confidence regions a desirable alternative to, or addition to, reporting test results whenever possible. Typically, a confidence region for a parameter is similar to an estimate plus or minus a measure of uncertainty, but it is statistically more refined. It provides insight into both the magnitude of the parameter and the reliability of the available information about it, measured in terms of precise, readily apprehended statistical properties. On the other hand, the most conclusive result possible from a test is rejection of a possibly "straw man" null hypothesis.

One caveat we must consider is that if the probability that $\theta_0 \in C(S)$, say $1 - \alpha(\theta_0)$, is not a constant $1 - \alpha$ for all θ_0, the interpretations above may need modification. In particular, they will more nearly hold for some kind of an average of $\alpha(\theta_0)$ than for a conservative nominal level or the exact level defined above as one minus the maximum of $\alpha(\theta_0)$. This distinction is not always negligible in practice. For instance, Fig. 6.3 shows, for $n = 5$, the probability $1 - \alpha(p_0)$ that the ordinary two-sided binomial confidence interval at nominal level 0.80 will include the true value p_0, as a function of p_0. The exact level is 0.83, but $1 - \alpha(p_0)$ is above 0.90 over most of the range of p_0, so both 0.80 and 0.83 seem to understate considerably the confidence one can have in the interval. We use a small n and large α here for simplicity, but the error rate would be overstated by a similar factor for smaller α and larger n.

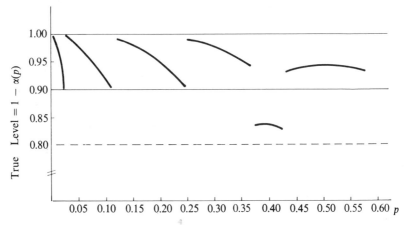

Figure 6.3   True confidence level for 80% confidence intervals for p when $n = 5$.

6.4 True Confidence Level

It is instructive to examine in some detail the calculation of the probability of
including the true p as a function of p in a binomial example. Recall that the
values given in Table 6.1 for lower and upper 90% binomial confidence limits
on p were each constructed to correspond to a one-tailed test at nominal level
0.10. The exact level of the test is a function of p. Hence so is the true level of the
confidence procedure, where, for any value of p, the true confidence level here
means the probability (computed under p) that the confidence region will
include p. For instance, looking at Table 6.1, we see that for $p = 0.4$, the two-
sided confidence interval will include p if $S = 1, 2$ or 3, but not otherwise. The
true confidence level then is the sum

$$\sum_{s=1}^{3} P_{0.4}(S = s) = P_{0.4}(S \leq 3) - P_{0.4}(S = 0) = 0.8352$$

from Table B. This can be calculated for any p, and Fig. 6.3 shows a graph of
the result.

Numerical values for selected p are shown in Table 6.2 under the heading
True Level. The true, two-sided confidence level can also be computed in two
pieces as follows. The probability that the interval includes p is one minus the
probability that it does not include p, and the latter will occur if and only if p is
either (a) smaller than the lower confidence limit, or (b) larger than the upper
confidence limit. But p is smaller than the lower limit if and only if, for that p,
the null hypothesis would be rejected by an upper-tailed test. This occurs if the
observed s satisfies $P_p(S \geq s) \leq 0.10$, that is, if s is in the upper 0.10 tail of the
distribution corresponding to p. Hence the probability of (a) is the largest
upper tail probability, not exceeding 0.10, of the distribution corresponding
to p. For $p = 0.4, n = 5$, for instance, we see in Table 6.1 that p is smaller than

Table 6.2 True Confidence Level for Confidence Intervals Corresponding to Equal-Tailed Binomial Tests at Nominal Level 0.20 when $n = 5$

p	0	$.021^-$	$.021^+$	.050	.100	$.112^-$	$.112^+$	.200	$.247^-$
Lower[a]	0.000	.000	.000	.000	.000	.000	.000	.000	.000
Upper[b]	0.000	.100	.004	.023	.081	.100	.012	.058	.100
Sum	0.000	.100	.004	.023	.081	.100	.012	.058	.100
True Level	1.000	.900	.996	.977	.919	.900	.988	.942	.900

p	0.247^+	.300	$.369^-$	$.369^+$	.400	$.416^-$	$.416^+$	.450	.500
Lower[a]	0.000	.000	.000	.100	.078	.069	.069	.050	.031
Upper[b]	0.015	.031	.065	.065	.087	.100	.012	.018	.031
Sum	0.015	.031	.065	.165	.165	.169	.081	.068	.062
True Level	0.985	.969	.935	.835	.835	.831	.919	.932	.938

[a] Probability upper confidence limit is smaller than p (equals largest lower tail probability not exceeding 0.10).
[b] Probability lower confidence limit is larger than p (equals largest upper tail probability not exceeding 0.10).

the lower confidence limit if and only if the observed s is 4 or 5, which has probability $P_{0.4}(S \geq 4) = 1 - 0.9130 = 0.0870$. This is indeed the largest upper binomial tail probability for $p = 0.4$ which does not exceed 0.10. Similarly, the probability of (b) is the largest lower binomial tail probability which does not exceed 0.10. For $p = 0.4$, this is $P_{0.4}(S \leq 0) = 0.0778$. Since (a) and (b) are mutually exclusive events, the true two-sided confidence level is $1 - (0.0870 + 0.0778) = 0.8352$, as found in the previous paragraph. Similar results for other values of $p \leq 0.5$ are also shown in Table 6.2 under the headings Upper and Lower respectively. The entries for values of p not included in Table B are found in the same way, but using more extensive tables of the binomial distribution or interpolation or a computer. Since the true level is not a continuous function of p, it is necessary to give special attention to the points of discontinuity, which are the possible confidence limits. A similar table for $p > 0.5$ can be obtained by changing the label p to $1 - p$ and interchanging the labels Lower and Upper in Table 6.2.

6.5 Including False Values and the Size of Confidence Regions

In this subsection we show that if a family of tests is powerful, the corresponding confidence regions will have high probability of excluding false values, and will also be small according to natural measures of size.[4] The exact sense in

[4] As used here, the word "size" is to be interpreted in a nontechnical sense, not to be confused with the previous technical definition as the level of a test. Since the technical term size is not used in this book, no difficulty in interpretation should occur.

which this is true is complicated by the fact that different tests may be powerful against different alternatives, and, correspondingly, different confidence regions have high probability of excluding false values and are of small size under different hypotheses. These ideas will be discussed briefly below. Further explanation is given by Pratt [1961] and Madansky [1962].

Suppose $C(s)$ is a confidence procedure corresponding to a family of tests with "acceptance" regions $A(\theta_0)$. Since $\theta_0 \in C(S)$ if and only if $S \in A(\theta_0)$, we have

$$P_{\theta_1}[\theta_0 \in C(S)] = P_{\theta_1}[S \in A(\theta_0)] \quad \text{for all } \theta_1. \tag{6.10}$$

For $\theta_1 = \theta_0$, this says that the probability of "accepting" the null hypothesis $\theta = \theta_0$ when it is true is equal to the probability that the confidence region will include the value θ_0 when θ_0 is the true value. This result has already been used in demonstrating that $1 - \alpha$ is the confidence level if all the tests have level α (establishing (6.9) from (6.8)). For $\theta_1 \neq \theta_0$, (6.10) says that the probability of a Type II error at θ_1 for the test of the null hypothesis $\theta = \theta_0$ equals the probability, when θ_1 is the true value, that the confidence region will include the *false* value θ_0. Thus, the ability of a confidence procedure to exclude false values is the same as the power of the corresponding tests, in this specific sense. This also leads to a correspondence between certain "optimum" properties of tests and confidence procedures. (See Sect. 7.3).

We consider next the *size* of a region, defined as its length (if it is an interval), its area (if it is a region in the plane), and in general its k-dimensional volume (if it is a region in k-dimensional space). It is perhaps more reasonable to be concerned with the probability of including false values than the size of a confidence region, since there is no merit in a small region if it is not even close to the true value. At the same time, it is natural to feel that a good confidence procedure would produce small regions. There turns out to be a direct connection between the size of the confidence region and the probability of including false values which implies that making either one small will tend to make the other small.

Since a confidence procedure does not ordinarily give a region of fixed size, let us consider the expected size of the confidence region. Consider, for instance, a two-sided confidence interval $[\theta_\ell, \theta_u]$ for a one-dimensional real parameter θ. The size is the length $\theta_u - \theta_\ell$, and the expected length is $E_{\theta_1}[\theta_u(S) - \theta_\ell(S)]$ if θ_1 is the true value. By Problem 41 we have

$$E_{\theta_1}[\theta_u(S) - \theta_\ell(S)] = \int P_{\theta_1}[\theta_\ell(S) \leq \theta \leq \theta_u(S)]d\theta \tag{6.11}$$

and the integral is unchanged if the integration is over all θ except the true value. In other words, the expected length is the integral of the probability of including false values. This is true of size generally (Problem 42). The essential condition is that the expected size and the probability of inclusion must be computed under the same distribution.

Similarly, suppose we are concerned only with the upper confidence bound

θ_u. Then we do not mind including false values smaller than the true value θ_1, but prefer to exclude those greater than θ_1. That is, it does not matter if $\theta_u > \theta$ for $\theta < \theta_1$, but we would like the probability that $\theta_u > \theta$ to be small for $\theta > \theta_1$. We would also like to overestimate θ_1 by as little as possible. In thise case the role of size is played by the "*excess*," defined as $\theta_u - \theta_1$ for $\theta_u > \theta_1$ and 0 for $\theta_u \leq \theta_1$. In particular, the expected "excess" equals the integral over all $\theta > \theta_1$ of the probability that θ_u exceeds θ. When this probability is small, the expected "excess is small, and conversely."

The foregoing statements imply that when confidence regions correspond to powerful tests, the confidence regions will have, first, high probability of excluding false values and, second, small expected size. The exact relationship between the properties of confidence regions and the relative emphasis placed on various alternatives in seeking powerful tests is subtle. It is common practice to choose tests which have desirable power functions without reference to confidence properties, and to use the corresponding confidence regions without investigating the confidence properties. The remarks above suggest that this practice will provide good confidence regions.

*6.6 Randomized Confidence Regions

In principle, application of the theory of randomized tests to confidence regions presents no difficulty. Given the observations, the probability that any point belongs to the confidence region equals the probability that the randomized test for the corresponding null hypothesis "accepts" it. For example, a lower-tailed randomized test at exact level $\frac{6}{64}$ in a binomial problem with 6 observations "accepts" the null hypothesis $p = 0.5$ with probability $\frac{1}{6}$ if one success occurs and always if more than one success occurs (see Sect. 5.1). The corresponding probability is easily computed for all values of p (see Problem 44). The upper confidence bound could then be reported in the form of a graph (for the particular outcome observed) showing the probability of "acceptance" for various values of p, according to a randomized test. This graph is monotonically decreasing (Problem 45). (A nonrandomized conservative upper confidence bound would be the point at which the graph reaches 0. The point at which the graph crosses 0.5 is the upper confidence bound corresponding to the nonrandomized test with exact level nearest $\frac{6}{64}$). To carry out the randomization, one could choose U uniformly distributed between 0 and 1 and take as the upper confidence bound the point at which the graph reaches the value U. For one-sided binomial tests at the same exact level for all p, this is always possible because the graph is monotonic. For two-sided randomized tests with maximum average power, it is not possible to choose the corresponding confidence region so that it is always an interval. (See also Pratt [1961].) In any case, the defining property of a confidence region could be satisfied by carrying out a separate randomization for each p, but the resulting region would be literally an unimaginable hodgepodge.

In practice, randomized confidence procedures are not used. Actually doing the randomization is, if anything, intuitively even less desirable here than in the context of hypothesis testing, and reporting what would happen if randomization were done is also more complicated.*

7 Properties of One-Tailed Procedures

7.1 Uniformly Most Powerful One-Tailed Tests

In Sect. 3 we discussed some optimum properties that point estimators might have in general, and do have in the binomial problem in particular. In developing tests of hypotheses for the binomial problem, we have so far relied mainly on intuition. Are there optimum properties which justify use of the one-tailed binomial test in a one-sided binomial problem? It does not go without proof, and indeed is not true in all problems, that a one-tailed test or procedure should necessarily be chosen when the null hypothesis is one-sided, or when the alternative is one-sided, or even when both are one-sided. It is true, however, in a wide class of problems arising in practice, in particular in all binomial problems, as we shall now show. (See Karlin and Rubin [1956] or Lehmann [1959] for a general theorem.)

The most important justification of one-tailed tests is a property of their power, which will be illustrated first in a special case. Suppose S is the number of successes in 10 independent Bernoulli trials, and we want a test of the null hypothesis $p \geq 0.6$ against the alternative $p = 0.5$ at the level 0.055. It will be shown below that, among tests of this null hypothesis at this level, the test which is most powerful against this simple alternative is the one which rejects the null hypothesis for $S \leq 3$ and "accepts" it otherwise. The same test is most powerful against the alternative that $p = p_1$ for any $p_1 < 0.6$. We summarize these facts by saying that the test is "uniformly most powerful" against the alternative $p < 0.6$. This means that, if the probability of rejection for any other test never exceeds 0.055 for $p \geq 0.6$, then it never exceeds the probability of rejection for this test (as in Fig. 4.1) for $p < 0.6$. Equivalently, if some other test has greater probability of rejection than this test for some $p < 0.6$, then it has probability of rejection greater than 0.055 for some $p \geq 0.6$.

In general, a test is the *uniformly most powerful* test against a certain alternative hypothesis if it simultaneously maximizes the power against every alternative distribution. The tests and alternatives considered may be restricted in some way. Thus a test is uniformly most powerful among tests of a specified class against a specified set of alternatives if no other test of the specified class has larger power against any alternative of the specified set. If the class of tests is not specified, then it is understood to be all tests of the same null hypothesis at the same level as the test in question (and based on the same observations, of course). A test may be uniformly most powerful against one alternative hypothesis but not against another. The test discussed in the previous paragraph, for example, is not uniformly most powerful against the alternative that S is not binomially distributed.

Now suppose that S is the number of successes in n Bernoulli trials with probability p of success on each trial. As defined in Sect. 5, a lower-tailed randomized test rejects always if $S < s_\ell$ and with probability ϕ_ℓ if $S = s_\ell$; it includes the lower-tailed nonrandomized tests ($\phi_\ell = 0$ or 1). The exact level of such a test for the simple null hypothesis $p = p_0$ is $P_{p_0}(S < s_\ell) + \phi_\ell P_{p_0}(S = s_\ell)$. A lower-tailed binomial test of the simple null hypothesis $p = p_0$ or the composite null hypothesis $p \geq p_0$ is uniformly most powerful at its exact level against alternatives below p_0. The result for $p = p_0$ is stated here as Theorem 7.1; it will be proved in Sect. 7.4.

Theorem 7.1. *A lower-tailed binomial test at exact level α of the simple null hypothesis $H_0 : p = p_0$ has maximum power against any alternative $p = p_1$ with $p_1 < p_0$, among all level α tests of H_0.*

The property in Theorem 7.1 also holds for the broader null hypothesis $p \geq p_0$ because the test has the same exact level for $p \geq p_0$, while any other test at level α for $p \geq p_0$ has level α for $p = p_0$, and therefore, by Theorem 7.1, cannot be more powerful against any alternative $p_1 < p_0$.

Of course, if a lower-tailed test has level α for the null hypothesis $p \geq p_0$, but its exact level is α_0 where $\alpha_0 < \alpha$, then there will be other tests at level α which are more powerful; however, they all must have exact level greater than α_0. As a matter of fact, slightly more is true. Any test whose power is greater than a lower-tailed test against any alternative $p_1 < p_0$ also has greater probability of a Type I error under every null distribution $p \geq p_0$.

Similar results hold for upper-tailed tests, and indeed follow simply by interchanging "success" and "failure."

*7.2 Admissibility and Completeness of One-Tailed Tests

The results of Sect. 7.1 imply that any lower-tailed test for the null hypothesis $p \geq p_0$ is *admissible*, meaning that any test with smaller probability of error under some distribution has greater probability of error under some other distribution. (The errors may be Type I or Type II, as the distributions may be null or alternative.) Briefly, any test which is better somewhere is worse somewhere else. If a test is not admissible, then it can be eliminated from consideration because there is another test at least as good everywhere, and better somewhere. This other test has power at least as great against every alternative and probability of a Type I error at least as small under every null distribution, and it either has greater power against some alternative, or else has smaller probability of a Type I error under some null distribution.[5] On the other hand, an admissible test cannot be excluded from consideration without adducing some further criterion, because any other test is either equivalent everywhere or inferior somewhere.

A further property of the lower-tailed tests is that they form a *complete* class of tests for the null hypothesis $p \geq p_0$. This means that, given any test, there is a lower-tailed test at least as good everywhere. Specifically, this is the lower-tailed test with the same probability of rejection under p_0. (The proof is given in Sect. 7.4.)

The definitions just given agree with the usual definitions of "admissible" and "complete" in decision theory.[6] Slightly different definitions are sometimes used in the theory of hypothesis testing. A test is sometimes called admissible if any test with larger power against some alternative has either smaller power against some other alternative or larger exact level. A class of tests is sometimes called complete if, given any test, there is a test in the class with exact level as small and power as large against every alternative. These alternative definitions are in accord with the emphasis on only the maximum of the probability of a Type I error in testing theory. They are not really in accord with common sense, however, and therefore will not be used here. It is possible for two tests, say A and B, to have the same level and the same probability of error everywhere except that test A has smaller probability of a Type I error than test B under some null distributions. Then A is better than B in common sense. By the alternative definitions, however, B might be admissible and might be that member of a complete class excluding A which was supposed to be as good as A.*

7.3 Confidence Procedures

The properties of a confidence procedure are analogous to the properties of the corresponding family of tests. Specifically, corresponding to the usual family of nonrandomized, lower-tailed binomial tests at conservative level α is the usual nonrandomized, upper confidence bound. Let $1 - \alpha(p)$ be the true confidence level of this procedure, the probability that the bound is at least p when p is the true value. For each p_0, among upper confidence bounds whose

---

[5] It does not immediately follow that all tests which are not admissible can be eliminated simultaneously, because one could imagine an infinite sequence of tests, each better than the one before but with no test better than all of them. This does not occur in practical testing problems, however. See, for instance, Lehmann [1959, Appendix 4].

[6] In restricting consideration to sufficient statistics, we have already adopted the view that when several procedures have the same probability of rejection for all p, they are equivalent, and only one of them need be considered. Our definition of "complete" reflects this view. Often "complete" is defined to require excluded procedures to be strictly inferior; what we call "complete" here is then called "essentially complete."

true level at p_0 is at least $1 - \alpha(p_0)$, this bound simultaneously maximizes the probability of falling below p_0 for all true values of p below p_0. Among bounds whose true level is at least as great everywhere, it simultaneously minimizes the expected "excess" over the true p whatever it may be (Sect. 6.5). (These properties do not hold if the true level is merely required to be at least $1 - \alpha$, rather than $1 - \alpha(p)$. Problem 46 gives an example of a level $1 - \alpha$, non-randomized upper confidence bound which is distinctly unreasonable but, under a particular p, has no greater probability of including any $p_0 > p$ than the usual bound, has smaller probability for some p_0, and has smaller expected "excess.")

Next consider the randomized, upper confidence bound corresponding to the family of randomized, lower-tailed binomial tests at exact level α (Sect. 6.6). Among upper confidence bounds at level $1 - \alpha$, whatever the true p may be, this bound simultaneously maximizes the probability of falling below p_0 for all $p_0 > p$ and minimizes the expected "excess" over p.

Finally, consider any confidence procedure corresponding to a family of one-tailed binomial tests. A property like that given above for the usual nonrandomized bound holds. It must, however, be suitably restated if the confidence region is too complicated to be given by a confidence bound. Whether it is or not depends on how the exact level $\alpha(p)$ varies with p. A confidence bound is obtained for the family of nonrandomized tests chosen to have exact level as close as possible to α (Problem 47) as well as for the two cases mentioned above.

*7.4 Proofs

Let $\alpha(p, \phi)$ be the probability that the null hypothesis $p = p_0$ will be rejected by the test ϕ when p is the true value. Given any $p_0, 0 < p_0 < 1$, and any $\alpha_0, 0 \le \alpha_0 < 1$, there is one and only one lower-tailed randomized test ϕ^* based on S for which $\alpha(p_0, \phi^*) = \alpha_0$. (This is easy to see by considering what happens as the lower tail is gradually augmented.) We will now prove that this test ϕ^* uniformly maximizes the probability of rejection $\alpha(p_1, \phi)$ for $p_1 < p_0$ and uniformly minimizes $\alpha(p_1, \phi)$ for $p_1 > p_0$, among tests ϕ for which $\alpha(p_0, \phi) = \alpha_0$. It will be left to the reader (Problems 48 and 49) to verify that all statements of Sect. 7.1–7.3 follow (with the help of hints given there and the relation between tests and confidence procedures, particularly the discussion of including false values in Sect. 6.5).

Let ϕ be any (randomized) test. As in Sect. 5, denote by $\phi(s)$ the probability that ϕ rejects the null hypothesis when a particular value s is observed. Then

$$\alpha(p_1, \phi) = E_{p_1}[\phi(S)] = \sum_s \phi(s) P_{p_1}(S = s). \qquad (7.1)$$

For $p_1 < p_0$ (H_0 false), we want to maximize the expression in (7.1), subject to the condition that

$$\alpha(p_0, \phi) = \sum_s \phi(s) P_{p_0}(S = s) = \alpha_0. \tag{7.2}$$

It is intuitively clear that we should choose $\phi(s) = 1$ for those s where the contribution to the sum (7.1) is greatest compared to the contribution to the sum (7.2). In other words, the maximizing ϕ will be of the form

$$\phi_0(s) = \begin{cases} 1 & \text{if } \lambda(s) > k \\ 0 & \text{if } \lambda(s) < k \end{cases} \tag{7.3}$$

where λ is the *likelihood ratio* given by

$$\lambda(s) = \frac{P_{p_1}(S = s)}{P_{p_0}(S = s)}.$$

The value of k and the value of $\phi_0(s)$ when the likelihood ratio equals k are chosen so that $\alpha(p_0, \phi) = \alpha$. Randomization will occur (that is, $0 < \phi_0(s) < 1$), if at all, only where $\lambda = k$.

Before proving this fact, we will first verify that (7.3) leads to a lower-tailed test. The likelihood ratio is

$$\lambda(s) = \frac{\binom{n}{s} p_1^s (1 - p_1)^{n-s}}{\binom{n}{s} p_0^s (1 - p_0)^{n-s}} = \left(\frac{1 - p_1}{1 - p_0}\right)^n \left(\frac{p_1}{p_0} \cdot \frac{1 - p_0}{1 - p_1}\right)^s. \tag{7.4}$$

The values of s for which (7.4) is larger than a constant k are simply those values of s which are less than a constant s_ℓ, because $(p_1/p_0)(1 - p_0)/(1 - p_1) < 1$ if $p_1 < p_0$. Thus ϕ_0 is indeed a lower-tailed test. The crucial property leading to this result is that the likelihood ratio is a monotone function of s.

If $p_1 > p_0$, one can prove that a lower-tailed test minimizes $\alpha(p_1, \phi)$ subject to $\alpha(p_0, \phi) = \alpha_0$ by a similar argument or by suitably using the result just proved.

The fact that (7.3) maximizes (7.1) subject to (7.2) is a special case of the following, well known theorem. Since the result does not depend on the binomial distribution, the theorem is stated for any two distributions P_0 and P_1 of any random variable S. E_0 and E_1 denote the expected value operations corresponding to the distributions P_0 and P_1.

Theorem 7.2 (Neyman–Pearson Fundamental Lemma). *Let P_0 and P_1 be distributions of a random variable S. Then $E_1[\phi(S)]$ is maximized, subject to the conditions*

(i) $E_0[\phi(S)] \leq \alpha_0$

and

(ii) $0 \leq \phi(s) \leq 1 \quad \text{for all } s,$

by any function ϕ of the form

(iii) $$\phi(s) = \begin{cases} 1 & \text{if the likelihood ratio} > k \\ 0 & \text{if the likelihood ratio} < k \end{cases}$$

provided the value of k and the values of $\phi(s)$ at those s where the likelihood ratio equals k are such that $E_0[\phi(S)] = \alpha_0$. Conversely, if ϕ maximizes $E_1[\phi(S)]$ subject to conditions (i) and (ii), then there is a constant k for which (iii) holds, and furthermore, $E_0[\phi(S)] = \alpha_0$ unless there is a set A such that $P_1(S \in A) = 1$ and $P_0(S \in A) < \alpha_0$.

If P_0 and P_1 are discrete, then the likelihood ratio is the ratio of the discrete frequency functions:

(iv) $$\text{likelihood ratio} = \frac{P_1(S = s)}{P_0(S = s)}.$$

If P_0 and P_1 have densities f_0 and f_1 respectively,[7] then

(v) $$\text{likelihood ratio} = \frac{f_1(s)}{f_0(s)}.$$

For the converse, (iii) is considered to hold if the set of s for which it fails has probability 0 under both P_0 and P_1.

PROOF (For more detail, see, e.g., Lehmann [1959]). Consideration of what happens as k increases from 0 to ∞ makes it clear that there exists a ϕ_0 of the form (iii) such that $E_0[\phi_0(S)] = \alpha_0$; that is, there is a k such that $E_0[\phi_0(S)] = \alpha_0$ if $\phi_0(s)$ satisfies (iii) and is defined suitably when the likelihood ratio equals k. Let ϕ_0 be so defined and suppose that ϕ satisfies (i) and (ii). Then, in the discrete case, we shall show that

$$E_1[\phi_0(S)] - E_1[\phi(S)] = \sum_s [\phi_0(s) - \phi(s)]P_1(S = s)$$

$$\geq \sum_s [\phi_0(s) - \phi(s)]kP_0(S = s)$$

$$= k\{E_0[\phi_0(s)] - E_0[\phi(S)]\}$$

$$\geq k\alpha_0 - k\alpha_0 = 0. \tag{7.5}$$

The first inequality holds term by term because $\phi_0(s) = 1 \geq \phi(s)$ where $P_1(S = s) > kP_0(S = s)$ and $\phi_0(s) = 0 \leq \phi(s)$ where $P_1(S = s) < kP_0(S = s)$. The inequality is strict unless ϕ also satisfies (iii). The second inequality holds because $E_0[\phi_0(S)] = \alpha_0 \geq E_0[\phi(S)]$ and $k \geq 0$. It is strict unless $k = 0$ or $E_0[\phi(S)] = \alpha_0$. It follows that $E_1[\phi(S)] \leq E_1[\phi_0(S)]$ for any ϕ satisfying (i) and (ii), thus proving the direct half of the theorem. If ϕ also maximizes $E_1[\phi(S)]$, then neither inequality is strict. It follows that ϕ satisfies

---

[7] Actually, this case covers any two distributions if densities with respect to an arbitrary measure are allowed. In fact, we may take $f_0 = dP_0/d(P_0 + P_1)$, $f_1 = dP_1/d(P_0 + P_1)$.

(iii), and that $E_0[\phi(S)] = \alpha_0$, except perhaps when $k = 0$. Thus the converse half of the theorem is proved except when $k = 0$, ϕ satisfies (iii), and $E_0[\phi(S)]$ $< \alpha_0$. In this case, the likelihood ratio is greater than k on the set A of all s where $P_1(S = s) > 0$; then $P_1(S \in A) = 1$, and $\phi(s) = 1$ for $s \in A$, so that $P_0(S \in A) \le E_0[\phi(S)] < \alpha_0$, satisfying the last clause of the theorem. $\qquad\square$

The proof for densities is similar and is left to the reader.*

8 Choice of Two-Tailed Procedures and Their Properties

8.1 Test Procedures

In the last section we found that in the one-sided binomial problem, a one-tailed randomized test based on S is uniformly most powerful against alternatives in the appropriate direction. For two-sided alternatives it is natural to use two-tailed tests, and later in this section, we shall show that no others should be considered. No test is uniformly most powerful against a two-sided alternative, however, since different one-tailed tests are most powerful against alternatives on the two sides. Hence, even if the level α is given, we need some further criterion to select among all two-tailed tests at level α.

One possible criterion for choice is called the *equal-tails criterion*. The idea is that a two-tailed test at level α should be a combination of two one-tailed tests, each at level $\alpha/2$. To make the exact levels equal, in most discrete problems, one of the one-tailed tests must ordinarily be randomized. (If the null distribution is symmetric, either both or neither must be randomized.) The usual nonrandomized two-tailed binomial test (Clopper and Pearson [1934]) discussed earlier in this book has equal nominal levels in the two tails. Specifically, it rejects at nominal level α when either one-tailed, non-randomized test at conservative level $\alpha/2$ would reject, and "accepts" otherwise.

The usual two-tailed binomial test is therefore conservative, as was illustrated in the context of confidence intervals by Fig. 6.3. In fact, it might be considered ultraconservative, since it would sometimes be possible to add a point to the rejection region without making the exact level greater than the nominal level. For the null distribution given in Table 4.1, for example, at level $\alpha = 0.06$, the usual test of H_0: $p = 0.6$ would reject only for $S = 0, 1, 2$, or 10. Either $S = 3$ or $S = 9$ could be added to the rejection region without raising the level above 0.06. The disadvantage of including such a point is that the level of one of the one-tailed tests would then exceed $\alpha/2$ so that the one-tailed and two-tailed procedures would not be simply related. Still, under the usual two-conclusion interpretation of tests, taken literally, or the first three-conclusion interpretation (Table 4.2), it seems unarguable that the point should be added when possible.

The primary advantage of using the criterion of equal tails is that it provides two-tailed tests which are simply related to one-tailed tests, and the corresponding confidence intervals have as endpoints the respective upper and lower confidence bounds, each at one-sided level $\alpha/2$ (see Sect. 8.2). It is difficult to internalize the inference if equal tails are not used. Nevertheless, in discrete problems, it is difficult to argue against eliminating at least ultra-conservatism if we consider the situation from a truly two-sided point of view, rather than as a combination of two one-sided problems. Perhaps, however, a truly two-sided view is rare. If, for instance, we adopt the second three-conclusion interpretation of tests (Table 4.3), the situation is different. Now to achieve a desired error probability α, the two tail probabilities α_1 and α_2 both must be less than or equal to the same value, although that value is α, not $\alpha/2$.

Another criterion which might be used to select among possible two-tailed tests is "unbiasedness." A (two-conclusion) test is called *unbiased* if the probability of rejecting the null hypothesis when it is false is at least as great as when it is true. In the binomial problem, for a specified null hypothesis $p = p_0$ and exact level α, there is one and only one unbiased two-tailed test (Problem 50). As we will show later in this section, this test is also *uniformly most powerful unbiased*, that is, uniformly most powerful among all unbiased tests at level α. Ordinarily, however, it is not equal-tailed; it is randomized at both the lower and upper critical values; and even adjusting α cannot eliminate randomization in both tails (Problem 50d).

If a nonrandomized test is desired, it will thus generally be impossible to find one which is unbiased, let alone uniformly most powerful unbiased. Then one would presumably choose the nonrandomized test which is in some sense most nearly unbiased. This sometimes gives results different from the usual equal-tailed, nonrandomized test. Thus, with or without randomization, the criterion of unbiasedness has the disadvantage of leading to two-tailed tests which are not simply related to one-tailed tests, as equal-tailed tests are.

While it is pleasant to find that a test chosen on other grounds is unbiased, unbiasedness is not really a satisfactory theoretical criterion for choosing among procedures anyway. For instance, a biased test might be considerably more powerful than an unbiased test except in a very small interval on one side of the null hypothesis, as in Fig. 8.1. In such a case, the biased test may be preferable to the unbiased one. To be sure, the criterion of equal tails is subject to the same criticism. For example, the unbiased test in Fig. 8.1 might be equal-tailed, while the presumably better test is not. Situations like Fig. 8.1 may be rare in practice, but the possibility makes clear that our fundamental preferences among power curves are not captured by either unbiasedness or equal tails. This criticism is less telling against equal tails than against unbiasedness, however, since the advantage claimed earlier for equal tails has no relation to power.

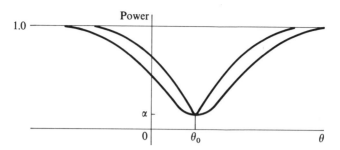

Figure 8.1   Two hypothetical power functions.

Another possible criterion for choosing critical regions is *minimum likelihood*, discussed in Sect. 4.5 in relation to *P*-values. By this procedure, the rejection region would be made up of those sample points which have the smallest probability under the null hypothesis, starting with the least probable and working up until the total probability in both tails combined is as large as possible without exceeding α.

To illustrate, we consider again the binomial problem with $n = 10$, H_0: $p = 0.6$, $\alpha = 0.10$; the point probabilities are given in Table 4.1. In order of increasing probability, the first five points entering the rejection region are $S = 0$, 1, 10, 2, and 9, and their total null probability is $0.000 + 0.002 + 0.006 + 0.010 + 0.040 = 0.058$. Adding the next point, $S = 3$, to the rejection region increases the rejection probability to $0.058 + 0.043 = 0.101$, which exceeds 0.100, so that $S = 3$ cannot be included for a test at level 0.10. By the minimum likelihood criterion then, the conservative, nonrandomized, two-tailed test of the null hypothesis $p = 0.6$ at the 0.10 level rejects when $S = 0, 1, 2, 9,$ or 10, and "accepts" otherwise. The test coincides with the usual equal-tailed test in this instance. At level 0.06, however, the two methods differ. The equal-tailed test rejects only for $S = 0, 1, 2,$ or 10, while the minimum likelihood test still rejects for $S = 0, 1, 2, 9,$ or 10. The randomized minimum likelihood procedure at exact level .10 rejects when $S = 0, 1, 2, 9,$ or 10 and with probability $0.042/0.043$ when $S = 3$, and differs from the randomized, equal-tailed test at the same exact level 0.10.

In any situation, a minimum likelihood test at exact level α which is based on a sufficient statistic S having a unique null distribution is most powerful against the alternative that all values of S are equally likely, among all tests at level α, randomized or not (Problem 52). In the binomial case, although there is no value of p under which all values of S are equally likely, they would be if p were a random variable having the uniform distribution on $(0, 1)$ (Problem 54). Thus the minimum likelihood test using the binomial distribution is most powerful against the alternative that p is uniformly distributed between 0 and 1. An equivalent property, which does not depend on regarding p as a random variable, is that the minimum likelihood procedure in the binomial case maximizes the average power, that is, the integral of the power over p,

$\int_0^1 P_p(p_0$ rejected$)dp$. Equivalently, it maximizes the area under the power curve, from 0 to 1. The power is not defined at the null hypothesis $p = p_0$ but the integral or area would not be changed by omitting this single value. Like the unbiased test, the test based on the minimum likelihood criterion has the disadvantage that it is not simply related to one-sided procedures. The advantage just described, however, precludes the conceivable sort of disadvantage of unbiasedness represented in Fig. 8.1. Average power is closer to our real concern than unbiasedness. Although reparameterization changes it, this corresponds to using a weighted average. In the Bayesian framework, the appropriate weights reflect the prior distribution and cost/benefit considerations or the "loss function."

In performing two-tailed tests, either in general or in binomial situations, statistical methodology does not clearly prescibe a particular procedure as optimum, whether it is randomized or nonrandomized. If one attempts to escape this problem by reporting two-tailed P-values, the difficulties discussed in Sect. 4.5 apply.

8.2 Confidence Procedures

If confidence regions are to be obtained from two-tailed tests, some procedure must be adopted to choose upper and lower critical values for any null hypothesis $p = p_0$ since an escape into P-values is not available. Most tables of confidence limits are constructed, following Clopper and Pearson [1934], to correspond to the usual equal-tailed, nonrandomized test of $p = p_0$. Then the upper and lower ends of the confidence region are the one-sided confidence bounds each at confidence level $1 - \alpha/2$, and for each p_0 the one-sided and two-sided confidence procedures are simply related.

Since the usual test is conservative, the corresponding confidence procedure is also conservative, as is illustrated in Sect. 6.4, and in fact even ultraconservative as explained in relation to testing in Sect. 8.1. However, when the advantages of the criterion of equal tails are sacrificed, it may be even more difficult to internalize the inference with confidence regions than with tests. Unequal tails can be quite misleading, especially if confidence is interpreted as if the Bayesian concept of subjective probability applied, that is, as the probability, after the observations are made, that the true value is in the confidence region (as explained in Sect. 6.3).

The confidence procedure corresponding to a family of tests, each of which is unbiased at exact level α, has the property that the probability of including the true value is at least $1 - \alpha$ for any true value, while the probability of including any particular false value is at most $1 - \alpha$ for any true value. Such a confidence procedure will be called *unbiased at confidence level* $1 - \alpha$.[8]

---

[8] If the tests are unbiased but have different exact levels for different members of the family, the corresponding confidence property is more complicated and will not be discussed here. (Problem 53 requests a statement of this property.)

Unbiasedness seems an even less important property of a confidence interval than of a test, and equal tails even more important.

The confidence regions which correspond to the minimum likelihood procedure, strangely enough, are not necessarily intervals. When only nonrandomized procedures are to be considered, however, it is apparently possible to modify the procedure so as to avoid this difficulty while preserving the property, discussed above, of maximizing the average power. For a non-randomized test, we have (Problem 55)

$$\text{average power} = \int_0^1 P_p(p_0 \text{ rejected})dp$$

$$= (\text{number of values of } S \text{ such that } p_0 \text{ rejected})/(n + 1).$$

(8.1)

This may be maximized by more than one level α test. Crow [1956] took advantage of this flexibility to modify the minimum likelihood procedure so that the corresponding confidence regions are intervals (for $n \leq 30$ and $\alpha = 0.10, 0.05,$ or 0.01). Crow made further modifications to shorten the intervals for S near 0 or n and to make the interval for a given S at level 0.01 contain that at level 0.05 and the interval at level 0.05 contain that at level 0.10.

Crow's confidence procedure, and any other confidence procedure corresponding to nonrandomized level α tests which maximize the number of values of S in each rejection region, have the following properties among nonrandomized procedures at level α. First, they minimize the average probability of including each point p_0, $\int_0^1 P_p(p_0 \text{ included})dp$, which can also be interpreted as the average probability of including the false value p_0. (In the integrand, if p is the true value and $p \neq p_0$, then p_0 is a false value; omitting $p = p_0$ from the region of integration has no effect.) Second, they also minimize both the average expected length of the confidence region and the ordinary, arithmetic average of the lengths of the $n + 1$ possible confidence regions, these being in fact equal (Problem 56). Specifically, for intervals these averages are

$$\int E_p[p_u(S) - p_\ell(S)]dp = \frac{1}{n + 1} \sum_{s=0}^n [p_u(s) - p_\ell(s)].$$

(8.2)

If, for some s, the confidence region $C(s)$ is not an interval, its generalized length (Borel measure) $\int_{C(s)} dp$ must be used here in place of the length $p_u(s) - p_\ell(s)$. (Since "p_0 included" is equivalent to "p_0 not rejected," the first property follows from (8.1) and the second from the relation

$$\sum_{s=0}^n \int_{C(s)} dp = \int_0^1 [\text{number of values of } s \text{ such that } p \in C(s)]dp,$$

(8.3)

whose proof is Problem 58.)

The solution of the corresponding problem for randomized procedures cannot be chosen so that the confidence region is always an interval, because there is flexibility in the choice of tests only at isolated values of p_0 [Pratt, 1961].

By a technique similar to that above, it is possible quite generally to minimize the average expected length of a confidence interval using any weight function in averaging. The essential step is to multiply (6.11) by the weight function and integrate with respect to θ_1.

*8.3 Completeness and Admissibility of Two-Conclusion Two-Tailed Tests

So far we have assumed, on an intuitive basis, that a two-sided binomial problem calls for a two-tailed test, rather than some more complicated type of test. This can be justified in terms of the concepts of completeness and admissibility introduced earlier in Sect. 7.2.

Consider first the two-conclusion interpretation in a two-sided binomial problem with null hypothesis $p = p_0$. It can be shown that the two-tailed tests form a *complete class* and somewhat more: Given any procedure which is not a two-tailed test, there is a two-tailed test with the same exact level and greater power everywhere, that is, the same probability of rejection when $p = p_0$ and greater probability of rejection for any $p \neq p_0$.[9] This justifies restricting consideration to two-tailed tests. Furthermore, none of these may be excluded from consideration without introducing some further criterion, because all two-tailed tests are *admissible*: Given any two-tailed test, no other test has power as great for every $p \neq p_0$ and at the same time exact level as small.

An interesting generalization also holds. The null hypothesis $p = p_0$ is sometimes used as an approximation to the null hypothesis that p lies in a specified interval containing p_0. That is, we might prefer to "accept" for p inside some interval, and to reject for p outside it. At the endpoints we may prefer "acceptance" or rejection, or we may be indifferent. (We assume that the endpoints of the interval are different and also that neither endpoint is 0 or 1. The degenerate case where they are equal is the situation discussed in the previous paragraph.) The previous result, that the two-tailed tests are admissible and form a complete class, holds here also. Explicitly, given any procedure which is not a two-tailed test, there is a two-tailed test which has greater probability than the given test of "accepting" when p is inside the interval and of rejecting when p is outside the interval (and the same probability as the given test of "accepting" and rejecting at the endpoints). Further,

---

[9] The definition of completeness requires only "at least as great." The statement is true only if one-tailed tests are included in the class of two-tailed tests, as we shall take them to be and as they are by the formal definition of two-tailed tests given above. In the binomial problem, we can take $s_\ell = 0, \phi_\ell = 0$ or $s_u = n, \phi_u = 1$.

given any two different two-tailed tests, each one has greater probability of making the correct decision than the other at some values of p. Thus we need consider only two-tailed tests, but none can be excluded from consideration without adducing some further criterion.

The facts of the last two paragraphs have been proved by Karlin [1955] for any strictly Pólya type 3 distribution. (The monotone likelihood ratio property mentioned in Sect. 7.4 is Pólya type 2. Pólya type 3 is a generalization.) Karlin's proof uses a fundamental theorem of game theory, that the class of all Bayes' procedures and their limits is complete, under certain compactness conditions. One could use instead the generalization of the Neyman–Pearson fundamental lemma to two side conditions. The proof of this generalization (see, for instance, Lehmann [1959]) is related to the usual proof of the fundamental theorem of game theory. With the help of completeness, a proof of admissibility is analogous to that given in Sect. 7.4.*

*8.4 Completeness and Admissibility of Three-Conclusion Two-Tailed Tests

Results similar to those of Sect. 8.3 hold for the various three-conclusion interpretations of two-sided problems. For definiteness (the discussion applies to the others also with trivial modifications), we consider the conclusions as (a) $p < p_0$, (b) "accept" the null hypothesis (that is, draw no conclusion), and (c) $p > p_0$. Suppose that when in fact $p < p_0$, we prefer (a) to (b) and (b) to (c); when $p = p_0$ we prefer (b) to either (a) or (c); and when $p > p_0$ we prefer (c) to (b) and (b) to (a). Then, given any procedure for reaching one of these conclusions, there is a two-tailed test which (in its three-conclusion interpretation) is at least as good for every p. Specifically, if α_1 equals the probability when $p = p_0$ of concluding $p < p_0$ and α_2 equals the probability when $p = p_0$ of concluding $p > p_0$ by the given procedure, then the two-tailed test combining the lower-tailed test at exact level α_1 and the upper-tailed test at exact level α_2 is at least as good as the given procedure, whatever the value of p. When $p < p_0$, its probability of concluding $p < p_0$ is at least as large, and its probability of concluding $p > p_0$ is at least as small, as that of the original procedure. (This follows from the results given in Sect. 7.) When $p > p_0$, the same statement holds with the inequalities reversed; and when $p = p_0$, the two procedures have the same probability of leading to each conclusion.

Thus the two-tailed tests form a complete class in the three-conclusion interpretation (with the natural preferences given above). Are all admissible in this interpretation, or might there now be one which is at least as good as another whatever the value of p and better for some p? The results given earlier for one-tailed tests imply immediately that, given any two-tailed test interpreted as a three-conclusion procedure, any procedure having greater probability under any $p > p_0$ of correctly concluding $p > p_0$ has also greater probability under all $p \le p_0$ of incorrectly concluding $p > p_0$. This statement also holds with the inequalities reversed. These results are not quite what we would like however, because we might prefer a procedure having somewhat smaller probability, under $p > p_0$, of correctly concluding $p > p_0$, if it also had a sufficiently smaller probability of incorrectly concluding $p < p_0$ (and therefore, of course, larger probability of "accepting" the null hypothesis). That is, a sufficient decrease in the probability of the least desirable conclusion $p < p_0$ might more than offset a decrease in the probability of the most desirable conclusion $p > p_0$.

In order to make the problem definite enough to investigate, let us suppose that the undesirability of a procedure, when p is the true value, is measured by weighting the probabilities of errors and adding. The weights may depend on p and are denoted by $L_j(p)$ for the conclusions $j = a, b$ and c. Specifically, then, when p is the true value, the undesirability is the sum

$$L_a(p)P_p(\text{test concludes } p < p_0)$$
$$+ L_b(p)P_p(\text{test "accepts"}) \qquad (8.4)$$
$$+ L_c(p)P_p(\text{test concludes } p > p_0).$$

In accordance with the preferences expressed earlier, we have

$$L_a(p) < L_b(p) < L_c(p) \quad \text{for } p < p_0 \qquad (8.5)$$

$$L_b(p_0) < L_a(p_0), \qquad L_b(p_0) < L_c(p_0) \qquad (8.6)$$

$$L_c(p) < L_b(p) < L_a(p) \quad \text{for } p > p_0. \qquad (8.7)$$

Beyond this, the L_j may be chosen almost arbitrarily and the statement we are about to make will still hold.

If undesirability is interpreted as the sum given in (8.4), and a mild further restriction is satisfied by the L_j, it can be proved (Karlin and Rubin [1956]; Karlin [1957b]; Problem 59) that any two-tailed test whose upper and lower critical values are not equal is admissible; that is, no other procedure is as desirable under all p and more desirable under some p. (For the sample sizes and null hypotheses that occur in practice, tests whose upper and lower critical values are equal have large probability of rejecting the null hypothesis when it is true and hence are never considered.)

These statements about the three-conclusion interpretation of two-sided binomial problems also hold if the null hypothesis $p = p_0$ is replaced by the null hypothesis that p lies in an interval containing p_0, as was done above for the two-conclusion interpretation.*

9 Appendices to Chapter 1

Straightforward tabulation of the binomial distribution involves three variables (n, p, and s) and leads to extremely bulky tables. (Table B is straightforward but very abbreviated and is not useful for in-between values, for large

n, or for finding confidence limits on p.) Straightforward machine computation is often inefficiently lengthy, and is sometimes infeasible because of overflow, underflow, or roundoff error. For these reasons, approximations are widely used. Two common approximating distributions are the Poisson (which requires a two-variable table) and the normal (which requires only a one-variable table). Simple limiting processes by which the binomial distribution leads to the Poisson and normal distributions will be developed in Appendix A. The accuracy of approximations can be vastly improved, however, by various kinds of transformations and adjustments. The appropriate computing forms of the resulting approximations may obscure their origins. The approximations suggested at the ends of Tables B and C are examples, but discussion of such matters would be out of place here.

Some basic aspects of limiting distributions generally will be discussed in Appendix B, primarily as background for Chap. 8.

Appendix A: Limits of the Binomial Distribution

A.1 Ordinary Poisson Approximation and Limit

A random variable S has a Poisson distribution with mean m if

$$P(S = s) = \frac{m^s e^{-m}}{s!} \quad \text{for } s = 0, 1, \ldots. \tag{9.1}$$

If S is binomial with parameters n and p, and n is large but np is moderate, then S has approximately a Poisson distribution with mean $m = np$, and (9.1) holds approximately. It follows that

$$P(s' \leq S \leq s'') \doteq \sum_{s=s'}^{s''} \frac{m^s e^{-m}}{s!} \tag{9.2}$$

where s' and s'' are integers, $0 \leq s' \leq s''$, and $\doteq$ denotes approximate equality (in absolute, not relative terms). If n is large but $n(1 - p)$ is moderate, then $n - S$ has approximately a Poisson distribution with mean $m = n(1 - p)$. This yields immediately an approximation to the distribution of S.

Precise limit statements corresponding to these approximations are as follows. Suppose S_n is binomial with parameters n and p, and p depends on n in such a way that $np \to m$ as $n \to \infty$. (The subscript has been added to S because n is no longer fixed and the distribution of S depends on n.) Then

$$P(S_n = s) \to \frac{m^s e^{-m}}{s!} \quad \text{as } n \to \infty, s = 0, 1, \ldots \tag{9.3}$$

$$P(s' \leq S_n \leq s'') \to \sum_{s=s'}^{s''} \frac{m^s e^{-m}}{s!} \quad \text{as } n \to \infty, 0 \leq s' \leq s'' \leq \infty. \tag{9.4}$$

In fact, for any set of integers A,

$$P(S_n \in A) \to \sum_{s \in A} \frac{m^s e^{-m}}{s!} \quad \text{as } n \to \infty. \quad (9.5)$$

Of course, these limit statements, although they have precise meanings, say nothing about when the limits are approximately reached.

*The statements in (9.3) and (9.4) are easily proved as follows. The exact probability function for S_n is

$$P(S_n = s) = \binom{n}{s} p^s (1 - p)^{n-s}$$

$$= \frac{1}{s!} \left[\frac{n}{n} \frac{n-1}{n} \cdots \frac{n-s+1}{n} \right] (np)^s [(1 - p)^{1/p}]^{np-sp}. \quad (9.6)$$

As $n \to \infty$, each factor in the first bracket approaches 1, $np \to m$, and therefore $p \to 0$; it follows that $(1 - p)^{1/p} \to e^{-1}$ and $sp \to 0$, which proves (9.3). Summing (9.3) over s gives (9.4) for $s'' < \infty$. Limits cannot be taken under infinite summation signs without further justification, but (9.4) now follows for $s'' = \infty$ as well, since

$$P(s' \le S_n) = 1 - P(S_n \le s' - 1)$$

$$\to 1 - \sum_{s=0}^{s'-1} \frac{m^s e^{-m}}{s!} = \sum_{s=s'}^{\infty} \frac{m^s e^{-m}}{s!}. \quad (9.7)$$

The proof of (9.5) will be given in Appendix B.*

A.2 Ordinary Normal Approximation and Limit

The density function of the standard normal distribution was given earlier in (2.3) as

$$\phi(z) = \frac{1}{\sqrt{2\pi}} e^{-z^2/2}. \quad (9.8)$$

In this book the symbol Φ will be reserved to denote the cumulative distribution function of the standard normal. Values of $\Phi(-z)$ are given in Table A for $z \ge 0$; by symmetry, $\Phi(z) = 1 - \Phi(-z)$. A random variable X is normal with mean μ and variance σ^2 if the standardized random variable, $(X - \mu)/\sigma$, has a standard normal distribution.

If S is binomial with parameters n and p, and np and $n(1 - p)$ are both large (that is, n is large and p is not too close to 0 or 1), then S is approximately normal with mean $\mu = np$ and variance $\sigma^2 = np(1 - p)$. In other words, the standardized random variable

$$\frac{S - \mu}{\sigma} = \frac{S - np}{\sqrt{np(1 - p)}} = \frac{(S/n) - p}{\sqrt{p(1 - p)/n}} = \left(\frac{S}{n} - p\right)\sqrt{\frac{n}{p(1 - p)}} \quad (9.9)$$

is approximately standard normal.

Since the normal distribution is continuous while the binomial is discrete, the approximation is generally improved by "correcting for continuity." This amounts to associating the binomial probability at each integer s with a continuous distribution on the interval between $s - \frac{1}{2}$ and $s + \frac{1}{2}$. Specifically, letting Z denote a standard normal random variable, $P(S = s)$ for an integer s can be approximated by either

$$P\left(\frac{s - \frac{1}{2} - \mu}{\sigma} \le Z \le \frac{s + \frac{1}{2} - \mu}{\sigma}\right) = \Phi\left(\frac{s + \frac{1}{2} - \mu}{\sigma}\right) - \Phi\left(\frac{s - \frac{1}{2} - \mu}{\sigma}\right)$$

or

$$\frac{1}{\sigma}\phi\left(\frac{s - \mu}{\sigma}\right), \quad (9.10)$$

where $\mu = np$, $\sigma = \sqrt{np(1 - p)}$. Further, for any integers s' and s'', $s' \le s''$, we have approximately

$$P(s' \le S \le s'') \doteq P\left(\frac{s' - \frac{1}{2} - \mu}{\sigma} \le Z \le \frac{s'' + \frac{1}{2} - \mu}{\sigma}\right)$$

$$= \Phi\left(\frac{s'' + \frac{1}{2} - \mu}{\sigma}\right) - \Phi\left(\frac{s' - \frac{1}{2} - \mu}{\sigma}\right). \quad (9.11)$$

The values of Φ are given by Table A. This approximation can be recommended only for extremely casual use, where ease of remembering and calculating are paramount. There is a normal approximation based on cube roots which is only a little more complicated but an order of magnitude more accurate. Another normal approximation based on logarithms is only slightly more complicated and is yet another order of magnitude more accurate. The latter is given at the end of Table B and the former is the basis of the approximate confidence bounds given at the end of Table C.

Precise limit statements corresponding to the approximations (9.10) and (9.11) are as follows. Suppose S_n is binomial with parameters n and p and p is fixed, $0 < p < 1$. If s_n is an integer depending on n in such a way that $(s_n - \mu)/\sigma$ approaches some number, say z, then

$$\frac{P(S_n = s_n)}{(1/\sigma)\phi(z)} \to 1 \quad \text{as } n \to \infty \quad (9.12)$$

and

$$P(S_n \le s_n) \to \Phi(z) \quad \text{as } n \to \infty. \quad (9.13)$$

Regardless of how s depends on n, we have

$$P(S_n \leq s) - \Phi\left(\frac{s - \mu}{\sigma}\right) \to 0 \quad \text{as } n \to \infty. \tag{9.14}$$

Replacing s in the second term of (9.14) by $s + \frac{1}{2}$ generally makes the left-hand side nearer 0 without affecting the limit; subtracting (9.14) for $s = s''$ from (9.14) for $s = s' - 1$ then leads to (9.11). Again, the limit statements have precise meanings but say nothing about when the limits are approximately attained. Further, (9.14) does not imply that the ratio of the approximate and true probabilities approaches one, and in fact it does not always.

PROOFS. The statement in (9.12) can be proved by straightforward but tedious calculation based on Stirling's formula, which says that

$$\frac{n!}{\sqrt{2\pi n}(n/e)^n} \to 1 \quad \text{as } n \to \infty.$$

Equation (9.13) can be proved from (9.12) by proving that (9.12) holds uniformly [Feller, 1968] or by another argument (see Appendix B), or, without (9.12), by applying the Central Limit Theorem for identically distributed random variables (Theorem 9.2). Equation (9.14) follows from (9.13) (Problem 60). $\qquad\qquad\square$

Appendix B: Convergence in Distribution and Asymptotic Distributions

In Appendix A, we showed that the Poisson and normal distributions may be used as approximations to the binomial distribution. Now we will discuss convergence of distributions and densities more generally.

B.1 *Convergence of Frequency Functions and Densities*

Suppose that for each n we have a sequence of random variables, say $\{X_n\}$, and a corresponding sequence of distributions. Suppose further that, as in the previous section, the random variables X_n are discrete and, for every real number x, the probability that $X_n = x$, say $f_n(x)$, approaches a limit, say $f(x)$, as $n \to \infty$. Then as $n \to \infty$

$$P(X_n = x) = f_n(x) \to f(x) \quad \text{for all } x.$$

It follows immediately that

$$P(X_n \in A) = \sum_{x \in A} f_n(x) \to \sum_{x \in A} f(x) \quad \text{for any finite set } A.$$

This is true for infinite sets as well if the limit f is a discrete frequency function. Specifically, we must have $\sum_x f(x) = 1$, which cannot be taken for granted. This added condition also holds when f is the Poisson frequency function, of course, and the proof of (9.4) for $s'' = \infty$ used it.

An analogous result holds for densities. These facts and others are given formally in the following theorem.

Theorem 9.1

(1) *If X_n has discrete frequency function f_n, $n = 1, 2, \ldots$, $f_n(x) \to f(x)$ for all x, and $\sum_x f(x) = 1$, then f is the discrete frequency function of a random variable X, and*

(i) $$\sum_x |f_n(x) - f(x)| \to 0 \tag{9.15}$$

(ii) $$P(X_n \in A) = \sum_{x \in A} f_n(x) \to P(X \in A) = \sum_{x \in A} f(x) \tag{9.16}$$

for every set A, and

(iii) $$E[h(X_n)] = \sum_x h(x) f_n(x) \to E[h(X)] = \sum_x h(x) f(x)$$

for every bounded function h.

(2) *If X_n has density f_n, $n = 1, 2, \ldots$, $f_n(x) \to f(x)$ for all x, and $\int f = 1$, then f is the density of a random variable X and*

(i′) $$\int |f_n - f| \to 0 \tag{9.17}$$

(ii′) $$P(X_n \in A) = \int_A f_n \to P(X \in A) = \int_A f \tag{9.18}$$

for every set A, and

(iii′) $$E[h(X_n)] = \int h f_n \to E[h(x)] = \int h f$$

for every bounded function h.[10]

(3) *In both (1) and (2), if X_n has cumulative distribution function F_n and X has cumulative distribution function F, then*

(iv) $$F_n(x) \to F(x) \quad \text{for all } x. \tag{9.19}$$

PROOF. (i) implies (ii) uniformly in A and (iii) uniformly in h for $|h| \le K$, while (ii) implies (iv). The same holds for (i′), (ii′), and (iii′). Hence we need prove only (i) and (i′). Since

$$f_n(x) + f(x) - |f_n(x) - f(x)| \ge 0 \quad \text{for all } x,$$

---

[10] Here, as elsewhere in this book, all sets and functions are assumed to be measurable without specific mention.

it follows (Problem 61) that

$$\sum_x \lim \inf [f_n(x) + f(x) - |f_n(x) - f(x)|]$$

$$\leq \lim \inf \sum_x [f_n(x) + f(x) - |f_n(x) - f(x)|].^{11} \qquad (9.20)$$

Since $f_n(x) \to f(x)$ and $\sum_x f(x) = 1$, this reduces to

$$2 \leq 2 - \lim \sup \sum_x |f_n(x) - f(x)| \quad \text{or} \quad \lim \sup \sum_x |f_n(x) - f(x)| \leq 0.$$

The statement in (i) follows. The proof of (i') is similar, using Fatou's Lemma, which says that

$$\int \lim \inf h_n \leq \lim \inf \int h_n \quad \text{for } h_n \geq 0.$$

Actually, the densities may be with respect to an arbitrary measure. With this understanding, the second case covers the first. (The method of proof used here appears in a more natural context in Young [1911]. See also Pratt [1960]. A somewhat different proof is given by Scheffé [1947].) □

The foregoing discussion does not apply directly to the approach of the binomial distribution to the normal, since the binomial distribution is discrete while the normal has a density. The discussion does apply indirectly, however. Specifically, suppose S_n is binomial with parameters n and p, and U is uniformly distributed between $-\frac{1}{2}$ and $\frac{1}{2}$. Then $S_n + U$ has a density; in fact, the density of $S_n + U$ is

$$g_n(y) = P(S_n = s) \qquad (9.21)$$

where s is the integer nearest y. (The definition for y half-way between adjacent integers is immaterial.) This density approaches 0 as $n \to \infty$ for y fixed, as would be expected since the variance of $S_n + U$ approaches ∞. Consider

$$X_n = \frac{S_n + U - \mu}{\sigma} \qquad (9.22)$$

where $\mu = np$ and $\sigma = \sqrt{np(1 - p)}$ as before. The density of X_n is

$$f_n(x) = \sigma g_n(\mu + x\sigma)$$
$$= \sigma P(S_n = s) \qquad (9.23)$$

where s is the integer nearest $\mu + x\sigma$. For this s, $(s - \mu)/\sigma \to x$, so that by (9.12)

$$\frac{f_n(x)}{f(x)} \to 1 \quad \text{for all } x \qquad (9.24)$$

---

[11] The abbreviation inf stands for infimum, which is the greatest lower bound. Similarly, sup denotes supremum or least upper bound. The infimum and supremum of a set of numbers always exist; either or both may be infinite.

where f is the standard normal density. Thus $f_n(x) \to f(x)$ for all x, and Theorem 9.1 applies.

It follows in particular that

$$F_n(x) = P(X_n \le x) \to F(x) \quad \text{for all } x \tag{9.25}$$

and hence (Problem 60) that (9.13) holds.

B.2 Convergence in Distribution

Suppose $X_1, X_2, \ldots$, and X are real- or vector-valued random variables with respective cumulative distribution functions $F_1, F_2, \ldots$, and F. Then the following conditions are equivalent (Problem 62).

(1) $F_n(x) \to F(x)$ for every x at which F is continuous.
(2) $E[h(X_n)] \to E[h(X)]$ for every bounded, continuous function h.
(3) $P(X_n \in A) \to P(X \in A)$ for every set A such that $P(X$ is a boundary point of $A) = 0$.
(4) $\liminf P(X_n \in A) \ge P(X \in A)$ for every open set A.
(5) $\limsup P(X_n \in A) \le P(X \in A)$ for every closed set A.

If these conditions hold (if one holds, they all do, since they are equivalent), then F_n is said *to converge in distribution* to F. Alternative terminology is that X_n converges in distribution to X, or X_n to F, or F_n to X; X_n or F_n is *asymptotically distributed* as X or F; F is the *limiting distribution* of X_n or F_n; etc.

Part of the definition of convergence in distribution is that the limit F should be a cumulative distribution function. It is possible for (1) to hold without F being a cumulative distribution function (Problem 63); then F_n does *not* converge in distribution to F (though the customary terminology is to say that it converges "weakly" to F). If (1) holds, F must satisfy the monotonicity properties of a cumulative distribution function; thus the further requirement is just that it should behave properly at $\pm \infty$, which amounts to the requirement that X be finite with probability one.

Conditions (1), (2), and (3) above are somewhat weaker than the corresponding statements of Theorem 9.1. Thus the hypotheses of Theorem 9.1 imply convergence in distribution, while convergence in distribution does not imply the conclusions (or hypotheses) of Theorem 9.1, even if all the distributions are discrete. (It does if all the distributions are concentrated on the same finite set of points. The proof is Problem 64.)

Notice also that the convergence in distribution of X_n to X does not imply that X_n is probably near X for n large. $X_1, X_2, \ldots, X$ might be independent; indeed, their joint distribution is not under discussion and need not exist. The convergence in distribution of X_n to X says only that the *distribution* of X_n is close to the *distribution* of X for n large, in a certain sense of close.

The most important limiting distribution in practice is the normal. A special terminology is convenient and often used, but must be handled with care, especially in mathematical arguments. A sequence of real random variables Z_n is said to be *asymptotically normal* with mean μ_n and standard deviation σ_n (or variance σ_n^2) if $0 < \sigma_n < \infty$ and $(Z_n - \mu_n)/\sigma_n$ has asymptotically a standard normal distribution. This terminology suggests approximating the distribution of Z_n by the normal distribution with mean μ_n and standard deviation σ_n. This is reasonable; it is the same as approximating the distribution of $(Z_n - \mu_n)/\sigma_n$ by the standard normal distribution, which is what the definition suggests, and what would have to be done anyway to make the standard normal tables apply. In mathematical arguments, however, the use of the special terminology can be misleading, because it suggests that the normal distribution with mean μ_n and standard deviation σ_n is in some way a limit, which of course is not true since it depends on n. Furthermore, if a parameter is involved and σ_n is of smaller order for some values of the parameter than others, usual statements and proofs may not apply to these values.

If X is normal, then $P(X$ is a boundary point of $A) = 0$ for ordinary sets A. Then, if X_n converges in distribution to X, by (3), we have $P(X_n \in A) \rightarrow P(X \in A)$ for ordinary sets A.

In particular, let A be the set of integers. Then $P(X \in A) = 0$ for any normal X. If S_n is binomial, $P(S_n \in A) = 1$, so that $P(S_n \in A)$ does not approach 0. This appears to contradict either (3) or the statement that S_n is asymptotically normal, but it actually does not. The point is that S_n does not converge in distribution, but rather that $(S_n - \mu_n)/\sigma_n$ does for suitable μ_n and σ_n; hence (3) applies to $(S_n - \mu_n)/\sigma_n$ rather than to S_n. This illustrates one way the special terminology can mislead.

B.3 *Two Central Limit Theorems*

The following theorems give two frequently applicable, convenient conditions for asymptotic normality. (Proofs are given in such texts as Cramér [1946], Doob [1953], Loève [1955], Fisz [1963]).

Theorem 9.2 (Central Limit Theorem for independent, identically distributed, real random variables). *If $X_1, X_2, \ldots$ are independent observations on a distribution with finite mean μ and finite variance σ^2, then*

$$\frac{\sum_1^n X_j - n\mu}{\sigma\sqrt{n}} \tag{9.26}$$

converges in distribution to the standard normal distribution; that is, $\sum_1^n X_j$ is asymptotically normal with mean $n\mu$ and variance $n\sigma^2$, and $\bar{X} = \sum_1^n X_j/n$ is asymptotically normal with mean μ and variance σ^2/n.

Theorem 9.3 (Liapounov Central Limit Theorem). *If $X_1, X_2, \ldots$ are independent real random variables with possibly different distributions, each having finite absolute moments of the order $2 + \delta$ for some number $\delta > 0$, and if*

$$\frac{\sum_1^n E(|X_j - \mu_j|^{2+\delta})}{(\sum_1^n \sigma_j^2)^{1+\delta/2}} \to 0 \quad as \ n \to \infty \tag{9.27}$$

where $\mu_j = E(X_j)$ and $\sigma_j^2 = \mathrm{var}(X_j)$, then

$$\frac{\sum_1^n (X_j - \mu_j)}{(\sum_1^n \sigma_j^2)^{1/2}} \tag{9.28}$$

converges in distribution to the standard normal distribution; that is, $\sum_1^n X_j$ is asymptotically normal with mean $\sum_1^n \mu_j$ and variance $\sum_1^n \sigma_j^2$, and $\bar{X} = \sum_1^n X_j/n$ is asymptotically normal with mean $\sum_1^n \mu_j/n$ and variance $\sum_1^n \sigma_j^2/n^2$.

Both of these theorems apply to the binomial distribution, since S can be represented as the sum of n random variables defined by $X_j = 1$ if the jth trial is a success, and $X_j = 0$ otherwise (Problem 65).

Actually, Theorem 9.3 applies whenever Theorem 9.2 does if the moment of order $2 + \delta$ is finite. In this case, for all j we have $\sigma_j^2 = \sigma^2$ and $E(|X_j - \mu_j|^{2+\delta}) = c$, say, so that the left-hand side of (9.27) becomes

$$\frac{nc}{(n\sigma^2)^{1+\delta/2}} = \frac{c}{n^{\delta/2}\sigma^{2+\delta}} \tag{9.29}$$

which indeed approaches 0 as $n \to \infty$. This illustrates the fact that the left-hand side of (9.27) has a tendency to approach 0 at the rate $n^{-\delta/2}$ for some $\delta > 0$, so the absolute moments $E(|X_j - \mu_j|^{2+\delta})$ must misbehave quite badly before (9.27) will fail.

PROBLEMS

1. Show that for any estimator $T_n(X_1, \ldots, X_n)$ of a parameter θ based on a sample of size n, if $\lim_{n\to\infty} E(T_n) = \theta$ and $\lim_{n\to\infty} \mathrm{var}(T_n) = 0$, then T_n is consistent for θ, that is, (3.5) holds.

2. Show that, for n Bernoulli trials, the probability that s successes occur on s specified trials is the same regardless of which s of the n trials are designated as successes.

3. Complete the proof of the equivalence of the three conditions for sufficiency given in Sect. 3.3
   (a) for the binomial distribution.
   (b) for an arbitrary frequency function.
   (c) for an arbitrary density function.

4. (a) Show that, if S is binomial with parameters n and p, then $S(n - S)/[n(n - 1)]$ is a minimum variance unbiased estimator of $p(1 - p)$.
   (b) Show that max $p(1 - p) = \frac{1}{4}$.
   (c) Show that the maximum value of the estimator in (a) exceeds $\frac{1}{4}$ by $1/4n$ for n odd, and by $1/(4n - 4)$ for n even.

5. (a) Show that the mean squared error of an estimator T of a parameter θ can be
        reduced to 0 for a particular value of θ by suitable choice of T.

   *(b) Suppose the same estimator has mean squared error 0 for two different values
        of θ. What unusual condition would follow for the distribution of the obser-
        vations?

*6. (a) Show that the nonrandomized estimator $T(S)$ defined in Sect. 3.4 has the same
        mean as the randomized estimator Y_S, and smaller variance unless they are
        equal with probability one.

   (b) Generalize (a) to any statistic S, sufficient or not, in any problem, and to any
        loss function $v(y, \theta)$ which is convex in the estimator y for each value of the
        parameter θ.

*7. Let X be any random variable. Show that
   (a) The points of positive probability are at most countable.
   (b) There exists a finite or countably infinite set A such that $P(X \in A) = 1$ if and
        only if the points of positive probability account for all of the probability.
        (Either of these conditions could therefore be used to define "discrete" for
        random variables or distributions.)

8. Give (real or hypothetical) examples of results of tests of binomial hypotheses
   which are
   (a) not statistically significant but are apparently practically significant.
   (b) statistically significant but are practically not significant.

9. Show that if a test of hypothesis has level 0.05, then it also has level 0.10.

10. Graph the power of the upper-tailed binomial test at level 0.10 of $H_0: p \le 0.6$ in
    the case where $n = 10$.

11. Show that any two tests which are equivalent have the same exact level and the
    same power against all alternatives, but not conversely.

12. Show that two test statistics are equivalent if they are strictly monotonically
    related.

13. Suppose that the rejection region of a lower-tailed binomial test is $S \le s_c$, and let
    $\alpha(p) = P_p(S \le s_c)$. Show that as p increases, $\alpha(p)$ decreases for any fixed n and s_c,
    so that $\alpha(p_0) > \alpha(p)$ for any $p > p_0$ and $\alpha(p_0) < \alpha(p)$ for any $p < p_0$. Hence, if the
    null hypothesis is $H_0: p \ge p_0$ and s_c is chosen in such a way that $\alpha(p_0) = \alpha$, then α is
    the maximum probability of a Type I error for all null distributions, and $1 - \alpha$ is the
    maximum probability of a Type II error for all alternative distributions, and both
    these maxima are achieved for the "least favorable case" $p = p_0$.

14. Suppose that 1 success is observed in a sequence of 6 Bernoulli trials. Is the sample
    result significant for the one-sided test of $H_0: p \ge 0.75$
    (a) at the 0.10 level?
    (b) at the 0.05 level?
    (c) at the 0.01 level?
    What is the level of just significance (critical level)?

15. Let p be the proportion of defective items in a certain manufacturing process. The
    hypothesis $p = 0.10$ is to be tested against the alternative $p > 0.10$ by the following
    procedure in a sample of size 10. "If there is no defective, the null hypothesis is

accepted; if there are two or more defectives the null hypothesis is rejected; if there is one defective, another sample of size 5 is taken. In this latter situation, the null hypothesis is accepted if there is no defective in the second sample, and it is rejected otherwise."

(a) Find the exact probability of a Type I error for this test procedure.

(b) Find the power of the test for the alternative distribution where $p \geq 0.20$.

(c) Graph the power curve as a function of p.

16. A manufacturing process ordinarily produces items at the rate of 5% defective. The process is considered "in control" if the percent defective does not exceed 10%.

(a) Find a procedure for testing $H_0: p \leq 0.05$ for a sample of size 20 and a significance level of 0.05.

(b) Find the power of this procedure when $p = 0.10$.

(c) A sample of size n is to be drawn to test the null hypothesis $H_0: p = 0.05$ against the alternative $p > 0.05$. Determine n so that the level is 0.10 and the power against $p = 0.10$ is 0.30.

17. Let p be the true proportion of voters who favor a school bond. Suppose we use a sample of size 100 to test $H_0: p \leq 0.50$ against the alternative $H_1: p > 0.50$, and 44 are in favor. Find the P-value. Does the test "accept" or reject at level 0.10?

18. A particular genetic trait occurs in all individuals in a certain population with probability either $\frac{1}{4}$ or $\frac{1}{6}$. It is desired to determine which probability applies to this population. If a sample of 400 is to be taken, construct a test at level 0.01 for the null hypothesis $H_0: p = \frac{1}{4}$.

(a) Find the power of this test.

(b) If 60 individuals in the sample have this genetic trait, what decision would you reach?

19. (a) Suppose that k individual tests of a null hypothesis H_0 are given, and their respective exact levels are $\alpha_1, \ldots, \alpha_k$. Show that the combined test, which rejects H_0 if and only if at least one of the given tests rejects H_0, has exact level $\alpha \leq \alpha^*$ where $\alpha^* = \alpha_1 + \cdots + \alpha_k$.

(b) Under what circumstances can we have $\alpha = \alpha^*$?

*(c) If the individual tests are independent under H_0, show that $\alpha^*/(1 + \alpha^*) < \alpha \leq \alpha^*$. (This implies, for instance, that α^* does not overstate α by more than 10% if $\alpha \leq 0.10$.) (Hint: Show that $1 - \alpha = \prod_{i=1}^{k} (1 - \alpha_i)$ and $1 + \alpha^* \leq \prod_{i=1}^{k} (1 + \alpha_i)$, and multiply.)

(d) If the individual tests have possibly different null hypotheses $H_{0i}, i = 1, \ldots, k$, show that (a) applies with $H_0 = \bigcap_{i=1}^{k} H_{0i}$, the intersection of the H_{0i}.

20. Define a two-tailed test $T(\alpha)$ at level α by combining two conservative one-tailed tests at level $\alpha/2$, as at (4.7) and (4.8). Let $\alpha^*(\alpha)$ be the exact level of this two-tailed test.

(a) Show that $\alpha^*(\alpha) \leq \alpha$.

(b) Under what circumstances will we have $\alpha^*(\alpha) = \alpha$? Does it matter whether the null hypothesis is simple? Does it matter whether the null distribution is continuous or symmetric?

(c) Under what circumstances will $T[\alpha^*(\alpha)] = T(\alpha)$?

(d) Under what circumstances will $\alpha^*[\alpha^*(\alpha)] = \alpha^*(\alpha)$?

21. Consider a two-tailed test for which "extreme" is defined by the one-tailed P-value of the test statistic. Assume the test statistic has a unique null distribution.
    (a) If this distribution is continuous, show that the two-tailed P-value is twice the one-tailed P-value.
    (b) Derive the two-tailed P-value (described in the text) in the discrete case.
    (c) Apply this definition to all possible outcomes in Table 4.1.
    (d) In Table 4.1, what test corresponds to this definition and what exact levels are possible?

22. Define a two-tailed P-value as the one-tailed P-value plus the nearest attainable probability from the other tail.
    (a) Apply this definition to all possible outcomes in Table 4.1.
    (b) In Table 4.1, what test corresponds to this definition and what exact levels are possible?
    (c) Compare these results with (c) and (d) of Problem 21.
    (d) Apply this definition to all possible outcomes for the following null distribution.

s	1	2	3	4	5	6
$P_{p_0}(S = s)$	0.13	0.22	0.14	0.08	0.18	0.25

    (e) Show that the P-values in (d) correspond to no test.
    (f) Invent a less pathological example, preferably unimodal, with the same property.

23. In a discrete case such as the binomial, show that the two-tailed P-value is a discontinuous function of the null hypothesis for all the procedures of Sect. 4.5 except doubling the one-tailed P-value.

24. Construct a table similar to Table 4.2 for the probabilities of erroneous conclusions in the three conclusion, two-tailed binomial tests defined below.
    (a) Observed: $S \le s_\ell$ $S \ge s_u$ $s_\ell < S < s_u$
        Conclusion: $p < p_0$ $p \ge p_0$ no conclusion
    (b) Observed: $S \le s_\ell$ $S \ge s_u$ $s_\ell < S < s_u$
        Conclusion: $p \le p_0$ $p \ge p_0$ no conclusion

25. In a binomial situation with H_0: $p = p_0$, let s'_ℓ, s''_ℓ, s'_u, s''_u be critical values at exact levels α'_1, α''_1, α'_2, and α''_2 respectively, where $\alpha'_1 < \alpha''_1$ and $\alpha'_2 < \alpha''_2$. Make a table like Table 4.2 for the test procedure below, and show that the maximum probability of error is $\max\{\alpha''_1, \alpha''_2, \alpha'_1 + \alpha'_2\}$.

Observed: $S \le s'_\ell$ $s'_\ell < S \le s''_\ell$ $s''_u \le S < s'_u$ $S \ge s'_u$ $s''_\ell < S < s''_u$

Conclusion: $p < p_0$ $p \le p_0$ $p \ge p_0$ $p > p_0$ no conclusion

26. Show algebraically that when $n = 6$ the power of the binomial test which rejects H_0: $p \ge 0.5$ when $S = 1$ only is greater than the power of the test which rejects when $S = 0$ only for all $p > \frac{1}{7}$.

27. Show that a lower-tailed randomized test with $\phi_c = 0$ or 1 for critical value s_c is equivalent to a lower-tailed non-randomized test with critical value $s_c - 1$ or s_c respectively.

28. Show that, for any exact level α, $0 \le \alpha \le 1$, there is exactly one lower-tailed randomized test based on a given statistic S if the distribution of S is uniquely determined under H_0 and all tests with the same critical function are considered the same.

29. Show that when $n = 6$, the binomial test which rejects either if there are no successes, or if there is just one success and it does not occur on the first trial, has exactly the same conditional probability of rejection given S for every p as does the lower-tailed randomized test based on S which rejects always if there are no successes and with probability $\frac{5}{6}$ if there is one success and it occurs on any trial. Hence, in particular, it has the same exact level ($\alpha = \frac{6}{64}$ for $p_0 = 0.5$) and the same power.

30. Show that for $n = 6$, $H_0: p \ge 0.5$, the most powerful (randomized) test at level α, for $\frac{1}{64} < \alpha < \frac{7}{64}$, is to reject always when no success is observed and with probability $(64\alpha - 1)/6$ when 1 success is observed.

31. Consider the binomial problem of testing the simple null hypothesis $H_0: p = 0.5$ against the simple alternative $p = 0.3$ when $n = 6$ using a lower-tailed test based on S, the observed number of successes in n trials. If we restrict consideration to non-randomized tests, there are 7 different critical values of S, and hence only 7 different exact levels α possible. For each of these, the corresponding probability of a Type II error β is easily found from Table B.
    (a) Plot these 7 pairs of values (α, β) on a graph to see how α and β interact.
    (b) If randomized tests are allowed, any exact α can be obtained. Find the randomized tests for some arbitrary values of α in between the exact values, and compute the corresponding values of β. Plot these points on the graph in (a).
    (c) Show that the points in (b) lie on the straight line segments which connect successive points in (a). Complete the (α, β) graph for randomized tests. If the nominal level is 0.10, the graph provides strong support for not choosing a conservative (nonrandomized) test in this situation, while if the nominal level is 0.05, the graph provides some support for using randomized tests in this case. What if the nominal level is 0.20?

*32. Consider a one-tailed test based on a statistic S whose distribution is uniquely determined under H_0. Show that the following hold under H_0:
    (a) If S is continuous, the P-value is uniformly distributed over $(0, 1)$.
    (b) If S is discrete, the c.d.f. of the P-value of a conservative test nowhere exceeds (is "stochastically smaller" than) that of the uniform distribution.
    (c) If S is discrete and a randomized test is carried out by rejecting for $U < \phi(x)$ where U is uniform on $(0, 1)$, then the corresponding P-value is uniformly distributed between the exact P-value and the next P-value.
    (d) This randomized P-value is uniformly distributed over $(0, 1)$.

*33. Let P' be the expectation of the P-value under a simple (or least favorable) null hypothesis. Then $P' - \frac{1}{2}$ is one possible measure of how conservative a test is. Evaluate this for the situation of Table 4.1 (binomial, $n = 10$, $H_0: p = 0.6$), for the three procedures discussed there.

34. Prove that if a confidence region has confidence level 0.99, then it also has level 0.95.

35. Verify the numerical values of the upper and lower 90% confidence limits shown in Table 6.1.

36. How do the regions $A(\theta_0)$ and $C(S)$ of Sect. 6.2 relate geometrically to the Clopper–Pearson charts?

37. Show that binomial confidence limits satisfy $p_\ell(s) = 1 - p_u(n - s)$.

38. Suppose that no successes occur in n Bernoulli trials with probability of success p. Find the one-sided upper confidence limit for p at an arbitrary level α using (a) the binomial distribution, (b) the Poisson approximation, and (c) the normal approximation. For $n = 4$, graph the upper limit as a function of α for each of the procedures (a), (b) and (c).

39. One of the large automobile manufacturers has received many complaints concerning brake failure in one of their current models. The cause was traced to a factory-defective part. This same part was found defective in six out of a group of sixteen cars inspected; these six cars were designated "unsafe."
    (a) Test the hypothesis that if this model is recalled for inspection, no more than 10% in this population will be designated "unsafe."
    (b) Find an upper confidence bound for the proportion "unsafe," with a level of 0.95.
    (c) Use the large-sample method to find an approximate upper 95% confidence bound.
    (d) Find a two-sided 95% confidence interval for the proportion of cars without the defective part.
    (e) What inference procedure seems most helpful to the company managers and why?
    (f) Which assumption for the binomial model is likely not be to satisfied in this example?

40. Show that the confidence region corresponding to the usual two-tailed binomial test (defined in Sect. 4.5) is an interval and that its endpoints are the confidence bounds (defined in Sect. 6.1) at level $1 - (\alpha/2)$.

41. Verify the result stated in (6.11) concerning the expected length of a confidence interval.

*42. Let $C(S)$ be a confidence region for a parameter θ, and let $V(S) = \int_{C(S)} d\theta$ be the size of $C(S)$. Denote by $Q(\theta)$ the probability that $C(S)$ includes θ under any fixed distribution of S, that is, $Q(\theta) = P[\theta \in C(S)]$. Show that the expected size is $E[V(S)] = \int Q(\theta)d\theta$. (Hint: This is just a change of order of integration in disguise.)

*43. What happens in Problem 42 if only a portion of the range of θ is considered (e.g., $\theta > \theta_0$ for θ one-dimensional)?

*44. For the randomized, lower-tailed binomial test of p at exact level α, show that the probability of "acceptance" when s is observed is $a(p, s)$ if $0 \le a(p, s) \le 1$, is 1 if $a(p, s) \ge 1$, and is 0 if $a(p, s) \le 0$, where

$$a(p, s) = \frac{P_p(S \le s) - \alpha}{P_p(S = s)}.$$

*45. Show that $a(p, s)$ as defined in Problem 44 is a decreasing function of p for fixed s.

*46. The table below gives the usual upper 90% confidence limits and some alternate limits for a binomial parameter p when $n = 6$. Note that the alternate limit is larger for $S = 0$ than for $S = 1$.
   (a) Show that the alternate procedure has confidence level 0.90.
   (b) Show that, when the true $p = 0.5$, the alternate limits have smaller probability of exceeding p_0 than the usual limits for $0.500 < p_0 < 0.510$ and the same probability for $p_0 > 0.510$.
   (c) Show that the alternate limits have smaller expected "excess" when $p = 0.5$.
   (d) What happens in (b) and (c) for other values of p?
   (e) Show that the "acceptance" region for $H_0: p = p_0$ corresponding to the alternate procedure is not an interval for $0.500 < p_0 < 0.510$.

S	0	1	2	3	4	5	6
Usual Limit	0.319	0.510	0.667	0.799	0.907	0.983	1.00
Alternate Limit	0.510	0.492	0.667	0.799	0.907	0.983	1.00

*47. Consider the family of nonrandomized, lower-tailed binomial tests which, for each null hypothesis value p, have exact level as near α as possible. Show that the corresponding confidence region is an interval.

*48. Using the facts stated in the first paragraph of Sect. 7.4, show that for a null hypothesis $p = p_0$ or $p \geq p_0$
   (a) a randomized, lower-tailed binomial test
      (i)   is uniformly most powerful at its exact level;
      (ii) uniformly minimizes the probability of rejection for $p > p_0$ among tests as powerful at p_1 for any $p_1 < p_0$;
      (iii) is admissible;
   (b) the class of all randomized, lower-tailed binomial tests is complete.

*49. (a) Show that, under any true value p_0, the usual, nonrandomized upper confidence bound for the binomial parameter p uniformly minimizes both
      (i) the probability of exceeding values of $p > p_0$,
      and
      (ii) the expected "excess" over p_0,
      among upper confidence bounds having no greater probability of falling below any true value p_0.
   (b) Show that the randomized upper confidence bound at exact level α for all p has the properties stated in (a).
   (c) Prove a similar result for any confidence procedure corresponding to a family of one-tailed binomial tests.

50. (a) Show that there is one and only one unbiased, two-tailed test for a given binomial null hypothesis $p = p_0 (0 < p_0 < 1)$ at a given level α.
   (b) Show that this test is not equal-tailed in general.
   (c) Show that, in general, it is randomized at both critical values.
   (d) Show that, in general, even adjusting α will not make the test nonrandomized.

51. Show that a one-tailed binomial test is unbiased, and hence that a one-sided confidence procedure is also unbiased.

52. Show that a minimum likelihood test based on a sufficient statistic S having a unique null distribution is most powerful against the alternative that S is uniformly distributed, among tests at the same exact level.

53. Suppose that an unbiased test of the binomial null hypothesis $p = p_0$ is given for each p_0, but the exact level $\alpha(p_0)$ varies with p_0. What property related to unbiasedness does the corresponding confidence procedure have?

54. Show that if p is distributed uniformly over $(0, 1)$ and, for given p, S is binomial with parameters n and p, then the marginal distribution of S is discrete uniform on the integers $0, 1, 2, \ldots, n$. (For a generalization, see Raiffa and Schlaifer [1961], pp. 237–241.)

55. (Continuation of Problem 54) Show that the average power of a nonrandomized test of a binomial null hypothesis $p = p_0$, that is, the integral of the power curve over p, equals the number of possible values of S which are in the rejection region divided by $n + 1$.

56. Demonstrate formula (8.2) for the average expected length of a binomial confidence region.

*57. Generalize formula (8.2) to binomial confidence regions which are not necessarily intervals.

58. (a) Suppose we have a nonrandomized confidence procedure for a binomial parameter p such that the region is always an interval. Show that we have $\sum_s L(s) = \int N(p)dp$, where $L(s)$ is the length of the confidence region for $S = s$ and $N(p)$ is the number of values of S for which the confidence region includes p.
    (b) Generalize (a) to regions which are not necessarily intervals.

*59. Show that any two-tailed binomial test whose upper and lower critical values are not equal is admissible in the three-conclusion interpretation with risk function (8.1), under the assumptions (8.4)–(8.7) and the assumption that $(p/q)[L_h(p) - L_c(p)]/[L_a(p) - L_b(p)]$ either approaches 0 as $p \to 0$ or approaches ∞ as $p \to 1$. Hint: Show that any such test is Bayes against some three-point prior distribution of p, and is unique Bayes except at the critical values.

60. Show that if $P(S_n \le s_n) \to \Phi(z)$ whenever $(s_n - \mu_n)/\sigma_n \to z$, as in (9.13), then $P(S_n \le s_n) - \Phi[(s_n - \mu_n)/\sigma_n] \to 0$ regardless of how s_n depends on n, as in (9.14).

61. Show that if $g_n(x) \ge 0$ for all x in some countable set B, then as $n \to \infty$, $\sum_{x \in B} \lim \inf g_n(x) \le \lim \inf \sum_{x \in B} g_i(x)$.

62. Show the equivalence of the conditions (1)–(5) for convergence in distribution given at the beginning of Sect. B.2.

63. (a) If X_n is normal with mean 0 and variance n, what is $\lim F_n(x)$?
    (b) If F is nondecreasing on $(-\infty, \infty)$ and $0 \le F \le 1$, construct a sequence of c.d.f.'s F_n such that $F_n(x) \to F(x)$ for every x at which F is continuous.

64. (a) Show that if a sequence of distributions on a finite set converges in distribution, then the conditions of Theorem 9.1(1) hold.
    (b) Give a counterexample for a countably infinite set.

65. Apply the Central Limit Theorems 9.2 and 9.3 to the binomial distribution.

CHAPTER 2

One-Sample and Paired-Sample Inferences Based on the Binomial Distribution

1 Introduction

The goal of statistical inference procedures is to use sample data to obtain information, albeit uncertain, about some larger population or data-generating process. The inferences may concern any aspect of a suitably defined population (or process) from which observations are obtained, for example, the form or shape of the probability distribution of some variable in the population, or any definable properties, characteristics or parameters of that distribution, or a comparison of some related aspects of two or more populations. Procedures are usually classified as nonparametric when some of their important properties hold even if only very general assumptions are made or hypothesized about the probability distribution of the observations. The word "distribution-free" is also frequently used in this context. We will not attempt to give an exact definition of "nonparametric" now or later, as it is only this general spirit, rather than any exact definition, which underlies the topics covered in this book.

In order to perform an inference in one-sample (or paired sample) problems using the methodology of parametric statistics, information about the specific form of the population must be postulated throughout or incorporated into the null hypothesis. The traditional parametric procedure then either postulates or hypothesizes a specific form of population, often the normal, and the inferences concern some population parameters, typically the mean or variance or both. The exact distribution theory of the statistic, and hence the probabilities of both types of errors in testing and the confidence level in estimation, depend on this population form. Such inference procedures may or may not be highly sensitive to the population form. If they are not, the procedure is said to be "robust." Robustness has been extensively studied. (See, for instance Bradley [1968, pp. 28–40] and references given there.)

A nonparametric procedure is specifically designed so that only very general characteristics of the relevant populations need be postulated or hypothesized, for example, that a distribution is symmetric about some specified point. The inference is then applicable to, and completely valid in, quite general situations. In the one-sample case, the inference concerns some definable property or aspect of the one population. For example, if symmetry is assumed, such an inference might concern the true value of the center of symmetry and its exact level may be the same for all symmetric populations. Symmetry is a much less restrictive assumption than normality. Alternatively, the inference may be an estimate or hypothesis test of the value of some other parameter in a general population. In short, a nonparametric procedure is designed so as to be perfectly robust in certain respects (usually the exact significance or confidence level) under some very general assumptions.

The remainder of this book will consider various situations where inferences can be made using nonparametric procedures, rather than studying post hoc the robustness of parametrically derived procedures. In this chapter the inferences will be based on the binomial probability distribution; however, they are valid for observations from general populations. The first type of inference to be covered relates to the value of a population percentile point like the median, the first quartile, etc. For data consisting of matched pairs of measurements, the same procedures are applicable for inferences concerning the population of differences of pairs. If the matched pair data are classificatory rather than quantitative, for example classified as either success or failure, inferences about the differences of pairs can also be made by similar procedures, but they merit separate discussion. Finally, we will discuss one-sample procedures for setting tolerance limits for the distribution from which observations are obtained.

2 Quantile Values

Most people are familiar with the terms percentiles, median, quartiles, etc., when used in relation to measurement data, for example, in reports of test scores. These are points which divide the measurements into two parts, with a specified percentage on either side. If a real random variable X has a continuous distribution with a positive density, the statement that the median or fiftieth percentile is equal to 10 means that 10 is the point having exactly one-half of the distribution of X below it and one-half above. This statement can be expressed in terms of probability as

$$P(X < 10) = 0.5, \qquad P(X > 10) = 0.5.$$

A similar probability statement could be made for other parameters of this type. For example, 10 is the 75th percentile point if $P(X < 10) = 0.75$, $P(X > 10) = 0.25$.

While such a definition of these order parameters is simple to understand and interpret, it is fully satisfactory only in those cases where the point so defined exists and is a unique number. In order to take care of all cases, we give the following more explicit definition. For any number p, where $0 \le p \le 1$, the *pth quantile*, or the *quantile of order p*, of the distribution of X is any value ξ_p which satisfies

$$P(X < \xi_p) \le p \le P(X \le \xi_p). \tag{2.1}$$

(This parameter may also be called the *pth fractile* or *fractile of order p*, the *p-point*, or the *(100p)th percentile point*.) Equation (2.1) says that the probability to the left of ξ_p is at most p if ξ_p is excluded, and at least p if ξ_p is included. An equivalent statement is that if ξ_p is excluded, the probability to the left is at most p and the probability to the right is at most $1 - p$.

We now investigate more specifically the implications of the definition in (2.1) for various types of distributions. We shall see that (a) ξ_p may be unique and belong to a unique p, (b) the possible values of ξ_p for given p may form a closed interval, or (c) ξ_p may have the same value for a closed interval of values of p. Thus p and ξ_p have a one-to-one relationship only in case (a). The three types of situation which lead to these possibilities are explained below and are illustrated in Fig. 2.1 as (a), (b) and (c) respectively.

(a) The simple case is where X has a strictly increasing, continuous c.d.f., as holds if it has a positive density. Then ξ_p exists, is unique and applies to

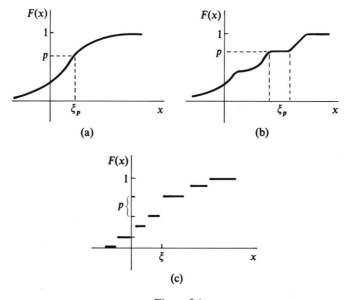

Figure 2.1

just one value of p. Thus there is a one-to-one relationship between p and ξ_p, and given ξ_p or p, the other one can be found as that unique number which satisfies $F(\xi_p) = p$.

(b) Suppose that F is not strictly increasing, that is, F is constant in some interval of positive length. If the value of F in the interval is p, every point in the interior of this interval has probability p to the left of it (and $1 - p$ to the right) and hence is a possible value of ξ_p. The endpoints of the interval also satisfy (2.1) (Problem 1). Thus the pth quantiles of the distribution form a closed interval of positive length. This nonuniqueness can easily be removed by some arbitrary convention, but this does not reduce the difficulties in inference procedures relating to a nonunique ξ_p. The most typical convention is to call the midpoint of the interval of values "the" pth quantile. We shall not follow this convention, although sometimes, as we have already done, we take the liberty of saying "the" pth quantile when "a" pth quantile would be more accurate.

(c) Suppose that F is discontinuous, as for a discrete distribution. At each discontinuity point ξ, there must be a jump in F. Then ξ is a pth quantile for a whole interval (endpoints included) of values of p, and this interval has positive length since, by (2.1), a given value ξ is the pth quantile for any p satisfying

$$P(X < \xi) \leq p \leq P(X \leq \xi). \tag{2.2}$$

This discussion shows that by our definition in (2.1), for any p there is at least one pth quantile ξ_p, and any ξ is a pth quantile for at least one value of p. In any case, the set of possible combinations of p and ξ_p is given by the c.d.f. with any vertical jumps filled in (Problem 2). Of course, a c.d.f. may exhibit more than one of (a)–(c) (Problem 4).

Since any quantile ξ_p is a population parameter, point estimates, tests of hypotheses, and confidence intervals for ξ_p are all of interest. The natural point estimate of ξ_p is a sample quantile of the same order p, but a precise definition of a sample quantile must be arbitrary if a unique value is desired. These problems do not arise if a confidence interval for ξ_p is obtained, and confidence intervals provide more information and are more useful in most situations anyway. For the most part, discussion of inferences about quantiles in this book will be restricted to confidence intervals and tests of hypotheses.

3 The One-Sample Sign Test for Quantile Values

3.1 Test Procedures

Let $X_1, \ldots, X_n$ be n independent observations drawn from the same distribution, and suppose we wish to test the null hypothesis that the median of this distribution is 0. Let us suppose that the point 0 does not have positive probability, that is, assume that $P(X_j = 0) = 0$. (The contrary case is more complicated and will be discussed in Sect. 6.) Then with probability 1, each observation is either positive (>0) or negative (<0), and $P(X_j < 0) = P(X_j > 0) = 0.5$ under the null hypothesis. We could then test the null hypothesis by counting the number S of negative (or positive) observations. Under the null hypothesis, S is binomial with parameter $p = 0.5$, and the tests discussed in Chap. 1 for this hypothesis may be applied to S as defined here.

An upper-tailed test, rejecting when S is larger than some critical value, that is, when there are too many negative observations, is appropriate against alternatives under which the probability of a negative observation is larger than 0.5, that is, alternatives with $P(X_j < 0) > 0.5$. Under such alternatives, the population distribution is more negative than under the null hypothesis in the sense that its median is negative instead of 0.

A lower-tailed test is appropriate against alternatives under which the population distribution has a positive median. A two-tailed test is appropriate when one is concerned with both alternatives.

Let F be the c.d.f. of the distribution from which the X_j are sampled. Since we are assuming that the point 0 does not have positive probability, $F(0) = P(X_j \leq 0) = P(X_j < 0)$, and the null hypothesis can be stated as H_0: $F(0) = 0.5$. Notice that an alternative distribution with $F(0) > 0.5$ is more negative, in the above sense, because if the probability of a negative observation exceeds 0.5, the population median must be negative. That is, loosely speaking, the larger F is, the more negative the population. This is illustrated in Fig. 3.1, for arbitrary c.d.f.'s F_1 and F_2, where $F_2(0) > F_1(0) = 0.5$ and the medians are related by $\xi_2 < \xi_1 = 0$.

The power of the tests above is, of course, just the power of whichever binomial test is used (lower-, upper-, or two-tailed) against the alternative $F(0) = p$ for some $p \neq 0.5$.

Of course, there is nothing special about the particular value 0 for the median. To test the null hypothesis that the distribution of every observation has median ξ_0, say, assuming that $P(X_j = \xi_0) = 0$ for all j, we would define S as the number of observations which are smaller than ξ_0. Under this null

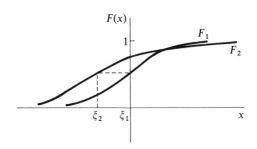

Figure 3.1

hypothesis S is again binomially distributed with parameter $p = 0.5$ and the tests previously discussed apply in exactly the same way.

Similarly, there is nothing special about the median rather than some other percent point. To test the null hypothesis that the distribution of every observation has p_0-point ξ_0, assuming that $P(X_j = \xi_0) = 0$ for all j, we would define S as the number of observations which are smaller than ξ_0. Under the null hypothesis, $P(X_j < \xi_0) = p_0$, so that S is binomially distributed with parameter $p = p_0$. Again the binomial tests may be employed.

These procedures are all frequently called "*sign tests*" because they use only the signs of $X_j - \xi_0$ and not the precise values of the observations.

While the independence of the observations is fundamental, the other assumptions can be relaxed some without affecting the validity of the test. If the observations are not identically distributed under the null hypothesis, the level of the test is unchanged provided that each observation is drawn from a distribution with p_0-point ξ_0. (For paired samples, the null hypothesis of median 0 may be justified by random assignment. See Sect. 7.) An upper-tailed (lower-tailed) test is appropriate against alternatives under which the population distributions of the observations all have negative (positive) medians. A two-tailed test is appropriate when one is concerned with both of these alternatives; this means either that the population distributions of the observations all have negative medians, or that they all have positive medians, but not that some have negative medians and some positive. Against strongly mixed alternatives of the last kind, one- and two-tailed sign tests may all be ineffective and other procedures would be desirable.

The situation is more complicated if the probabilities $p_j = P(X_j < \xi_0)$ are allowed to differ under the null hypothesis. A "random effects" model under which the p_j are independently sampled from a population with mean p reduces immediately to the original situation provided p is the parameter of interest. A simple result for the "fixed effects" model is that the one-tailed tests are valid for either $H_0: p_j \leq p_0$ for all j or $H_0: p_j \geq p_0$ for all j, as appropriate. A much deeper result for the fixed effects model is that if one is interested in the parameter $p = \sum_j p_j/n$, then a one- or two-tailed sign test is valid for the null hypothesis $p = p_0$ provided the level is small enough (up to approximately 0.5) so that the "acceptance" region includes the integers just above and below np_0 (np_0 itself if it is an integer). This "main effect" p may well not be the parameter of interest, however. Indeed its definition even depends on the sample size unless the experiment or sample is specifically designed to avoid this. Unfortunately, for other null hypotheses the sign test may be invalid or its validity not strictly proved.

The power against alternatives under which the observations are not identically distributed can be calculated from the binomial distribution as before if p_j is the same for all j, or if the random effects model mentioned above holds. Under a fixed effects model with differing p_j, the power might be approximated by treating S as binomial with parameter $p = \sum_j p_j/n$. This overestimates the power whenever the "acceptance" region includes the integers just above and below np, that is, whenever $p \geq (s_\ell + 1)/n$ and/or $p \leq (s_u - 1)/n$ where s_ℓ and s_u are the lower and upper critical values respectively. It underestimates the power of a one-tailed test for $p \leq s_\ell/n$ or $p \geq s_u/n$ as relevant (where the power exceeds 0.5, approximately). The power of a two-tailed test can be underestimated in the same region by simply ignoring the (ordinarily negligible) probability of rejection in the "wrong" tail and using the one-tailed lower bound for the "correct" tail. These results for the "fixed effects" model are due to Hoeffding [1956].

3.2 "Optimum" Properties

The one-tailed sign tests are, in a sense which will now be described, the best possible. Certain of the two-tailed tests have similar but less strong properties, and all are admissible.

Assume once more that the observations $X_1, \ldots, X_n$ are independent and identically distributed and we wish to test the null hypothesis that the median ξ of the population distribution is a particular number ξ_0. Assume also that $P(X_j = \xi_0) = 0$ under both null and alternative hypotheses. Consider the class of tests based on S, the number of observations smaller than ξ_0. Since S is binomial, there is a level α test based on S which is uniformly most powerful against one-sided alternatives, namely the appropriate one-tailed binomial test at exact level α. This test is just the one-tailed sign test at exact level α.

What if one considers not only tests based on S, but also tests which make further use of the original observations? One might think that a better test could be produced by taking into account how far above and below ξ_0 the observations fall, rather than just how many fall above and how many below. This is not possible, however, as long as one insists that the test have level α for *every* population distribution with median ξ_0. Compared to any other such test, the one-tailed sign test at exact level α which rejects when S is too small has greater power than any other test against every alternative distribution under which the median exceeds ξ_0. A similar statement holds for the test rejecting in the other tail and alternatives on the other side. In other words, a one-tailed sign test at exact level α is uniformly most powerful against the appropriate one-sided alternative, among all level α tests based on $X_1, \ldots, X_n$ for the null hypothesis that the median is ξ_0.

For two-sided alternatives there is, of course, no uniformly most powerful test of the null hypothesis that the median is ξ_0. Suppose, however, that we consider only unbiased tests, that is, tests which reject with probability at least as great under every alternative as under the null hypothesis. Among these tests, the unbiased sign test (which is equal-tailed) is uniformly most powerful.

The symmetry of the situation suggests another way of restricting the class of tests to be considered with a two-sided alternative. Suppose we had observed not $X_1, \ldots, X_n$ but $Y_1, \ldots, Y_n$, where Y_j is the same distance from ξ_0 as X_j but on the other side (that is, $Y_j = 2\xi_0 - X_j$). If the X_j satisfy the null hypothesis, so do the Y_j; if they satisfy the alternative, so do the Y_j. Hence, in the absence of other considerations, it seems equally reasonable to apply a test to the Y's as to the X's, and it would be unpleasant if different outcomes resulted. This suggests requiring that a test be symmetric in the sense that applying it to the Y's always gives the same outcome as applying it to the X's. Among such tests also, the equal-tailed sign test is uniformly most powerful.

Every two-tailed sign test is admissible (in the two-conclusion interpretation of two-tailed tests); that is, any test having greater power at some alternative has either smaller power at some other alternative or greater probability of rejection under some distribution of the null hypothesis.

The restriction to identically distributed observations can be relaxed without affecting any of the properties above. That is, the results hold for alternatives under which $X_1, \ldots, X_n$ are independent, with $P(X_j < \xi_0)$ the same for all j, but are not necessarily identically distributed, provided the null hypothesis is similarly enlarged.

If "p_0-point" is substituted for the hypothesized median value ξ_0 throughout, the foregoing properties continue to hold, except that an unbiased test is not equal-tailed when $p_0 \neq 0.5$, and the discussion of symmetry no longer applies. In summary, the "optimum" properties of the sign tests are as follows.

If $X_1, \ldots, X_n$ are independent and identically distributed with $P(X_j = \xi_0) = 0$, then among tests of the null hypothesis $P(X_j < \xi_0) = p_0$:

(a) A one-tailed sign test is uniformly most powerful against the appropriate one-sided alternative;

(b) Any two-tailed sign test is admissible;

(c) A two-tailed, unbiased sign test is uniformly most powerful against the two-sided alternative $P(X_j < \xi_0) \neq p_0$ among unbiased tests and, when $p_0 = 0.5$, among symmetric tests.

If $X_1, \ldots, X_n$ are not necessarily identically distributed but are independent with $P(X_j < \xi_0) = P(X_j \leq \xi_0) = p$ for all j, then the same statements apply to the null hypothesis $p = p_0$ and the alternative $p \neq p_0$.

The proof of the foregoing statements, which will be given in Sect. 3.3, depends on the fact that the null hypotheses are very broad and are satisfied by some peculiar distributions, like the density represented by the dotted line in Fig. 3.2(b). If one is willing to test a more restrictive null hypothesis, there could well be tests which are more powerful, at least against some alternatives. For instance, the hypothesis that the median is ξ_0 might be replaced by the hypothesis that the distribution is symmetric around ξ_0. Nonparametric tests of this null hypothesis will be discussed in Chap. 3 and 4.

For $p_0 \neq 0.5$, restrictions of the null hypothesis that the p_0-point is ξ_0 have been studied only for parametric situations. For instance, the hypothesis might be that the observations come from a normal population with p_0-point

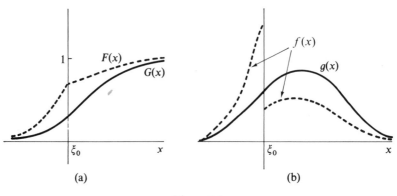

Figure 3.2

ξ_0. Procedures based on such assumptions are outside the scope of this book, but it is relevant to mention that considerable risk accompanies their apparent advantages. The risks and advantages are easily seen in terms of the estimators involved. With no distribution assumption, the true population probability below ξ_0, say p, would be estimated by S/n, where S is the number of observations below ξ_0 in the sample. With the assumption of a normal population, the probability below ξ_0 is $p' = \Phi[(\xi_0 - \mu)/\sigma]$, where μ and σ are the population mean and standard deviation and Φ is the standard normal c.d.f. Intuition would lead one to use $\hat{p} = \Phi[(\xi_0 - \bar{X})/s]$ as an estimator for p', where $\bar{X}$ and s are the sample mean and standard deviation. (This estimator $\hat{p}$ is slightly biased for p' in normal populations, but can be adjusted to be unbiased and in fact minimum variance unbiased [Ellison, 1964].) Under normality, $\hat{p}$ (or $\hat{p}$ adjusted) is a much better estimator than S/n. However, a departure from normality which looks minor can easily lead to an important difference between p' and the true proportion p and therefore very poor properties for $\hat{p}$. (Typical goodness-of-fit tests of the normality assumption will throw almost no light on this crucial question.) In fact, such information as the sample provides about p beyond the value of S/n is, in common sense, relatively little and difficult to extract. The advantage of the estimator $\hat{p}$ over S/n relies most heavily on the assumed normal shape when ξ_0 is in the extreme tails, and hence these reservations about $\hat{p}$ apply most strongly when p is close to 0 or 1, which is unfortunately just when the advantage of $\hat{p}$ is also greatest. (See also Sect. 3.1 of Chap. 8.)

Why do similar reservations not apply to nonparametric procedures based on symmetry? Because symmetry may well be more plausible than normality, and the effect of departures from the assumption are less serious. In fact, nonparametric procedures based on symmetry are often used to make inferences about the location of the "center" of the population distribution; a departure from symmetry will require that this "center" be defined somehow, and the definition implicit in the procedure used may be satisfactory. In contrast, an inference about p is typically made in a situation where the proportion of the population beyond some critical point is of interest (a tolerance limit, for instance). Then it will not be satisfactory if the inference is really about some other parameter, like p' above, which may differ appreciably from p under mild departures from the assumptions.

*3.3 Proofs

We will demonstrate the optimum properties summarized above for the sign tests only in the case where $X_1, \ldots, X_n$ are independent and identically distributed with $P(X_j = \xi_0) = 0$. We will prove that the one-tailed, and unbiased two-tailed, sign tests of the null hypothesis $P(X_j < \xi_0) = p_0$ are, respectively, uniformly most powerful, and uniformly most powerful unbiased (and, when $p_0 = 0.5$, uniformly most powerful symmetric) against the appropriate alternatives. The proofs of admissibility and the extension to observations with possibly different distributions are requested in Problems 11 and 12.

Consider any alternative distribution G with $P_G(X_j < \xi_0) = P_G(X_j \leq \xi_0) = p_G \neq p_0$. For any p, define F as a distribution with probability p to the left of ξ_0 and the same conditional distribution as G on each side of ξ_0. (To simplify notation, the dependence of F on p will not be indicated.) Specifically, let

$$
F(x) = \begin{cases} \dfrac{p}{p_G} G(x) & \text{for } x \leq \xi_0 \\[2ex] 1 - \dfrac{1-p}{1-p_G} [1 - G(x)] & \text{for } x \geq \xi_0. \end{cases}
$$

Then $P_F(X_j < \xi_0) = P_F(X_j \leq \xi_0) = p$. Also, the conditional distribution of X_j, given that $X_j < \xi_0$, is the same under F as G, as is the conditional distribution given that $X_j > \xi_0$. The family of distributions F includes G (when $p = p_G$) and a null distribution F_0 (when $p = p_0$). The point of the definition is that F_0 is the null distribution "most like" the given alternative G, or "least favorable" for testing against G, as we shall see.

Figure 3.2(a) illustrates possible c.d.f.'s of the type defined by F and G; notice the abrupt change of slope in F at ξ_0. The corresponding densities f and g are shown in Fig. 3.2(b), where f and g are related by

$$
f(x) = \begin{cases} \dfrac{p}{p_G} g(x) & \text{for } x < \xi_0 \\[2ex] \dfrac{1-p}{1-p_G} g(x) & \text{for } x > \xi_0. \end{cases}
$$

Thus f is the same as g on each side of ξ_0 except for a scale factor, but f has ξ_0 as a quantile of order p.

If $X_1, \ldots, X_n$ are independent and identically distributed according to one of the distributions F, then S, the number of observations below ξ_0, is a sufficient statistic (Problem 13) and is binomially distributed with parameters n and p. It follows that the appropriate one-tailed, level α sign test is most powerful against G among level α tests of the null hypothesis F_0. It is therefore most powerful against G among level α tests of the original null hypothesis that $P(X_j < \xi_0) = p_0$ because (a) it has level α for the original null hypothesis, while (b) any level α test of the original null hypothesis is also a level α test of F_0 and hence cannot be more powerful against G. Since G was arbitrary, a one-tailed sign test is uniformly most powerful against alternatives on the appropriate side.

Similarly, the unbiased, level α, two-tailed sign test is most powerful against G among level α tests of the null hypothesis F_0 which are unbiased against the alternatives F for which $p \neq p_0$. It is, therefore (Problem 14), most powerful against G among level α, unbiased tests for the original problem, and hence, uniformly most powerful.

When $p_0 = 0.5$, the symmetric, level α, two-tailed sign test is most powerful against G among symmetric, level α tests of the null hypothesis F_0. It is, therefore (Problem 15), uniformly most powerful among level α, symmetric tests for the original problem.

The above proof contains, in effect, a proof of the following theorem, which will be useful later in this book. The formal proof is requested in Problem 16.

Theorem 3.1. *Let H_0 be a null hypothesis and suppose that H_0' is contained in H_0.*

(a) *A test is most powerful against an alternative G among level α tests of H_0 if it has level α for H_0 and is most powerful against G among level α tests of H_0'.*

(b) *A test is most powerful against G among level α tests of H_0 which are unbiased against H_1 if it has level α for H_0, is unbiased against H_1, and is most powerful against G among level α tests of H_0' which are unbiased against H_1', where H_1' is contained in H_1.*

(c) *A test is most powerful against G among level α symmetric tests of H_0 if it is symmetric, has level α for H_0, and is most powerful against G among symmetric level α tests of H_0'.*

(d) *The property in (c) holds if the requirement of symmetry is replaced by any requirement that does not depend on H_0.**

4 Confidence Procedures Based on the Sign Test

Assume that $X_1, \ldots, X_n$ are independent observations, identically and continuously distributed. Then, with probability one, no two observations are equal. We have discussed the sign test procedure for hypotheses of the form $P(X_j < \xi) = p$. We shall now derive two types of confidence procedures which correspond to these tests. First, confidence intervals are derived for the true value of p when ξ is fixed, and second, confidence intervals are derived for the true value of ξ when p is fixed, i.e., for the quantile ξ_p. The first requires only brief mention. The second involves the concept of "order statistics," which merit some discussion in their own right.

The confidence bounds for $p = P(X_j < \xi)$ for a fixed ξ which correspond to the sign test are obtained by direct application of the standard binomial confidence procedures (Sect. 6 of Chap. 1) to S, the number of observations below ξ, since the test is based on S and since S is binomial with parameters n and p. Many properties of these confidence bounds follow immediately from the previous section and the results of Chap. 1.

The confidence bounds for a quantile ξ_p, such as the median, which correspond to the sign test turn out to be certain *order statistics* of the sample, which are defined as follows. For a set of observations $X_1, \ldots, X_n$ which are all different, let $X_{(1)}$ denote the smallest of the set, $X_{(2)}$ the second smallest, etc., and $X_{(n)}$ the largest. Then, since we assumed that there are no ties, we have $X_{(1)} < X_{(2)} < \cdots < X_{(n)}$, and these are the original observations after arrangement in increasing order of magnitude. They are collectively called the order statistics of the sample, and $X_{(r)}$, for $1 \le r \le n$, is called the rth order statistic. Note that the sign of an observation is considered in determining its size; for instance, $-3, -2, 0, 1$ are arranged in order of size. (If there are ties, we can only require that $X_{(1)} \le X_{(2)} \le \cdots \le X_{(n)}$.)

The property of order statistics which is of immediate relevance is that, for any ξ, we have $X_{(r)} < \xi$ if and only if at least r of the n observations are less than ξ. Consider now a one-tailed sign test at level α with critical value s_ℓ in the lower tail of S. This test "accepts" the null hypothesis $\xi_p = \xi$ if and only if at least $s_\ell + 1$ of the observations are smaller than ξ. This is equivalent to having $X_{(s_\ell + 1)} < \xi$, by the property just stated. Therefore $X_{(s_\ell + 1)}$ is the level $1 - \alpha$ lower confidence bound for ξ_p corresponding to the sign test.

For example, a lower-tailed binomial test of $H_0: p = 0.5$ with $n = 18$ observations rejects at level 0.05 if there are 5 or fewer successes (observations smaller than the hypothesized median). Therefore $X_{(6)}$, the sixth smallest observation, is a lower 95% confidence bound for the population median for any set of 18 observations. Since from Table B the exact level of this test is 0.0481, the exact confidence level is 0.9519.

An upper confidence bound is found similarly. If s_u is the critical value for an upper-tailed, level α binomial test with parameters n and p, then the (s_u)th smallest observation is the level $1 - \alpha$ upper confidence bound for ξ_p since the sign test rejects $\xi_p = \xi$ if and only if at least s_u observations are smaller than ξ. For instance, 13 is the critical value for a level 0.05, upper-tailed test of $p = 0.5$ with $n = 18$. Therefore $X_{(13)}$, the 13th smallest observation, is a 95% upper confidence bound for the median, with exact confidence coefficient again 0.9519.

The interval between the lower confidence bound at level $1 - \alpha_1$ and the upper confidence bound at level $1 - \alpha_2$ is, of course, a confidence interval for the quantile ξ_p at level $1 - \alpha_1 - \alpha_2$. This corresponds to a two-tailed sign test with lower and upper tail probabilities α_1 and α_2.

For $p = 0.5$, an equal-tailed test will have $s_u = n - s_\ell$, by symmetry. The corresponding confidence interval for the median is then the interval between the $(s_\ell + 1)$th and the $(n - s_\ell)$th smallest observations. This interval has the symmetry one would expect, inasmuch as the lower endpoints is the $(s_\ell + 1)$th smallest observation and the upper endpoint is the $(s_\ell + 1)$th from the largest. (The jth smallest observation is the $(n + 1 - j)$th from the largest, and $n + 1 - j = n + 1 - s_u = s_\ell + 1$ here.) For instance, with 18 observations, the interval between the 6th and 13th smallest observations is a 90 % confidence interval for the median, with exact level $1 - 2(0.0481) = 0.9038$. There is no such symmetry for $p \neq 0.5$, that is, for quantiles other than the median.

If the quantile is not unique, there is an interval of values for the quantile as explained earlier in Sect. 2. Then the confidence procedures just described are equally valid for any determination of the quantile. In fact, the probability is at least $1 - \alpha$ that the lower $1 - \alpha$ confidence bound will fall at or below the lower end of the interval, and similarly that the upper bound will be at or above the upper end (Problem 19).

Whether or not the confidence interval is taken to include its endpoints makes no practical difference in any ordinary application, and it makes no theoretical difference as long as the population distribution is continuous, that is, has no points of positive probability. The error rate for a discontinuous population will be at most that for a continuous one if the endpoints are included in the confidence interval, and at least that for a continuous one if they are excluded (Problem 20, or, more generally, see Sect. 6 of Chap. 3).

If the observations $X_1, \ldots, X_n$ are independent but are not identically distributed, the procedures above remain valid provided that the quantile ξ_p is identical for all populations from which the observations are drawn. If the quantiles are not unique, the confidence intervals are valid for every ξ_p which is a quantile of order p for all the populations.

The properties of these confidence procedures correspond to those of the tests from which they were obtained, which were developed in Sect. 3. Here we shall say only that one can hope to do better only by making further assumptions. Alternative confidence procedures for the median based on the assumption of a symmetric population are discussed in Chap. 3, especially Sect. 4. For other quantiles, the only available alternative procedures are based on parametric assumptions and are sensitive to deviations from these assumptions, as discussed earlier.

We have developed the procedures for finding confidence bounds and confidence intervals for quantiles as those corresponding to the sign test. Since the end points are order statistics of the sample observations, these confidence regions can also be developed independently of the sign test using the principles and properties of order statistics. Thus we will now give a brief introduction to some properties of order statistics and show that the confidence regions developed above can be obtained directly from them.

Order statistics are particularly useful in nonparametric statistics, partly because of the properties of an important device called the *probability integral transformation*. If X has a continuous c.d.f. F, then the transformed random variable $U = F(X)$ has a uniform distribution on the interval $(0, 1)$. Furthermore, if $X_{(r)}$ for $1 \leq r \leq n$ are the order statistics of a sample of n from any distribution with continuous c.d.f., then the transformed random variables $U_{(r)} = F(X_{(r)})$ for $1 \leq r \leq n$ are the order statistics of a sample of n from the uniform distribution on the interval $(0, 1)$, and therefore the $U_{(r)}$ are distribution-free. These and other properties of order statistics are Problems 21, 29–36.

Consider the rth order statistic for a sample of n from any distribution F. Then since $X_{(r)} \leq x$ if and only if at least r of the n observations are less than or equal to x, the c.d.f. of $X_{(r)}$ is the upper tail of the binomial distribution with parameters n and $F(x)$, that is

$$P(X_{(r)} \leq x) = \sum_{k=r}^{n} \binom{n}{k} [F(x)]^k [1 - F(x)]^{n-k}. \qquad (4.1)$$

Now suppose that F is continuous and x is its quantile of order p so that $F(\xi_p) = p$. Then (4.1) can be written as

$$P(X_{(r)} \leq \xi_p) = P(X_{(r)} < \xi_p) = \sum_{k=r}^{n} \binom{n}{k} p^k (1 - p)^{n-k}. \qquad (4.2)$$

Thus a lower confidence bound for ξ_p at any desired (attainable) level $1 - \alpha$ can be found by setting the right-hand side of (4.2) equal to $1 - \alpha$, that is, by determining that value of r which has upper tail probability $1 - \alpha$ in the binomial distribution with parameters n and p. The lower bound at conservative level $1 - \alpha$ is $X_{(r)}$ for the smallest integer r which makes the right-hand side of (4.2) at least $1 - \alpha$. An upper confidence bound for ξ_p is found similarly. Further, we have (Problem 34)

$$P(X_{(r)} < \xi_p < X_{(v)}) = P(X_{(r)} < \xi_p) - P(X_{(v)} \leq \xi_p)$$

$$= \sum_{k=r}^{v-1} \binom{n}{k} p^k (1 - p)^{n-k}. \qquad (4.3)$$

Thus we obtain a confidence interval for ξ_p at any desired confidence level $1 - \alpha$ by suitable choice of r and v, using the binomial distribution.

The result in (4.2) can also be written in the form of the beta distribution (Problem 30) as

$$P(X_{(r)} < \xi_p) = \int_0^p [B(r, n - r + 1)]^{-1} u^{r-1} (1 - u)^{n-r} du \qquad (4.4)$$

where

$$B(r, v) = \frac{(r + v - 1)!}{(r - 1)! (v - 1)!}. \qquad (4.5)$$

As a final comment, we note that the event $X_{(r)} < \xi_p$ is equivalent to $F(X_{(r)}) < p$. Applying the probability integral transformation, the latter inequality can be replaced by $U_{(r)} < p$ where $U_{(r)}$ is the rth order statistic from a uniform distribution on $(0, 1)$. Since the density of $U_{(r)}$ is the integrand of (4.4) (Problem 30), this observation provides a direct verification of the expression given in (4.4).

5 Interpolation between Attainable Levels

When the possible values of α are discrete, as above, there may be no (non-randomized) confidence bound at exactly the level desired. One might then be content with a nearby level which is attainable. If not, a bound at approximately the desired level may be obtained by interpolating linearly between the two neighboring attainable levels. That is, if W_1 and W_2 are two lower (or two upper) confidence bounds at levels $1 - \alpha_1$ and $1 - \alpha_2$ respectively, with $\alpha_1 < \alpha < \alpha_2$, then a lower (or upper) confidence bound with level approximately $1 - \alpha$ (and certainly between $1 - \alpha_1$ and $1 - \alpha_2$) is

$$\frac{(\alpha_2 - \alpha)W_1 + (\alpha - \alpha_1)W_2}{\alpha_2 - \alpha_1}. \tag{5.1}$$

In the example of the previous section where $n = 18$, a lower confidence bound was desired for the median at level 0.95. The relevant lower-tailed rejection regions of S for $p = 0.5$ are $S \leq 5$ at the level 0.0481 and $S \leq 6$ at level 0.1189, and the corresponding lower confidence bounds on ξ_p are $X_{(6)}$ and $X_{(7)}$, with levels 0.9519 and 0.8811 respectively. Using (5.1), a lower confidence bound on ξ_p at level approximately 0.95 is

$$\frac{(0.0689 X_{(6)} + 0.0019 X_{(7)})}{0.0708}.$$

The approximation in (5.1) is based on simple linear interpolation and could be used with almost any confidence procedure. Since an hypothesis test can be performed by comparing the hypothesized value to the confidence interval endpoints, a confidence procedure at level approximately $1 - \alpha$ yields directly a test at level approximately α.

The accuracy of this approximation can be given algebraically in the following, interesting case. Suppose that for any n, two neighboring order statistics $X_{(i)}$ and $X_{(i+1)}$ constitute two lower confidence bounds for some quantile ξ at levels say $1 - \alpha_1$ and $1 - \alpha_2$ respectively. If $(X_{(i)} + X_{(i+1)})/2$ is used for the confidence bound on ξ, then the level is approximately $1 - (\alpha_1 + \alpha_2)/2$ according to linear interpolation as in (5.1). However, when ξ is the median and the population is symmetric about ξ, the methodology explained in Sect. 7.3 of Chap. 3 will show that $(X_{(i)} + X_{(i+1)})/2$ is a lower confidence bound at the true level $1 - \alpha$, where $\alpha = (1 - i/n)\alpha_1 + (i/n)\alpha_2$

(see Problem 49, Chap. 3). Thus the weights for the true α are $(1 - i/n, i/n)$, while the approximation above uses $(1/2, 1/2)$. For α fixed, $i/n \to 1/2$ as $n \to \infty$, but at typical levels and in samples small enough to make discreteness severe and interpolation really interesting, i/n is not close to $1/2$ and the approximation in (5.1) is disappointingly inaccurate. For instance, for α near 0.05 the exact values of i/n are 0.28 when $n = 9$ and 0.32 when $n = 17$ (Problem 39). These two sample sizes were chosen since the attainable levels for testing $p = 0.5$ are approximately equidistant from 0.05 (see Table B).

6 The Sign Test with Zero Differences

6.1 Discussion of Procedures

Suppose that $X_1, \ldots, X_n$ are independent and identically distributed observations and we wish to test the null hypothesis that the distribution has p_0-point ξ_0, i.e., p_0 th quantile ξ_0, as in Sect. 3. If one or more of the observations is exactly equal to ξ_0, the test has not yet been defined. Such observations are called "*zero differences*" or sometimes "*ties.*" They can arise theoretically only if the population distribution has positive probability at ξ_0, and the appropriate procedure depends on how we wish to treat such populations. In practice, they can always arise because of limitations on precision in measurement.

If a test procedure is of interest simply as a method of obtaining a confidence interval for the p-point, the problem of zero differences generally need not be resolved because it affects only whether or not the endpoints are included in the interval (see the discussion in Sect. 4).

However, if one really wants to perform a test for its own sake, or to obtain confidence intervals for a population probability p for fixed ξ, then any of the following parameters may be of interest:

(a) the true probability at or below ξ_0, that is $p_\le = P(X_j \le \xi_0)$;
(b) the true probability below ξ_0, that is, $p_< = P(X_j < \xi_0)$;
(c) the p_0-point ξ_{p_0} as defined in Sect. 2;
(d) the difference between the probability below and the probability above ξ_0, that is, $P(X_j > \xi_0) - P(X_j < \xi_0) = p_> - p_<$.

The following facts are relevant to these respective cases.

(a) The number $S_\le$ of observations at or below ξ_0 has a binomial distribution with parameters n and $p_\le$. Tests of hypotheses and confidence procedures for $p_\le$ can thus be based on $S_\le$.

(b) The number $S_<$ of observations below ξ_0 has a binomial distribution with parameters n and $p_<$. Tests of hypotheses and confidence procedures for $p_<$ can thus be based on $S_<$.

(c) Since ξ_0 is a p_0-point if and only if $p_< \le p_0 \le p_\le$, by definition (2.1), tests concerning the p_0-point ξ_0 can be obtained from (a) and (b). Specifically, the null hypothesis $p_< \le p_0$ is equivalent to $\xi_p \ge \xi_0$ and the alternative $p_< > p_0$ is equivalent to $\xi_{p_0} < \xi_0$ (Problem 40. In each case, the largest ξ_{p_0} is to be used if ξ_{p_0} is not unique.) Similarly, the null hypothesis $p_\le \ge p_0$ is equivalent to $\xi_{p_0} \le \xi_0$ and the alternative $p_\le < p_0$ to $\xi_{p_0} > \xi_0$ (Problem 40. Here the smallest ξ_{p_0} is to be used if ξ_{p_0} is not unique.) The relevant one-tailed tests reject these null hypotheses for $S_< \ge s_u$ and $S_\le < s_\ell$ respectively, where s_ℓ and s_u are the lower and upper critical values of binomial tests of the null hypothesis $p = p_0$. A test of the null hypothesis that ξ_0 is a p_0-point against two-sided alternatives therefore rejects if either $S_\le \le s_\ell$ or $S_< \ge s_u$. Note that the precise definition of p_0-point is (unpleasantly) important here.

(d) Frequently one is concerned with whether the proportion of the population below ξ_0 is smaller than, equal to, or larger than the proportion above ξ_0, that is, with the relation between $p_< = P(X_j < \xi_0)$ and $p_> = P(X_j > \xi_0)$. Then those observations equal to ξ_0 seem irrelevant to the desired inference. This suggests that the inference be based on only those observations which do not equal ξ_0. If there are N observations different from ξ_0 and $S_<$ is the number smaller, then the distribution of $S_<$, given the value of N, is binomial with parameters N and p, where p is defined by

$$p = \frac{p_<}{p_< + p_>}. \tag{6.1}$$

It is obvious that $p = 0.5$ when $p_< = p_>$, and that p is larger or smaller than 0.5 according as $p_<$ is larger or smaller than $p_>$. A test of the null hypothesis $p_< = p_>$ (or equivalently that $p = 0.5$), against either one- or two-sided alternatives, can then be based on $S_<$. This amounts to omitting from the sample those observations which equal ξ_0 and applying to the remaining observations a test of Sect. 3.1 for the null hypothesis that ξ_0 is the median, using the reduced sample size. Any test of this type will be called a *conditional sign test*, because it is conditional on the value of N.

The parameter p may itself be of interest, as may the quantity

$$2p - 1 = \frac{p_< - p_>}{p_< + p_>}. \tag{6.2}$$

Confidence bounds and other inferences about these parameters can be obtained from the conditional distribution of $S_<$ given N, which, as already mentioned, is binomial with parameters N and p.

However, for a test of the null hypothesis that $p_< - p_>$ is some value other than 0, the parameter p in (6.1) is not determined unless $p_< + p_>$ is a constant. Therefore, exact methods of testing this hypothesis and hence of setting confidence bounds for $p_< - p_>$ are not easily developed. With large samples, approximate methods can be used, based on the fact that $S_< - S_>$ is approximately normally distributed with mean $n(p_< - p_>)$ and variance $n[p_< + p_> - (p_< - p_>)^2]$, provided that this variance is not too small (Problem 42). The estimated variance $S_< + S_> - [(S_< - S_>)^2/n]$ can be used in place of the unknown true variance. It may be appropriate to incorporate a correction for continuity in the amount of $\frac{1}{2}$ to $S_< - S_>$, although the appropriate correction would be 1 rather than $\frac{1}{2}$ in the case $p_< + p_> = 1$, when $S_< + S_> = n$ with certainty (Problem 42).

6.2 Conditional Properties of Conditional Sign Tests

Since the procedures and properties of tests concerning cases (a), (b) and (c) of the previous subsection have already been discussed in Sect. 3, the remainder of this section will be devoted to those tests appropriate for case (d), specifically, the conditional sign tests just described. We first consider a direct argument in favor of performing a test for the null hypothesis $H_0 : p_< = p_>$ conditionally on N, the number of observations which are not equal to ξ_0.

The argument is as follows. The number N does not pertain to the matter under test but is, in effect, the sample size, because the $n - N$ observations which equal ξ_0 (the "ties") are irrelevant. Accordingly, whatever test is performed, the effective sample size N and the properties of the test for that given N should be reported. Failing to do so would be tantamount to using a procedure involving a sample size which is random but reporting its overall properties for any size sample instead of its properties for the sample size actually used. For example, suppose a sample is taken, of size n_1 or n_2, where the probability of each sample size is $\frac{1}{2}$. Suppose further that a test at level 0.01 is made when the sample size is n_1 and a test at level 0.09 when it is n_2. The overall level of this procedure is 0.05, but it would be misleading to report simply that a test had been made at level 0.05, withholding the information about whether in a particular instance the level was really 0.01 or 0.09. This is not an argument against varying the level in this way, but only an argument for quoting the level and sample size actually used.

Such an argument for a conditional procedure can be very compelling. On the other hand, there are situations where the conditional argument leads to chaos at best. It is not always possible to condition on everything one might like. Worse yet, the conditional argument, together with some apparently harmless assumptions, leads to the radical conclusion that tail probabilities are irrelevant and inferences should be based only on the probability of the actual sample under the various hypotheses (the likelihood) [Birnbaum, 1962]. Thus conditioning poses fundamental problems for orthodox inference methods; these problems have no satisfactory resolution entirely within the frequency interpretation of probability. Even the radical conclusion is in accord with Bayesian and likelihood philosophies, however.

Accepting the argument above for conditional procedures, at least as applied to N in the situation currently under discussion, we are led to consider the conditional properties, for given N, of tests in this situation. These properties all relate to the conditional distribution of $X_1, \ldots, X_n$, given N. This can be derived for any null or alternative distribution and hence the conditional probability, given N, of rejecting the null hypothesis can be computed for any test. The *conditional level* of the test is defined as the maximum of this conditional probability over null distributions of $X_1, \ldots, X_n$. Similarly, the *conditional power* against an alternative distribution is the conditional probability under that alternative, given N, of rejecting the null hypothesis. A test is called *conditionally unbiased* against an alternative hypothesis if its conditional power against each alternative distribution is at least equal to its conditional level. Some conditional properties imply corresponding unconditional properties (Problem 44).

Conditionally on N, the present situation is no different from that discussed in earlier sections. Therefore, a *one-tailed, level α, conditional sign test*, that is, a one-tailed level α sign test, applied to the N observations different from ξ_0, has uniformly greatest conditional power against the appropriate one-sided alternative among tests at conditional level α. Further, the *equal-tailed, level α, conditional sign test*, that is, the equal-tailed level α sign test applied to the N observations different from ξ_0, has uniformly greatest conditional power against any alternative with $p_< \neq p_>$, among tests at conditional level α which are conditionally unbiased against the alternative hypothesis $p_< \neq p_>$, or symmetric in the sense defined in Sect. 3.2 (or both).

In summary, when we assume that ties may occur theoretically, that is, when $P(X_j = \xi_0) > 0$, the conditional sign tests have conditional properties corresponding to the properties of the ordinary sign tests for the situation where we assume $P(X_j = \xi_0) = 0$. The correspondence carries over to other properties which are not specifically discussed above, for instance to the situation when $X_1, \ldots, X_n$ are not identically distributed. The point is that if one accepts the argument for conditioning on N, then *all* of the reasons for using sign tests when ties have probability 0 are equally valid for using conditional sign tests when ties have positive probability. Of course, the remarks in Sect. 3.2 continue to apply, and improvement is possible even when ties have positive probability if one is willing to make an additional assumption, such as symmetry.

The next subsection presents another argument leading to conditional sign tests by invoking (unconditional) unbiasedness. This unbiasedness argument leads to conditional tests at the same conditional level for each N, however, which the direct argument for conditional procedures does not require. In the absence of a convincing and practical method for choosing significance levels, this difference loses importance. As regards P-values, if a one-tailed P-value is to be reported, either argument suggests finding the conditional P-value, that is, the conditional probability given N, under the null hypothesis, of an outcome as extreme as or more extreme than that obtained.

6.3 Unconditional Properties of Conditional Sign Tests

In the situation under discussion, suppose we do not insist upon a conditional test. If we seek instead one which is *unconditionally unbiased*, we will again be led to a conditional sign test, with the slight difference just mentioned. In fact, the following two statements are true.

(a) A one-tailed, level α, conditional sign test is, from an unconditional point of view (that is, unconditionally), uniformly most powerful against the appropriate one-sided alternative, $p_< < p_>$ or $p_< > p_>$, among level α tests which are unbiased against this alternative.

(b) An equal-tailed, level α, conditional sign test is uniformly most powerful against the two-sided alternative $p_< \neq p_>$, among level α tests which are unbiased against this alternative.

*6.4 Proof for One-Sided Alternatives

Consider the one-sided alternative $p_< < p_>$, that is, $P(X_j < \xi_0) < P(X_j > \xi_0)$. Let G be any distribution satisfying this alternative hypothesis. In order to adapt the method of Sect. 3.3 to the present situation, for any $p_<$ and $p_>$, we define a distribution F as follows: $P_F(X_j < \xi_0) = p_<$; the conditional distribution of X_j, given that $X_j < \xi_0$, is the same under F as under G; $P_F(X_j = \xi_0) = 1 - p_< - p_>$; $P_F(X_j > \xi_0) = p_>$; and the conditional distribution of X_j, given that $X_j > \xi_0$, is the same under F as under G (Problem 43a). Then the family $\mathscr{F}$ of distributions F includes G and it includes a null distribution for each value of $p_< = p_>$ (Problem 43b).

For notational convenience we let $S = S_<$, the number of observations below ξ_0. Then for this family of distributions, the statistics S and N are jointly sufficient (Problem 43c). We shall show that any unbiased test based on S and N is conditional. We already know that, among tests at conditional level α, a one-tailed conditional sign test at level α has uniformly greatest conditional power against the appropriate one-sided alternative. The desired conclusion, (a) of Sect. 6.3, then follows (Problem 44c).

It remains to show that any unbiased test $\phi(S, N)$ is conditional. We note first that,

$$E_F[\phi(S, N)] \leq \alpha \tag{6.3}$$

for all null distributions F while

$$E_K[\phi(S, N)] \geq \alpha \tag{6.4}$$

for all alternative distributions K, by unbiasedness. Second, every null distribution F is a limit of alternative distributions (e.g., the alternatives $K = [(m - 1)F + G]/m$, or alternatives in the family $\mathscr{F}$ for which $p_<$ and $p_<$ approach their values under F). It follows that

$$E_F[\phi(S, N)] = \alpha \tag{6.5}$$

for all null distributions F.

Consider now the null distributions of the family $\mathscr{F}$ above. For this subfamily, N is a sufficient statistic (Problem 43d), and hence the conditional probability of rejection given N is a function of N alone, say $\alpha(N)$. That is, for $p_< = p_> = r/2$ say,

$$E_r[\phi(S, N)|N] = \alpha(N) \tag{6.6}$$

where $\alpha(N)$ does not depend on r. Since (6.5) holds for all null distributions, taking the expected value of both sides of (6.6) gives

$$E_r[\alpha(N)] = \alpha \quad \text{for all } r. \tag{6.7}$$

Now N, the number of observations not equal to ξ_0, is binomially distributed with parameters n and $r = p_< + p_>$. Since the family of binomial distributions is complete (Chap. 1, Sect. 3.4) it follows that $\alpha(N) = \alpha$ for all N, that is, the test is conditional. This is all that remained to be proved.   □

Remarks

The type of argument employed in the previous two paragraphs often applies. It is summarized in the following theorem, whose proof is requested in Problem 45.

Theorem 6.1. *Any unbiased test at level α has probability exactly α of rejection on the common boundary K of the null hypothesis H_0 and the alternative H_1. If T is a complete sufficient statistic for K, then any unbiased test at level α has conditional level exactly α for all distributions of K, conditional on T. Hence if T is a complete sufficient statistic for H_0, and if $H_0 = K$, that is, H_0 is contained in the boundary of H_1, then any unbiased test at level α is a conditional test, conditional on T.*

"Complete" could be replaced by "boundedly complete" throughout, meaning that there exists no bounded, nontrivial, unbiased estimator of 0 (cf. Chap. 1, Sect. 3.4). The "common boundary" means those distributions which are limits of both null distributions and alternative distributions. The relevant definition of limit here is that $F_n \to F$ if $E_{F_n}[\phi(x)] \to E_F[\phi(x)]$ for all bounded functions ϕ, though stronger definitions typically hold also.*

*6.5 Proof for Two-Sided Alternatives

A level α test of the null hypothesis $p_< = p_>$ which is unbiased against $p_< \neq p_>$ must have conditional level α given N, as follows from either corresponding one-sided statement, but it need not be conditionally unbiased (Problem 46). Accordingly, in contrast to the one-sided case (Sect. 6.4), the fact that the equal-tailed, level α, conditional sign test is uniformly most powerful among conditionally unbiased tests does not imply directly that it is uniformly most powerful among unconditionally unbiased tests. To prove that it is, let G be any alternative and consider its family $\mathscr{F}$ of distributions F as defined in Sect. 6.4. We will prove that, among tests having level α for the null distributions of the family $\mathscr{F}$ and unbiased against the alternatives of the family $\mathscr{F}$, the equal-tailed, level α, conditional sign test is uniformly most powerful against these alternatives, and, in particular, is most powerful against G. Since it is in fact a level α, unbiased test for the original, more inclusive, null and alternative hypotheses, it is, among such tests also, most powerful against G (Theorem 3.1) and thus uniformly most powerful, since G was arbitrary.

Now restrict the problem to the family $\mathscr{F}$. For this family, a sufficient statistic is (S, N). The distribution of (S, N) may be described as follows. N is binomial with parameters n and $r = p_< + p_>$, while given N, S is binomial with parameters N and $p = p_</r$, as at (6.1).

We seek a test $\phi(S, N)$ of the null hypothesis $p = 0.5$ (that is, $p_< = p_>$), which is unbiased against the alternative $p \neq 0.5$ (that is, $p_< \neq p_>$). Let

$$\alpha(r, p) = E_{r, p}[\phi(S, N)] \tag{6.8}$$

be the power (the level, when $p = 0.5$) of the test ϕ. Let

$$\alpha(p|N) = E_{r, p}[\phi(S, N)|N] \tag{6.9}$$

be the conditional power (level, when $p = 0.5$) of ϕ given N, which is a function of p and N alone, not depending on r, because the conditional distribution of S given N is a function of p and N alone. If ϕ is unbiased at level α, then

$$\alpha(r, 0.5) \leq \alpha,$$
$$\alpha(r, p) \geq \alpha \quad \text{for } p \neq 0.5. \tag{6.10}$$

It follows, as we saw in the one-sided case, that $\alpha(r, 0.5) = \alpha$ and that the conditional level $\alpha(0.5|N) = \alpha$. It also follows (Problem 47) that

$$\frac{\partial \alpha(r, p)}{\partial p} = 0 \quad \text{at } p = 0.5. \tag{6.11}$$

Now (Problem 47) it is also true that

$$\alpha(r, p) = E_{r, p}[\phi(p|N)]. \tag{6.12}$$

Since the distribution of N depends on r alone, we may differentiate with respect to p under the expectation (Problem 47), obtaining

$$\frac{\partial \alpha(r, p)}{\partial p} = E_{r, p}\left[\frac{\partial \alpha(p|N)}{\partial p}\right] \tag{6.13}$$

and then, by (6.11),

$$E_{r, p}\left[\frac{\partial \alpha(p|N)}{\partial p}\right] = 0 \quad \text{at } p = 0.5. \tag{6.14}$$

Since the family of distributions of N is complete, it follows that

$$\frac{\partial \alpha(p \mid N)}{\partial p} = 0 \quad \text{at } p = 0.5. \tag{6.15}$$

We have now proved that if $\phi(S, N)$ is an unbiased test of $p = 0.5$ against $p \neq 0.5$, then it must be a conditional test and its conditional power must have derivative 0 at $p = 0.5$. Recall that S is conditionally binomial with parameters N and p. As stated in Sect. 8.3, Chap. 1, only two-tailed tests are admissible for this situation. The only such tests whose power has derivative 0 at 0.5 are equal-tailed. Hence the equal-tailed, level α, conditional test is the only admissible, unbiased test based on (S, N). Since (S, N) is a sufficient statistic when the problem is restricted to the family $\mathscr{F}$, this test, which is the equal-tailed, level α, conditional sign test, is therefore uniformly most powerful unbiased for the restricted problem. As mentioned initially, it follows that this test is uniformly most powerful unbiased for the original problem.   □

Notice that S, $N - S$, and $n - N$ are the cell frequencies in a sample of n where the three cells have respective probabilities, $p_<$, $p_>$, and $1 - p_< - p_>$. Thus the restricted problem reduces by sufficiency to a trinomial problem. The main part of the foregoing proof was essentially a proof that, in a trinomial problem, when testing equality of two cell probabilities against a two-sided alternative, a uniformly most powerful unbiased procedure is to omit those observations falling outside the cells of interest and apply the natural, equal-tailed binomial test to the reduced sample.

The argument involved showing that an unbiased test is conditional (which is true even for one-sided alternatives), and that its conditional power has derivative 0 at the null hypothesis. It was essential at (6.13) that the distribution of N depend on r alone, and in going from (6.14) to (6.15) that the conditional power depend on p alone. A most powerful, unbiased test can be derived by this method for any exponential family if the null hypothesis specifies the value of one of the "natural" parameters of the family [Lehmann, 1959, Sect. 4.4]. The trinomial distributions form an exponential family. In place of the fact that only two-tailed tests are admissible, Lehmann uses a generalization of the Neyman–Pearson fundamental lemma to two side conditions. See also Sect. 8.3, Chap. 1.*

7 Paired Observations

Frequently in practice measurements or observations occur in pairs; the two members of a pair might be treated and untreated, or male and female, or math score and reading score, etc. While the pairs themselves may be independent, the members of a pair are related in some way. This relationship within pairs may be present because of the nature of the problem, or may be artificially imposed by design, as when experimental units are matched according to some criterion. The units or pairs may be drawn randomly from some population of interest, and the assignment within pairs may be random. If not, additional assumptions may be needed, depending on the type of inference desired.

For example, suppose that a random sample of individuals is drawn, and a pair of observations is obtained for each individual, like one before and one after some treatment. Then each individual acts as his own "control." Under the assumption (not to be treated lightly) that there is no time-related change other than the treatment, one can estimate the effect of the treatment, and with smaller sampling variability than if the controls were chosen independently of the treated individuals. If instead each individual receives two treatments, such as a headache remedy administered on two completely separate occasions, it may be possible to assign the treatments to the occasions randomly for each individual. For comparing the two treatments in terms of some measure of effectiveness, this provides a similar advantage in efficiency without requiring such strong assumptions. One of the treatments could, of course, be a placebo or other control treatment.

More generally, suppose that the units to be observed are formed into pairs in some way, either naturally or according to some relevant criterion, and observations are made on both members of each pair. A pair here might be two siblings, two litter mates, a husband-and-wife couple, two different sides of a leaf, two different but similar schools, etc., or one individual at two times, as above. If the matching is such that the members of a pair would tend to respond similarly if treated alike, random variation within pairs is reduced and the nonrandom variation is easier to observe. If a difference is then observed between two treatments, the difference can be attributed to the effect of the treatments rather than to random differences between units with more assurance than could an equal difference observed in a situation without matching.

If, within each pair, one unit is selected at random to receive a certain treatment, and the other unit receives a second treatment (or serves as a control), we have a *matched-pair experiment*. If the pairs themselves are independently drawn from some population of pairs, we have a *simple random sample* of pairs. Another possibility is to draw a simple random sample of individuals and then form pairs within this sample. If either type of randomization is lacking, as in the before-after example above, it is especially important to consider the assumptions required for the type of inference being made.

In an analysis of paired observations, it is technically improper and ordinarily disadvantageous to disregard the pairing. A convenient approach to taking advantage of the pairing usually results if the measurements on the two members of each pair are subtracted and the analysis is performed on the resulting sample of differences. To a great extent, this procedure effectively reduces a paired-sample problem to a one-sample problem, but the assumptions which are appropriate for the sample of differences depend on what assumptions are appropriate for the pairs.

For example, suppose that each pair consists of a control measurement V and a treatment measurement W. If the treatment has absolutely no effect, it is often natural to assume that the measurements are permutable within pairs, that is, (V_i, W_i) has the same distribution as (W_i, V_i) for each i. (This holds, for instance, in a matched-pair experiment, where one unit of each pair is chosen at random to be a control.) Under this assumption, and independence between pairs, the null hypothesis of no treatment effect can be tested by applying the methods of Sect. 3.1 (median equal to zero) to the differences $X_i = W_i - V_i$. To discuss point estimation or confidence intervals, however, we need to make some assumption about the treatment effect if there is one. A strong assumption would be that the treatment has the same effect on every unit, specifically, that any treated unit has a value larger by an amount β than it would have had if untreated. Under this assumption and the earlier one, $(V_i, W_i - \beta)$ and $(W_i - \beta, V_i)$ have the same distribution, and therefore (Problem 48) the treatment effect β is the median of the differences $X_i = W_i - V_i$, for each i. If also the pairs are independent, then the methods of Sect. 4 can be used to find a confidence interval for the common population median of the differences X_i, which is here a confidence interval for the treatment effect. If the treatment effect varies from unit to unit, however, the situation is more complicated, and confidence intervals obtained in this way may be invalid. Even with simple random sampling, the median of the differences X_i need not equal the difference of the medians of W_i and V_i. Furthermore, in a matched-pair experiment, if the treatment has no effect on the median of the distribution for any unit, but its dispersion increases in direct relation to the median, then the differences X_i are typically skewed to the right and their medians are typically negative but can also be positive. See also Problems 49–52, Sect. 2, Chap. 3, and Sect. 9, Chap. 8.

When the relevant assumptions are satisfied by the differences of pairs of observations, all of the procedures and properties of the one-sample sign test discussed in Sects. 3, 4, and 5 are equally applicable to the set of differences of the matched or paired observations. Hence, further discussion of such techniques will not be given here. We note, however, that the methods explained later in Chaps. 3 and 4 may be particularly appropriate for differences because completely arbitrary distributions are less plausible for differences than for raw measurements, and the assumption of symmetry in particular is more plausible for differences (if the measurements are continuous or nearly so).

8 Comparing Proportions Using Paired Observations

We will now discuss a common and important situation which reduces to an application of the type of inference described in (d) of Sect. 6.1 which concerned the relation between the parameters $p_<$ and $p_>$ in the presence of ties. The present situation is simply described as a set of paired observations where each observation is dichotomous and hence can be recorded as either 0 or 1. Then the difference of variables in any pair can only be $-1, 0$ or $+1$, so that many ties are likely. The analysis follows that of the sign test with ties, but it will be worthwhile to discuss this application and its interpretation, especially in relation to other similar situations. In this section we will present the test procedure, its interpretation and properties, and some related matters including a model discussed by Cox.

Suppose that the observations occur in pairs, each pair consisting of a unit, element, or measurement of Type I and one of Type II. For instance, as in Sect. 7, each pair might contain a control unit (Type I) and a treated unit (Type II), or a measurement before (Type I) and after (Type II), or the two types might be husband and wife, two sides of a leaf, etc. Suppose further that each observation merely measures the presence or absence of some characteristic or response. It is natural to represent this by scores of 1 and 0 respectively. Then we have essentially the same situation of matched or paired observations as described in Sect. 7, except that here the variables are "indicator" functions since each observation is a dichotomous measurement, being 1 or 0. Some actual examples of this type of situation are:

(a) Some soldiers were asked whether they thought the war with the Japanese would last over a year, given a lecture on the difficulty of fighting the Japanese, and then asked the same question again [McNemar, 1947].
(b) Specimens were taken from the throats of persons suspected of having diphtheria and grown on each of four media. Suppose two media were to be compared with respect to the probability of growth taking place [Cochran, 1950].
(c) Two drugs were tried on a number of patients, each drug on each patient, to see which was more likely to cause nausea [Mosteller, 1952].

In the matched-pair situation described in Sect. 7, we were concerned with inference about the population mean or median of the difference between the two variables measured on each pair. The analogous comparison here is between the population proportion p_1 of Type I scores which are 1 and the proportion p_{II} of Type II scores which are 1, since these proportions are the population means (expected values) of the two types of score, which we will call score on I and score on II. Note that each observation is classified as either Type I or Type II and as either score 1 or score 0. For the examples mentioned above, we might designate the types and scores as follows:

(a) Type I is before the lecture; Type II is after. Score 1 is yes; score 0 is no.
(b) Type I and II are the two media. Score 1 if growth takes place; score 0 if not.
(c) Type I and II are the drugs. Score 1 if nausea results; score 0 if not.

In what follows, we shall assume that a simple random sample of n is drawn and two observations are made on each member, one of Type I and one of Type II, and each observation is scored as either 1 or 0. Thus we have

n independent, identically distributed pairs of dichotomous observations. We will discuss making inferences about the difference $p_{II} - p_I$ of the population proportions of the two types of scores which are 1. (The same analysis can be applied to a comparative experiment with nonrandom pairs as long as the elements within a pair are randomly assigned to be either Type I or Type II, and we are concerned only with testing a null hypothesis such as no treatment effect whatever exists. See also Sect. 8.7.)

8.1 Test Procedure

In the situation under discussion here, the data might be recorded using the format of Table 8.1. For each pair, we subtract the score on I from the score on II to obtain differences which are either $+1$, -1, or 0. While there may be any number n of pairs observed, there are only four categories of response, that is four distinguishable pairs of scores, and these are listed in the table. The last column shows the four symbols which we shall use to designate the number of pairs observed in each of the four categories.

Now suppose that the null hypothesis of primary interest is that the probability of a score of 1 on I is the same as the probability of a score of 1 on II, that is $p_I = p_{II}$. The difference $p_{II} - p_I$ is equal to the probability of a positive difference score II $-$ I, minus the probability of a negative difference score II $-$ I (Problem 53). That is, the null hypothesis $p_{II} - p_I = 0$ is equivalent to the hypothesis that the difference scores of $+1$ and -1 are equally likely to occur in the population. Accordingly, the test suggested in (d) of Sect. 6.1 applied to the numbers A, B, C, and D in Table 8.1 is appropriate in this situation. (The number B corresponds to $S_<$ in Sect. 6.1.) The $A + D$ zero difference scores are ignored, and under the null hypothesis, given $B + C = N$, the distribution of B is binomial with parameters N and $p = \frac{1}{2}$. Hence the sign test with zero differences or ties, which was introduced in Sect. 6, can be used to test this hypothesis. The properties and interpretation of this test, which will be called here a *test for equality of proportions* based on paired or matched observations (it is also frequently called the McNemar test) will be discussed later in this section. An example is given in Sect. 8.3.

Table 8.1

Score		Difference Score	Response Category	Observed Number
I	II	II $-$ I	of Pair	in Category
1	1	0	1	A
1	0	-1	2	B
0	1	1	3	C
0	0	0	4	D

8.2 Alternative Presentations

The relevant data of the situation under discussion can be presented in another way, as shown in Table 8.2. Each observed pair falls into one of the cells of this table, and the total number in each cell is recorded using the symbols A, B, C and D defined as in Table 8.1.

Table 8.2 is a "double dichotomy", one kind of 2×2 (contingency) table. It may appear that the usual tests, namely, Fisher's exact test (Sect. 3.2, Chap. 5) and the chi-square (approximate) test of independence or "no association," are applicable. However, the hypothesis of interest here, $p_I = p_{II}$, is not the usual one and cannot be tested by these procedures. Independence of the row and column characteristics in a 2×2 table, here I and II, is equivalent to equality of the population (unmatched) proportions within the two columns, or within the two rows, which is not our present concern. In our situation, association of the Types designated by I and II is presumably present, whether by necessity or by design, and to a high degree. We wish to take advantage of the association, not to test its existence.

For instance, consider situation (a) at the beginning of Sect. 8. A soldier who is optimistic about the war before the lecture is more likely to be optimistic after the lecture than one who is pessimistic beforehand. Our concern is whether the net result of such a lecture is to make more soldiers pessimistic than were before, that is, whether the lecture can be expected to change more optimists to pessimists than vice versa. Those soldiers whose point of view does not change as a result of the lecture are, in a sense, irrelevant to the point under discussion (see also Sect. 6). For this reason, the test suggested here is based on the values of B and C alone. (See also Sect. 8.4.)

In this situation then, there is association between Types I and II because of the matching, but we are interested in the equality of the proportions p_I and p_{II}. These would be estimated from Table 8.2 by $(A + B)/(A + B + C + D)$ and $(A + C)/(A + B + C + D)$ respectively, the respective proportions in the first row and the first column. The difference $p_{II} - p_I$ is then estimated by $(C - B)/(A + B + C + D)$, which is to be compared to 0. This is equivalent to comparing the proportions in the lower left and upper right cells of Table 8.2, and also to comparing the proportion in the second row with the proportion in the second column.

Table 8.2

		Score on II	
		1	0
Score on I	1	A	B
	0	C	D

Table 8.3

		I	II
Score	1	A + B	A + C
	0	C + D	B + D

Table 8.3 shows another alternative method of presenting the data. With this 2×2 format, the quantity we are interested in is the difference between the proportions in the two columns, since this is equal to $(C - B)/(A + B + C + D)$. It may therefore appear that Fisher's exact test and the chi-square test of "no association" are now applicable, but again they cannot be used, in this case because the assumptions are not satisfied. The quantities in the two columns labeled I and II are not independent. In fact, the numbers or proportions here refer to matched pairs, and each pair appears twice in Table 8.3, once in each column. An adjustment can be made, but it leads either to the test already suggested or to a large-sample approximation to that test (Stuart [1957]).

Of course, the format for presentation of the data is largely a matter of taste and is irrelevant to proper analysis, provided that the situation is correctly understood. Table 8.1 is quite clear, but is less compact than might be desired. Tables 8.2 and 8.3, although compact, might lead to misinterpretation. In addition, since Table 8.3 gives only the marginal totals of Table 8.2, it alone does not contain sufficient information for application of the appropriate test which requires knowledge of at least B and $B + C = N$. Table 8.2 is the more common, but the format of Table 8.1 generalizes more easily to more than two types of unit or measurement when each observation is still recorded as 0 or 1. This generalization is equivalent to 0–1 observations occurring as k-tuples rather than as matched pairs, and the usual test procedure is Cochran's Q Test (Problem 57).

8.3 Example

Twenty married couples were selected at random from a large population and each person was asked privately whether he would prefer that a week's summer vacation for the family be spent in the mountains or at the beach, all other factors being equal. The subjects were told to ignore factors such as cost and distance so that their preference would reflect only their assessment of the pleasure derived by the family from the two kinds of vacation. The preferences expressed are shown in the pairs below, with B denoting beach and M mountains and the husband's view of family preference always the first member of the pair:

$(M, B), (M, B), (M, M), (B, B), (M, B), (B, M), (B, M), (M, M), (M, B), (B, B),$
$(B, M), (M, B), (B, B), (M, B), (M, B), (M, M), (M, M), (B, B), (M, B), (M, M).$

Table 8.4

Preference Category of (H, W)	Number
(M, M)	5
(M, B)	8
(B, M)	3
(B, B)	4

The purpose of the study was to determine whether views of family preference for vacation are largely influenced by sex, and hence a possible source of serious disharmony between husband and wife. Specifically, we wish to determine whether a married man's view of family preference differs systematically or only randomly from his wife's view.

We first present the data in each of the formats that were described in Sect. 8.2. The frequencies of occurrence for the four response categories of pairs are easily counted. The results shown in Tables 8.4 and 8.5 are examples of the general format of Tables 8.1 and 8.2 respectively. Table 8.6 is analogous to Table 8.3, and it is clear here that the entries in the two columns are not independent, because each couple appears twice in the table.

Suppose we wish to test the null hypothesis that the probability that the husband responds mountains while the wife responds beach is equal to the reverse type of disagreement, that is,

$$P(M, B) = P(B, M). \tag{8.1}$$

If we add $P(M, M)$ to both sides of (8.1), the left-hand side is simply the probability that the husband responds mountains since

$$P(M, B) + P(M, M) = P[M, B \text{ or } M] = P_H(M)$$

say, while the right-hand side of (8.1) similarly becomes the probability $P_W(M)$ that the wife responds mountains. Hence the null hypothesis can also be stated as either

$$P_H(M) = P_W(M) \quad \text{or} \quad P_H(B) = P_W(B),$$

which may be easier to interpret than (8.1). If H is Type I and M is score 1, then $P_H(M) = P_W(M)$ represents $p_I = p_{II}$ here. The ordinary binomial test

Table 8.5

| | Wife's Preference | |
	M	B
Husband's M	5	8
Preference B	3	4

Table 8.6

		H	W
Preference	M	13	8
	B	7	12

with parameters $N = 11$, $p = 0.5$ is appropriate. From Table B, given 11 disagreements, the probability of obtaining 3 or less pairs of category (B, M) is $P(S \leq 3) = 0.1133$; this of course equals the probability of 8 or more pairs of category (M, B). The two-tailed P-value is then $2(0.1133) = 0.2266$.

8.4 Interpretation of the Test Results

The interpretation of the results of a before versus after test of any kind bears close scrutiny. Suppose, for example, that the same characteristic is measured before (I) and after (II) a treatment, and by a one-sided, level α test for equality of proportions based on matched observations there are significantly more 1's after the treatment than before. Then, if the units constitute a random sample from some population, the inference, at level α, is that if all elements in the population had been treated, there would have been more 1's after treatment than before. However, the inference that the population would have changed in this direction if treated does not automatically justify the inference that the treatment would have changed the population. The observations in themselves provide no information about what would have happened in the absence of the treatment. In order to make an inference about the treatment, it is necessary either to assume that the proportion of 1's in the population would not have changed in the absence of treatment or to run an additional control experiment (Problem 60).

In example (a) at the beginning of Sect. 8, for instance, it is reasonable to assume that the soldiers would not have changed their opinions in the absence of a lecture. Hence the effect, if real, may reasonably be attributed to the lecture and the circumstances surrounding it. Of course, it is conceivable that a dull lecture on any topic would have made the soldiers pessimistic about a speedy end to the war. In example (c) if the drugs were given in the same order to every patient, an apparent difference between the drugs might be due to a time effect. If the order was randomized for each patient independently, this difficulty would be obviated. Actually, in this experiment exactly half of the patients in the sample were chosen at random and given drug I first while the rest were given drug II first. This alters the null distribution of the test statistic if there is a time effect, but may be more powerful inasmuch as it balances out the time effect by design rather than by randomization. The test for equality of matched proportions will still be approximately "valid" as long as the time effect is not too large and is "conservative" in any case (Problem 61). An exact test is given in Problem 106c of Chap. 5. See also Problem 9c and text in Chap. 5.

Even if the effect is attributable to the treatment, the inference is limited to the population sampled; the effect might be quite different on a population with a different proportion of 1's initially. Sometimes, of course, one might be willing to assume that the effect on different populations would be in the same direction, especially if there is a continuous measurement underlying the dichotomous one. For instance, the soldiers could have been asked how long they thought the war would last. The reader may ponder to what extent the effectiveness of the lecture on one population of soldiers guarantees its effectiveness on another, such as a population more pessimistic initially, or more experienced in fighting the Japanese, or made up of officers, or Marines, etc.

With continuous measurements, one is often willing to assume or able to verify that the treatment effect is approximately additive, that is, approximately the same on different populations. One possible corresponding assumption for the present situation is discussed in Sect. 8.7. Such assumptions, however, do not justify transferring inferences to different populations without assuming that the effect on different populations would be in the same direction. In fact, they amount to assuming this and more.

If, initially, substantially more units score 1 than 0, there are many more units available to change from 1 to 0 than from 0 to 1. One might argue that this puts the treatment at a disadvantage (if 1 is better than 0). This does not invalidate an inference about the effect of the treatment on the population, but merely reflects the danger of tampering with the status quo if it is good. It would be undesirable to give everyone polio vaccine unless the chance of preventing polio in people who would otherwise have contracted it is much greater than the chance of causing polio in people who would otherwise have escaped it. This again emphasizes the possible danger in transferring the inference to another population. A polio vaccine could be very advantageous for children of the most susceptible age in a city undergoing an epidemic, but very disadvantageous for use by everyone in the country. Some consider yellow fever shots undesirable in the U.S. but desirable in some other countries.

Another difficulty in interpretation, also by no means restricted to situations involving matched proportions, is that it may not be clear what population, if any, was actually sampled. For example, if the soldiers present at the lecture constitute the population, and all of them were questioned both before and after, one has a complete census of the population. To make an inference about the effect of giving similar lectures to various groups at various times, allowance must be made for the variation of the lectures, of which one has a sample of only one, and for the fact that the group of soldiers is not a sample of various soldiers at various times. Even if we were willing to regard the soldiers as an independent, random sample of some sort before the lecture, their responses afterwards are certainly not independent because of the effect they have on one another as they listen to the same lecture. In view of its sampling assumption, the test under discussion here may be only partially relevant to the questions of interest. It may nevertheless be the most nearly relevant procedure which is readily available. In situations where two treatments (one may be a control) are compared by randomizing over matched pairs, there is usually less problem in interpretation.

8.5 Properties of the Test

Some properties of the sign test with ties described earlier in Sect. 6 carry over to the present situation. We assume throughout that the only data available are the numbers A, B, C, and D in the four response categories for a simple random sample of pairs, and that the null hypothesis is $p_I = p_{II}$, with no further restriction on the probabilities.

Then, specifically, the one-tailed test as applied in this section has uniformly greatest conditional power against the appropriate one-sided alternative among tests at its conditional level, where "conditional" here means "given $B + C$." The equal-tailed test has uniformly greatest conditional power against any alternative among tests at its conditional level which are conditionally unbiased against the alternative that the probability p_I of scoring 1 on I differs from the probability p_{II} of scoring 1 on II. The one-tailed, level α, conditional test is, from an unconditional point of view, uniformly most powerful against the appropriate one-sided alternative $p_I < p_{II}$ or $p_I > p_{II}$, among level α tests which are unbiased against this alternative. The equal-tailed, level α, conditional test is, from an unconditional point of view, uniformly most powerful against the alternative $p_I \neq p_{II}$, among level α tests which are unbiased against this alternative. Proofs are requested in Problem 63.

8.6 Other Inferences

In the situation under discussion, inferences other than a test for equality of paired proportions may also be of interest. Some of these will be discussed in this subsection. We continue to use the notation introduced in Sect. 8.1, that p_I and p_{II} are the proportions of pairs in the population scoring 1 on I and II respectively. We will also now be referring to the joint classification of observations on the basis of scores of both Types; hence it will be convenient to introduce the notation p_{ij}, for $i = 0, 1, j = 0, 1$, to denote the proportion of pairs in the population with Type I score i and Type II score j. For example, p_{01} is the true proportion scoring 0 on I and 1 on II. Thus p_{11}, p_{10}, p_{01} and p_{00} denote the true proportions of the population of pairs corresponding to the observed numbers A, B, C, and D respectively in Tables 8.1–8.3. Notice that $p_I = p_{10} + p_{11}$ and $p_{II} = p_{01} + p_{11}$.

The test already described in Sect. 8.1 was for $H_0: p_{II} - p_I = 0$, or equivalently $p_{01} - p_{10} = 0$; under this null hypothesis, given $B + C = N$, the test statistic B follows the binomial distribution with parameters N and $p = \frac{1}{2}$. The difference $p_{01} - p_{10}$ is relevant for comparison of the proportions of kinds of "disagreements" between scores for the two Types, or kinds of "switches." This parameter corresponds to the difference parameter $p_< - p_>$ (see Section 6.1) discussed in the context of the conditional sign test with ties in Sect. 6, so that the test and confidence interval procedures discussed there are relevant here also.

In the present context and notation, the parameter p defined in Eq. (6.1) can be expressed as $p = p_{10}/(p_{10} + p_{01})$. It represents the conditional probability of a score of 1 on I given a disagreement between scores for the two types. Since B is the number of observations scoring 1 on I and 0 on II, we know that, given $B + C = N$, B follows the binomial distribution with parameters N and p. (Recall that B corresponds to the number $S_<$ defined in Sect. 6.1.) Hence the usual binomial procedures are appropriate for tests of hypotheses and confidence intervals for this p and also for $(1/p) - 1 = p_{01}/p_{10}$ (Problem 64a). In the present situation, the primary advantage of the parameter $(1/p) - 1$ seems to be its adaptability to simple inference techniques, but it will be given a useful interpretation in the next subsection.

Another quantity which might be of interest is the true proportion of observations scoring 1 on II among those scoring 1 on I, or $p_{11}/p_I = p_{11}/(p_{10} + p_{11})$. This quantity is the conditional probability of a score of 1 on II given a score of 1 on I. Alternatively, the conditional probability of a score of 1 on II given a score of 0 on I might be of interest. Inferences about these quantities can appropriately be based on the binomial distribution, and the procedures are easily developed (Problem 64b).

We have discussed inferences concerning the "disagreements" between scores on the two categories; now what about the "agreements?" For example, one might be interested in testing the null hypothesis that the probability of Type I and II scoring the same does not depend on the score of Type I. This condition reduces successively to

$$P(\text{same}|1 \text{ on I}) = P(\text{same})$$

$$p_{11}/(p_{10} + p_{11}) = p_{11} + p_{00}$$

$$p_{11}(1 - p_{10} - p_{11} - p_{00}) = p_{00}p_{10}$$

$$p_{11}p_{01} = p_{00}p_{10}, \tag{8.2}$$

and the result is identical if we start with either of the relations $P(\text{same}|0 \text{ on I}) = P(\text{same})$ or $P(\text{same}|0 \text{ on I}) = P(\text{same}|1 \text{ on I})$. If the data are represented in a new 2×2 table using the format of Table 8.7, it is clear that the usual contingency table test of independence (of score on I and sameness) is appropriate for the null hypothesis in (8.2). This is of course equivalent to a test of equality of proportions within the rows of Table 8.7, or

$$p_{11}/(p_{10} + p_{11}) = p_{00}/(p_{00} + p_{01}) \tag{8.3}$$

which in the present context says $P(\text{same}|1 \text{ on I}) = P(\text{same}|0 \text{ on I})$. Another equivalent way of stating the null hypothesis in (8.2) is as an equality of odds, or

$$p_{11}/p_{10} = p_{00}/p_{01} \tag{8.4}$$

which says here that the odds for "same given 1 on I" are equal to the odds for "same given 0 on I," or

$$P(\text{same}|1 \text{ on } I)/P(\text{different}|1 \text{ on } I) = P(\text{same}|0 \text{ on } I)/P(\text{different}|0 \text{ on } I).$$

Table 8.7   Table 8.8

		II	
		same	different
I	1	A	B
	0	D	C

		I	
		same	different
II	1	A	C
	0	D	B

A test of the null hypothesis that the probability of Types I and II scoring the same does not depend on the score of Type II can also be performed using a test of independence (of score on II and sameness), or a test of equality of proportions within the rows of the new table shown as Table 8.8. This hypothesis is not the same as (8.2)–(8.4), but is equivalent (Problem 65) to

$$p_{11}p_{10} = p_{01}p_{00}. \tag{8.5}$$

These hypotheses of independence of sameness and score on I or II are not as easy to interpret as they may seem. If there are many more scores of 1 than 0 on II, then it is easier to be the same given 1 on I than given 0 on I. (Compare Sect. 8.4.) It may further exemplify the difficulty of interpretation of independence in Table 8.7, and Table 8.8, to remark that independence in both implies that $p_{11} = p_{00}$ and $p_{10} = p_{01}$, except in the degenerate case where either $P(\text{same}) = 0$ or $P(\text{different}) = 0$ (Problem 66).

8.7 Cox Model

In Sects. 8.1–8.6, the essential assumption was that the observed pairs constitute a simple random sample from some population of pairs. In Sect. 7, we indicated that in the case of matched pairs of continuous observations an alternative assumption is often made. This is that the observations of a given pair are random (as in a matched-pair experiment), while the pairs themselves have "fixed effects" and need not be random at all. Each observation then reflects the fixed effect of the pair to which it belongs, as well as the effect of the treatment (or Type). An analogous assumption for paired dichotomous observations has been discussed by D. R. Cox (1958c). We explain it here in the context of the drug example mentioned in (c) at the beginning of Sect. 8, where two drugs are tried on a group of patients, each drug once on each patient. Consider the group of patients as fixed, and suppose that, for patient i, drug I causes nausea with probability $p_{\text{I}i}$ and drug II causes nausea with probability $p_{\text{II}i}$. Note that the randomness is now associated with different possible outcomes on a given patient, rather than with the choice of a patient from the population. Suppose also that the outcomes of separate trials on the same patient are independent (as trials on different patients would be). If each drug is tried once on each patient, then the probability p_{11i} that patient i scores 1 (nausea) on each drug is, by the independence assumption,

$$p_{11i} = p_{1i}p_{IIi}. \tag{8.6}$$

Similarly, with obvious definitions, we have

$$p_{10i} = p_{1i}(1 - p_{IIi}) \tag{8.7}$$

$$p_{01i} = (1 - p_{1i})p_{IIi} \tag{8.8}$$

$$p_{00i} = (1 - p_{1i})(1 - p_{IIi}). \tag{8.9}$$

Suppose now that we make the additional assumption that the drug effect is constant, in the sense that, for all patients, the odds for nausea under drug II are the same multiple θ of the odds for nausea under drug I. In symbols, this assumption is

$$\frac{p_{IIi}}{1 - p_{IIi}} = \theta \frac{p_{1i}}{1 - p_{1i}} \quad \text{for all } i. \tag{8.10}$$

Notice that when there is no difference between the drugs we have $\theta = 1$. With the assumptions (8.6)–(8.9), θ in (8.10) is given by

$$\theta = \frac{(1 - p_{1i})p_{IIi}}{p_{1i}(1 - p_{IIi})} = \frac{p_{01i}}{p_{10i}} \quad \text{for all } i. \tag{8.11}$$

The fact that θ is the same for all i means that every patient has the same conditional probability of a score of 1 on drug I given that he scored 1 on exactly one of the drugs. Specifically, this conditional probability is

$$\frac{p_{10i}}{p_{10i} + p_{01i}} = \frac{1}{\theta + 1} \quad \text{for all } i. \tag{8.12}$$

Under these assumptions, then, it is again true that, given $B + C = N$, B is binomial with parameters N and $p = 1/(\theta + 1)$. Hence the usual binomial procedures can be applied to obtain inferences about p, and hence also about $\theta = (1/p) - 1$. Here, however, the underlying assumptions are much stronger, and as a result θ has an interpretation not previously available. In particular, Eq. (8.10) implies that for every patient the same drug has the higher probability of causing nausea, and indeed by the same amount in the sense of multiplying the odds by the same factor.

This model for dichotomous observations corresponds to an additive model for continuous observations with a pair effect α_i and a treatment difference β, as can be seen by taking logarithms in (8.10). (Here $\log p_{1i}/(1 - p_{1i})$ plays the role of α_i while $\log \theta$ plays the role of β.) This is a special case of an adaptation of the general linear or regression model to dichotomous dependent variables (Cox [1958b, 1970]) where the logarithm of the odds is assumed to be a linear function, with unknown coefficients, of some "independent" variables whose values are known (and which may be design or dummy variables).

The properties of the binomial tests discussed in Sect. 8.5 carry over to the present situation in the following form (Problem 67). Regarding the p_{1i} as nuisance parameters (the p_{11i} are then functions of θ and the p_{1i}), one-tailed binomial tests for p are uniformly most powerful unbiased tests for θ against one-sided alternatives, and the unbiased two-tailed binomial tests for p are uniformly most powerful unbiased tests for θ against two-sided alternatives.

9 Tolerance Regions

One-sample procedures based on the binomial distribution are also useful for obtaining tolerance regions. The methodology can be viewed as a generalization of the procedure for constructing confidence limits for the median or any specified p-point (quantile) of a distribution. Because of difficulties analogous to that of defining a unique p-point for discrete distributions, it is convenient to assume throughout this section that the relevant distribution is continuous.

Recall that the median of a population is the 50 % point of the distribution, or the point such that 50 % of the population lies below it. Let X^* be an upper 95 % confidence bound for the population median. This means that X^* is obtained in such a way that it has probability 0.95 of exceeding the population median. It follows that X^* has probability 0.95 of exceeding at least 50 % of the population. Equivalently, the region to the left of X^* has probability 0.95 of covering (including) at least 50 % of the population. This is perhaps the simplest example of a "tolerance region," more specifically, a "50 % tolerance region" at the "confidence" level 0.95.

In this section we will define tolerance regions exactly, mention some practical situations where they might be useful, and explain a simple method of constructing them from a random sample. Then we will discuss their usefulness for description and prediction, pointing out some difficulties in the interpretation of tolerance regions. Finally we will generalize the construction procedures. The question of what would be a "good" or "best" tolerance procedure will not be discussed.

9.1 Definition

For any fixed region R of a given population, we define the *coverage* of R as the proportion of the population which lies in R, that is, the proportion of the population covered by R. In random variable terminology, the coverage of R is

$$C(R) = P(X \in R) \tag{9.1}$$

where X is drawn at random from the population.

Suppose that, for some purpose, we would ideally like to find a region with coverage 0.5, that is, a region including 50% of the population. Lacking special knowledge about the population distribution, we cannot accomplish this exactly. We might be willing, instead, to define a region (depending on a sample) so that there is probability 0.95 that it will have coverage at least 0.5. This would perhaps sound difficult to do, had an example not been given above.

In general, a *tolerance region* is a random region having a specified probability, say $1 - \alpha$, that its coverage is at least a specified value, say c. Various names are given to $1 - \alpha$ and c in the literature. We shall call $1 - \alpha$ the *confidence level* and c the *tolerance proportion*, the latter because in some situations it is the minimum proportion of the population which it is considered tolerable to cover. We shall also speak of a "c tolerance region with confidence $1 - \alpha$." Regions which have this property under essentially no restrictions on the population are sometimes called "nonparametric tolerance regions," to distinguish them from "parametric tolerance regions," which have the required property as long as the population belongs to some specified parametric family, but not in general otherwise. Only nonparametric tolerance regions will be discussed here.

9.2 Practical Uses

In nonspecific settings, tolerance regions are often suggested for the purpose of describing the underlying population or for predicting future observations. For these purposes, however, a tolerance region at a conventional confidence level like 0.95 is of doubtful value. For description, for instance, the difficulty of interpreting a tolerance region is analogous to the difficulty of interpreting a confidence bound by itself as an estimator. Such difficulties, and possible remedies, will be explained further later, in Sects. 9.4 and 9.5.

The specific context in which tolerance regions are most often employed is that of production processes, since then it is natural to be concerned with whether the items produced are meeting specifications or measuring within some design tolerances, such as 100 ohms $\pm$ 10 ohms. If certain deviations of various characteristics from designated values are specified in advance as tolerable, it is easy to make nonparametric inferences about the proportion of the population in this region of tolerable values (Problem 68). This is not the type of tolerance region defined above, however; we shall be concerned here not with prespecified tolerance limits or regions of tolerable values, but rather with finding a tolerance region, based on a sample, such that a prespecified proportion of the population will be covered by that region with a preselected level of confidence. Because these regions are based on a sample, they are often called "statistical" tolerance regions. We will not repeat the adjective in the discussion to follow, since all tolerance regions here will be statistical.

These tolerance regions are also frequently used in connection with a production process. Suppose that no particular specifications are of special interest and one merely wishes to establish a rule for keeping tabs on the process. The process might be watched carefully for a limited period in order to collect sample data to use in finding a tolerance region having, for example, tolerance proportion 0.99 and confidence level 0.95. Thereafter, a cause of trouble is sought only when a sampled observation falls outside the tolerance region thus established. Whatever limits are set on the basis of the first sample, there will be some long run proportion of trouble-shooting even when the process is in control. The probability is 0.95 that the first sample will set tolerance limits such that trouble-shooting will be required at most 1 % of the time in the long run, if the process stays "in control," that is, does not change. Of course, more complicated conditions for trouble-shooting might well be used in practice (Problem 69).

As another possible use of tolerance limits in production processes, suppose that a producer wants to make some kind of money-back guarantee that his output will lie in a certain range, and he does not care exactly what the range is as long as it is not too much wider than necessary. Of course, he wants to be reasonably sure that no more than a very small proportion of his production will fall outside the guaranteed range. If the guaranteed range is a 0.995 tolerance interval with confidence 0.99 say, then the probability is 0.99 that, in the long run, at least 99.5 % of the production will satisfy the guarantee and the probability is only 0.01 that as much as 0.5 % will fail to satisfy it. This assumes no change in the process.

As an example from another field, consider setting norms for a physiological measurement, say the level of cholesterol in the blood.[1] Suppose that ideally one would like to be able to say that 98 % of normal people have between a and b milligrams of cholesterol per milliliter of blood as measured in a particular way. However, limits must be set on the basis of measurements on a finite sample of normal people. The endpoints of a 0.98 tolerance interval with confidence 0.95 might be chosen as limits of the norm, that is, the "normal range." (For example, the normal range of total serum cholesterol in adults is 150–250 mg/100 ml of blood.) Then the probability is 0.95 that the limits will include at least 98 % of normal people, and the probability is only 0.05 that more than 2 % of normal people will fall outside.

In each of these situations, once the concept of tolerance regions and the difficulties of interpretation discussed in the next two subsections are clearly understood, serious objections to tolerance regions as a solution to the real problem at hand will come readily to mind. To improve upon them substantially, however, is not so simple and requires consideration of aspects of each individual problem which were not even touched on above. Further, the relevant information on these aspects may be difficult or impossible to

---

[1] The authors are indebted to Frederick Mosteller for a discussion of uses of tolerance regions, and in particular for suggesting this example.

find. For example, clearly relevant but hard to estimate are the costs of the various possible acts in each problem, both when the production process is unchanged or the people are normal as regards cholesterol level, and when any of a variety of possible alternatives holds.

9.3 Construction of Tolerance Regions: Wilks' Method

Assume that $X_1, \ldots, X_n$ are independent observations on the same distribution, with c.d.f. F. Let $X_{(1)}, \ldots, X_{(n)}$ denote the order statistics of these observations. Assume that F is continuous, so that, with probability one, there are no ties and a unique ordering $X_{(1)} < X_{(2)} < \cdots < X_{(n)}$ exists. Let C_k be the coverage of the interval between $X_{(k-1)}$ and $X_{(k)}$. Then by (9.1) we have for $k = 2, \ldots, n$,

$$C_k = F(X_{(k)}) - F(X_{(k-1)}). \tag{9.2}$$

(Since F was assumed continuous, the coverage is the same whether the endpoints are included in the interval or not. If a specific statement were required, we would assume that right (upper) endpoints are included, and left (lower) endpoints are not.) We further define

$$C_1 = F(X_{(1)}) \quad \text{and} \quad C_{n+1} = 1 - F(X_{(n)})$$

as the coverage of the interval below $X_{(1)}$ and the interval above $X_{(n)}$ respectively. The definition in (9.2) applies also to these two intervals once we define $X_{(0)} = -\infty$ and $X_{(n+1)} = \infty$.

We now have $n + 1$ coverages, $C_1, C_2, \ldots, C_{n+1}$, corresponding to the $n + 1$ intervals into which the n sample points divide the real line. These $n + 1$ coverages are random variables, and their joint distribution has a number of interesting properties (Problems 70 and 71). A property which provides an easy method of construction of tolerance regions is the following. Let $i_1, \ldots, i_s$ be any s different integers between 1 and $n + 1$ inclusive; then the sum C of the corresponding coverages,

$$C = C_{i_1} + C_{i_2} + \cdots + C_{i_s},$$

has the same distribution as the sth smallest observation in a sample of n from the uniform distribution on the interval $(0, 1)$ (Problem 71g), namely

$$P(C \geq c) = n\binom{n-1}{s-1} \int_c^1 u^{s-1}(1-u)^{n-s} \, du \tag{9.3}$$

$$= \sum_{k=0}^{s-1} \binom{n}{k} c^k (1-c)^{n-k}. \tag{9.4}$$

Notice that the distribution depends on s and n only, not on which s integers are chosen, nor on the distribution from which the sample was drawn, as long as F is continuous. Notice also that C has a beta distribution by (9.3) and that (9.4) is a left-tail binomial probability.

Thus we have the following simple method of constructing tolerance regions. Select s distinct integers between 1 and $n + 1$ inclusive, and for each integer i selected, include the interval between $X_{(i-1)}$ and $X_{(i)}$ in the tolerance region. This gives a tolerance region with tolerance proportion c and confidence level $1 - \alpha$ if s is the critical value for an upper-tailed, level α binomial test of the null hypothesis $p = c$ based on a sample of size n. Equivalently, $1 - \alpha = P(C \geq c)$ is the probability of $s - 1$ or less successes, and $\alpha = P(C < c)$ is the probability of s or more successes, under the binomial distribution with parameters n and $p = c$. Thus, given the sample size n and the tolerance proportion c, the confidence level corresponding to each value of s can be obtained from a binomial table with this n and $p = c$.

If the s integers chosen are $1, 2, \ldots, s$, then the tolerance region obtained by this method is simply the interval with endpoints $-\infty$ and $X_{(s)}$, where $X_{(s)}$ is the upper confidence bound for the c-point ξ_c of the distribution, as explained in Sect. 4. In this sense, tolerance regions are a generalization of confidence bounds for quantile points, and the probabilities obtained previously continue to apply in this case, as indicated in connection with the median at the beginning of the section.

A more usual procedure is to choose s integers, $k + 1, k + 2, \ldots, k + s$ where $1 \leq k \leq k + s \leq n$. Then the tolerance region is the interval with endpoints $X_{(k)}$ and $X_{(k+s)}$, both finite. Note that, in accordance with what has already been said, the confidence level associated with this interval, when regarded as a tolerance interval, depends only on s and not on k. (This is not true when the interval is regarded as a confidence interval for a quantile point. See Problem 72.)

Sometimes it is convenient to consider not the number s of included intervals but instead the number m of excluded intervals. The procedure then would be to select m distinct integers between 1 and $n + 1$ inclusive, delete the interval between $X_{(i-1)}$ and $X_{(i)}$ for every selected integer i, and let the tolerance region consist of the remainder of the real line. Then (Problem 71h) the confidence level $1 - \alpha$ is the probability of m or more successes and α is the probability of $m - 1$ or less successes, under the binomial distribution with parameters n and $p = 1 - c$ (*not* $p = c$). Again, given n and c, the confidence level $1 - \alpha$ corresponding to each value of m can be obtained from a binomial table with this n and $p = c$. If the excluded intervals lie at the extremes of the sample, then the resulting tolerance region is again an interval.

To decide on the sample size n, one might select m, c, and α and choose n just large enough so that the tolerance region omitting m intervals has tolerance proportion c and confidence level at least $1 - \alpha$. Determining the required n is fairly easy using binomial tables. One might, for example, plan to omit just two intervals, the ones to the left and right of all sample values, and use as a tolerance region the interval with endpoints $X_{(1)}$ and $X_{(n)}$, that is, the entire range of the sample; this fixes $m = 2$. If the k leftmost and l rightmost intervals are omitted, so that the tolerance region is the interval whose endpoints are the kth smallest and lth largest observations $X_{(k)}$ and

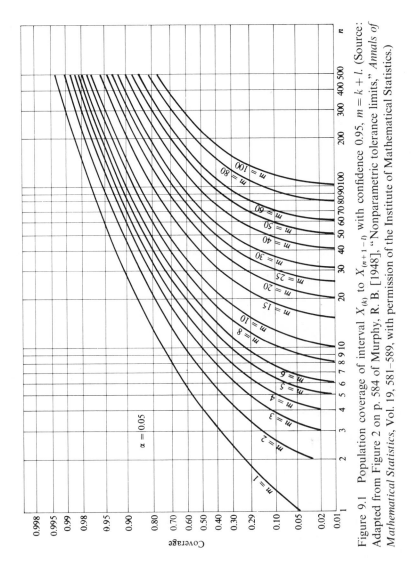

Figure 9.1 Population coverage of interval $X_{(k)}$ to $X_{(n+1-l)}$ with confidence 0.95, $m = k + l$. (Source: Adapted from Figure 2 on p. 584 of Murphy, R. B. [1948], "Nonparametric tolerance limits," *Annals of Mathematical Statistics*, Vol. 19, 581–589, with permission of the Institute of Mathematical Statistics.)

$X_{(n+1-l)}$, then $m = k + l$. If values of c and α are also selected, then n is the smallest value for which the probability of m or more successes exceeds $1 - \alpha$, under the binomial distribution with parameters n and $p = 1 - c$.

Murphy [1948] gives graphs of c versus n for various values of m and $1 - \alpha = 0.90, 0.95$, and 0.99. These are somewhat easier to use than binomial tables for some purposes. They are essentially equivalent to graphs of binomial confidence limits or percent points of the beta distribution (as, for instance, Fig. 6.2, Chap. 1), arranged in a certain way (Problem 73). Figure 9.1 reproduces the graph for $1 - \alpha = 0.95$; m denotes the number of intervals omitted, as explained in the previous paragraph.

9.4 Tolerance Regions for Description

As a descriptive device, a tolerance region with a high confidence level is deceptive, especially if it is based on a small amount of data. The reason can be explained as follows. Consider, for instance, a 0.65 tolerance region with confidence 0.95. If, as in the previous subsection, there is no probability that the coverage is exactly 0.65, then such a region has probability 0.95 of covering more than 65% of the population and probability only 0.05 of covering less than 65%. Then "typically" the coverage will be more than 65%, and substantially more if the region is based on little data. This is clearly reflected in Fig. 9.2, which shows the distribution of the actual coverage C of a 0.65 tolerance region with confidence 0.95 based on 20 independent observations and obtained (Problem 74) by the method described above. It is as though the tolerance proportion 0.65 were a lower 0.95 confidence bound for the actual coverage, except that what is random is the region rather than its tolerance proportion.

Another aspect of the same phenomenon is that a tolerance region having a particular tolerance proportion at a specified confidence level has at the same time other tolerance proportions at other confidence levels. For example, the region just mentioned, based on 20 observations, is not only a 0.65 tolerance region with confidence 0.95, but also a 0.50 tolerance region with confidence 0.999, a 0.75 tolerance region with confidence 0.775, etc. These and other combinations of tolerance proportion and confidence level can also be read from Fig. 9.2 (Problem 74).

The relation between the tolerance proportion and confidence level depends on the sample size (for any reasonable method of obtaining tolerance regions). Thus it is impossible to know, let alone to apprehend, just how conservative a tolerance region is merely from knowledge of the tolerance proportion and confidence level.

The difficulties just mentioned can be seen clearly in the special case mentioned earlier when the tolerance region is the interval from $-\infty$ to $X_{(s)}$, and $X_{(s)}$ is an upper confidence limit at level $1 - \alpha$ for the population c-point ξ_c. Then the use of the tolerance region for description amounts to

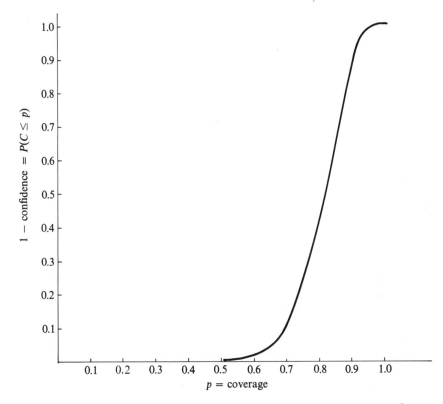

Figure 9.2   Distribution of actual coverage of a 0.65 tolerance region with confidence 0.95 when $n = 20$.

the use of the upper confidence limit $X_{(s)}$ as a sample descriptor of ξ_c. However, a single confidence limit at a typical level would not ordinarily be considered a descriptor in this sense, and would be a very lopsided descriptive device. For example, an upper 95 % confidence limit for a parameter, say the median, is not by itself very descriptive of what one knows about the parameter, since the limit could be any distance from the parameter.

Some methods of avoiding this deceptiveness when using tolerance regions as a descriptive device are listed below.

(1) Use the confidence level 0.50. Then the tolerance region is analogous to a median unbiased estimator.[2]
(2) State several combinations of the tolerance proportion and confidence level for the region given, for instance, the tolerance proportions corresponding to the confidence levels 0.05, 0.25, 0.50, 0.75, and 0.95.

[2] An estimator T is called median unbiased for a parameter θ if T has median θ for any allowed distribution. A significant fact about median unbiasedness is that $h(T)$ is median unbiased for $h(\theta)$ if T is median unbiased for θ and h is monotone (see also van der Vaart [1961]). This property does not hold for ordinary (mean) unbiasedness.

(3) Instead of giving any tolerance proportion and confidence level, give the expected coverage, as defined in the next subsection. This is somewhat analogous to unbiased estimation.

(4) Give two regions, one a tolerance region as defined already, and the other what might be called an inner tolerance region with the same tolerance proportion c and confidence level $1 - \alpha$. This is analogous to giving both upper and lower confidence bounds. By an inner tolerance region is meant a region having probability $1 - \alpha$ that its coverage is at most c. It can be chosen to lie inside the ordinary tolerance region (as long as $\alpha < 0.5$). Its complement is a tolerance region with tolerance proportion $1 - c$ and the same confidence level $1 - \alpha$.

9.5 Tolerance Regions for Prediction

The prediction problem we shall discuss is as follows. A sequence of observations is to be made, and after the first n observations, we want to predict whether the $(n + 1)$th observation will lie in some region. The first n observations may be used in constructing this region, and it should provide adequate probability of correct prediction. More precisely, given $n + 1$ independent, identically distributed random variables $X_1, \ldots, X_n, X_{n+1}$, construct a region depending only on $X_1, \ldots, X_n$. What is the probability that X_{n+1} will lie in the region?

The question is still not precise, and there are two natural ways to define this probability. One is conditional on $X_1, \ldots, X_n$, and then the probability is just the coverage of the region. That is,

$$P[X_{n+1} \in R(X_1, \ldots, X_n) | X_1, \ldots, X_n] = C[R(X_1, \ldots, X_n)] \quad (9.5)$$

where R is the region. The other is unconditional, and then the probability is the expected coverage of the region, or

$$P[X_{n+1} \in R(X_1, \ldots, X_{n+1})] = E\{C[R(X_1, \ldots, X_n)]\}. \quad (9.6)$$

The definition in (9.5) is the tolerance proportion or coverage we have been discussing in the previous subsections. This probability is relevant to an infinite number of predictions, being the proportion of correct predictions if each of the future observations $X_{n+1}, X_{n+2}, \ldots$ is predicted separately to lie in $R(X_1, \ldots, X_n)$. While this proportion is unknown, depending on the unknown population distribution, we can make confidence statements about it if R is a tolerance region.

The probability in (9.6) is the probability, before any observations are taken, that the particular prediction $X_{n+1} \in R(X_1, \ldots, X_n)$ will prove correct. For tolerance regions of the type we have discussed, this probability does not depend on the unknown population distribution, and is simply $s/(n + 1)$ (Problem 71), where s is the number of intervals $X_{(i-1)}$ to $X_{(i)}$ in the region R. For example, before any observations are taken, the probability that the $(n + 1)$th observation will lie within the entire range $X_{(1)}$ to $X_{(n)}$ of the first n is $(n - 1)/(n + 1)$, and the probability it will lie between any two successive observations $X_{(i-1)}$ and $X_{(i)}$ is $1/(n + 1)$.

If an ordinary tolerance region at a high confidence level is used for the region R, then the probability in (9.6) will be larger, and may be much larger, than the tolerance proportion c, since the actual coverage $C(X_1, \ldots, X_n)$ exceeds the tolerance proportion with probability equal to the confidence level chosen. This illustrates in another way the difficulty of making a simple interpretation of a tolerance region as a descriptor.

9.6 More General Construction Procedures

The procedure described in Sect. 9.3 for constructing tolerance regions can be generalized in a number of ways. As an aid to understanding, we shall proceed informally and one step at a time.

Suppose first that we have a sample of bivariate observations rather than univariate observations, and we seek a tolerance region in two-dimensional rather than one-dimensional space. Of course, we could look only at the first coordinate of each observation and construct a univariate tolerance region based on these n univariate observations. The corresponding (equivalent) region in the plane would then be a bivariate tolerance region. For instance, if the univariate region is an interval I, then the corresponding bivariate region would be simply the vertical band whose intersection with the horizontal axis is I.

A more interesting possibility would be to look at some real-valued function ϕ other than the first coordinate of the bivariate observations, which we denote by X_i. Suppose that $Z = \phi(X)$ has a continuous c.d.f. G. Define a tolerance region S in Z-space based on the order statistics $Z_{(1)}, \ldots, Z_{(n)}$ of the sample of values $Z_i = \phi(X_i)$. Let R be the corresponding region in X-space, or formally, $R = \{x : \phi(x) \in S\}$. Then R has the same coverage for X that S has for Z and hence is a tolerance region in X-space with the same tolerance proportion and confidence level. For example, if $\phi(X)$ is the distance of the point X from the origin, or the length of the vector X, then the Z_i are the distances of the X_i from the origin. If the Z tolerance region S is the interval from $Z_{(k)}$ to $Z_{(k+s)}$, then the corresponding X tolerance region R is the ring consisting of those points x whose distance from the origin is between $Z_{(k)}$ and $Z_{(k+s)}$. If some other well-behaved function ϕ had been used in place of distance, the boundaries of R would still be the two contours where ϕ has the values $Z_{(k)}$ and $Z_{(k+s)}$.

The same method can be applied to any kind of X-space whatever, by letting ϕ be a real-valued function on this space. We shall require only that $\phi(X)$ have a continuous distribution G (this avoids the difficulties of discreteness). To carry the ideas a bit further, let $Z_{(1)}, \ldots, Z_{(n)}$ again be the order statistics of the sample of values $Z_i = \phi(X_i)$. The $Z_{(i)}$ separate the Z-space (the real line) into $n + 1$ intervals. Let $R_1, \ldots, R_{n+1}$ be the corresponding regions in X-space, that is, R_i is the X-region where $\phi(X)$ is between $Z_{(i-1)}$ and $Z_{(i)}$. Specifically, $R_i = \{x \colon Z_{(i-1)} < \phi(x) \le Z_{(i)}\}$ where $Z_{(0)} = -\infty$ and $Z_{(n+1)} = \infty$ as before. The coverage C_i of R_i is the probability under the X distribution in the region R_i, which is the probability under the Z distribution in the interval between $Z_{(i-1)}$ and $Z_{(i)}$, which is $G(Z_{(i)}) - G(Z_{(i-1)})$. The joint distribution of these coverages is therefore the same as in Sect. 9.3, and in particular, the union of any s of the regions R_i is a tolerance region having a coverage C whose distribution is given by (9.3) and (9.4). The indices $i_1, \ldots, i_s$ of the included regions are to be selected in advance, of course.

Regions R_i whose coverages have the same joint distribution as in Sect. 9.3 are called "statistically equivalent blocks," the equivalence being that any permutation of the coverages has the same joint distribution as any other. By generalizing the procedure for constructing statistically equivalent blocks, we can obtain more general tolerance procedures. We shall proceed further in this way.

Instead of using the same function throughout, we could use a sequence of functions $\phi_1, \phi_2, \ldots, \phi_n$. One way to do so is as follows. First, let R_1 be the region where $\phi_1(X)$ is smaller than the smallest value $\phi_1(X_i)$ observed. Next, remove the minimizing X_i from the sample and R_1 from the X-space, and apply the same procedure to the remaining sample and the remaining portion of X-space, using ϕ_2 in place of ϕ_1. And so on. At each stage, the remaining X's are a sample from the original distribution except restricted to the remaining portion of the X-space; therefore, the conditional distribution of the coverage of the next region to be removed given the coverages of the regions already removed is the same no matter what function ϕ_j is used next and hence is the same as when all functions ϕ_j are the same. Therefore, the regions $R_1, R_2, \ldots, R_{n+1}$ obtained are again statistically equivalent blocks.

The first step above may be thought of as using ϕ_1 to cut the X-space into two regions, one consisting of one block and one consisting of $n - 1$ blocks not yet subdivided. The second step then uses ϕ_2 to cut the latter region into two regions, of 1 and $n - 2$ blocks, etc. Geometrically, the successive cuts are along contours of $\phi_1, \phi_2, \ldots, \phi_n$. Instead of cutting off one block at each step, however, we could choose some arbitrary split. In this case, the first step is to choose an integer r_1 between 1 and n, find the r_1th from the smallest of the values $\phi_1(X_i)$, $i = 1, 2, \ldots, n$, and cut the X-space into two regions according to whether $\phi_1(x)$ is smaller or larger than this r_1th value. We then have one region containing $r_1 - 1$ X's and still to be subdivided into r_1 blocks, and a second region containing $n - r_1$ X's and still to be subdivided into $n - r_1 + 1$ blocks; the remaining X, say $X_{[1]}$, is the borderline value through which the first cut passes. The second step is to cut one of these two regions, using the function ϕ_2 and another arbitrary integer. After n steps, n cuts have been made, all the X_i have been used, and there are $n + 1$ regions, each fully subdivided, i.e., consisting of a single block. These $n + 1$ regions are again statistically equivalent blocks, by a slight generalization of the previous reasoning.

An even further generalization is to let the function and/or integer used in the second step depend on the borderline value $X_{[1]}$ through which the first cut passes. Similarly, the function and integer used in any later step may depend on the borderline X-values (and functions and integers) used in earlier steps. They cannot, however, depend on the X_i other than the borderline values in earlier steps. Since all the X_i must be examined to determine the borderline values, it is essential for preventing any dependence on other than borderline values that the procedure be fully specified in advance (or perhaps that the borderline values be determined by a second party or a computer without revealing any other values).

One other point should be clarified. "Statistical equivalence" of course does not mean that a tolerance region could be formed from the s smallest blocks obtained, for instance. The indices of the blocks to be included must be specified in advance, and the indexing of the blocks must be carried out in such a way that statistical equivalence holds. (Some relaxation of the former is possible, but adjustment of the indexing process can accomplish the same thing.) The requirement on the indexing is that an appropriate number of indices must be assigned to each region at every step, and more particularly that, before each step, from the indices already assigned to the region about to be cut, an appropriate number must be selected and assigned to each of the two regions which will result from the cut. What this amounts to for the tolerance procedure is that, at every step, in each region, the number of blocks which will ultimately belong to the tolerance region must be specified, and more particularly that, before each step, the number of blocks which will ultimately belong to the tolerance region must be specified for each of the two regions which will result from the next cut. Again, dependence is permitted on previous borderline X-values but not on other X_i.

We shall not attempt to state fully and formally the most general procedure obtainable along the foregoing lines for either statistically equivalent blocks or tolerance regions. Such a statement is bound to be very cumbersome and therefore difficult to understand and use in checking the validity of a proposed procedure. It is probably easier to understand the ideas and validate procedures directly on the basis of this understanding.

The generalizations above are applicable even in univariate situations. For example, given a univariate sample $X_1, \ldots, X_n$, one might remove three blocks, leaving a tolerance region of $n - 2$ blocks, as follows. First, make cuts at the smallest and largest observations, $X_{(1)}$ and $X_{(n)}$, and remove the two blocks $(-\infty, X_{(1)})$ and $(X_{(n)}, \infty)$. (This is equivalent to using the cutting function $\phi(x) = x$, or any strictly monotone function thereof, for each of the first two cuts.) Let

$$\phi_3(x) = \left| x - \frac{1}{2}(X_{(1)} + X_{(n)}) \right|,$$

and make the third cut at the largest value of $\phi_3(X_{(i)})$, $i = 2, \ldots, n - 1$. Then the tolerance region will be the interval from

$$X_{(1)} + X_{(n)} - X_{(n-1)} \text{ to } X_{(n-1)} \quad \text{if } X_{(2)} - X_{(1)} > X_{(n)} - X_{(n-1)}$$
$$X_{(2)} \text{ to } X_{(n)} - X_{(2)} + X_{(1)} \quad \text{otherwise}$$

(Problem 76). The rationale for such a procedure might be that excluding three blocks would more nearly permit the desired tolerance proportion and confidence level than excluding two or four, and this is a more symmetric procedure for excluding three blocks than using always $X_{(1)}$ to $X_{(n-1)}$ or always $X_{(2)}$ to $X_{(n)}$.

More elaborate univariate procedures, including this as a special case, are discussed by Walsh (1962b). The possibilities of generalization seem to have been opened up in Scheffé and Tukey (1945) and Tukey (1947). Further discussion and references are given by Fraser (1957) and Guttman (1970).

PROBLEMS

1. Show that, if $F(x) = p$ for $a < x < b$, then a and b are both pth quantiles of F.

2. Show that ξ is a pth quantile of F if and only if the point (ξ, p) lies on the graph of F with any vertical jumps filled in.

3. Show that ξ is a pth quantile of a distribution if and only if, when ξ is included, the left tail probability is at least p and the right tail probability is at least $1 - p$.

4. (a) Sketch seven c.d.f.'s to exhibit the seven possible combinations of cases (a)–(c) of Sect. 2.
   (b) Which combinations are possible for
   (i) discrete distributions?
   (ii) distributions with densities?

5. What is the relation between the quantiles and the inverse function of a c.d.f.?

6. An estimator is called median unbiased for a parameter if its median is that parameter. Show that, for odd sample sizes, the sample median is a median unbiased estimator of the population median.

7. (a) If ξ_p is a pth quantile of a distribution on a finite set, show that either p or ξ_p is not unique. Relate this to cases (a)–(c) of Sect. 2.
   (b) For what countably infinite sets does (a) hold?

8. A constant problem in manufacturing plants is machine breakdowns. Time consumed while the machine is being repaired is called "down time." Both the expense of repair and amount of down time need to be kept small for efficient and profitable production. The managers of a plant are considering replacing the type of machine currently in use. The decision should depend on the costs of purchase and changeover, in addition to a comparison of down time. To get data as input for this decision, twenty machines of the new type are leased and observed for a fixed period of time. Only one machine had a breakdown. The probability of a breakdown is known to be 0.10 for the machine previously used. Does the replacement machine seem to have a smaller probability of breakdown? Find the P-value.

9. It is frequently claimed that working together on a common project makes people like each other more. A sociologist ran an experiment to test the null hypothesis that no systematic change in people's friendliness occurs through joint participation, against the alternative that people become friendlier. Twenty-five pairs of individuals were selected at random, each pair was observed together in several situations for one week, and notes made on their relationship. After each pair had worked together on a single project for one week, their relationship was observed again. Fifteen pairs were noted as having a friendlier relationship after the project. Perform a statistical test, and comment on its appropriateness and limitations.

*10. Let $X_1, X_2, \ldots, X_n$ be independent, identically distributed m-vectors and let $p = P(X_j \in A)$ for A a given set in m-space. Find a uniformly most powerful test of the null hypothesis $H_0: p \leq p_0$ against the alternative $H_1: p > p_0$ [Lehmann, 1959, p. 93].

*11. Prove that any two-tailed sign test is admissible if $X_1, \ldots, X_n$ are independent and identically distributed with $P(X_j = \xi_0) = 0$, where ξ_0 is the hypothesized median value.

*12. Prove the optimum properties of the sign test for $H_0: p = p_0$ given in Sect. 3.2 for the case where the observations are not necessarily identically distributed but are independent with $P(X_j < \xi_0) = P(X_j \leq \xi_0) = p$ for all j.

*13. Prove that S, the number of observations below ξ_0, is a sufficient statistic for p if $X_1, \ldots, X_n$ are independent and identically distributed with distribution F belonging to the family defined in Sect. 3.3, paragraph 2.

*14. Show that, if $X_1, X_2, \ldots, X_n$ are independent and identically distributed with $P(X_j < \xi_0) = P(X_j \leq \xi_0) = p$, then the unbiased, two-tailed sign test of the null hypothesis $H_0: p = p_0$ is uniformly most powerful unbiased against $H_1: p \neq p_0$.

*15. Show that, in Problem 14, the two-tailed, symmetric, level α sign test is most powerful among all symmetric, level α tests.

*16. Prove Theorem 3.1, which gives conditions under which a test is most powerful.

17. An automobile manufacturer wishes to design a certain new model such that the front-seat headroom is sufficient for all but the tallest 5% of male drivers. A random sample of 100 male drivers is taken. The heights of the 9 tallest are as follows:

$$70.1, \ 72.3, \ 71.9, \ 70.5, \ 73.4, \ 76.1, \ 74.5, \ 70.9, \ 75.8.$$

(a) Find a 90% two-sided confidence interval for the 95th percent point of the population of male drivers.
(b) Former studies by the Federal government have shown that the 95th percentile point for height of U.S. males is 70.2 inches. Does this result appear to be valid now and for the population of male drivers?

18. A sample of 100 names was drawn from the registered voters in Appaloosa County and sent questionnaires regarding a proposed taxation bill. Of the 75 usable returns, 50 were in favor of the bill. Find a 95% confidence interval for the true proportion of registered voters in favor of the bill. What assumption are you making about the unusable returns?

19. Suppose that a quantile of order p is not unique for fixed p, that is, ξ_p is any value in the closed interval $[\xi_p', \xi_p'']$ for some $\xi_p' < \xi_p''$. Show that if lower and upper confidence bounds, say L and U, are determined by the usual sign test procedure, each at level $1 - \alpha$, then

$$P(L \le \xi_p') \ge 1 - \alpha \quad \text{and} \quad P(U \ge \xi_p'') \ge 1 - \alpha.$$

20. (a) Suppose that L is a lower confidence bound for a pth quantile ξ_p constructed by the usual sign test procedure at exact level $1 - \alpha$ for a continuous population. Show that if the population is discontinuous, the lower confidence bound has at least the indicated probability of falling *at or below* ξ_p, and at most the indicated probability of falling *strictly below* ξ_p, that is

$$P(L \le \xi_p) \ge 1 - \alpha \ge P(L < \xi_p).$$

(b) Show that a corresponding statement holds for the upper confidence bound.

(c) Show that the two-sided confidence interval which includes its endpoints has at least the indicated probability of covering ξ_p (and hence at most the indicated error rate), while the interval excluding its endpoints has at most the indicated probability of covering ξ_p.

21. Show that
(a) If X has a continuous, strictly increasing c.d.f. F, then $Y = F(X)$ has a uniform distribution over $(0, 1)$. (This is the fundamental property of the "probability integral transformation" $F(X)$, named thus presumably because F is the integral of the density of X when it has one.)

(b) Conversely, if Y has a uniform distribution over $(0, 1)$ and F is a continuous, strictly increasing c.d.f., then $X = F^{-1}(Y)$ has c.d.f. F.

(c) In (b), if F is any c.d.f. whatever and X is any quantile of order Y in F, then X has c.d.f. F.

22. Suppose that four independent, dichotomous trials are observed, with true probability p_j of success on the jth trial, $j = 1, 2, 3, 4$. Let $p = \sum p_j/4$. If the upper confidence limit for p is taken to be 0.44, 0.68, 0.85, 0.97, and 1.00 respectively when there are 0, 1, 2, 3, and 4 successes, graph the true confidence level as a function of p for each of the following situations:

(a) $p_1 = p_2 = p_3 = p_4 = p$

(b) $p_1 = p_2 = p - 0.1, p_3 = p_4 = p + 0.1$

(c) $p_1 = 0, p_2 = p_3 = p_4 = 4p/3$

(d) $p_1 = 1, p_2 = p_3 = p_4 = (4p - 1)/3$.

Be sure to study what happens in the neighborhood of $p = 0.44, 0.68, 0.85, 0.97$; a few other values of p will suffice. (Notice that the procedure is conservative. See Hoeffding [1956].)

23. Consider "fixed effects" models with $p_j = P(X_j < \xi_0)$ arbitrary except that $p = \sum_j p_j/n$ is given. Show that, for sign tests with critical values s_c' and s_c'':

(a) If $p \ge (s_c' + 1)/n$ and/or $p \le (s_c'' - 1)/n$ as appropriate, then there exist values of p_j for which the probability of "acceptance" is 1 (and hence the power is 0 if $p \ne p_0$).

(b) If $p \le s_c'/n$ or $p \ge s_c''/n$, then there exist values of p_j for which the probability of rejection is 1.

24. Let $X_{(r)}$ denote the rth from the smallest in a random sample of size 5 from any continuous population with ξ_p the quantile of order p. Evaluate the following probabilities:

(a) $P(X_{(1)} < \xi_{0.50} < X_{(5)})$

(b) $P(X_{(1)} < \xi_{0.25} < X_{(3)})$

(c) $P(X_{(4)} < \xi_{0.80} < X_{(5)})$.

25. If $X_{(1)}$ and $X_{(n)}$ are respectively the smallest and largest observations in a random sample of size n from any continuous distribution F with median $\xi_{0.50}$, find the smallest value of n such that

(a) $P(X_{(1)} < \xi_{0.50} < X_{(n)}) \geq 0.95$

(b) $P[F(X_{(n)}) - F(X_{(1)}) \geq 0.50] \geq 0.95$.

26. Let V denote the proportion of the population lying between the smallest and largest observations in a random sample of size n from any continuous population. Find the mean and variance of V.

27. In a random sample of size n from any continuous population F, the interval $(X_{(r)}, X_{(n-r+1)})$ for any $r < n/2$ gives a level $1 - \alpha$ confidence interval for the median of F. Show that α can be written as

$$\alpha = (0.5)^{n-1} \sum_{k=0}^{r-1} \binom{n}{k} = 2n \binom{n-1}{r-1} \int_0^{0.50} x^{n-r}(1-x)^{r-1}\, dx.$$

28. Show that the exact confidence level of a confidence interval for the median with endpoints the second smallest and second largest observation is equal to

$$1 - (n+1)/2^{n-1}$$

for a random sample of n observations from any continuous population.

29. Show that if $X_{(1)} < \cdots < X_{(n)}$ are the order statistics of an independent random sample from a continuous distribution with density f, then the joint density of the order statistics is

$$n! \prod_{i=1}^{n} f(x_i) \quad \text{for } x_1 < x_2 < \cdots < x_n.$$

For example, the density of the normal order statistics is

$$n!(2\pi\sigma^2)^{-n/2} e^{-\Sigma_j(x_j - \mu)^2/2\sigma^2} \quad \text{for } x_1 < x_2 < \cdots < x_n.$$

30. Let $X_{(r)}$ be the rth order statistic of a random sample of size n from a population with continuous c.d.f. F.

(a) Differentiate (4.1) to show that the marginal density of $X_{(r)}$ is

$$g(x) = r \binom{n}{r} [F(x)]^{r-1} [1 - F(x)]^{n-r} f(x).$$

(b) Show that the c.d.f. of the density in (a) can be written as

$$G(t) = P(X_{(r)} \leq t) = \int_0^{F(t)} [B(r, n-r+1)]^{-1} u^{r-1}(1-u)^{n-r}\, du,$$

and hence this binomial sum is equivalent to the incomplete beta c.d.f. above.

(c) Integrate (a) by parts repeatedly to obtain the binomial form in (4.1).

31. By considering $P(X_{(r)} > t/n)$ in the binomial form given in (4.1), find the asymptotic distribution of $X_{(r)}$ for r fixed and $n \to \infty$ if
    (a) F is the uniform distribution on $(0, 1)$.
    (b) F is an arbitrary continuous c.d.f.

32. Let $X_{(n)}$ denote the largest value in a random sample of size n from the population with density function f.
    (a) Show that $\lim_{n \to \infty} P(n^{-1} X_{(n)} \le x) = \exp(-\alpha/\pi x)$ if $f(x) = \alpha/[\pi(\alpha^2 + x^2)]$ (Cauchy).
    (b) Show that $\lim_{n \to \infty} P(n^{-2} X_{(n)} \le x) = \exp(-2\sqrt{2/\pi x})$ if $f(x) = (\alpha/\sqrt{2\pi})x^{-3/2} \exp(-\alpha^2/2x)$ for $x \ge 0$.

33. Show that the joint density of two order statistics $X_{(r)}, X_{(v)}, 1 \le r < v \le n$, of an independent random sample from a population with continuous c.d.f. F, is

$$g(x, y) = n(n-1)\binom{n-2}{r-1, v-r-1, n-v}[F(x)]^{r-1}[F(y) - F(x)]^{v-r-1}$$

$$[1 - F(y)]^{n-v} f(x) f(y) \quad \text{for } x < y$$

where $\binom{m}{r, s, t} = m!/r!s!t!$ for $r + s + t = m$.

34. For any two order statistics $X_{(r)}$ and $X_{(v)}$ of an independent random sample from a population with continuous c.d.f. F, where $r < v$, show that the event $X_{(r)} < \xi_p$ occurs if and only if either $X_{(r)} < \xi_p < X_{(v)}$ or $\xi_p \ge X_{(v)}$. Since these latter two events are mutually exclusive, it follows that

$$P(X_{(r)} < \xi_p < X_{(v)}) = P(X_{(r)} < \xi_p) - P(X_{(v)} \le \xi_p).$$

This result can be expressed in binomial form as in (4.3), or, by Problem 30(b), it can be written as the difference

$$\int_0^p [B(r, n - r + 1)]^{-1} u^{r-1}(1 - u)^{n-r} \, du$$

$$- \int_0^p [B(v, n - v + 1)]^{-1} u^{v-1}(1 - u)^{n-v} \, du.$$

35. Let $X_{(1)} < \cdots < X_{(n)}$ be order statistics of a random sample of size n from the exponential density $f(x) = e^{-x}, x \ge 0$.
    (a) Show that $X_{(r)}$ and $X_{(v)} - X_{(r)}$ are independent for any $r < v$.
    (b) Find the distribution of $X_{(r+1)} - X_{(r)}$.
    (c) Interpret the significance of these results if the sample arose from a life test of n items with exponential lifetimes.

36. Find the density and c.d.f. of the range, $X_{(n)} - X_{(1)}$, of a random sample size n from any continuous population.

37. Suppose we want to use a random sample of size n to find a level 0.95 confidence interval for θ in the density $f(x) = \exp[-(x - \theta)]$ for $x > \theta$. Since the smallest observation $X_{(1)}$ is a sufficient statistic for θ (and also its maximum likelihood estimator), some functions of $X_{(1)}$ would be a natural choice for the confidence bounds. If $X_{(1)}$ is the upper confidence bound, find that lower confidence bound $g(X_{(1)})$ which gives a two-sided level of 0.95.

38. Suppose we have a normal population with unknown mean and median ξ and known variance σ^2 and we require a test of the null hypothesis $H_0: \xi = \xi_0$ against the simple alternative $\xi = \xi_1$, where $\xi_1 > \xi_0$, such that the level is α and the power is $1 - \beta$. Let α and β be fixed while $(\xi_1 - \xi_0)/\sigma = \delta$ approaches 0.

   We consider two different tests for the situation described above. Test A is the appropriate normal theory test for a sample of size n_A, that is, the test based on $Z = \sqrt{n_A}(\bar{X} - \xi_0)/\sigma$ with a right-tail critical value z_α from Table A. Test B is the sign test for a sample of size n_B, that is, the test based on S, the number of observations smaller than ξ_0.

   (a) Obtain a general expression for the sample size n_A required by the normal theory test.

   (b) Obtain an approximate expression for the sample size n_B required by the sign test, by approximating the binomial distribution by the normal distribution.

   (c) The limit of the ratio n_A/n_B as $\delta \to 0$ is known as the asymptotic efficiency of the sign test relative to the optimum normal theory test. Using (a) and (b), show that the asymptotic relative efficiency equals $2/\pi$. This example and others will be discussed at length in Chap. 8.

39. (a) Show that the value of i/n such that $X_{(i)}$ is a lower confidence bound for the median at conservative level 0.95 is 0.28 when $n = 9$, and 0.32 when $n = 17$. What is the true level in each case?

   (b) Find the true level of $X_{(i+1)}$ as a lower confidence bound for the median when $i/n = 0.28$, $n = 9$ and when $i/n = 0.32$, $n = 17$.

   (c) Since the true levels in (a) and (b) are approximately equidistant from 0.95 for each n, one might consider using $(X_{(i)} + X_{(i+1)})/2$ as a lower confidence bound for the median. If the population is symmetric, what is the true level of this bound when $n = 9$, and when $n = 17$?

40. Show that the following pairs of hypotheses are equivalent:

   (a) $\max \xi_{p_0} < \xi_0$ and $p_< > p_0$

   (b) $\min \xi_{p_0} \le \xi_0$ and $p_\le \ge p_0$

   (c) $\min \xi_{p_0} > \xi_0$ and $p_\le < p_0$

   (d) $\max \xi_{p_0} \ge \xi_0$ and $p_< \le p_0$.

   Here $\max \xi_{p_0}$ and $\min \xi_{p_0}$ denote the maximum and minimum possible values of ξ_{p_0} as required if ξ_{p_0} is not unique.

41. Use the results of Problem 40 to show that $\xi_{p_0} = \xi_0$ for all p_0 points if and only if $p_< \le p_0 \le p_\le$, and thus that the two-sided test that rejects if $S_\le \le s_\ell$ and also if $S_< \ge s_u$ is appropriate for a two-sided test of the null hypothesis that ξ_0 is a p_0-point when one or more observations is equal to ξ_0.

42. (a) Show that

$$E(S_< - S_>) = n(p_< - p_>)$$

$$\text{var}(S_< - S_>) = n[p_< + p_> - (p_< - p_>)^2].$$

   (b) Show that if $p_< + p_> = 1$ then $S_< - S_>$ is even with probability 1 for n even, odd for n odd.

(c) Show that $S_< - S_>$ is asymptotically normal with the mean and variance given in (a), even if the parameters $p_<$ and $p_>$ depend on n, provided that this variance approaches infinity.

(d) Derive approximate tests and confidence bounds for $p_< - p_>$ and show that their levels are asymptotically valid.

43. (a) Given any c.d.f. G and any positive $p_<$, $p_>$ with $p_< + p_> \le 1$, show that there is exactly one c.d.f. F having the same conditional distributions as G on each side of ξ_0, $P_F(X < \xi_0) = p_<$, and $P_F(X > \xi_0) = p_>$, and express F algebraically in terms of G, $p_<$, and $p_>$.

(b) Show that the family of such distributions F includes G and includes a distribution with $p_< = p_>$ for each $p_<$, $0 < p_< < \frac{1}{2}$.

(c) Show that the statistics S and N are jointly sufficient for this family, where S and N are the numbers of observations in a sample which are respectively $< \xi_0$ and $\le \xi_0$.

(d) Show that, for the subfamily of such distributions F with $p_< = p_>$, the statistic N is sufficient.

44. (a) Show that a conditional test at conditional level α is an unconditional test at level α.

(b) Show that a conditionally unbiased test is unconditionally unbiased.

(c) Show that if all unbiased tests against a certain alternative are conditional and if a certain conditional test has uniformly greatest conditional power, then this test is uniformly most powerful unbiased against this alternative.

*45. Prove Theorem 6.1 relating unbiased and conditional tests.

46. In the situation and notation of Sect. 6, for $n = 4$, let $\phi(S_<, N)$ be defined by the accompanying table, with $\phi(S_<, N) = 0$ elsewhere

N	0	1	2	3	4
$S_<$	0	0, 1	0, 2	1, 2	0, 4
ϕ	$\frac{1}{8}$	$\frac{1}{8}$	$\frac{1}{4}$	$\frac{1}{6}$	1

(a) Show that the test ϕ has conditional level $\frac{1}{8}$ for each N.

(b) Show that ϕ is biased conditional on $N = 3$.

(c) Show that ϕ has power

$$\tfrac{1}{8}(p_0 + p_1) + \tfrac{1}{4}p_2[p^2 + (1 - p)^2] + \tfrac{1}{2}p_3 p(1 - p) + p_4[p^4 + (1 - p)^4]$$

where $p_m = P(N = m) = \binom{4}{m}r^m(1 - r)^{4-m}$ with $r = p_< + p_>$ and $p = p_</r$.

*(d) Show that the conditional power for $N = 4$ exceeds that for $N = 2$, i.e., $p^4 + (1 - p)^4 > \frac{1}{4}[p^2 + (1 - p)^2]$ for $p \ne \frac{1}{2}$.

(e) Show that the conditional power for $N = 2$ and that for $N = 3$ average to $\frac{1}{8}$, i.e., $\frac{1}{8}[p^2 + (1 - p)^2] + \frac{1}{4}p(1 - p) = \frac{1}{8}$.

*(f) Show that $p_3 \le p_2 + p_4$.

*(g) Use these facts to show that ϕ is unconditionally unbiased.

*47. Fill in the details of the proof that the equal-tailed conditional sign test is uniformly most powerful unbiased when ξ_0 has positive probability by showing the results stated in (6.11)–(6.13).

48. Suppose that a pair of control and treatment measurements (V, W) would be permutable if the treatment had no effect, while the effect of the treatment is to add a constant amount β to W. Show that the median of the difference $X = W - V$ is β.

49. Suppose that a confidence interval for the median is obtained by applying the methods of Section 4 to the differences $X_i = W_i - V_i$ of independent, identically distributed pairs (V_i, W_i). Show that the confidence procedure is valid for the "shift" θ which makes $P(W_i - \theta < V_i)$ and $P(W_i - \theta > V_i)$ both no larger than 0.5. (In the continuous case, $W_i - \theta$ is equally likely to be less than or greater than V_i. Nevertheless, θ need not equal the difference between the medians of W_i and V_i; see the following problems and Sect. 2 of Chapter 3.)

50. This problem and the next one show that the population medians of the treatment-control differences in matched-pair experiments may be all positive or all negative even though the treatment has no effect on the mean or median for any unit. If the dispersion is larger for larger-valued units, and if the treatment accentuates this, then the treatment-control differences typically have negative medians although they may have positive medians. The following problem gives a similar result for skewness.

   Suppose that each unit possesses a "unit effect" u and would yield the observed value $U = u$ if untreated but $T = u + \tau(u)Z$ if treated, where $\tau(u)Z$ is a "random error," independent from unit to unit, with positive scale factor $\tau(u)$. Suppose that a pair of units is given whose unit effects are u' and u'', with $u' < u''$, say; then one of the two units is chosen at random for treatment. Let X be the treatment-control difference observed, and let $\delta = u'' - u'$, $\tau' = \tau(u')$, and $\tau'' = \tau(u'')$.

   (a) Show that $P(X \le x) = \frac{1}{2}P(\delta + \tau''Z \le x) + \frac{1}{2}P(-\delta + \tau'Z \le x)$.
   (b) Suppose that $P(Z = -1) = P(Z = 1) = 0.5$. Show that the possible medians of X are all positive if $\tau'' < \delta < \tau'$, all negative if $\tau' < \delta < \tau''$, and include 0 otherwise. [Hint: $P(X = -\delta - \tau') = P(X = -\delta + \tau') = P(X = \delta - \tau'') = P(X = \delta + \tau'') = 0.25$.]
   (c) Suppose that $\tau(u)$ is an increasing function of u and that Z is symmetrically distributed about zero, that is, $P(Z \ge z) = P(Z \le -z)$ for all z. Show that $P(X < 0) \ge 0.5$ with equality holding if and only if $P(\delta/\tau'' \le Z \le \delta/\tau') = 0$. (In the case of inequality, the median of X must be negative.)

51. Suppose that the situation of the previous problem holds except that U is distributed as $u + v(u)Z$ with $v(u) > 0$, and the pairs (U, T) are independent from unit to unit. (For any one unit, only the marginal distributions of U and T matter. Beyond this their joint distribution is irrelevant.) Let $v' = v(u')$ and $v'' = v(u'')$.

   (a) Show that $P(X \le x) = \frac{1}{2}P(\tau''Z'' - v'Z' \le x - \delta) + \frac{1}{2}P(\tau'Z' - v''Z'' \le x + \delta)$ where Z' and Z'' are independently distributed as Z.
   (b) Show that $P(X < 0) > 0.5$ if Z is normal with mean 0 and the treatment variance minus the control variance, $\tau^2(u) - v^2(u)$, is an increasing function of the unit effect u
   *(c) Show that if Z is uniformly distributed over the interval $[-R\ R]$, $\tau'' - v'' \ge \tau' - v'$, and $(v''\tau'' - v'\tau')(\tau'\tau'' - v'v'') > 0$, then $P(X < 0) \ge 0.5$ with equality if and only if $R \le \delta/(v' + \tau'')$.
   (d) Show that τ and v satisfy the conditions of (b) and (c) if $\tau(u) - v(u)$ is a positive, increasing function of u and $v(u)$ is nondecreasing.

(e) Show that $P(X \le 0) = \frac{4}{9}$ if $P(Z = z) = \frac{1}{3}$ for $z = -1, 0, 1$ and $v' < \delta < v'' < \tau' + \delta$ and $\delta < \tau' < \tau'' < v' + \delta$.

If $v(u)$ and $\tau(u) - v(u)$ are increasing functions of u, parts (b)–(d) illustrate that the median of X will typically be negative, although it may be positive, even for symmetric Z, since these conditions do not preclude the conditions of part (e).

52. In the situation of the previous two problems, replace the assumptions about U and T by $E(U) = E(T) = u$ (so the treatment has no effect on the mean), var$(U) = v^2(u)$ and var$(T) = \tau^2(u)$ are finite, and $E(U - u)^3 = E(T - u)^3 = 0$, with (U, T) still independent from unit to unit. Show that $E(X) = 0$ and $E(X^3) = 1.5\delta(\tau''^2 - v''^2 - \tau'^2 + v'^2)$. Therefore the distribution of X is skewed to the right if $\tau^2(u) - v^2(u)$ is an increasing function of u, even though U and T have no skewness.

53. In the paired situation of Sect. 8, define a difference score for each pair as the score on II minus the score on I. Show that the probability of a positive difference score minus the probability of a negative difference score equals $p_{11} - p_1$, where p_1 and p_{11} denote the probabilities of a score of 1 on I and II respectively.

54. Suppose that one male and one female chick are selected at random from each of 10 litters and inoculated with an organism which is thought to produce an equal chance of life (1) or death (0) for every chick. The death occurs within 24 hours if at all, and the organism has no effect on life after the first 24 hours. Test the data below to investigate whether the sexes differ in their response to the organism.

Litter	Male Response	Female Response
1	1	0
2	0	0
3	1	1
4	1	0
5	0	1
6	0	0
7	0	0
8	0	1
9	1	0
10	1	0

55. Prior to a nationally televised series of debates between the presidential candidates of the two major parties, a random sample of 100 persons were asked their preference between these candidates. Sixty-three persons favored the Democrat at this time. After the debate the same 100 people were asked their preference again, and now seventy-two favored the Democratic candidate. Of these, 12 had previously stated a preference for the Republican candidate. Test the null hypothesis that the voting preferences were not significantly changed after the debate. Can you say anything about the effect of the debate?

56. In a study to determine whether constant exposure of children to violence on TV affects their tendency toward violent behavior and possibly crime, disturbance, etc., a group of 100 matched pairs of children were randomly selected. The pairs

were formed in the population by matching the children as well as possible as regards home environment, genetic factors, intelligence, parental attitudes, etc., so that differences in other factors influencing aggressive behavior could be minimized. In each pair, one child was randomly selected to view routinely the most violent shows, while the other was permitted to watch only "acceptably nonviolent" shows. Psychologists rated each child's tendency toward aggression, both before and after the experiment, and noted whether the children exhibited more (1) or the same (or less) (0) tendency to aggression. The numbers of children in these rating groups are shown below. Analyze the data to test

(a) The hypothesis that the proportions of children exhibiting more tendency to aggression is the same regardless of what kind of TV shows they watch.

(b) The hypothesis that the proportion of children "switching" their aggression pattern in each direction is the same. Find a confidence interval at level 0.90 for the difference of these proportions. Comment on the meaning of this difference.

Score of Children Exposed to Violence	Nonviolence	Number of Pairs
1	1	18
1	0	43
0	1	8
0	0	31

57. Suppose that an achievement test is divided into k parts, all covering the same type of achievement but using different methods of investigation. All parts of the test are given to each of r subjects and each is given a score of pass or fail on each part of the test. The data might be recorded as follows, where 1 represents pass and 0 represents fail.

Subject	Score on part 1	Score on part 2	...	Score on part k	R
1	1	0	...	1	R_1
2	1	1	...	0	R_2
⋮	⋮	⋮		⋮	⋮
r	0	1	...	0	R_r
	C_1	C_2	...	C_k	N

The symbols $R_1, \ldots, R_r$ and $C_1, \ldots, C_k$ represent the row and column totals respectively, so that R_i is the number of parts passed by subject i and C_j is the number of subjects who passed part j. N is the total number of parts passed by all subjects.

Let p_{ij} denote the probability that the score of the ith subject on the jth part of the test will be a 1 (pass). Suppose we are interested in the null hypothesis

$$H_0: p_{i1} = p_{i2} = \cdots = p_{ik} \quad \text{for } i = 1, \ldots, r.$$

This says that the probability of a pass is the same on all parts of the test for each subject, or that the parts are of equal difficulty. The Cochran Q test statistic is defined as

$$Q = k(k-1)\sum_1^k (C_j - N/k)^2 \Big/ \sum_1^r R_i(k - R_i).$$

(a) Give a rationale for the use of the quantity Q to test the hypothesis of interest. What is the appropriate tail for rejection?

(b) While the exact null distribution of Q conditional on $R_1, \ldots, R_r$ can be generated by enumeration, it is time-consuming and difficult to tabulate. Hence a large sample approximation is usually used instead unless r is quite small, specifically the chi square distribution with $(k-1)$ degrees of freedom. Give a justification of this approximation along the following lines. Within each column, the observations are Bernoulli trials. Hence for each j, C_j follows the binomial distribution with mean $\sum_{i=1}^r p_{ij}$ and variance $\sum_{i=1}^r p_{ij}(1 - p_{ij})$. Under the null hypothesis, this mean and variance can be estimated from the sample data as N/k and $\sum_{i=1}^r (R_i/k)[1 - (R_i/k)]$, but this latter estimate is improved by multiplying by the correction factor $k/(k-1)$. Standardized variables are then squared and summed to obtain Q. While the C_j are not independent, as r increases the C_j approach independence. One degree of freedom is lost for the estimation procedure.

(c) Show that the test statistic Q above can be written in the equivalent form

$$Q = (k-1)\left(k \sum_1^k C_j^2 - N^2\right)\Big/\left(kN - \sum_1^r R_i^2\right)$$

which is easier for calculation.

(d) Show that when $k = 2$, the test statistic Q reduces to

$$Q = (C - B)^2/(C + B)$$

where C and B are defined as in Table 8.1, that is, C is the number of $(0, 1)$ pairs and B is the number of $(1, 0)$ pairs. Hence the test statistic Q when $k = 2$ is equivalent to the test statistic for equality of paired proportions presented in Sect. 8.1.

58. Forty-six subjects were each given drugs A, B and C and observed as having a favorable (1) or unfavorable (0) response to each. The results reported in Grizzle et al. [1969, p. 494] are shown in the table below.

Response to Drug			Number of
A	B	C	Subjects
1	1	1	6
1	1	0	16
1	0	1	2
0	1	1	2
1	0	0	4
0	1	0	4
0	0	1	6
0	0	0	6

Test the null hypothesis that the drugs are equally effective.

59. Four different surgical procedures for a duodenal ulcer are A (drainage and vagotomy), B (25% resection and vagotomy), C (50% resection and vagotomy) and D (75% resection). Each procedure was used for a fixed period of time in each of 15 different hospitals, and an overall clinical evaluation made of the severity of the "dumping syndrome," an undesirable aftereffect of surgery for duodenal ulcer. The overall evaluation was made as simply "not severe" (0) or "present to at least some degree" (1). Analyze the results below for any significant difference between aftereffects of the four surgical procedures.

Hospitals	Surgical Procedure			
	A	B	C	D
1, 7, 8, 11	1	0	1	0
2, 3, 13	0	0	1	0
4, 10	1	1	0	1
5, 12	1	1	0	0
6	0	1	0	1
9, 14, 15	1	0	0	0

60. Suppose that for a group of n matched pairs of individuals, one member of each pair is selected to receive a treatment for a certain period while the other serves as a control (is untreated or given a placebo). Each individual is measured both before and after the treatment period so that there are a total of $4n$ observations. Indicate how this can be reduced to a matched pair situation while making use of both before-after and treatment-control information. If all $4n$ observations measure only the presence or absence of response, can the procedures of Sect. 8.1 be applied? (Hint: How many different response categories are there for the differences?)

*61. Suppose that half of a given group of patients is selected at random to receive Drug I at a certain time and Drug II at a later time, while the remaining patients receive Drug II first and Drug I second. On each occasion, the characteristic measured is dichtomous. Suppose further that the two drugs are known to have exactly the same effect on all patients but there is a time effect. Show that a test for equality of proportions using matched observations ignoring the time effect is conservative. (Of course one would expect to obtain better power from a suitable test which takes into account the time effect. See also the end of Sect. 3.1.)

62. Suppose that the randomization in Problem 61 is carried out within pairs of patients rather than over the whole group of patients. How does the situation then relate to that of Problem 60?

63. Prove the test properties stated in Sect. 8.5.

64. In the paired dichotomy situation and notation of Sect. 8.6, derive tests and confidence procedures for
   (a) the parameter $\lambda = p_{01}/p_{10}$,
   (b) the parameter $\gamma = p_{11}/p_1$.

65. In the paired dichotomy situation and notation of Sect. 8.6, show that independence of score on II and sameness is equivalent to $p_{11}p_{10} = p_{01}p_{00}$.

66. In the paired dichotomy situation and notation of Sect. 8.6, show that if sameness is independent of score on I and of score on II, then either $P(\text{same}) = 0$ or $P(\text{different}) = 0$ or $p_{11} = p_{00}$ and $p_{10} = p_{01}$.

67. Show that, under the model of Sect. 8.7 for paired dichotomous observations, a one-sided test of $p = 0.5$ as described in Sect. 8.1 (or $\theta = \theta_0$ as described in Sect. 8.6) is uniformly most powerful against a one-sided alternative. Show also that the related unbiased tests are uniformly most powerful unbiased against the alternative $p \neq p_0$.

68. (a) Let c be the coverage of a specified (nonrandom) region R. Suggest nonparametric tests and confidence procedures for c.
    (b) What optimum properties would you expect your tests in (a) to have?
    (c) Show that the optimum properties you stated in (b) hold.

69. Suppose a tolerance region with tolerance proportion 0.90 and confidence level 0.95 is set up for a production process. Thereafter, pairs of items are observed, and trouble-shooting is undertaken whenever both fall outside the tolerance region. What can you say about the amount of trouble-shooting required when the process is in control?

70. Let $C_1, \ldots, C_{n+1}$ be the coverages of a sample of n from a continuous c.d.f. F. Let $U_s = \sum_1^s C_i, s = 1, \ldots, n$.
    (a) What is the joint density of $U_1, \ldots, U_n$? Of $C_1, \ldots, C_n$?
    (b) What is the joint density of U_i, U_j for $i \neq j$? Of C_i, C_j for $i \neq j$?

71. Let $C_1, \ldots, C_{n+1}$ be the coverages of a sample of n from a continuous c.d.f. F, and $U_s = \sum_1^s C_i$ for $s = 1, \ldots, n$. Show that the following properties hold.
    (a) The joint distribution of $C_1, \ldots, C_{n+1}$ does not depend on F.
    (b) The random variables U_s are the order statistics of a sample of n from the uniform distribution on $(0, 1)$.
    (c) Given $U_1, \ldots, U_{s-1}, U_{s+1}, \ldots, U_n$, the conditional distribution of U_s is uniform on the interval (U_{s-1}, U_{s+1}).
    (d) Any permutation of $C_1, \ldots, C_{n+1}$ has the same joint distribution as $C_1, \ldots, C_{n+1}$.
    (e) $E(C_i) = 1/(n + 1)$ for all i.
    (f) The correlation of C_i and C_j is $-1/n$ for all i, j with $i \neq j$. (Hint: This can be shown without integration.)
    (g) The sum of any s coverages is distributed like U_s (which has the beta distribution (9.3) by Problem 30).
    (h) If m intervals $(X_{(i-1)}, X_{(i)})$ are excluded, the probability that the coverage of the remaining region is at least c is the probability of m or more successes under the binomial distribution with parameters n and $p = 1 - c$.
    (i) The conditional distribution of $U_1, \ldots, U_s$ given U_{s+1} is that of the order statistics of a sample of s from the uniform distribution on $(0, U_{s+1})$.
    (j) The random variables $C_i' = C_i/U_{s+1} = C_i/(C_1 + \cdots + C_{s+1}), i = 1, \ldots, s + 1$ are jointly distributed like the coverages of a sample of s from a continuous distribution.
    (k) Given U_3, U_7, the conditional distribution of $U_1, U_2, U_4, U_5, U_6, U_8, \ldots, U_n$ $(n > 7)$ is that of the order statistics of three independent samples, one of two observations from the uniform distribution on $(0, U_3)$, one of three observations from the uniform distribution on (U_3, U_7), and one of $n - 7$ observations

from the uniform distribution on $(U_7, 1)$. (Note that this generalizes to any set of given U's.)

(l) The coverages C_i have the same joint distribution as the random variables $Z_i/(Z_1 + \cdots + Z_{n+1})$ where the Z_i are independently distributed with density e^{-z}, $z \geq 0$.

72. In a sample of size n from a continuous distribution, let I_1 be the interval between the smallest and largest observation and I_2 be the interval between $-\infty$ and the next-to-largest observation. Show that

(a) I_1 and I_2 are both tolerance regions with tolerance proportion 0.5 and confidence level $1 - (n + 1)/2^n$.

(b) I_1 is a confidence interval for the median with confidence level $1 - 1/2^{n-1}$.

(c) I_2 is a confidence interval for the median with confidence level $1 - (n + 1)/2^n$.

(d) I_1 and I_2 are tolerance regions with tolerance proportion 0.75 and confidence level $1 - (n + 3)3^{n-1}/4^n$.

(e) I_1 is a confidence interval for the upper quartile with confidence level $1 - (1 + 3^n)/4^n$.

(f) I_2 is a confidence interval for the upper quartile with confidence level $1 - (n + 3)3^{n-1}/4^n$.

73. (a) Show that Murphy's graphs (Fig. 9.1) give the upper confidence limit for a binomial parameter p, as a function of n, for a fixed number of successes m and confidence level $1 - \alpha$.

(b) To what extent do you agree with the accompanying estimates of ease and accuracy of using various types of tables and graphs for various purposes? Make reasonable assumptions about the grids of values employed, etc. The notation is that $1 - \alpha$ is the probability of r or more successes in n binomial trials with parameter p.

	Tables		Graphs	
	Direct binomial	Fisher–Yates	Clopper–Pearson	Murphy
Size of table or graph	large	small	medium	medium
Given r, n, p, find α (binomial tail probability)	low accuracy usually easy, high accuracy often hard		essentially impossible	
Given r, n, α, find p (binomial confidence limit)	low accuracy usually easy, high accuracy usually hard	easy and accurate	low accuracy usually easy, high accuracy impossible	
Given p, n, α, find r (binomial critical value)	usually easy, accurate	hard, accurate	easy usually accurate	
Given p, r, α, find n	usually easy, accurate	hard, accurate	medium difficulty usually accurate	easy

74. (a) Verify that the coverage of a 0.65 tolerance region with confidence 0.95 based on 20 observations has the distribution graphed in Fig. 9.2.
    (b) Show that the same region has coverage c with confidence $1 - \alpha$ for any values c and $1 - \alpha$ related according to the graph.

75. How could a tolerance region be defined for all sample sizes $n \geq 10$ so that the relationship between the tolerance proportion and the confidence level is the same for all n?

76. (a) Show that the region obtained from the cutting functions $\phi_1(x) = x, \phi_2(x) = x$, $\phi_3(x) = |x - \frac{1}{2}(X_{(1)} + X_{(n)})|$ is a valid univariate tolerance region.
    (b) Which of the generalizations discussed in Sect. 9.6 are adequate to cover this case, and how do they do so?

CHAPTER 3

One-Sample and Paired-Sample Inferences Based on Signed Ranks

1 Introduction

In Chap. 2 we saw that the sign test is the best possible test (in strong senses of "best") at level α for a null hypothesis which is as inclusive as the statement "the observations are a random sample from a population with median 0 (or ξ_0)." It certainly seems as though better use could be made of the observations by taking their magnitudes into account. However, since the sign test is optimum at level α for this inclusive set of null distributions, any procedure which considers magnitudes would be a better test at level α only for some smaller and more restrictive set of null distributions. Such procedures are of special relevance if the restricted set is the one of interest anyway, or if the "restriction" of the null hypothesis can reasonably be assumed as a part of the model, so that the restricted hypothesis essentially amounts to the unrestricted one above. Furthermore, their exact levels may vary only slightly under the kinds of departure from assumptions which are likely, and their increased power may well be worth the small price. Recall that, in principle, level and power considerations should always be balanced off.

This chapter will present and discuss tests and confidence procedures suggested by the more restrictive hypothesis that the observations are a random sample from a population which is symmetric with median 0 (or any other value). (Symmetry will be defined precisely below.) Other related hypotheses will also be considered. All of the inference procedures presented are based on what are called the "signed ranks" of the observations, and the primary emphasis is on the well-known Wilcoxon signed-rank test. All of the discussion here will be in the context of observations in a single random sample. However, as in the last chapter, these procedures may be performed on the differences of paired observations, like treatment-control differences

145

in matched pairs, without any change in techniques or properties as long as it is remembered that the relevant distribution is the distribution of differences of pairs.

2 The Symmetry Assumption or Hypothesis

We noted in Chap. 1 that the normal distribution is symmetric about its mean, and that the binomial distribution with $p = 0.5$ is symmetric about $n/2$, but an explicit definition of symmetry was not given. A random variable X is said to have *a symmetric distribution* or to be *symmetrically distributed* about the *center of symmetry* μ if

$$P(X \le \mu - x) = P(X \ge \mu + x) \quad \text{for all } x. \tag{2.1}$$

This can be written in terms of the c.d.f. F of X as

$$F(\mu - x) = 1 - F(\mu + x) + P(X = \mu + x) \quad \text{for all } x. \tag{2.2}$$

Equivalently, a distribution with discrete frequency function or density f is symmetric about μ if and only if (Problem 1)

$$f(\mu - x) = f(\mu + x) \quad \text{for all } x. \tag{2.3}$$

If there is symmetry, there must be a center of symmetry, μ, and this point is both the mean (if it exists) and a median of the distribution (Problem 3). It may be noted that (2.1) holds for all x if and only if it holds for all $x > 0$, and that the two (nonstrict) inequalities therein may be replaced by two strict inequalities (Problem 2).

Symmetry can be viewed in several alternative ways. Some common conditions, each of which is equivalent to the statement that X is symmetrically distributed about μ (Problem 4), are listed below.

(a) $X - \mu$ is symmetrically distributed about 0.
(b) $X - \mu$ and $\mu - X$ are identically distributed.
(c) Given that $|X - \mu| = x$, for any $x \ne 0$ the sign of $X - \mu$ is equally likely to be positive or negative.

The null hypothesis of interest in this chapter is that the observations are a random sample from a population symmetric about a given value μ, that is, the observations are independently, identically and symmetrically distributed about μ, where μ is specified. Since μ is then the median, the sign test of Chap. 2 also applies here, but the additional assumption of symmetry permits "better" tests. These tests have corresponding confidence procedures for the center of symmetry.

The procedures developed in this chapter are tests and confidence intervals for location, where the location parameter is the center of symmetry. In this sense, they are nonparametric analogs of the classical (normal-theory) tests and confidence intervals for the mean. Even when the symmetry assumption is relaxed in certain ways, the procedures of this chapter remain valid, that is, they retain their level of significance or confidence.

In paired-sample applications with observation pairs (V, W), the symmetry property must hold for the distribution of differences $W - V$. The difference between the individual population medians of V and W is not necessarily a center of symmetry, or even a median, of the difference $W - V$ (Problem 6). It is a center of symmetry when V and W are independent and related by translation, or when their joint distribution is symmetric (permutable, exchangeable) in the sense that (V, W) has the same distribution as (W, V) (Problems 7 and 8). This means that in a treatment-control experiment, for example, randomization of treatment and control within pairs makes a test of the null hypothesis of 0 center of symmetry valid as a test of the null hypothesis that the treatment has no effect on any unit. (See Sect. 7, Chap. 2 and Sect. 9, Chap. 8 for more complete discussion.)

3 The Wilcoxon Signed-Rank Test

In this section we present the Wilcoxon signed-rank test for an hypothesized center of symmetry. The test and its properties will be developed first under the assumption that the observations $X_1, \ldots, X_n$ are an independent, random sample from a population (distribution) which is continuous and symmetric about μ. In Sect. 3.5, we will see that both the assumptions of independence and of identical distributions can be relaxed in certain ways without affecting the level, and hence the validity, of the test. The confidence procedures related to this test will then be discussed in Sect. 4. The continuity assumption ensures that $P(X_i = \mu) = 0$ for all i and that $P(X_i = X_j) = 0$ for all $i \neq j$ so that the effects of "zeros" and "ties" need not be considered until later, in Sect. 6.

3.1 Test Procedure and Exact Null Distribution Theory

Suppose the observations X_j are ranked in order of absolute value, using positive integer ranks with 1 for the smallest $|X_j|$ and n for the largest. (While it is not necessary, it may be more convenient to rearrange the X_j in order of absolute value to determine these ranks.) The *signed rank* of an observation is the rank of its absolute value with a plus or minus sign attached; this sign is $+$ if the observation is positive, and $-$ if the observation is negative. For example, consider the following data.[1]

X_j	49	−67	8	16	6	23	28	41	14	29	56	24	75	60	−48
rank of $|X_j|$	11	14	2	4	1	5	7	9	3	8	12	6	15	13	10
signed rank	11	−14	2	4	1	5	7	9	3	8	12	6	15	13	−10

The *signed-rank sum* T is defined as the sum of the signed ranks. It may be expressed as the *positive-rank sum* T^+ minus the *negative-rank sum* T^-, where T^+ and T^- are the sums of the ranks of the positive and negative observations respectively. (Note that, as defined, T^- is positive.) Thus, in this example, we have

$$T^+ = 11 + 2 + \cdots + 13 = 96, \qquad T^- = 14 + 10 = 24,$$

$$T = T^+ - T^- = 72.$$

Each of these three statistics determines the other two by a linear relation, so that only the most convenient one need be computed. Specifically, because $T^+ + T^-$ equals the sum of all the ranks, or $1 + 2 + \cdots + n = n(n + 1)/2$, we have

$$T^+ = n(n + 1)/2 - T^- = n(n + 1)/4 + T/2 \tag{3.1}$$

$$T^- = n(n + 1)/2 - T^+ = n(n + 1)/4 - T/2 \tag{3.2}$$

$$T = T^+ - T^- = n(n + 1)/2 - 2T^- = 2T^+ - n(n + 1)/2. \tag{3.3}$$

Any of these three statistics may be called the *Wilcoxon signed-rank statistic*, as the test may be based on any one of them.

Under the null hypothesis that the population center of symmetry μ is zero, the signs of the signed ranks are independent and equally likely to be positive or negative (Problem 9). This fact determines the null distribution of each of the statistics. The null distribution of T^+ is the same as that of T^- (Problem 10). The probability of a particular value of T^+, say, is

$$P(T^+ = t) = u_n(t)/2^n \tag{3.4}$$

where $u_n(t)$ is the number of ways that plus and minus signs can be attached to the first n integers $1, 2, \ldots, n$ such that the sum of the integers with positive signs equals t. Equivalently, $u_n(t)$ is the number of subsets of the first n

---

[1] The X_j are differences, in eighths of an inch, between heights of cross- and self-fertilized plants of the same pair, as given by Fisher [1966 and earlier editions]. The original experiment was done by Darwin. Fisher's discussion is extensive and interesting, including pertinent quotations from Darwin and Galton. After applying the ordinary t test to these data, obtaining $t = 2.148$ and a one-tailed probability of 0.02485, Fisher introduces the method of Sect. 2.1, Chap. 4 and obtains a one-tailed probability of 0.02634.

integers whose sum is t. The values of $u_n(t)$ and the probabilities in (3.4) may be easily generated using recursive techniques (Problem 11).

The possible values of T^+ (and of T^-) range from 0 to $n(n + 1)/2$. The mean and variance (Problem 14) under the null hypothesis are

$$E(T^+) = E(T^-) = n(n + 1)/4 \qquad (3.5)$$

$$\text{var}(T^+) = \text{var}(T^-) = n(n + 1)(2n + 1)/24 = (2n + 1)E(T^+)/6. \qquad (3.6)$$

From these results and the relation (3.3), we obtain

$$E(T) = 0 \qquad (3.7)$$

$$\text{var}(T) = 4\,\text{var}(T^+) = n(n + 1)(2n + 1)/6. \qquad (3.8)$$

The null distributions of T^+, T^- and T are all symmetric about their respective means (Problem 15).

The left-tail cumulative probabilities from (3.4), that is $P(T^+ \le t) = P(T^- \le t)$, are given in Table D for all different integer values of t not exceeding $n(n + 1)/4$ (probabilities not exceeding 0.5), for all sample sizes $n \le 20$. The tables in Harter and Owen [1970] give these probabilities for all $n \le 50$. We now describe how these tables are used to perform the test.

If the population center of symmetry μ is positive, one would anticipate more positive signs than negative, and hence more positive signed ranks among the observations, at the higher ranks as well as the lower. Such an outcome would give larger values of T, and hence of T^+. This suggests a test rejecting the null hypothesis if T is too large, that is, when T falls at or above its critical value, which is the upper α-point of its null distribution for a test at level α. By (3.1)–(3.3), this is equivalent to rejecting if T^+ is too large, and also to rejecting if T^- is too small. Since Table D gives only the lower-tailed cumulative null distribution of T^+ or T^-, the convenient rule for rejection based on this table, for the alternative $\mu > 0$, is T^- less than or equal to its critical value. The table entry for an observed value of T^- is the one-tailed (left-tailed) P-value according to the Wilcoxon signed-rank test.

The corresponding test against the alternative $\mu < 0$ rejects if T is too small (i.e., highly negative), which is equivalent to T^+ too small (or T^- too large). Hence the appropriate P-value here, the probability that T^+ is less than or equal to an observed value, is again found from Table D as a left-tailed probability.

An equal-tailed test against the two-sided alternative $\mu \ne 0$ rejects at level 2α if either of the foregoing tests rejects at level α. Since the table gives left-tailed probabilities only, it should be entered with the smaller of T^+ and T^- for an equal-tailed test, and the P-value is twice the tabulated value.

In the example above we found $T^- = 24$. According to Table D, under the null hypothesis the probability that $T^- \le 24$ is 0.0206. A one-tailed Wilcoxon test in this direction would then reject at level 0.025 but not at 0.020; an equal-tailed test would have P-value 0.0412 and would reject at level 0.05 but not at 0.04.

In order to test the null hypothesis that the center of symmetry is $\mu = \mu_0$ for any value of μ_0 other than 0, the analogous procedure is to subtract μ_0 from every X_j and then proceed to find the signed ranks as before; that is, the Wilcoxon test is applied to $X_1 - \mu_0, \ldots, X_n - \mu_0$ instead of $X_1, \ldots, X_n$. The corresponding confidence bounds on μ will be discussed in Sect. 4.

A different representation of the Wilcoxon signed-rank statistics will be convenient later, although it is not convenient for hypothesis testing. We define a *Walsh average* as the average of any two observations X_i, X_j, that is

$$(X_i + X_j)/2 \quad \text{for } 1 \le i \le j \le n. \tag{3.9}$$

Note that each observation is itself a Walsh average where $i = j$. From a sample of n then, we obtain $n(n + 1)/2$ Walsh averages, $n(n - 1)/2$ where $i < j$ and n where $i = j$. The sign of $(X_i + X_j)/2$ is equal to the sign of either X_i or X_j, whichever is larger in absolute value. The theorem below follows easily from this observation (Problem 18).

Theorem 3.1. *The positive-rank sum T^+ and negative-rank sum T^- are respectively the number of positive and negative Walsh averages $(X_i + X_j)/2$, $1 \le i \le j \le n$.*

Equivalent to this result is a method of expressing the Wilcoxon signed-rank statistics as functions of the indicator variables defined by

$$T_{ij} = \begin{cases} 1 & \text{if } X_i + X_j > 0 \\ 0 & \text{otherwise.} \end{cases} \tag{3.10}$$

Specifically, we can write (3.1) as

$$T^+ = \sum_{i \le j} \sum T_{ij}, \tag{3.11}$$

and T^- and T can be similarly represented.

3.2 Asymptotic Null Distribution Theory

For n large, T, T^+, and T^- are approximately normal under the null hypothesis, with the means and variances given in (3.5)–(3.8). (Asymptotic normality was discussed in Sect. 9, Chap. 1.) The normal approximation may be used for sample sizes outside the range of Table D. The procedure for finding approximate normal tail probabilities based on T^+ or T^- is shown at the bottom of this table. When using T, the mean and variance in (3.7) and (3.8) are appropriate. A continuity correction may be incorporated, in the amount of $\frac{1}{2}$ for T^+ or T^- and 1 for T, since T takes on only alternate integer values (Problem 19). However, the amount of the continuity correction is small (Problem 20) and in fact it reduces the accuracy of the approximation for small tail probabilities. The approximation without continuity correction is very good for $P = 0.025$, and the approximation with continuity correction is very good for $P = 0.05$. In general, for very small P, the correction is in the opposite direction from the error in the approximation, and of much smaller magnitude. For details and an extensive investigation of the accuracy of critical values based on the normal approximation, see McCornack [1965] and Claypool and Holbert [1975]. Some simple approximations involving the t distribution are investigated in Iman [1974].

In the example of Sect. 3.1, with $n = 15$, the mean and variance of T under the null hypothesis are 0 and 1240 respectively. The approximate normal deviate corresponding to $P(T \geq 72)$ is thus

$$\frac{72}{\sqrt{1240}} = 2.045 \quad \text{without correction for continuity}$$

$$\frac{71}{\sqrt{1240}} = 2.016 \quad \text{with correction for continuity.}$$

From Table A we find the corresponding tail probabilities are 0.0204 and 0.0219 respectively. For comparison, the exact tail probability is 0.0206.

The accuracy of the normal approximation can be improved by using an Edgeworth expansion. Taking terms to order $1/n$, the Edgeworth approximation gives the same degree of accuracy with $n = 15$ as the normal approximation with $n = 100$ [Fellingham and Stoker, 1964].

Formally, we may say that under the null hypothesis, T, T^+, and T^- are asymptotically normal (see Sect. 9, Chap. 1) with the mean and variance given by (3.5)–(3.8). This follows immediately for T^+ and T^- once it is proved for T (Problem 21). To prove it for T, note that T has the same distribution as $\sum_1^n R_j$ where $R_1, R_2, \ldots$ are independent with $P(R_j = j) = P(R_j = -j) = 1/2$ under the null hypothesis. It is easily verified (Problem 22) that the Liapounov criterion for the Central Limit Theorem (Theorem 9.3 of Chap. 1) is satisfied.

3.3 Large Sample Power

For n large, the distributions of T^+, T^- and T are approximately normal under alternative hypotheses as well as in the null case. Hence, once the appropriate means and variances under alternative distributions are calculated, a large-sample approximation to the power of the Wilcoxon signed-rank test can be computed. Asymptotic normality will not be proved here, but the mean and variance of T^+ are found in the theorem below.

Theorem 3.2. Let $X_1, \ldots, X_n$ be a random sample from some continuous population. Then

$$E(T^+) = np_1 + n(n-1)p_2/2, \tag{3.12}$$

where p_1 is the probability that a single X is positive and p_2 is the probability that the sum of two independent X's is positive, that is,

$$p_1 = P(X_j > 0) \quad \text{for all } j, \tag{3.13}$$

$$p_2 = P(X_i + X_j > 0) \quad \text{for all } i \neq j. \tag{3.14}$$

The variance is

$$\mathrm{var}(T^+) = np_1(1 - p_1) + n(n - 1)[p_2(1 - p_2)/2 + 2(p_3 - p_1 p_2)]$$
$$+ n(n - 1)(n - 2)(p_4 - p_2^2), \tag{3.15}$$

where

$$p_3 = P(X_i > 0 \text{ and } X_i + X_j > 0) \quad \text{for all } i \neq j, \tag{3.16}$$

$$p_4 = P(X_i + X_j > 0 \text{ and } X_i + X_k > 0) \quad \text{for all distinct } i, j, k. \tag{3.17}$$

PROOF. By using the representation of T^+ in (3.11), its mean and variance are easily expressed in terms of moments of the T_{ij} defined in (3.10) as

$$E(T^+) = nE(T_{ii}) + n(n - 1)E(T_{ij})/2, \tag{3.18}$$

$$\mathrm{var}(T^+) = \sum_{i \leq j} \sum \sum_{h \leq k} \sum \mathrm{cov}(T_{ij}, T_{hk})$$

$$= n\,\mathrm{var}(T_{ii}) + \binom{n}{2}\mathrm{var}(T_{ij}) + 2n(n - 1)\mathrm{cov}(T_{ii}, T_{ik})$$

$$+ 2n\binom{n-1}{2}\mathrm{cov}(T_{ij}, T_{ik}) + \binom{n}{4}\mathrm{cov}(T_{ij}, T_{hk}), \tag{3.19}$$

where h, i, j and k are all different.

Using the probabilities defined in (3.13), (3.14), (3.16) and (3.17), and the fact that X_i, X_j are independent for all $i \neq j$, gives the moments of the T_{ij} (Problem 24) as

$$E(T_{ii}) = p_1 \quad \text{for all } i$$

$$E(T_{ij}) = p_2 \quad \text{for all } i \neq j$$

$$\mathrm{var}(T_{ii}) = p_1 - p_1^2 = p_1(1 - p_1) \quad \text{for all } i$$

$$\mathrm{var}(T_{ij}) = p_2 - p_2^2 = p_2(1 - p_2) \quad \text{for all } i \neq j$$

$$\mathrm{cov}(T_{ii}, T_{ik}) = p_3 - p_1 p_2 \quad \text{for all } i \neq k$$

$$\mathrm{cov}(T_{ij}, T_{ik}) = p_4 - p_2^2 \quad \text{for distinct } i, j, k$$

$$\mathrm{cov}(T_{ij}, T_{hk}) = 0 \quad \text{for distinct } i, j, h, k.$$

The results in (3.12) and (3.15) follow once these moments are substituted in (3.18) and (3.19). $\square$

The moments in Theorem 3.2 hold for a random sample from any continuous population. If the population is symmetric about zero, as under the null hypothesis that the population center of symmetry is zero, the probabilities defined in Theorem 3.2 have the values (Problem 25)

$$p_1 = \tfrac{1}{2}, \qquad p_2 = \tfrac{1}{2}, \qquad p_3 = \tfrac{3}{8}, \qquad p_4 = \tfrac{1}{3}.$$

3.4 Consistency

A test is called *consistent* against a particular alternative if the power of the test against the alternative approaches 1 as the sample size approaches infinity. Of course, the test must be defined for each sample size. Thus consistency is, strictly speaking, a property of a sequence of tests, one test for each sample size. (In some contexts it might be desirable to let the alternative also depend on the sample size.)

The Wilcoxon signed-rank test is consistent against any alternative distribution, even nonsymmetric, for which the probability is *not* $\tfrac{1}{2}$ that the sum of two independent observations is positive, that is, for which

$$p_2 = P(X_1 + X_j > 0) \neq \tfrac{1}{2} \quad \text{for } i \neq j.$$

(Note that the value of the parameter p_1 does not determine consistency here, while that parameter alone determines consistency and even power for the sign test (Sect. 3.2, Chap. 2).) More precisely, consider a two-tailed Wilcoxon signed-rank test with level bounded away from 0 as $n \to \infty$, and a fixed continuous alternative distribution F for which $p_2 \neq \tfrac{1}{2}$. We show below that if $X_1, \ldots, X_n$ are independently distributed according to F for each n, then the probability of rejection approaches 1 as $n \to \infty$. Similarly, the one-tailed test is consistent against alternatives with p_2 on the appropriate side of $\tfrac{1}{2}$.

* In order to prove this, we first show a fact of some interest in itself, namely that if $X_1, \ldots, X_n$ are independently, identically distributed (according to any distribution whatever), then $2T^+/n^2$ is a *consistent estimator* of p_2, or equivalently, *converges in probability* to p_2. By the definition given in Eq. (3.5) of Chap. 1, this means that, for every $\varepsilon > 0$,

$$P\left(\left| \frac{2T^+}{n^2} - p_2 \right| > \varepsilon \right) \to 0 \quad \text{as } n \to \infty. \tag{3.20}$$

In this specified sense, the distribution of $2T^+/n^2$ becomes concentrated near p_2 as $n \to \infty$. Note that the distribution of $2T^+/n^2$ depends on n through the numerator as well as the denominator, since the distribution of T^+ itself depends on n (although the notation does not explicitly express this fact). Note also that a more natural estimator of p_2 is $T_0^+ = \sum\sum_{i<j} T_{ij}/\binom{n}{2}$, which is unbiased for p_2 (Problem 39). (See also Sect. 5.)

PROOF. Letting $Z_n = 2T^+/n^2$, by (3.12) and (3.15) we have

$$E(Z_n) \to p_2 \quad \text{and} \quad \text{var}(Z_n) \to 0 \quad \text{as } n \to \infty.$$

A simple application of Chebyshev's inequality then shows (Problem 26) that (3.20) holds, that is, Z_n converges in probability to p_2, so that Z_n is a consistent estimator of p_2.

It remains to show that the equal-tailed Wilcoxon signed-rank test at level α is consistent against alternatives for which $p_2 \neq \frac{1}{2}$. Under the null hypothesis, we have $p_2 = \frac{1}{2}$; therefore, by (3.20), for any $\varepsilon > 0$, we have for sufficiently large n

$$P_0(|Z_n - \tfrac{1}{2}| > \varepsilon) < \alpha/2.$$

This implies that the upper and lower $(\alpha/2)$-points of the null distribution of Z_n lie between $\frac{1}{2} - \varepsilon$ and $\frac{1}{2} + \varepsilon$. Since Z_n is just a multiple of T^+, it follows that the test rejects the null hypothesis at least for $|Z_n - \frac{1}{2}| > \varepsilon$, provided n is sufficiently large.

Consider a particular alternative with $p_2 \neq \frac{1}{2}$, and let $\varepsilon = |p_2 - \frac{1}{2}|/2$. Then $|Z_n - \frac{1}{2}| > \varepsilon$ whenever $|Z_n - p_2| \leq \varepsilon$, so that for sufficiently large n the test rejects at least for $|Z_n - p_2| \leq \varepsilon$. But the probability of this event approaches 1 under the alternative, by (3.20), which proves that the test is consistent.   □

The type of proof used here works quite generally to show that consistent estimators lead to consistent tests. Consider a parameter θ, an estimator which is consistent for θ, and a test based on that estimator as a test statistic. Suppose that (a) θ has one value θ_0 under the null hypothesis and a different value under the alternative, (b) the test rejects in the appropriate tail of the distribution of the estimator, (c) the probability of rejection under the null hypothesis is bounded away from 0, and (d) either the distribution of the estimator under the null hypothesis is completely determined, or the estimator is uniformly consistent under the null hypothesis. ($\hat{\theta}_n$ is a *uniformly consistent estimator* of θ_0 under H_0 if $\hat{\theta}_n$ converges uniformly in probability to θ_0 under H_0, that is for every $\varepsilon > 0$, $P_0(|\hat{\theta}_n - \theta_0| > \varepsilon) \to 0$ as $n \to \infty$, uniformly in the distributions of H_0.)

The first step in the proof is to observe that the test must, for sufficiently large n, reject whenever the estimator falls more than ε away from the value to which it converges in probability under the null hypothesis (more than ε away in one direction if the test is one-tailed). The second step is to consider an alternative under which the estimator converges in probability to a different value, and to observe that, under such an alternative, the probability approaches 1 that the estimator will lie within ε of this different value, and hence, if ε is small enough, more than ε away from the null hypothesis value, whence rejection occurs.

Note that it may be necessary to redefine the test statistic so that it becomes a consistent estimator of something. For instance, we considered above $Z_n = 2T^+/n^2$ instead of T^+. Tails of the distribution of T^+ correspond to tails of the distribution of Z_n, but Z_n converges in probability to p_2, while the distribution of T^+ moves further and further to the right with variance increasing as $n \to \infty$.

A similar approach to a general proof could be based on the test statistic standardized under the null hypothesis (Problem 29). Under the alternative, the variance of this statistic is bounded but the mean approaches ∞ as $n \to \infty$.*

3.5 Weakening the Assumptions

We have been assuming that $X_1, \ldots, X_n$ are independent and identically distributed, with a continuous and symmetric common distribution. The Wilcoxon test procedures were developed using only the fact that the signs of the signed ranks are independent and are equally likely to be positive or negative. The level of the test will therefore be unaffected if, in particular, the X_j are independent with continuous distributions symmetric about μ_0, even if their distributions are not the same (Problem 30). (The level of the corresponding confidence procedure to be discussed in Sect. 4 is preserved if the X_j are independent with continuous distributions possibly different, but all symmetric about the same μ.)

Even independence is not required for the validity of the Wilcoxon test, provided that the conditional distribution of each X_j given the others is symmetric about μ (Problem 31). It is valid, for example, under the null hypothesis that the treatment actually has no effect when the X_j are the treatment-control differences in a matched-pairs experiment, as long as the controls are chosen independently and at random, one from each pair (Problem 32).

If the continuity assumption is relaxed, then there is positive probability of a zero or tie, and the test has not yet been defined. This situation is discussed in some detail in Sect. 6.

For the one-tailed test, even the assumption of symmetry can be relaxed. Specifically, suppose the X_j are independent and continuously distributed, not necessarily identically. Consider the one-tailed test at level α for the null hypothesis of symmetry about the origin with rejection region in the lower tail of the negative-rank sum. It can be shown (Problem 35) that this test rejects with probability at most α if

$$P(X_j < -x) \geq P(X_j > x) \quad \text{for all } x \geq 0 \text{ and all } j \qquad (3.21)$$

and with probability at least α if

$$P(X_j < -x) \leq P(X_j > x) \quad \text{for all } x \geq 0 \text{ and all } j. \qquad (3.22)$$

Under (3.21), the probability in any left tail of the distribution of X_j exceeds or equals the probability in the corresponding right tail, so that the distribution of X_j is "to the left of or equal to" ("stochastically smaller" than) a distribution symmetric about 0 (Problem 33). Equation (3.22) means the opposite—the distribution of X_j is "stochastically larger" than a distribution symmetric about zero. Since the probability of rejection is at most α when (3.21) holds, the null hypothesis could be broadened to include (3.21) without affecting the significance level. Similarly, it is natural to broaden the alternative hypothesis to include (3.22); since the probability of rejection is at least α when (3.22) holds, the test is by definition "unbiased" against (3.22). This is the least we would hope for in broadening the alternative. (Of course, any particular distribution whatever could be included in the alternative hypothesis without affecting the validity of the test in the sense that the significance level, which depends only on the null hypothesis, would be preserved. However, we would not want to add an "alternative" distribution under which the probability of rejection is less than α.)

The statement above, that the probability of rejection is at most α under (3.21) and at least α under (3.22), follows (Problem 35) from a fact which is of interest in itself. Suppose a (nonrandomized) test ϕ is "increasing" in the sense that, if ϕ rejects at $X_1, \ldots, X_n$ and any X_j is increased, ϕ still rejects. (The upper-tailed Wilcoxon test obviously has this property.) Then the probability that ϕ will reject increases (not necessarily strictly) when the distribution of X_j is moved to the right ("stochastically increased"), that is, when the c.d.f. F_j of X_j is replaced by G_j where $G_j(x) \leq F_j(x)$ for all x. (In such a case, we say that G_j is *stochastically larger* than F_j.) Formally, for randomized tests ϕ, we have the following theorem.

Theorem 3.3. *If $X_1, \ldots, X_n$ are independent, X_j has c.d.f. F_j, and $\phi(X_1, \ldots, X_n)$ is a randomized test function which is increasing in each X_j, then the probability of rejection*

$$\alpha(F_1, \ldots, F_n) = E_{F_1, \ldots, F_n}[\phi(X_1, \ldots, X_n)] \qquad (3.23)$$

is a decreasing function of the F_j, that is,

$$\alpha(F_1, \ldots, F_n) \leq \alpha(G_1, \ldots, G_n) \text{ if } F_j(x) \geq G_j(x) \text{ for all } x \text{ and } j. \quad (3.24)$$

The same holds if the words increasing and decreasing are interchanged and the first inequality of (3.24) is reversed. (The theorem does not depend on the fact that ϕ is a test, but it will be applied only to tests.)

PROOF. Suppose that $X_1, \ldots, X_n$ are independent and X_j has c.d.f. F_j. Then there exist independent random variables $Y_1, \ldots, Y_n$, such that Y_j has c.d.f. G_j and $P(Y_j \geq X_j) = 1$ for each j (Problem 34). It follows that, for ϕ increasing in each argument,

$$\alpha(G_1, \ldots, G_n) = E[\phi(Y_1, \ldots, Y_n)]$$
$$\geq E[\phi(X_1, \ldots, X_n)] = \alpha(F_1, \ldots, F_n).$$

For ϕ decreasing, the inequality is reversed. $\qquad \square$

4 Confidence Procedures Based on the Wilcoxon Signed-Rank Test

Suppose that we have a random sample $X_1, \ldots, X_n$ from a continuous, symmetric distribution and we want to find the confidence bounds for the center of symmetry μ which correspond to the Wilcoxon signed-rank test. The confidence region consists of those values of μ which would be "accepted" if the usual Wilcoxon test for zero center of symmetry were applied to the set $X_1 - \mu, \ldots, X_n - \mu$. We could proceed by trial and error, testing various values of μ to see which ones lead to "acceptance." However, the interpretation of the Wilcoxon test statistic as a function of Walsh averages $(X_i + X_j)/2$ provides a more direct and convenient method of determining the corresponding confidence bounds.

In Theorem 3.1, we found that the test statistics T^+ and T^- are the number of positive and negative Walsh averages respectively. Suppose that μ is subtracted from every X_j. Since

$$(X_i - \mu + X_j - \mu)/2 < 0 \quad \text{if and only if} \quad (X_i + X_j)/2 < \mu,$$

the number of negative Walsh averages among the $(X_j - \mu)$ equals the number of Walsh averages less than μ for the original X_j. Therefore, the Wilcoxon test with rejection region $T^- \leq k$, when applied to the observations $(X_j - \mu)$, would reject or accept according as μ is smaller than or larger than the $(k + 1)$th smallest Walsh average $(X_i + X_j)/2$, counting from smallest to largest in order of algebraic (not absolute) value. That is, the $(k + 1)$th Walsh average from the smallest is a lower confidence bound for μ with a one-sided α equal to the null probability that $T^- \leq k$. Similarly, the $(k + 1)$th from the largest Walsh average is an upper confidence bound at the same level. Hence, the endpoints of any confidence region based on the Wilcoxon signed-rank procedure are order statistics of the Walsh averages.

All of the Walsh averages can be easily generated and ordered (sorted) by computer. However, there is also a convenient graphical procedure for finding the confidence bounds as Walsh averages. Plot each X_j value on a horizontal line, the "X axis," as in Fig. 4.1 for the data $-1, 2, 3, 4, 5, 6, 9, 13$ with $n = 8$. On one side of the X axis (either above or below), draw two rays emanating from each X_j, as in the diagram. All rays should make equal angles with the X axis. Then the Walsh averages are exactly the horizontal coordinates of the intersections of the rays. The points on the X axis must be included, as they correspond to the n Walsh averages where $i = j$, that is, the original observations. The $(k + 1)$th smallest Walsh average can be identified by counting from the left in the diagram. Its value may be read from the graph or, if greater accuracy is desired, calculated as the corresponding $(X_i + X_j)/2$. In the latter case, it may be necessary to calculate more than one Walsh average to determine exactly which is the $(k + 1)$th. Continuing the example, for $n = 8$ we find from Table D that $P(T^- \leq 3) = 0.0195$,

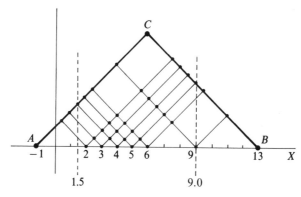

Figure 4.1   Adapted from Gibbons, Jean D. [1971], *Nonparametric Statistical Inference*, New York: McGraw-Hill, p. 118, Fig. 3.1. With Permission of the publisher and author.

so that $k = 3$ corresponds to a one-tailed test at exact level 0.0195. The 4th smallest Walsh average is therefore a lower confidence bound for μ with confidence coefficient 0.9805, and this number can be read off Fig. 4.1 as 1.5, or identified and calculated as $(-1 + 4)/2 = 1.5$. Similarly, an upper confidence bound with the same confidence coefficient is 9.0, and therefore the two-sided confidence interval, $1.5 < \mu < 9.0$, has confidence coefficient $1 - 2(0.0195) = 0.961$.

For $k \leq 2$, the situation becomes especially simple. The critical values $k = 0, 1, 2$ give one-sided levels $\alpha = 2^{-n}, 2(2^{-n}), 3(2^{-n})$ respectively for any $n > k$. The corresponding confidence bounds are (Problem 36) $X_{(1)}, (X_{(1)} + X_{(2)})/2$, and $\min\{X_{(2)}, (X_{(1)} + X_{(3)})/2\}$ respectively, where $X_{(i)}$ is the ith order statistic, that is the ith from the smallest of the X_j in order of algebraic (not absolute) value. For $k \geq 3$, this sort of description rapidly becomes more complicated (Problem 37).

5 A Modified Wilcoxon Procedure

In computing the rank sums, suppose that the ranks $1, 2, \ldots, n$ are replaced by the numbers $0, 1, \ldots, n - 1$, which we will call *modified ranks*. Denote the new ranks sums by T_0, T_0^+, T_0^-. As an example, for the data in Sect. 3.1, the following results are obtained.

X_j	49	−67	8	16	6	23	28	41	14	29	56	24	75	60	−48
modified rank of $|X_j|$	10	13	1	3	0	4	6	8	2	7	11	5	14	12	9
modified signed rank	10	−13	1	3	0	4	6	8	2	7	11	5	14	12	−9

$$T_0^+ = 83, \qquad T_0^- = 22, \qquad T_0 = 61, \qquad n = 15.$$

Under the null hypothesis of symmetry about 0, with either the original or weakened assumptions, T_0^- has the same distribution in a sample of size n as does T^- in a sample of size $n - 1$ (Problem 38). Accordingly, Table D applies to T_0^-, and similarly to T_0^+, provided that $(n - 1)$ is used in place of n. In the example above, then, the one-tailed P-value for the lower tail of T_0^- is 0.0290 for the modified test, since this is the probability from Table D that $T^- \leq 22$ for 14 observations.

With $n - 1$ in place of n, the formulas in (3.5) and (3.6) for the mean and variance under the null hypothesis apply to T_0^+, as does the asymptotic normality discussed in Section 3.2.

T_0^+ is the number of positive Walsh averages *excluding* those with $i = j$, that is, the number of positive $(X_i + X_j)/2$, for $i < j$ (Problem 39). Therefore, the graphical method of obtaining confidence bounds works for the modified procedure with slight changes. Specifically, the points on the X axis must now be excluded, and of course the critical value k is now that for T_0^+ rather than T^+.

Under alternative distributions, we have (Problem 40)

$$E(T_0^+) = n(n - 1)p_2/2 \tag{5.1}$$

$$\text{var}(T_0^+) = n(n - 1)(n - 2)(p_4 - p_2^2) + n(n - 1)p_2(1 - p_2)/2 \tag{5.2}$$

where p_2 and p_4 are defined by (3.14) and (3.17) respectively. Although we will not prove it here, T_0^+ is approximately normal with this mean and variance. Thus we can compute the approximate power of a test based on T_0^+. Since the dominant terms of (5.1) and (5.2) are the same as those of the corresponding moments of T^+ given in (3.12) and (3.15), it follows that T_0^+ and T^+ are asymptotically equivalent, in a sense which is not trivial to make precise (see Chap. 8). In particular, the consistency results for T^+ in Sect. 3.4 apply to T_0^+.

A natural estimator of p_2, the probability of a positive Walsh average, is $2T_0^+/n(n - 1)$. When the X_j are a random sample from some continuous population, this estimator is unbiased, by (5.1), and consistent, as just remarked. If the class of possible distributions is broad enough—all continuous distributions, for instance—then no other unbiased estimator has variance as small (Problem 41). On the basis of asymptotic normality, an approximate confidence interval for p_2 could also be obtained from

$$|T_0^+ - n(n - 1)p_2/2| \leq z\sqrt{\text{var}(T_0^+)} \tag{5.3}$$

where z is an appropriate standard normal deviate. From (5.2) we see that $\text{var}(T_0^+)$ is a function of p_4 as well as p_2; hence p_4 must be eliminated before (5.3) will yield confidence limits on p_2.

We could estimate p_4 from the data as the proportion of triples i, j, k, all distinct, for which $X_i + X_j$ and $X_i + X_k$ are both positive; alternatively, p_4 could be replaced by the upper bound p_2 (Problem 42). In either case, the right-hand side of (5.3) still depends on p_2. This inequality could then be solved for p_2 (Problem 43). Another possibility is to estimate $\text{var}(T_0^+)$ as given by (5.2), using one of the foregoing methods for p_4 and the estimate $2T_0^+/n(n-1)$ for p_2, and substitute this estimate of $\text{var}(T_0^+)$ on the right-hand side of (5.3) (Problem 44).

The statistic T_0^+ is thus especially natural and useful for estimation of p_2. There are two further reasons for introducing the modified Wilcoxon procedure. The first is that it illustrates the inevitable arbitrariness involved in the choice of a test or confidence procedure. Neither T nor T_0 has any definitive theoretical or practical advantage over the other, and they are equally easy to use. In the example of Sect. 3.1 with $n = 15$ we get a one-tailed P-value of 0.0206 from T and 0.0290 from T_0, and lower and upper confidence bounds of 10 and 39 each at the one-tailed level 0.0535 from T, 7 and 39 each at the one-tailed level 0.0520 from T_0 (Problem 45). There is no reason to say that one set of values is any better than the other, and both methods are reasonable. A single "correct" or "best" method of statistical analysis does not exist in anything like the present framework.

The other reason for introducing the modified procedure is that the choice of exact levels is thereby increased. It may happen that T provides a nonrandomized procedure at nearly the level desired but T_0 does not, or vice versa. For instance, suppose the nominal level selected is 0.05 for $n = 10$. From Table D, we see that T_0 provides a test at the one-sided level 0.0488 while the one-sided levels for T which are nearest 0.05 are 0.0420 and 0.0527. If there is some real reason for insisting on one particular level and a randomized procedure is undesirable, the fact that T and T_0 provide tests at different natural levels may be grounds for choice between them in a particular case. Of course, with either T or T_0, the approximate method of interpolating confidence limits between those at attainable levels may be used. (This was explained generally in Sect. 5, Chap. 2.) The accuracy of such interpolation has not been investigated extensively, but results given in Problem 49 and quoted in Chap. 2 for interpolation midway between any two successive order statistics, and results given in Problem 50 for interpolation between the two smallest (or largest) Walsh averages, suggest that, especially when discreteness matters most, the accuracy is often quite poor but interpolation tends to be conservative.

6 Zeros and Ties

6.1 Introduction

An observation which equals 0 is called a "zero;" its sign has not been defined. Two or more observations which have the same absolute value (zero or not) are called a "tie;" their ranks (in order of absolute value) have not been defined. As a result, if zeros or ties occur, the sum of the signed ranks cannot be determined without some further specification. We have avoided this difficulty so far by assuming continuous distributions; this ensures that the probability of a zero or a tie is equal to zero. In practice, it is necessary to deal with zeros and ties because measurements are not always sufficiently precise to avoid them, even with theoretically continuous distributions, and discontinuous distributions do occur. (The modified Wilcoxon procedure of Sect. 5 assigns rank 0 to the smallest observation in absolute value, whether it is a zero or not. Therefore, a single zero creates no problem in this procedure. Otherwise, the discussion below applies to it also, with the necessary changes.)

Of course, the Walsh averages are well defined even with zeros or ties, although some of them may now be tied. The confidence procedures given earlier in Sect. 4 depend only on the values of certain order statistics of the set of Walsh averages, and these values are well defined for any data (and easily found by the graphical procedure or some other technique). It can be shown that, if k is defined as before and found from Table D, the $(k + 1)$th smallest average, say L, is still a lower confidence bound for the population center of symmetry μ in the present situation. However, the exact statement for the confidence property is now more delicate. The lower confidence bound L has at least the indicated probability of falling *at or below* μ, and at most the indicated probability of falling *strictly below* μ. That is

$$P(L \leq \mu) \geq 1 - \alpha \geq P(L < \mu), \tag{6.1}$$

where $1 - \alpha$ is the exact confidence level in the continuous case (Problem 51; see also Problems 35 and 109). A corresponding statement holds for the upper confidence bound. Accordingly, a two-sided confidence interval including its endpoints has at least the indicated probability of covering μ, while the interval excluding its endpoints has at most the indicated probability of covering μ. Thus the confidence procedures (one- and two-sided) of Sect. 4 are still applicable. The only question is whether or not to include the end-points in the interval; ordinarily this does not matter and the question need not be resolved.

Since the confidence procedures are still applicable, a test of hypothesis about μ at level α can be performed by finding the corresponding confidence bound(s) at level $1 - \alpha$. If μ_0 is not an endpoint of this region, the test "accepts" or rejects according as μ_0 is inside or outside the region. If μ_0 is an endpoint, it may be sufficient to state this fact and not actually carry the test procedure further (or to be "conservative" and "accept" the null hypothesis. As we shall see, this is equivalent to the "conservative" procedure of rejecting μ_0 if and only if resolving all ambiguities in the definition of signed ranks in favor of "acceptance" would still lead to rejection.) It may be desirable to proceed further, however, in order to report an accurate P-value or to obtain more nearly the desired significance level, and to avoid the loss of power resulting from "conservatism," which is serious in situations where many zeros or ties are likely. The following subsections will discuss various methods of handling ties and zeros. For some of these methods, the null distribution of the Wilcoxon test statistic as given in Table D can still be applied, while others produce a different null distribution so that ordinary tables cannot be used. Furthermore, some other surprising traps and anomalies can arise. See Pratt [1959] for a more complete discussion than is given here.

6.2 Obtaining the Signed Ranks

There are several different methods of obtaining signed ranks for observations which include zeros and/or ties. We will now describe each of these methods briefly. In the next three subsections, we illustrate the resulting test procedures and discuss certain properties of them.

Ties may occur in only nonzero observations, in only zero observations, or in both simultaneously. Zeros may occur with or without ties. We will consider each relevant case starting with nonzero ties.

The three basic methods of handling nonzero ties which we shall discuss are called (a) the average rank method, (b) breaking the ties randomly, and (c) breaking the ties "conservatively."

The *average rank* (or *midrank*) method assigns to each member of a group of tied observations the simple average of the ranks they would have if they were not tied. In general, for a set of tied observations larger than the $(k - 1)$th largest and smaller than the $(l + 1)$th largest observations, that is, tied values in rank positions k through l, the average rank is $(k + l)/2$, which is always an integer or half integer, and this rank is given to each of the observations in this tied set. This approach gives a unique set of ranks, and tied observations are given tied ranks. Since the Wilcoxon statistic uses the signed ranks for the absolute values of the observations, the possible ranks are averaged for sets of tied *absolute* values of the observations and then signs are attached to these resulting average ranks. (Examples will be given shortly.)

Methods which *break the ties* give each observation, including the tied values, a distinct integer rank, that is, as if there were no ties. Two methods of doing so are the "random" method and the "conservative" method. In the present context, with tied absolute values at ranks k through l, if m of these belong to negative observations, the random method would attach the m minus signs to a sample of size m drawn at random without replacement from the integers k through l. The "conservative" method would attach the m minus signs to the smallest m of these integers when testing for rejection in the direction of smaller values of μ, the largest m in the other direction, thus breaking all ties in favor of "acceptance." Other methods of breaking ties will not be considered here.

For zeros, which may occur singly or as ties, there are analogs of each of the foregoing methods. Still another method is to discard the zeros from the sample before ranking the remaining observations. This last method will be called the *reduced sample procedure*. For the signed-rank test, Wilcoxon

[1945] recommended this practice, with the ordinary test then being applied to the reduced sample if there are no nonzero ties. However, this leads to a surprising anomaly, which will be discussed later. Further, one might argue purely intuitively that the ambiguity about the signed ranks of the zero observations is irrelevant to the ranks of the nonzero observations, and hence these ranks should not be changed by discarding the zeros. Nevertheless, if the zeros are retained, they must also be given signed ranks. One possibility is to give each zero a signed rank of zero; we call this the *signed-rank zero procedure*. The signed-rank zero is actually an average rank, since it is the average of the two signed ranks which could be assigned to any zero, regardless of what method is used to obtain the unsigned rank for the zero.

Tiebreaking procedures, on the other hand, would assign the signed ranks ± 1, ± 2, etc., to the zeros, choosing the signs either randomly (independently with equal probability) or "conservatively" (all $+$ signs when testing for rejection in the direction of smaller values of μ, all $-$ signs in the other direction).

If both nonzero ties and one or more zeros are present, it would be possible to use any one of the procedures for zeros in conjunction with any one of the procedures for nonzero ties. With any procedure used for nonzero ties, however, it is natural to use either the corresponding procedure for zeros or the reduced sample procedure. When the "conservative" procedure is viewed as inadequate, we recommend that the signed-rank zero method be used in conjunction with the average rank procedure, for reasons indicated later in Sects. 6.4 and 6.5.

6.3 Test Procedures

We illustrate the basic methods of handling ties, with zeros also present, for the following data (arranged in order of absolute value):

$$0,\ 0,\ -1,\ -2,\ 2,\ 2,\ 3. \tag{6.2}$$

(a) The assignment of ranks to all observations by the *average rank method, in combination with the signed-rank zero procedure*, is illustrated in Table 6.1 for the data in (6.2). By the average rank and signed-rank zero methods, the values of the Wilcoxon signed-rank statistics are $T^+ = 17$, $T^- = 8$, $T = 9$. Note that the relationships between T^+, T^- and T, as given in (3.1)–(3.3), must be modified; if v zeros are present, then $v(v + 1)$ must be subtracted from $n(n + 1)$ throughout (Problem 52d).

Table 6.1

X_j	0	0	-1	-2	2	2	3
Possible ranks	1, 2	1, 2	3	4, 5, 6	4, 5, 6	4, 5, 6	7
Average rank	1.5	1.5	3	5	5	5	7
Signed rank	0	0	-3	-5	5	5	7

The null distributions of these test statistics are not as given in Table D, as is obvious since the ranks with signs are not the first n integers. An exact test can, however, be performed conditionally on the number of zeros present and the ranks assigned to the nonzero observations. Under the null hypothesis of symmetry about 0, given the pattern of zeros and ties (and even the absolute values) present, the conditional distribution of the positive and negative rank sums by the average rank method is determined by the fact that all assignments of signs to the average ranks of the nonzero observations are equally likely.

For the data in (6.2), given the absolute values present, including two zeros, three observations tied at ranks 4–6, and two untied observations at ranks 3 and 7, there are 2^5 equally likely assignments of $+$ and $-$ signs to the relevant average ranks, which are 3, 5, 5, 5, and 7. We enumerate these assignments and calculate T^- as follows.

Negative Ranks	T^-	Number of Cases
none	0	1
3	3	1
5	5	3
7	7	1
3, 5	8	3
3, 7	10	1
5, 5	10	3
5, 7	12	3
3, 5, 5	13	3
3, 5, 7	15	3
5, 5, 5	15	1
5, 5, 7	17	3
3, 5, 5, 5	18	1
3, 5, 5, 7	20	3
5, 5, 5, 7	22	1
3, 5, 5, 5, 7	25	1

(T^+ and T^- are equivalent conditional test statistics, since $T^+ = 25 - T^-$ given two zeros.) The conditional null distribution of T^- is then as follows.

t	0	3	5	7	8	10	12	13	15	17	18	20	22	25
$2^5 P(T^- = t)$	1	1	3	1	3	4	3	3	4	3	1	3	1	1
$2^5 P(T^- \leq t)$	1	2	5	6	9	13	16	19	23	26	27	30	31	32

In the sample observed, we had $T^- = 8$. The probability of a value at least as small as that observed, that is, the P-value, is

$$P(T^- \leq 8) = \tfrac{9}{32} = 0.28.$$

The probability of a value smaller than that observed, the next P-value, is

$$P(T^- < 8) = P(T^- \le 7) = \tfrac{6}{32} = 0.19.$$

When there are relatively many ties, as here, the normal approximation may be very inaccurate. The relevant mean and variance are easily obtained and the distribution is symmetric (Problem 52), but the uneven spacing of the possible values of T^- makes correction for continuity difficult and of doubtful value. Further, general lumpiness of the distribution removes any possibility of high accuracy, even when a continuity correction is used. However, when enumeration of the null distribution is too laborious, one could use the Monte Carlo method (simulation) described further in Sect. 2.5, Chap. 4. A table of critical values for the case when there are zeros but no nonzero ties is given in Rahe [1974].

(b) If a *tiebreaking procedure* is applied to the data in (6.2) *without omitting the zeros*, there are 12 different possible signed-rank assignments for the possible ranks shown in Table 6.1, obtained as follows. The two zeros, with possible ranks 1 and 2, could be given signed ranks either 1 and 2, or -1 and 2, or 1 and -2, or -1 and -2. The three observations with absolute value 2 must each be assigned one of the ranks 4, 5, and 6. When signs are attached, one of these ranks must be negative since one of the 2's was negative. Thus the signed rank associated with -2 is either -4, -5, or -6, and the two observations which are $+2$ have the remaining two ranks with positive sign. The observation -1 has signed rank -3 and the observation 3 has signed rank 7 in any case.

As a result of these possibilities, breaking the ties in this set of observations could lead to any one of $4(3) = 12$ sets of signed ranks. In all 12 cases, the negative-rank sums are between 7 and 12 inclusive. The two methods of breaking the ties which produce (i) the smallest, and (ii) the largest, negative-rank sum are shown in Table 6.2. There are in addition two cases with $T^- = 8$, three with $T^- = 9$, three with $T^- = 10$, and two with $T^- = 11$, as shown in Table 6.3 (Problem 53).

Table 6.2

X_j	0	0	-1	-2	2	2	3	T^-
Signed ranks (i)	1	2	-3	-4	5	6	7	7
Signed ranks (ii)	-1	-2	-3	-6	5	4	7	12

Table 6.3 Results of Breaking the Ties

Number of cases	1	2	3	3	2	1
T^-	7	8	9	10	11	12
Wilcoxon tail probability	0.1484	0.1875	0.2344	0.2891	0.3438	0.4063

The *random method of breaking ties* selects one of the possible resolutions of ties by using some supplementary random experiment which ensures that each possible set of signed-rank assignments is equally likely to be chosen. Because this preserves the usual null distribution of the Wilcoxon signed-rank statistic (Problem 54), the usual critical value can then be used, or the *P*-value can be found from Table D. Thus one of the columns of Table 6.3 would be selected, with probability proportional to the number of cases.

Instead of actually breaking the ties at random and using standard tables, we might report the probability for which doing so would lead to rejection, as in reporting $\phi(x)$ for a randomized test (Sect. 5.2, Chap. 1). For example, for the data of (6.2), if we were using the critical value $t = 8$, three of the twelve possible ways of breaking ties would lead to rejection. If the ties were broken at random, therefore, there would be a $\frac{3}{12} = 0.25$ chance of rejection. Instead of actually randomizing, one could report this probability. (For *P*-values one could in principle report similarly that randomizing would give $P = 0.1484$ with probability $\frac{1}{12}$, $P = 0.1875$ with probability $\frac{2}{12}$, etc., or some summary of this information.)

The "*conservative*" method of breaking ties, when rejecting for small values of T^-, would assign negative signs to the two zeros of the data (6.2), and assign rank 6 rather than 4 or 5 to the negative observation -2, so as to maximize T^-, as in the bottom line of Table 6.2. The "conservative" value of T^- would therefore be $T^- = 12$ and the "conservative" *P*-value from Table 6.3 is 0.4063. This means that all methods of breaking ties would give $T^- \leq 12$ and would reject at any critical value of 12 or more and hence at any level $\alpha \geq 0.4063$.

The other side of the conservative coin is that no method of breaking ties would give a value $T^- < 7$. Hence all methods of breaking ties lead to "acceptance" for any critical value of 6 or less. For critical values between 7 and 11 inclusive, however, the conclusion is indeterminate, and the ordinary Wilcoxon test would reject by one resolution of ties but not by another.

Approaching the hypothesis testing problem from a confidence interval viewpoint sheds some light on the interpretation of results when ties are broken in all possible ways. The Wilcoxon signed-rank test for the null hypothesis $\mu = 0$ should presumably "accept" $\mu = 0$ when 0 is inside and reject when 0 is outside the usual confidence interval defined by the order statistics of the Walsh averages. By Problem 55, the ordinary signed-rank test leads to rejection no matter how the ties are broken if and only if 0 is outside the confidence interval and not an endpoint of it. The same holds for "acceptance" and "inside." The remaining possibility, that 0 is an endpoint of the confidence region, occurs when and only when the ordinary Wilcoxon test would reject by one method of breaking the ties but not by another. Hence, instead of resolving the indeterminacy as to whether the test "accepts" or rejects $\mu = 0$, one might simply state that 0 is an endpoint of the corresponding confidence interval, as suggested earlier. For the data in (6.2), we have seen that indeterminacy occurs for the critical values 7–11, and it is easily verified that the corresponding confidence bounds, the Walsh averages at ranks 8–12, are all 0 (Problem 56).

Note that these comments apply only to tiebreaking procedures, meaning that the ordinary Wilcoxon test is used after the ties are broken. They unfortunately do not apply to the average rank procedure, which, as we shall see, may give a conclusion opposite to that based on a tiebreaking procedure even when the latter is determinate. Though inconvenient, this is not a telling objection to the average rank procedure, since there is nothing sacrosanct about the Wilcoxon procedure itself, even in the absence of ties.

(c) A *reduced sample procedure* would omit the two zeros in the data of (6.2), leaving a sample of size 5 with a three-way nonzero tie. The tie could be handled by any of the methods described above. We shall not illustrate this procedure here. However, note that it can disagree with tiebreaking in the complete sample, and has a still more objectionable property which will be described shortly.

6.4 Warnings and Anomalies: Examples

Nonzero ties

For a given pattern of zeros and ties (strictly, for given absolute values), if the average rank procedure would be used in some cases, it is to be used in all cases, even those where tiebreaking is unambiguous. Furthermore, it may lead to the opposite conclusion. Thus, it is not a valid shortcut to use tiebreaking when it is unambiguous and average ranks when tiebreaking is indeterminate. Similar comments would presumably apply to other procedures for handling zeros and ties, in the absence of proof to the contrary.

To illustrate the difficulty, consider the following sample, in which the tied observations all have the same sign:

$$1, 1, 1, 1, 2, 3, -4, 5. \tag{6.3}$$

Any method of breaking the ties gives the same signed ranks, namely

$$1, 2, 3, 4, 5, 6, -7, 8. \tag{6.4}$$

Thus, one is tempted to apply the Wilcoxon test to these signed ranks without further ado. The null probability of a negative-rank sum of less than 7 is $14/2^8$ while that of 7 or less is $19/2^8$ (0.0547 and 0.0742 respectively, from Table D). Hence, when the ties are broken in any way, (6.3) would be judged not significant at any one-sided level $\alpha \leq 0.0547$ and significant at any level $\alpha \geq 0.0742$. (For $0.0547 < \alpha < 0.0742$, the exact level α is unavailable, but such values of α are not required for the present discussion.)

Now if the average rank procedure is used on the sample in (6.3), the signed ranks are

$$2.5, 2.5, 2.5, 2.5, 5, 6, -7, 8, \tag{6.5}$$

and the test statistic still has the value $T^- = 7$. The left tail of the null distribution of T^- given the ties at ranks 1, 2, 3, and 4 is shown below (Problem 57).

t	0	2.5	5	6	7	7.5	8	8.5	9.5
$2^8 P(T^- = t)$	1	4	7	1	1	8	1	4	4

Hence, by the average-rank procedure, $P(T^- \leq 7) = 14/2^8 = 0.0547$, and the sample in (6.3) would be judged significant at the level 0.0547, even though it is not significant at this level when the ties are broken, no matter how they are broken. For $\alpha = 0.0547$, the two methods reach opposite conclusions. (This is true for all α in some interval including 0.0547, but this interval depends on what is done when the exact level α is not available. Furthermore, similar disagreement in the other direction is also possible (Problem 58).)

In terms of P-values, the exact P-value is 0.0547 by the average-rank procedure, while it is 0.0742 by the Wilcoxon test no matter how the ties are broken. In terms of confidence bounds, for $\alpha = 0.0547$, the lower confidence bound is 1 by the average-rank procedure but -0.5 by the usual procedure. Thus the two procedures give bounds with opposite signs (and 0 is not an endpoint of either confidence interval).

There is no contradiction here, but there is a warning; namely, it is not valid to use a tiebreaking procedure when tiebreaking is unambiguous, not even when all tied observations have the same sign, if one would have used another procedure (such as average ranks) in other cases (with the same absolute values).

Consider, for example, a sample with the same absolute values as (6.3), but different observed signs, as in (6.6):

$$-1, \ -1, \ -1, \ 1, \ 2, \ 3, \ 4, \ 5. \tag{6.6}$$

Using the average rank procedure, we find $T^- = 7.5$. The null distribution of T^- for the average rank procedure is the same as that given above for sample (6.2), so that $P(T^- \leq 7.5) = 22/2^8 = 0.0859$. But this computation of the null distribution of T^- by the average rank procedure for sample (6.6) assumes that the signed ranks for sample (6.3) are those of (6.5), not those of (6.4) which result from breaking the ties. Thus, if we would use the average rank procedure for the sample (6.6), we must also use it for the sample (6.3), even though tiebreaking would be easier and unambiguous for (6.3). The alternative would be to use tiebreaking in both samples (and others with the same absolute values). When the ties in (6.6) are broken, the possible values of T^- are $6 \leq T^- \leq 9$ with corresponding P-values from Table D as 0.0547 and 0.1250 at the extremes. If this degree of ambiguity is too great, one might prefer the average rank procedure, but one must make this decision in advance, without knowing the signs and hence without knowing whether the actual sample is (6.3), or (6.6), or some other sample with the same absolute values.

Zeros

In the case of zeros, if there are no nonzero ties, it can be shown that the signed-rank zero procedure gives the same conclusion as tiebreaking whenever the latter is unambiguous (Problem 59). The reduced sample procedure, however, may exhibit strange behavior in this and other respects, as can be illustrated by applying it to the 13 observations

$$0, 2, 3, 4, 6, 7, 8, 9, 11, 14, 15, 17, -18. \qquad (6.7)$$

Dropping the zero before ranking leaves 12 observations with a negative-rank sum of 12, which is not significant at any one-sided level $\alpha \leq 55/2^{12} = 0.0134$ and is significant at any $\alpha \geq 70/2^{12} = 0.0171$, these being the null probabilities of less than 12 and 12 or less, respectively. On the other hand, tiebreaking in favor of the null hypothesis, assigning 0 the signed rank -1, would result in 13 observations with negative ranks 1 and 13 and a negative-rank sum of 14, which is significant at $\alpha = 109/2^{13} = 0.0133$, the null probability of 14 or less. Thus, for $0.0133 \leq \alpha \leq 0.0134$, the reduced sample procedure disagrees with tiebreaking even though tiebreaking is unambiguous (and, as before, disagreement occurs for a wider range of α which, however, depends on what is done when the exact level α is not available). The *P*-value is 0.0171 by the reduced sample procedure, but it cannot exceed 0.0133 if the zero is retained, no matter what sign is given to the zero. These results are comparable to those above for the average rank procedure for nonzero ties. When we examine confidence regions, however, an anomaly appears which is more striking and disturbing than before.

For $\alpha = 0.0133$, the usual lower confidence bound is 1. What is the confidence region by the reduced sample procedure? It contains $\mu = 0$, since the reduced sample procedure would accept this hypothesis at this level. If an amount μ between 0 and 1 is subtracted from every observation, there will be no zero or tie, so the usual procedure will be used, and will reject the value μ. This is already strange—the sample is not significant as it stands but becomes significant in the *positive* direction if every observation is *reduced* by the same small amount. Correspondingly, the confidence region is not an interval; it contains the point $\mu = 0$, excludes all other values $\mu < 1$, and contains all $\mu > 1$. Thus it is an interval plus an exterior point. (Strictly speaking, for those integer and half-integer values of μ where nonzero ties occur, the procedure has not been defined, but the statement holds for the average rank procedure and for any tiebreaking procedure.)

It is also possible for the reduced sample procedure to judge a sample significant in the positive direction, yet not significant when every observation is increased by the same small amount, corresponding to a confidence region which is an interval with an interior point removed (Problem 61).

Thus the reduced sample procedure is not only inconsistent with tiebreaking, but also inconsistent with itself, in the sense that shifting the sample in one way may shift the conclusion the opposite way. The signed-rank zero and average rank procedures are self-consistent in this sense.

6.5 Comparison of Procedures

In this subsection we will conclude our discussion of various methods of handling zeros and ties. We shall list some requirements which seem intuitively desirable to avoid anomalies, and then check the various methods against them. Of course the most important consideration in general is power, but this is not the main focus of the present discussion. All the methods are variants of the Wilcoxon procedure, and if one were seriously trying to improve upon the power of the Wilcoxon test in some respect, one would probably be led to a different procedure altogether. Ease of application is also a consideration, but this depends heavily on individual circumstances and hence will not be focused on either.

We have in mind alternatives involving primarily changes in location rather than shape. Accordingly, the following requirements seem intuitively desirable.

(i) A significantly positive sample shall not become insignificant nor an insignificant sample significantly negative when (a) some observations are increased, or (b) all observations are increased by equal amounts. (Requirement (b) is weaker than (a). It is included because it seems even more desirable than (a) and because some procedures will satisfy (b) and not (a).)

(ii) Those values of the center of symmetry μ which would be "accepted" if tested shall form an interval. (This says that the corresponding confidence region shall be an interval, and is equivalent to (i)(b) (Problem 62).)

(iii) A sample shall be judged significantly positive if it is significantly positive however the ties are broken; similarly for significantly negative and not significant. (This is implied by (i)(a) but not by (i)(b).)

Consider first the methods of handling zeros. The data in (6.7) show that the reduced sample procedure satisfies none of the conditions above, no matter how nonzero ties are handled if they are present. These factors are the primary justification for our recommendation against this procedure in Sect. 6.2, even though the ordinary Wilcoxon tables can be used if there happen to be no nonzero ties, once n is reduced appropriately. They also supplant considerations of power for the kinds of situation we have in mind, although Conover [1973] exemplifies situations where the reduced sample procedure is both more and less powerful than the signed-rank zero procedure.

The signed-rank zero procedure, in the absence of nonzero ties, satisfies all of the above conditions (Problem 59).

The average rank procedure (in conjunction with the signed-rank zero procedure) satisfies (i)(b) and (ii) but not (i)(a) and (iii) (Problem 60). This procedure presumably gives better power, at least in any ordinary situation, than breaking ties either "conservatively" or randomly. The regular tables do not apply, and the null distribution must be generated for the set of average ranks observed.

The "conservative" procedure, that is, breaking the ties and choosing the value of the statistic least favorable to rejection, satisfies all of these conditions. However, the true level α is unknown and may be much less than the nominal level, and considerable loss of power may result, especially if many ties are likely.

Breaking the ties at random, whether by actually doing so and using standard tables or by reporting the probability with which doing so would lead to rejection, satisfies all of the above conditions and also the following version of (i) (Problem 63), which is stronger for randomized test procedures.

(i′) The probability that a sample is judged significantly positive shall not decrease, nor the probability that it is judged significantly negative increase, when (a) some observations are increased, or (b) all observations are increased by equal amounts.

Breaking the ties at random permits use of the regular tables but then the analysis depends on an irrelevant randomization. Imposing this extraneous randomness in an artificial way is unpleasant in itself, and presumably greater power could be achieved without it [see also Putter, 1955]. The unpleasantness can be somewhat mitigated, but not entirely eliminated, by reporting instead the probability with which breaking the ties would lead to rejection for the sample at hand, rather than actually breaking the ties in one particular, randomly chosen way. This, however, requires additional calculation.

However reasonable the properties above may seem in general, in particular cases larger observations may not be greater evidence of positivity in the population (Problem 64). Even the normal-theory t statistic does not satisfy (i)(a) and may decrease when some observations are increased, since this affects the sample variance as well as the mean (Problem 65). Nevertheless, in the absence of information about the underlying distribution, as in the present nonparametric context, the conditions appear desirable intuitively.

Because the average rank procedure does not satisfy (i)(b) and (iii), it is all the more tempting to resort to it only when tiebreaking is indeterminate, i.e., only for μ_0 at the end of the usual confidence interval. Unfortunately there seems to be no easy way to do this and preserve the level α. Accordingly our recommendation is to use the "conservative" procedure if it is not too conservative in view of the type of inference desired and the extent of zeros and ties expected. (If one is testing a null hypothesis, and not forming a confidence interval, one may look at the absolute values present before deciding, but not at the signs.) Otherwise, we recommend the average rank procedure in conjunction with the signed-rank zero procedure.

7 Other Signed-Rank Procedures

Now we return to the situation where the random sample, $X_1, \ldots, X_n$, is drawn from a continuous population (so that ties need not be considered). For tests of the null hypothesis that the population is symmetric about 0, the Wilcoxon signed-rank procedure uses the ranks $1, 2, \ldots, n$ in place of the absolute values of the observations. In Sect. 7.1 we consider procedures which employ some other set of constants $c_1, c_2, \ldots, c_n$ in place of the absolute values. The sum of the signed constants is the test statistic.

In Sect. 7.2, we will see that these and all other (permutation-invariant) signed-rank tests depend, like the Wilcoxon test, only on the signs of the Walsh averages. The Walsh averages therefore again determine the boundary points of the corresponding confidence regions for the center of symmetry. These regions are discussed in Sect. 7.3. Some particular tests and confidence procedures involving Walsh averages are presented in Sect. 7.4.

7.1 Sums of Signed Constants

Suppose the observations X_j are replaced in order of the ranks of their absolute values by the constants $c_1, c_2, \ldots, c_n$, and that the sign of the corresponding X is attached to each constant. The result is called a *signed constant*, corresponding to the term signed rank defined in Sect. 3.1. To be more specific, suppose that X_j has rank k in order of absolute magnitude; then its signed constant is $+ c_k$ if X_j is positive and $- c_k$ if X_j is negative. For the data in Sect. 3.1, this gives the following results.

X_j	49	-67	8	16	6	23	28	41	14	29	56	24	75	60	-48
$k = $ rank of $|X_j|$	11	14	2	4	1	5	7	9	3	8	12	6	15	13	10
signed c_k	c_{11}	$-c_{14}$	c_2	c_4	c_1	c_5	c_7	c_9	c_3	c_8	c_{12}	c_6	c_{15}	c_{13}	$-c_{10}$

Analogously to the Wilcoxon signed rank statistic T, we define a statistic which is the sum of these signed constants. Tests based on the sum of the signed c_k's are equivalent to tests based on the sum of the negative c_k's, or the sum of the positive c_k's (Problem 79).

Under the null hypothesis that the population is symmetric about 0, all assignments of signs are equally likely. This fact determines the null distribution of the test statistic. Therefore, tables could be generated for any particular set of constants c_k, although not as easily as in Problem 11 for the Wilcoxon case unless the c_k for different n satisfy some special relationships.

If the hypothesized center of symmetry is some value other than 0, say μ_0, the foregoing test can be applied to the $X_j - \mu_0$. The set of values of μ_0 which would be accepted by this procedure forms a confidence region for an assumed center of symmetry μ. The confidence bounds are Walsh averages, but for arbitrary c_k they are determined differently from the Wilcoxon procedure, as we shall see in Sect. 7.3.

For $c_k = k$, the sum of signed-constants procedure is identical to the Wilcoxon signed-rank procedure. For $c_k = k - 1$, it is the modified Wilcoxon procedure of Sect. 5. For $c_k = 1$, we have the sign test of Chap. 2, and the corresponding confidence bounds, which are order statistics. Other possibilities for c_k arise naturally (in Sect. 9, for instance). However, even if appropriate tables are available, these other tests (and especially the corresponding confidence procedures) are at least somewhat more difficult to use than those just mentioned. In the absence of tables, the normal approximation could be used (Problem 80), but it has only limited, though perhaps adequate, accuracy in small samples. In large samples, where the normal approximation is more accurate, confidence limits may be preferable to tests but especially difficult to obtain. With more effort, the null distribution for any particular set of constants c_k can be obtained by enumeration, or approximated by Monte Carlo methods (simulation). These approximations will be discussed in the next chapter, in connection with "observation-randomization" procedures (where tabulation is impossible). Here, we need only note that the distribution can be determined or at least approximated well. In some problems, the advantages of these procedures may warrant the extra effort in analysis.

For any signed-rank test and corresponding confidence procedure, the assumptions made in the introduction to this section can be relaxed as in the first two paragraphs of Sect. 3.5. In some circumstances, the continuity assumption can also be relaxed so that ties have positive probability; in particular, tests based on sums of signed constants and corresponding confidence bounds are conservative for discrete distributions if the c_k satisfy $0 \leq c_1 \leq c_2 \leq \cdots \leq c_n$ (Problem 109). For one-tailed signed-rank tests, the assumption of symmetry can also be relaxed as in Sect. 3.5, provided the test satisfies the condition of Theorem 3.3. This condition is satisfied, in particular, by any one-tailed test based on a sum of signed constants c_k such that $0 \leq c_1 \leq \cdots \leq c_n$ (Problem 83).

7.2 Signed Ranks and Walsh Averages

The tests based on sums of signed constants c_k depend only on the signed ranks of the X_j; if X_j has signed rank $-k$, the signed constant is $-c_k$, and if X_j has signed rank $+k$, the signed constant is $+c_k$. Tests depending only on the signed ranks, including those in this general class and many more, are called *signed-rank tests*.

We have already seen that the Wilcoxon signed-rank test depends only on the signs of the Walsh averages (because the positive-rank sum is the number of positive Walsh averages, by Theorem 3.1), and that the corresponding confidence limits are order statistics of the Walsh averages. We will see in this subsection that all signed-rank tests are in practice equivalent to tests depending only on the signs of the Walsh averages, and in the next subsection that the corresponding confidence limits are always Walsh averages.

The exact statement of the relationship for tests is given in the following theorem.

Theorem 7.1. *The signed ranks determine the signs of the Walsh averages, so any test depending only on the signs of the Walsh averages is a signed-rank test. Conversely, the signs of the Walsh averages determine the signed ranks except possibly for the order in which they occur. Therefore, any signed-rank test which does not take into account the order of the signed ranks depends only on the signs of the Walsh averages.*

The proof of this theorem is similar to that of Theorem 3.1 (Problem 81). To illustrate the point about order, consider a sample X_1, X_2, X_3 whose Walsh averages have the signs given below.

i	j	sign of $(X_i + X_j)/2$
1	1	+
1	2	−
1	3	+
2	2	−
2	3	−
3	3	+

Clearly, X_1 and X_3 are positive, and X_2 is negative with larger absolute value than either X_1 or X_3. However, there is no way to determine whether the respective signed ranks of X_1, X_2, X_3 are 2, −3, 1 or 1, −3, 2. Thus the signs of the Walsh averages determine the signed ranks collectively, but they do not completely determine their order.

Of course, intuitively, there is no reason to care about order anyway in the situations of concern here, or to think that order is relevant. The tests we have been considering do not take the order of the X_j into account, that is, they are invariant under permutations of the X_j (see Sect. 8.1). For a permutation-invariant test, Theorem 7.1 says that it is a signed-rank test if and only if it depends only on the signs of the Walsh averages.

7.3 Confidence Bounds Corresponding to Signed-Rank Tests

Given a sample $X_1, \ldots, X_n$ from a continuous, symmetric distribution, we now want to find the confidence region for the center of symmetry μ which corresponds to some (particular) signed-rank test. The region is, of course, the set of values of μ which would be accepted if the test were applied to the values $X_j - \mu$. As μ varies, the signed ranks of the $X_j - \mu$ will change only when $\mu = (X_i + X_j)/2$ for some i, j, and hence the outcome of a signed-rank test for μ will change only at these values of μ (Problem 84. For permutation-invariant tests this is also a consequence of Theorem 7.1.) It follows that the boundary points of the confidence region corresponding to any signed-rank test are Walsh averages of the original sample. (The region will be an interval provided the test satisfies condition (i)(b) of Sect. 6.5. This holds for a test based on a sum of signed c_k's provided $c_{k+1} \geq c_k \geq 0$ for all k (Problem 85).)

As a result, the confidence limits corresponding to signed-rank tests are always Walsh averages. However, except in the Wilcoxon case, the confidence limit at a given level does not always have the same rank among the Walsh averages, and the ordered Walsh averages have different confidence levels in different samples. For certain tests, such as the sign test and others mentioned in the next subsection, the relevant Walsh average can be easily identified. In general it cannot, but the following trial and error procedure seems likely to identify it fairly quickly in most cases.

Consider the Walsh averages arranged in order of algebraic size. Let $T(\mu)$ be the value of the test statistic for the hypothesized value μ. For a signed-rank test, $T(\mu)$ is constant for μ in each interval between adjacent Walsh averages. Suppose, as is usual, that $T(\mu)$ is a monotone function of μ, so that its values in successive intervals are increasing (or decreasing) throughout $-\infty < \mu < \infty$. Start the search with some Walsh average, such as the Wilcoxon confidence bound or a Walsh average close to the normal-theory bound or to some other approximate bound appropriate to the test being used. Find the value of the test statistic $T(\mu)$ for μ just below the starting point (i.e., between it and the next smaller Walsh average). Move to the next smaller or greater Walsh average depending on whether $T(\mu)$ is smaller than or greater than the critical value of the test. Continue (in the same direction) until the value of $T(\mu)$ equals the critical value, or until successive values bracket it. The Walsh average which separates rejectable from "acceptable" values of $T(\mu)$ is the confidence bound sought. It may be helpful to rank beforehand all Walsh averages in what seems to be the relevant range. At each step, at most two signed ranks will change, and it may be easier to calculate the change in $T(\mu)$ than to recalculate $T(\mu)$ from scratch (Problem 84; see also Bauer [1972]).

7.4 Procedures Involving a Small Number of Walsh Averages

If, as for the Wilcoxon confidence intervals, a procedure involves ranking all the Walsh averages, at least implicitly, then it will automatically be permutation invariant (independent of the order of the observations), and it is immaterial whether the Walsh averages are defined in terms of the original observations X_j or the sample order statistics $X_{(j)}$. Confidence procedures for the center of symmetry which are simpler and still permutation invariant, can, however, be obtained from the sample order statistics by using a small number of Walsh averages of the form $(X_{(i)} + X_{(j)})/2$. We cannot choose a completely arbitrary function of the Walsh averages; for validity under the assumption that the population is symmetric, the corresponding test should be a signed-rank test and hence should depend only on the signs of the Walsh averages (Theorem 7.1).

The simplest case would be to use a single Walsh average $(X_{(i)} + X_{(j)})/2$. For $i = j$, the confidence limit is simply an order statistic and the procedure corresponds to a sign test, as discussed in Chap. 2. For $i < j$, the confidence limit corresponds to a sign test on the $n + i - j$ observations which are largest in absolute value (Problem 86; [Noether, 1973]).

With two Walsh averages there are more possibilities, and a more difficult probability problem must be solved to obtain the confidence level, but for sample sizes $n \leq 15$, procedures have been worked out by Walsh [1949a,b]. Specifically, his tests are of the form reject the null hypothesis $\mu = 0$ (or $\mu \leq 0$) in favor of the alternative $\mu > 0$ if both $(X_{(i)} + X_{(j)})/2$ and $(X_{(k)} + X_{(l)})/2$ are positive, and "accept" otherwise, where the four indices i, j, k, and l are chosen, not necessarily all different, to give the desired level. The corresponding lower confidence bound is

$$\min[(X_{(i)} + X_{(j)})/2, (X_{(k)} + X_{(l)})/2]. \tag{7.1}$$

A lower-tailed test and corresponding upper confidence bound can be obtained similarly, and combining one-sided procedures gives two-sided procedures. For $4 \leq n \leq 15$, Walsh [1949a,b] gives a table of tests of this form with one-tailed levels near the conventional values 0.005, 0.01, 0.025, 0.05, and two-tailed levels twice these values. He does not define a particular method of choosing i, j, k, and l for $n > 15$.

The Wilcoxon procedures with the critical value 0, 1, or 2 are of this type. Specifically, for any $n \geq 3$ (Problem 36),

$$T^- = 0 \quad \text{if and only if } X_{(1)} > 0 \tag{7.2}$$

$$T^- \leq 1 \quad \text{if and only if } (X_{(1)} + X_{(2)})/2 > 0 \tag{7.3}$$

$$T^- \leq 2 \quad \text{if and only if } \min[X_{(2)}, (X_{(1)} + X_{(3)})/2] > 0. \tag{7.4}$$

As a result, fourteen of the procedures in Walsh's table are equivalent to Wilcoxon procedures, while the remaining thirty are not. The modified Wilcoxon procedure of Sect. 5 satisfies similar relations (Problem 47), but happens to be equivalent to a procedure in Walsh's table only when it is also equivalent to an ordinary Wilcoxon procedure ($T_0^- = 0$ if and only if $T^- \leq 1$).

The levels available using these Walsh procedures are discrete and do not include exactly the conventional levels. However, Walsh [1949a] gives modified procedures which have a conventional level under normality and bounded level under the original symmetry assumption. The modification is made as follows. In place of $X_{(i)}$ in (7.1), substitute $[aX_{(h)} + (1 - a)X_{(i)}]$ for some $h < i$, $0 < a < 1$. This substitution gives a lower confidence bound of

$$\min\{[aX_{(h)} + (1 - a)X_{(i)} + X_{(j)}]/2, (X_{(k)} + X_{(l)})/2\}. \tag{7.5}$$

Since $X_{(h)} \leq X_{(i)}$, we have always

$$X_{(h)} \leq aX_{(h)} + (1 - a)X_{(i)} \leq X_{(i)}. \tag{7.6}$$

Hence the confidence level of (7.5) is between the confidence level of (7.1) and the confidence level that (7.1) would have if $X_{(i)}$ were replaced by $X_{(h)}$. The exact level depends on the form of the distribution sampled. By appropriately choosing the value of a, the exact level under normal distributions can be adjusted to the desired value. The calculation involved will not be discussed here. When $n = 5$, for instance, one procedure of this type has level $\alpha = 0.05$ when the X_j are independently, identically, normally distributed, and has level between 0.031 and 0.062 when the observations are independently, continuously distributed and symmetric about a common center of symmetry μ.

8 Invariance and Signed-Rank Procedures

The concept of permutation invariance came up briefly in Sect. 7.2 where we noted that a permutation-invariant test is a signed-rank test if and only if it depends only on the signs of the Walsh averages. In this section we define this concept in more detail and for more general one-sample procedures and discuss the justifications for restricting consideration to permutation-invariant procedures. Procedures which are invariant under other classes of transformations are also frequently desirable. Accordingly, we will go on to show that a strictly increasing, odd transformation on a set of observations does not change their signed ranks, and conversely that the only procedures which are invariant under all such transformations are signed-rank procedures. This provides a possible justification for the use of signed-rank tests when this type of invariance is considered important (but not for the corresponding confidence procedures, as will be explained).

8.1 Permutation Invariance

A procedure $\phi(X_1, \ldots, X_n)$ is called *permutation invariant* in $X_1, \ldots, X_n$ if it is unchanged by permutations of the X_j. In other words, ϕ is not changed if the order of the X's is changed, so that

$$\phi(X_1, \ldots, X_n) = \phi(X_{\pi_1}, \ldots, X_{\pi_n}) \tag{8.1}$$

for every permutation $\pi_1, \ldots, \pi_n$ of $1, \ldots, n$.

A procedure is permutation invariant if and only if it is a function of the order statistics $X_{(1)}, \ldots, X_{(n)}$ alone. The order statistics form a sufficient statistic if the X_j are independent and identically distributed under all contemplated distributions. (For test procedures, this must hold under both alternative and null hypotheses.) Accordingly, for independent and identically distributed observations, it follows from the properties of sufficiency

(Sect. 3.3, Chap. 1) that given any procedure, an equivalent, permutation-invariant procedure exists. This procedure (possibly randomized) is based on the order statistics alone and has exactly the same operating characteristics as the given procedure based on the observations and their order. Thus, for independently, identically distributed observations, restricting consideration to permutation-invariant procedures is clearly justified because nothing can be gained by looking beyond them.

If the observations may have different distributions, such a strong justification of permutation invariance is not applicable because the order statistics no longer form a sufficient statistic. However, another argument may apply if it still seems unreasonable to take the order of the X_j into account. If $X_1, \ldots, X_n$ provide intuitively the same relevant information as any permutation $X_{\pi_1}, \ldots, X_{\pi_n}$, then a "reasonable" procedure ϕ would satisfy $\phi = \phi_\pi$, where ϕ and ϕ_π denote the left- and right-hand sides of (8.1) respectively. The argument, which we will call "invoking *the principle of invariance* (for permutations)," asserts that a procedure ϕ satisfying $\phi = \phi_\pi$ "should" be used, or at least will be. It is just an assumption, and is based on a rationale which may or may not apply in any particular problem.

To carry the argument a little further, suppose we have a procedure ϕ where $\phi \neq \phi_\pi$. If permuting the X_j would not affect our judgment of the situation, then we would be indifferent between ϕ and ϕ_π for all permutations π. Accordingly, we should also be indifferent between ϕ and the procedure ψ which consists of choosing a permutation π at random and using ϕ_π. Since the procedure ψ is permutation invariant (Problem 93), corresponding to any procedure ϕ there is a permutation-invariant procedure ψ which seems equally desirable. In this sense, nothing is lost by restricting consideration to permutation-invariant procedures. This argument applies to procedures for testing, estimation, or anything else.

Now we discuss this argument from a slightly different point of view, using the context of testing for convenience.

For any joint distribution F of $X_1, \ldots, X_n$ and any permutation $\pi_1, \ldots, \pi_n$, let F_π be the joint distribution of $X_{\pi_1}, \ldots, X_{\pi_n}$. Suppose that for any null or alternative distribution F, F_π is also a null or alternative distribution respectively. Then the level of a test ϕ_π under a null distribution F is the same as that of a test ϕ under the null distribution F_π, and the power of ϕ_π against an alternative F is the same as that of ϕ against the alternative F_π. The power function of ϕ_π is then the same as that of ϕ, except for a permutation of the points in the space on which the power function is defined (the space of alternative distributions). In addition, the exact (overall maximum) level of ϕ_π is the same as the exact level of ϕ. If a power function remains equally desirable when the alternative distributions are permuted in this way, and the same is true for the level and the null distributions, then there is no preference between ϕ and ϕ_π. If $\phi \neq \phi_\pi$, the choice of one procedure over the other must be arbitrary; a "reasonable" test ϕ will then satisfy $\phi = \phi_\pi$. If this holds for all permutations π, a "reasonable" test will be permutation invariant.

Consider the procedure ψ above which randomly selects a permutation π and uses ϕ_π. Then $\psi = \sum_\pi \phi_\pi / n!$. If each ϕ_π is as desirable as ϕ, then ψ is as desirable as ϕ and is also a permutation-invariant test. One specific sense in which ψ is as desirable as ϕ is that for any F, ψ and ϕ have the same average power over the distributions obtainable from F by permutation, that is

$$\frac{1}{n!} \sum_\pi E_{F_\pi}[\psi(X_1, \ldots, X_n)] = \frac{1}{n!} \sum_\pi E_{F_\pi}[\phi(X_1, \ldots, X_n)]. \tag{8.2}$$

When power is of interest only through such averages, the power of ψ is as good as that of ϕ. It also follows from (8.2) that a suitable permutation invariant test will have certain properties if any test has them. For instance, if there exists a test which is uniformly most powerful, or uniformly most powerful unbiased, then there is a permutation-invariant test which has the same property.

The original argument was that in situations where we feel that any rearrangement $X_{\pi_1}, \ldots, X_{\pi_n}$ provides the same information as $X_1, \ldots, X_n$, we will want to use a procedure that treats them alike, that is, a permutation-invariant procedure. The argument stated above in the context of testing changes the emphasis slightly, by saying that if our attitudes toward F and F_π are the same for all F and π, then any "reasonable" procedure is permutation invariant. Two ways in which our attitudes toward F and F_π might differ should be distinguished. First, we might consider an alternative F more likely than F_π, and therefore prefer high power against F and low power against F_π to the reverse. Alternatively, the consequences of an error might be more severe under F than F_π, so that the power against F is again more important than the power against F_π. In either case, it might be quite reasonable to prefer some procedure which is not permutation invariant to any procedure which is. In a more formal framework where such things can be discussed explicitly, it is appropriate to say that we will be led to permutation-invariant procedures if the prior distribution and loss structure are both permutation invariant, but generally not otherwise. In any framework, the mere fact that permutations carry null into null and alternative into alternative distributions is a necessary condition, but by no means a sufficient reason, to invoke the principle of invariance.

8.2 Invariance under Increasing, Odd Transformations

In the previous subsection, the notion of invariance was discussed specifically in the context of transformations which are permutations of the observations. We gave two possible reasons for requiring a permutation-invariant procedure, namely, the fact that the order statistics are sufficient when the observations are independent and identically distributed, and the "principle of invariance." Permutation invariance is, however, a property of all procedures which are ordinarily considered for the situation at hand. One justification for restricting consideration to signed-rank procedures is based on another kind of invariance. The notion of invariance applies very generally, with similar rationale and limitations, and the "principle of invariance" can be invoked for any suitable class of transformations. In this subsection we consider a large class of transformations which leads to a much reduced and very useful class of invariant procedures, in particular, to signed-rank tests.

Suppose, for convenience, that $X_1, \ldots, X_n$ are independent and identically distributed, and that we are testing the null hypothesis that the distribution is symmetric about 0 against the alternative that it is not. Consider the class of transformations defined by all strictly increasing, odd functions g, where odd means that

$$g(-x) = -g(x) \quad \text{for all } x. \tag{8.3}$$

Then if $X_1, \ldots, X_n$ satisfy the null hypothesis, so also do $g(X_1), \ldots, g(X_n)$, and similarly for the alternative hypothesis. The transformation in (8.3) then carries null distributions into null distributions and alternative distributions into alternative distributions (Problem 97). Accordingly, we could "invoke the principle of invariance," that is, require that the test treat $X_1, \ldots, X_n$ and $g(X_1), \ldots, g(X_n)$ in the same way. If this is required for all strictly increasing, odd functions g, then any two sets of observations with the same signed ranks must be treated alike, because any set of observations can be carried into any other set with the same signed ranks by such a function g (Problem 98). In short, the signed-rank tests are the only tests which, for these hypotheses, are invariant under all strictly increasing, odd transformations g.

The signed-rank tests are also invariant under this class of transformations for some other null and alternative hypotheses we have been considering in this chapter. The argument above applies to any hypotheses for which strictly increasing, odd transformations carry null distributions into null distributions and alternatives into alternatives. This restriction is satisfied for the null hypothesis that the X_j are independent with possibly different distributions but all symmetric about 0, and for null and alternative hypotheses of the form $P(X_j < 0) > P(X_j > 0)$ or the reverse (as in Sect. 6.1, Chap. 2), etc. However, it does not hold for alternatives under which the X_j are symmetrically distributed about a value other than 0 (Problem 101).

Similarly, the confidence procedures for the center of symmetry μ which correspond to signed-rank tests are not justifiable by this invariance argument, because they are not invariant under all strictly increasing, odd transformations. They are not themselves signed-rank procedures, that is, they are not functions of the signed ranks of the original observations. The relevant transformations are different for different μ.

The argument for invariance under the class of transformations in (8.3) is far less compelling than the argument for permutation invariance. On general grounds, when the class of transformations is too large, as it is here, it may not be possible to average over it as was done in (8.2). When this occurs, some test may have optimum properties which no invariant test has. It can even happen that there is a uniformly most powerful invariant test which is seriously inadmissible, that is, there exists a noninvariant test which is uniformly as good and considerably better in a large region [Lehmann, 1959, p. 231]. Thus it would seem that requiring invariance can lead (though here it does not) to the use of a highly inferior procedure when the class of transformations is too large. This does not fundamentally invalidate the "principle of invariance" however, because one's attitude is never literally invariant in such a situation. The point is that if one is going to use an invariant procedure when one has an only approximately invariant attitude, then one should make sure that there is no non-invariant procedure available which is significantly better.

This brings us to the real reason why the argument for invariance is not compelling in the present situation. One might not want to treat $X_1, \ldots,$ X_n and $g(X_1), \ldots, g(X_n)$ alike in all instances. For an extreme example, there is a strictly increasing, odd function g which carries the observations

$$-1.2, \ -1.1, \ -1.0 \ -0.9, \ 7.6, \ 16.7, \ 24.1, \ 42.9, \ 51.0, \ 83.4$$

into

$$-12, \ -11, \ -10, \ -9, \ 13, \ 14, \ 15, \ 16, \ 17, \ 18.$$

However, one does not feel compelled to regard the two samples as providing the same evidence concerning the null hypothesis of symmetry about 0; the first might well be considered much less compatible with this null hypothesis than the second. (For this it is not necessary to have measurements on an "interval scale." While a difference of one unit might have varying or indefinite meaning throughout the measurement scale, a difference of 20 might always be bigger than a difference of 1.)

If such a discrepancy is at all likely, one might prefer not to use a signed-rank test. However, ordinarily one is quite content to treat alike practically all samples which have the same signed ranks, with exceptions having small probability of occurring. Then presumably little or nothing is lost by using a signed-rank test, and there are indeed some very good tests of this type. By the "principle of invariance," we can thus justify restricting consideration to the class of signed-rank tests. Of course, the choice of which signed-rank test remains.

9 Locally Most Powerful Signed-Rank Tests

This section is concerned with most powerful signed-rank tests. Surprisingly, problems which admit uniformly most powerful tests do not generally admit uniformly most powerful signed-rank tests. We can, however, find that signed-rank test which is most powerful against any alternative distribution, and in particular against alternatives of various kinds which are "close to" the null distribution. We assume that the observations are independently, identically and continuously distributed so that ties have probability zero. Also, by sufficiency, we can ignore the order of the observations and restrict consideration to permutation-invariant tests (Sect. 8.1). If we did not, they would result anyway. This and some other points which arise here will be discussed more fully in Chap. 5.

For a sample of size n, there are 2^n possible assignments of signs to the ranks $1, \ldots, n$, and accordingly 2^n sets of signed ranks $r_1, \ldots, r_n$, where $r_j = \pm j$. We assume as usual that all 2^n possible sets of signed ranks are equally likely under the null hypothesis. By the Neyman-Pearson Lemma (Theorem 7.1 of Chap. 1), it follows (Problem 101) that among signed-rank tests at level α, the most powerful test against the alternative F rejects if the probability under F of the observed set of signed ranks is greater than a constant k, and "accepts" if it is less than k, where k and the probability of rejection at k are chosen to make the level exactly α. Letting $P_F(r_1, \ldots, r_n)$ be the probability under F of signed ranks $r_1, \ldots, r_n$, we may express this test as

$$\text{reject if } P_F(r_1, \ldots, r_n) > k$$
$$\text{"accept" if } P_F(r_1, \ldots, r_n) < k. \tag{9.1}$$

The most powerful signed-rank test against the alternative F depends, of course, on F. Even if we restrict consideration to normal alternatives with positive mean μ, the most powerful signed-rank test depends on μ. If we consider only *small* positive μ, however, we will find that there is a certain signed-rank test which, among signed-rank tests, is uniformly most powerful against normal alternatives with sufficiently small, positive μ. Such a test is called *locally most powerful* against normal alternatives with $\mu > 0$.

More generally, consider a one-parameter family of distributions F_θ with densities f_θ, and suppose that $\theta = 0$ satisfies the null hypothesis of symmetry about 0, that is, $f_0(-x) = f_0(x)$. Let s_j denote the sign of r_j. Then we can write the probability under F_θ of signed ranks $r_1, \ldots, r_n$ (Problem 102) as

$$P_\theta(r_1, \ldots, r_n) = n! \int \cdots \int_{0 < y_1 < \cdots < y_n < \infty} \prod_1^n f_\theta(s_i y_i) \, dy_1 \cdots dy_n. \tag{9.2}$$

Assume it is legitimate to differentiate (9.2) under the integral sign, and let

$$h(x) = \frac{\partial}{\partial \theta} [\ln f_\theta(x)] \Big|_{\theta = 0} = \frac{1}{f_0(x)} \frac{\partial}{\partial \theta} [f_\theta(x)] \Big|_{\theta = 0}. \tag{9.3}$$

Then the derivative of (9.2) at $\theta = 0$ is

$$
n! \int \cdots \int_{0 < y_1 < \cdots < y_n < \infty} \sum_1^n h(s_j y_j) \prod_1^n f_0(s_i y_i)\, dy_1 \cdots dy_n = \sum_1^n E_0[h(s_j |X|_{(j)})]/2^n
$$

(9.4)

where $|X|_{(1)} < \cdots < |X|_{(n)}$ in (9.4) are the absolute values of a sample of n from F_0, arranged in order of size. Expanding $P_\theta(r_1, \ldots, r_n)$ in a Taylor's series about $\theta = 0$ and using (9.4) for its derivative, we have

$$
P_\theta(r_1, \ldots, r_n) = 2^{-n}\left\{ 1 + \theta \sum_1^n E_0[h(s_j |X|_{(j)})] + \text{smaller order terms} \right\}.
$$

(9.5)

Substituting this in (9.1), it follows that, for sufficiently small $\theta > 0$, the most powerful signed-rank test has the following form:

$$
\text{reject if } \sum_1^n E_0[h(s_j |X|_{(j)})] > k
$$
$$
\text{"accept" if } \sum_1^n E_0[h(s_j |X_{(j)}|)] < k.
$$

(9.6)

This is equivalent to a test based on a sum of signed constants c_j (Problem 104) where

$$
c_j = E_0[h(|X|_{(j)}) - h(-|X|_{(j)})].
$$

(9.7)

It follows that a test of the form

$$
\text{reject if } \sum_1^n s_j c_j \geq k
$$
$$
\text{"accept" otherwise}
$$

(9.8)

is locally most powerful against F_θ, $\theta > 0$, among signed-rank tests at the same level α. If the level α desired is not attainable with a test of the form (9.8), then randomization may be necessary at k, and if also more than one set of signed ranks gives the critical value of $\sum_1^n s_j c_j$, then higher order terms in (9.5) will be required to determine the locally most powerful test (but not to maximize the derivative of the power at $\theta = 0$).

Similar statements hold for $\theta < 0$, with rejection when $\sum_1^n s_j c_j$ is too small.

For normal alternatives with mean $\mu = \theta$ and fixed variance σ^2, from (9.3) we have (Problem 105) $h(x) = x/\sigma^2$ and thus

$$
c_j = 2E(|X|_{(j)})/\sigma^2
$$

(9.9)

where $|X|_{(1)} < \cdots < |X|_{(n)}$ are the ordered absolute values of a sample of n from the normal distribution with mean 0 and variance σ^2. A test of the form (9.8) based on these c_j is equivalent to one based on

$$
c_j^* = E(|Z|_{(j)})
$$

(9.10)

where $|Z|_{(1)} < \cdots < |Z|_{(n)}$ are the ordered absolute values of a sample of n from the standard normal distribution. Accordingly, this test is the locally most powerful signed-rank test against normal alternatives with positive mean. The corresponding lower-tailed test is similarly the locally most powerful signed-rank test against normal alternatives with negative mean.

The test with the c_j in (9.9) (or, equivalently (9.10)), is frequently referred to as the Fraser (normal scores) test since it was derived by Fraser [1957a]. Additional properties, as well as the values of the scores in (9.10) and the critical values, are given in Klotz [1963]. This test is asymptotically equivalent to a test with "inverse normal scores" as constants, that is, $c_j = \Phi^{-1}(j/(n + 1))$, where $\Phi(x)$ is the standard normal c.d.f. The values of these scores are more readily accessible than those in (9.10); e.g., Fisher and Yates [1963] and van der Waerden and Nievergelt [1956]. This test is mentioned in van Eeden [1963].

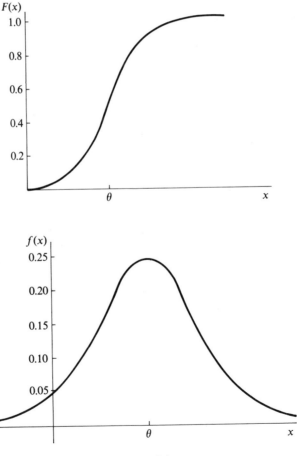

Figure 9.1

Consider next the logistic distribution with

$$F_\theta(x) = \frac{1}{1 + e^{-(x-\theta)}}, \quad f_\theta(x) = \frac{e^{-(x-\theta)}}{[1 + e^{-(x-\theta)}]^2}. \qquad (9.11)$$

This distribution is shown in Fig. 9.1; it is very close to a normal distribution. For the logistic distribution, we have (Problem 105) $h(x) = 2F_0(x) - 1$ and

$$c_j = 2j/(n+1). \qquad (9.12)$$

The tests of the form (9.8) based on the c_j in (9.12) are exactly the upper-tailed (nonrandomized) Wilcoxon signed-rank tests, which are therefore locally most powerful among signed-rank tests against the logistic alternative distributions in (9.11) where $\theta > 0$. The lower-tailed test has a similar property for $\theta < 0$. An arbitrary scale factor σ would not alter this property.

The argument leading to (9.8) shows that any locally most powerful signed-rank test is a one-tailed test based on a sum of signed constants, subject to the qualification following (9.8). It can also be shown that any test of the form (9.8) is locally most powerful against some alternative (Problem 96, Chap. 5). A more difficult problem is to determine which c_j provide locally most powerful tests for some restricted alternatives like $F_\theta(x) = F_0(x - \theta)$ or $F_\theta(x) < F_0(x)$ for $\theta > 0$. This problem will not be discussed here, but see Problem 103, Chap. 5.

PROBLEMS

1. Show that a distribution with discrete frequency function f is symmetric as defined by (2.1) or (2.2) if and only if f satisfies (2.3).

2. (a) Show that the condition in (2.1) is equivalent to the same condition with strict inequalities, that is, $P(X < \mu - x) = P(X > \mu + x)$ for all x.
   (b) Show that (2.1) holds for all x if and only if it holds for all $x > 0$.

3. Suppose that X is symmetrically distributed about μ. Show that
   (a) μ is a median of X.
   (b) μ is the mean of X (provided it exists).
   (c) μ is the mode of X if X is unimodal.

4. Show that each of the symmetry conditions given as (a), (b), and (c) in Sect. 2 is equivalent to the condition that X is symmetrically distributed about μ.

*5. Show that a distribution is symmetric about 0 if and only if its characteristic function is real everywhere. (Hint: Each is equivalent to the statement that X and $-X$ have the same distribution.)

6. Let V and W have the joint density

$$f(v, w) = \frac{1}{2} \quad \text{for} \begin{cases} -1 \le w \le v \le 1, v - w \le 1; \\ 0 \le v + 1 \le w \le 1. \end{cases}$$

   (a) Show that the marginal distributions of V and W are both the uniform density over $(-1, 1)$, and hence the medians of V and W are each equal to 0.

(b) Show that $P(W < V) = \frac{3}{4}$, which implies that the median of the population of differences $X = W - V$ must be negative and hence cannot equal the difference between the medians of W and V.

(c) Show that the density of the difference $X = W - V$ is

$$f(x) = \begin{cases} (2 + x)/2 & \text{for } -1 < x \le 0 \\ (2 - x)/2 & \text{for } 1 < x \le 2. \end{cases}$$

This distribution is not symmetric and does not have median 0. It has a unique median $-2 + \sqrt{3}$.

7. Suppose that V and W are independent and $X = W - V$. Show that the following properties hold. Note that (a)–(c) give conditions under which the difference X is symmetrically distributed about a median which equals the difference of the medians of W and V, while (d) shows that, even if X is symmetrically distributed, its center of symmetry may not be equal to the difference of the medians of W and V.

   (a) If V and W are symmetrically distributed about μ and λ respectively, then X is symmetrically distributed about $\lambda - \mu$.

   (b) If V and W are identically distributed, then X is symmetrically distributed about 0.

   (c) If W has the same distribution as $V + \theta$ for some "shift" θ, then X is symmetrically distributed about θ.

   *(d) If V and W have any two distributions of the following family, then X is symmetrically distributed about 0, even though the medians of V and W may differ. The family of distributions is discrete with frequency functions f_θ given by

$$f_\theta(1) = 2\theta, \qquad f_\theta(2) = \tfrac{2}{3} - 3\theta, \qquad f_\theta(3) = \tfrac{1}{3}, \qquad f_\theta(4) = \theta,$$

   all for $0 \le \theta \le \frac{2}{9}$. The median is uniquely 2 for $\theta < \frac{1}{6}$, uniquely 3 for $\theta > \frac{1}{6}$. What is the median for $\theta = \frac{1}{6}$?

8. Define the difference $X = W - V$ where W and V need not be independent. The example in Problem 6 shows that in general, the median of X need not be equal to the difference of the medians of W and V even if the marginal distributions of V and W are identical and symmetric. Show that in the following situations, the median of the difference is equal to the difference of the individual medians, by showing first the results stated.

   (a) If (V, W) has the same distribution as either (W, V) or $(-V, -W)$, then X is symmetrically distributed about 0 (and the medians of V and W are equal).

   (b) If the distributions of V, W, and X are each symmetric (with finite means), then the center of symmetry of X is equal to the difference of the centers of symmetry of W and V.

9. For a set of n independent observations from a population which is continuous and symmetric about zero, show that the signs of the signed ranks are mutually independent and each is equally likely to be positive or negative. Show that the signed ranks themselves are dependent if the original order of the observations is retained.

10. Show that T^+ and T^- have identical null distributions.

11. (a) If $u_n(t)$ denotes the number of subsets of the first n integers whose sum is equal to t, show that

$$u_n(t) = u_{n-1}(t - n) + u_{n-1}(t)$$

for all $t = 0, 1, \ldots, n(n + 1)/2$ and all positive n, with the following initial and boundary conditions:

$$u_n(t) = 0 \quad \text{for all } t < 0$$
$$u_0(0) = 1$$
$$u_0(t) = 0 \quad \text{for all } t \neq 0$$
$$u_n(t) = 0 \quad \text{for } t > n(n + 1)/2.$$

This provides a simple recursive method for generating the frequencies of values of T^+ and hence the null probability function $p_n(t) = P(T^+ = t)$ for samples of size n using

$$2p_n(t) = p_{n-1}(t - n) + p_{n-1}(t).$$

(b) What change is required in order to generate directly the null cumulative probabilities $F_n(t) = P(T^+ \leq t)$?

12. Use the recursive method developed in Problem 11 to generate the complete null distribution of T^+ for all $n \leq 4$. Check your results against Table D.

*13. Define $u_n(t)$ as in Problem 11 and let $u(t)$ be the number of subsets of all positive integers whose sum is equal to t. Show that
   (a) $u_n(t) = u(t)$ for $t \leq n$.
   (b) The number of subsets of all positive integers with sum t and maximum m is $u_{m-1}(t - m)$.
   (c) $u_n(t) = u(t) - \sum_{i=0}^{\infty} u_{n+i} (t - n - 1 - i)$, where the sum actually terminates because the summand vanishes for $i > t - n - 1$. (Hint: The terms in the sum count the subsets with sum t and maximum $n + 1 + i$.)
   (d) $u_n(t) = u(t) - u^{(1)}(t - n - 1)$ for $t \leq 2n + 1$, where $u^{(1)}(s) = \sum_{i=0}^{\infty} u(s - i)$.
   (e) $u_n(t) = u(t) - u^{(1)}(t - n - 1) + \sum_{i=0}^{\infty} \sum_{j=0}^{\infty} u_{n+i+j}(t - 2n - 2 - 2i - j)$.
   (f) $u_n(t) = u(t) - u^{(1)}(t - n - 1) + u^{(2)}(t - 2n - 2)$ for $t \leq 3n + 2$, where $u^{(2)}(s) = \sum_{i=0}^{\infty} u^{(1)}(s - 2i)$.
   (g) $u_n(t) = \sum_{k=0}^{\infty} (-1)^k u^{(k)}(t - kn - k)$ where $u^{(0)}(s) = u(s)$, and

$$u^{(k)}(s) = \sum_{i=0}^{\infty} u^{(k-1)}(s - ki).$$

Both sums terminate because $u^{(k)}(s) = 0$ for $s < 0$. (For a complete, formal proof, it may be convenient to introduce $u_n^{(k)}(t) = \sum_{i=0}^{\infty} u_{n+i}^{(k-1)}(t - ki)$ and prove by induction on k that $u_n^{(k)}(t) = u^{(k)}(t) - u_n^{(k+1)}(t - n - 1)$.)
   (h) All of the foregoing equalities hold for $U_n(t) = \sum_{i=0}^{t} u_n(i) =$ the number of subsets of $\{1, 2, \ldots, n\}$ with sum at most t, if $u^{(k)}$ is replaced by $U^{(k)}$ where $U^{(k)}(t) = \sum_{i=0}^{\infty} U^{(k-1)}(t - ki)$ and $U^{(0)}(t) = U(t) = \sum_{i=0}^{t} u(i) = u^{(1)}(t)$. How is (b) interpreted in this case?
   (i) The null probability function and c.d.f. of T^+ satisfy

$$P_n(T^+ = t) = 2^{-n} \sum_{k=0}^{\infty} (-1)^k u^{(k)}(t - kn - k),$$

$$P_n(T^+ \leq t) = 2^{-n} \sum_{k=0}^{\infty} (-1)^k U^{(k)}(t - kn - k).$$

Note: Instead of tabulating the null distribution directly, one could tabulate the functions $u^{(k)}$ (for point probabilities) or $U^{(k)}$ (for tail probabilities). The total number of lower tail probabilities less than 0.5 for sample sizes $1, \dots, n$ is $[[n(n + 1)(n + 2)/12]] - [[(n + 2)/4]]$ where $[[x]]$ denotes the largest integer not exceeding x. The number of function-values $U^{(k)}(t)$ required to cover the same range is $[[(n + 4)/4]]\{[[n(n + 1)/4]] - [[n/4]](n + 1)/2\}$. For large n, this is $\frac{3}{8}$ as large a number, and the values of t covered are covered for all sample sizes. The tabulation required could be reduced still further by omitting $U^{(k)}$ for alternate values of k (odd or even) and using $U^{(k-1)}(s) = U^{(k)}(s) - U^{(k)}(s - k)$. Tables of the functions $u^{(k)}$ and $U^{(k)}$ are easily generated for successive k from their definitions and a table of u. The function u is a well-studied partition function and is tabled in National Bureau of Standards [1964, Table 24.5] where further references may be found. It can be generated recursively (without need for $u_n(t)$) from the nontrivial relation

$$u(t) = \sum_{k=1}^{\infty} (-1)^{k-1} \left[u\left(t - \frac{3k^2 - k}{2}\right) + u\left(t - \frac{3k^2 + k}{2}\right) \right] + s(t)$$

where $u(t) = 0$ for $t < 0$, $u(0) = 1$, and $s(t) = (-1)^r$ if $t = 3r^2 \pm r$ for some integer r, $s(t) = 0$ otherwise. Alternatively, for tail probabilities, $U(t)$ can be generated recursively from

$$U(t) = \sum_{k=1}^{\infty} (-1)^{k-1} \left[U\left(t - \frac{3k^2 - k}{2}\right) + U\left(t - \frac{3k^2 + k}{2}\right) \right] + S(t)$$

where $U(t) = 0$ for $t < 0$, $U(0) = 1$, and $S(t) = (-1)^r$ if $3r^2 + r \leq n < 3(r + 1)^2 - r - 1$ for some integer r, $S(t) = 0$ otherwise. This relation follows from the previous one.

14. Derive the mean and variance of T as given in (3.7) and (3.8) by using the fact that under the relevant null hypothesis, T is a sum of the first n integers with factors $+1$ and -1 attached at random.

15. Show that T^+, T^-, and T have symmetric null distributions.

16. Show that for samples with n nonzero observations and no ties, the null probability distribution of T^+ can be written as

$$P(T^+ = t) = \sum_{k=0}^{n} \binom{n}{k} 2^{-n} P(T^+ = t \mid S = k)$$

where S denotes the number of positive observations. This representation might be useful for systematic generation of the null distribution of the signed-rank statistic. Further, $P(T^+ = t \mid S = k)$ is a null probability for the Wilcoxon two-sample statistic (covered in Chap. 5) where the positive observations are interpreted as from one sample and the negative from another. Problem 17 gives further insight into the relationship between the one-sample and two-sample statistics.

17. Let $D_1, \dots, D_N$ be a sample of N nonzero observations and define X_i or Y_i for each i by

$$D_i = \begin{cases} X_i & \text{if } D_i > 0 \\ -Y_i & \text{if } D_i < 0. \end{cases}$$

Assume there are m X values and n Y values, where $m + n = N$.

(a) Show that the signed-rank test statistic T^+ calculated for these D_i is equal to the sum of the ranks of the X observations in the combined ordered sample of m X's and n Y's.

(b) Show that $T^+ - T^-$ is the sum of the ranks of the X's minus the sum of ranks of the Y's in the combined ordered sample. The sum of the ranks of the X's is the test criterion for the Wilcoxon two-sample statistic to be discussed in Sect. 4, Chap. 5. Show that T^+ might be used to test the null hypothesis that the distributions of X and Y are identical, and relate this to a test based on T^+ for the null hypothesis that the center of symmetry of the D population equals 0.

18. (a) Prove Theorem 3.1, concerning the relation between Walsh averages and signed ranks.

(b) Show that, if there are no zeros or ties, $T = T^+ - T^-$ can be written

$$\sum\sum_{1 \leq i \leq j \leq n} \text{sign}(X_i + X_j)$$

where

$$\text{sign}(X) = \begin{cases} 1 & \text{if } X > 0 \\ -1 & \text{if } X < 0. \end{cases}$$

19. (a) Show that the possible values of $T = T^+ - T^-$ are alternate integers between $-n(n + 1)/2$ and $n(n + 1)/2$ inclusive.

(b) For what values of n are the possible values of T even?

20. (a) Show that the continuity correction to the approximate normal deviate for the Wilcoxon test is $[6/n(n + 1)(2n + 1)]^{1/2}$.

(b) Show that this correction is less than 0.02 if (and only if) $n \geq 20$, less than 0.01 if (and only if) $n \geq 31$.

(c) Show that the corresponding values for the sign test (of an hypothesized median) are $1/n^{1/2}$, 2501, and 10001.

*21. Show that T^+ and T^- are asymptotically normal by using the fact that T is asymptotically normal.

*22. Show that $\sum_1^n R_j$, as defined in Sect. 3.2, satisfies the Liapounov criterion.

23. Show that the standardized statistics $[T^+ - E(T^+)]/\sqrt{\text{var}(T^+)}$ and $[T - E(T)]/\sqrt{\text{var}(T)}$ are identical in value as long as the means and variances are calculated under the same distribution.

24. Verify the moments of the T_{ij}, which are given in the proof of Theorem 3.2.

25. Under the null hypothesis of symmetry about 0,

(a) Show that the probabilities defined in Theorem 3.2 have the values

$$p_1 = \tfrac{1}{2}, \qquad p_2 = \tfrac{1}{2}, \qquad p_3 = \tfrac{3}{8}, \qquad p_4 = \tfrac{1}{3}.$$

(b) Verify that the expressions given in (3.12) and (3.15) for the mean and variance of T^+ reduce correctly to (3.5) and (3.6).

26. Use the method of Problem 1 of Chap. 1 to show that $2T^+/n^2$ is a consistent estimator of p_2.

27. (a) Consider the sign test of Chap. 2 for an hypothesized median based on an independent random sample. Against what alternatives is it consistent?

    (b) Give an example of an alternative against which the sign test is consistent but the Wilcoxon test is not, and vice versa.

28. Show that the Wilcoxon test based on an independent random sample from a symmetric distribution is consistent against shift alternatives.

*29. Suppose that $|E(Z_n)| < B$ and $\text{var}(Z_n) \leq B$ for all n and all null distributions. Use Chebyshev's inequality to show that an equal-tailed test based on Z_n is consistent against any alternative under which $|E(Z_n)| \to \infty$ and $\text{var}(Z_n)$ is bounded.

30. Show that if $X_1, \ldots, X_n$ are independent with distributions which are continuous and symmetric about μ_0, then a Wilcoxon test for center of symmetry μ_0 has the same level irrespective of whether the distributions are identical or not.

31. If the conditional distribution of X_1 given $X_2, \ldots, X_n$ is symmetric about 0, show that the conditional probability that $X_1 > 0$ equals the conditional probability that $X_1 < 0$ given the signs of $X_2, \ldots, X_n$.

32. Consider a matched-pairs experiment with the null hypothesis that the treatment actually has no effect. Show that randomization validates the null distribution of the Wilcoxon test statistic defined on treatment-control differences.

33. (a) Let X have continuous c.d.f. F and let $G(x) = \frac{1}{2} + \frac{1}{2}[F(x) - F(-x)]$. Show that G is the c.d.f. of a symmetric distribution and is stochastically larger than F if and only if $P(X < -x) \geq P(X > x)$ for all $x \geq 0$.

    (b) Generalize (a) to allow discrete distributions.

*34. (a) If X has c.d.f. F and $F(x) \geq G(x)$ for all x, show that there exists a random variable Y with c.d.f. G such that $P(Y \geq X) = 1$.

    (b) If $X_1, \ldots, X_n$ are independent, X_j has c.d.f. F_j and $F_j(x) \geq G_j(x)$ for all x and j, show that there exist independent random variables Y_j such that Y_j has c.d.f. G_j and $P(Y_j \geq X_j) = 1$ for all j.

35. Use Theorem 3.3 to show that a suitable one-tailed Wilcoxon test rejects with probability at most α under (3.21) and at least α under (3.22).

*36. Let $X_1, \ldots, X_n$ be independent observations on a continuous distribution which is symmetric about μ.

    (a) Show that for any $n \geq 3$, we have

    $$P(X_{(1)} > \mu) = 2^{-n}$$

    $$P[(X_{(1)} + X_{(2)})/2 > \mu] = 2(2^{-n})$$

    $$P[X_{(2)} > \mu \text{ and } (X_{(1)} + X_{(3)})/2 > \mu] = 3(2^{-n}).$$

    These results give lower confidence bounds at the respective confidence levels $1 - 2^{-n}, 1 - 2^{-(n-1)}, 1 - 3(2^{-n})$ for any sample of size $n \geq 3$.

    (b) Show that these confidence bounds correspond to the Wilcoxon test with critical values 0, 1, 2 respectively.

*37. Consider the Walsh averages $W_{ij} = (X_{(i)} + X_{(j)})/2$, for $i \leq j$, defined in terms of the order statistics $X_{(1)}, \ldots, X_{(n)}$ of a set of n observations.

(a) Note that always $W_{11} \le W_{12} \le$ all other W_{ij}. What other inequalities always hold between W_{ij} with $i \le j \le 4$?

(b) Recall that the three smallest W_{ij} are W_{11}, W_{12}, and $\min\{W_{22}, W_{13}\}$. Which W_{ij} can be fourth smallest for some data sets? (There are three possibilities.)

(c) Show that the minimum possible rank of W_{ij} among the Walsh averages is $\frac{1}{2}i(2j - i + 1)$.

(d) Show that the maximum possible rank of W_{1j} among the Walsh averages is $\frac{1}{2}j(j - 1) + 1$. What is the maximum possible rank of W_{ij} for $i \ge 2$?

(e) Which W_{ij} can be fifth smallest for some data sets (four possibilities)? Sixth smallest (six possibilities)?

(f) Show that the fourth smallest W_{ij} is $\min\{W_{14}, \max(W_{22}, W_{13})\} = \max\{W_{13}, \min(W_{22}, W_{14})\}$.

(g) Show that the fifth smallest W_{ij} is $\min\{W_{15}, W_{23}, \max(W_{14}, W_{22})\} = \max\{\min(W_{14}, W_{23}), \min(W_{15}, W_{22})\}$.

(h) Show that the fourth and fifth smallest W_{ij} are confidence bounds corresponding to one-sided Wilcoxon tests at level $\alpha = 5/2^n$ and $7/2^n$ for $n \ge 5$.

38. Show that the modified Wilcoxon test statistic T_0^- of Sect. 5 has the same null distribution in a sample of size n as T^- in a sample of size $n - 1$.

39. For a sample with neither zeros nor ties, show that
   (a) $T^+ = T_0^+ + S$ where S is the number of positive observations. (This result relates the Wilcoxon statistic, the modified Wilcoxon statistic and the sign test statistic.)
   (b) T_0^+ is the number of positive Walsh averages $(X_i + X_j)/2$ with $i < j$.

40. Verify the expressions given in (5.1) and (5.2) for the mean and variance of T_0^+.

41. Show that $2T_0^+/n(n - 1)$ is the minimum variance unbiased estimator of p_2 for the family of all continuous distributions.

42. With the definitions of Sect. 3.3, show that $p_4 \le p_2$ for all distributions.

*43. (a) Show that the suggestions following (5.3) lead to inequalities of the form $(\hat{p}_2 - p_2)^2 \le C + 2Bp_2 - Ap_2^2$ where $\hat{p}_2 = 2T_0^+/n(n - 1)$ and A, B, C are nonnegative constants.
   (b) Show that the corresponding confidence region is an interval with endpoints
   $$\{\hat{p}_2 + B \pm [(\hat{p}_2 + B)^2 - (\hat{p}_2 - C)(1 + A)]^{1/2}\}/\{1 + A\},$$
   except that it is empty if the quantity in brackets is negative (which is impossible if p_4 is replaced by the upper bound p_2 and extremely unlikely if it is estimated as described following (5.3)).

*44. Assuming that the asymptotic distribution of $[T_0^+ - E(T_0^+)]/\sqrt{\operatorname{var}(T_0^+)}$ is standard normal, show that $P\{|T_0^+ - n(n - 1)p_2/2| \le z\sqrt{V}\} \to 1 - 2\alpha$ if z is the upper α point of the standard normal distribution and V is an estimator of $\operatorname{var}(T_0^+)$ satisfying $V/n^3 \to p_4 - p_2^2$ in probability. This provides an asymptotic confidence interval for p_2.

45. For the data in Sect. 2, verify the P-values, confidence bounds and confidence levels given in Sect. 5 for the Wilcoxon and modified Wilcoxon procedures.

46. Show that $T_0^- = 0$ if and only if $T^- \le 1$. Thus the modified Wilcoxon test with critical value 0 is equivalent to the ordinary Wilcoxon test with critical value 1. (Otherwise the tests are not equivalent.)

*47. Modifying Problem 37, consider only those Walsh averages W_{ij} with $i < j$.
   (a) Show that the smallest three W_{ij} with $i < j$ are W_{12}, W_{13} and $\min\{W_{23}, W_{14}\}$.
   (b) Show that the minimum and maximum possible ranks of the W_{ij} among those with $i < j$ are the same as those of $W_{i,j-1}$ in Problem 37.
   (c) Show that the formulas for the ordered W_{ij} in Problem 37 apply here if the second subscript is increased by 1 throughout. For instance, (a) is so related to 37(b). Similarly 37(d) gives that the fourth smallest W_{ij} among those with $i < j$ is $\min\{W_{15}, \max(W_{23}, W_{14})\} = \max\{W_{14}, \min(W_{23}, W_{25})\}$.
   (d) Show that the first five ordered W_{ij} with $i < j$ are confidence bounds corresponding to one-sided modified Wilcoxon tests at levels $2/2^n$, $4/2^n$, $6/2^n$, $10/2^n$, and $14/2^n$ respectively for $n \geq 6$. In particular, $T_0^- \leq 0$, 1, or 2 respectively if and only if $0 < W_{12}$, W_{13}, or $\min\{W_{23}, W_{14}\}$.

48. For the data in Sect. 2 use procedures corresponding to the tests based on T^+ and T_0^+ to find upper and lower confidence bounds for μ, each at level approximately 0.025, by applying the methods of interpolation between attainable levels (explained in Sect. 5, Chap. 2).

*49. In order to investigate the effect of interpolating halfway between two adjacent order statistics of a random sample to find a confidence bound for the population median μ, note that the true level of the interpolated confidence bound is

$$P[(X_{(i)} + X_{(i+1)})/2 > \mu] = P(X_i > \mu) + pP(X_{(i)} \leq \mu < X_{(i+1)})$$
$$= (1 - p)P(X_i > \mu) + pP(X_{(i+1)} > \mu)$$

where

$$p = P(X_{(i+1)} - \mu > \mu - X_{(i)}|X_{(i)} \leq \mu < X_{(i+1)}).$$

Linear interpolation approximates p by $\frac{1}{2}$. Show that, for a continuous, symmetric population,

$$p = P(R_1 = -1|S^- = i) = i/n$$

where R_1 is the first signed rank and S^- is the number negative among the values $X_i - \mu$. Thus, linear interpolation overestimates the error probability (for one-tail probabilities below 0.5).

*50. (a) In order to investigate the effect of interpolating between the two smallest (or largest) Wilcoxon confidence limits, show that for n observations on a density f,

$$P(r) = P[(1 - r)X_{(1)} + r(X_{(1)} + X_{(2)})/2 > 0]$$

$$= n(n - 1) \iint f(x)f(y)[1 - F(y)]^{n-2} \, dy$$

where the region of integration is $(1 - r/2)x + ry/2 > 0$, $x < y$. Show that for $0 \leq r \leq 1$, this reduces to

$$n(n - 1) \int_0^\infty f(y)\left[F(y) - F\left(\frac{-ry}{2 - r}\right)\right][1 - F(y)]^{n-2} \, dy$$

$$= \begin{cases} [1 - F(0)]^n & \text{for } r = 0 \\ 2/2^n & \text{for } r = 1 \text{ and } f \text{ symmetric about } 0. \end{cases}$$

The accuracy of linear interpolation depends on how linear the integral is, and hence on the behavior of $F[-ry/(2 - r)]$, as a function of r, $0 \le r \le 1$.

(b) Show that, for the uniform distribution, $F[-ry/(2 - r)]$ is a concave function of r in the relevant range, and hence $P(r)$ is convex.

(c) Show that, for the standard normal distribution, $F[-ry/(2 - r)]$ is a concave function of r for $0 \le r \le 2 - y^2$ and hence for $0 \le r \le 1$ and $0 \le y \le 1$. (Values of $y > 1$ contribute relatively little to the integral above, since both the first and last terms in the integrand decrease rapidly as y increases above 1.) These results suggest that the tail probability $P(r)$ tends to be convex in r and hence to be overestimated by linear interpolation.

51. Show that, if L is the $(k + 1)$th smallest Walsh average of a sample from a distribution which is symmetric about μ, then $P(L \le \mu) \ge 1 - \alpha \ge P(L < \mu)$ where $1 - \alpha$ is the exact confidence level in the continuous case. (Hint: What confidence region corresponds to the randomization method of breaking ties?)

52. (a) Show that the null mean and variance of T^+, based on n nonzero observations with ties and calculated using the average rank procedure, conditional on the ties observed, are

$$E(T^+) = n(n + 1)/4$$
$$\text{var}(T^+) = [n(n + 1)(2n + 1) - \sum t(t^2 - 1)/2]/24.$$

where t is the number of observations tied for any given rank, and the sum is over all distinct sets of tied ranks for any $t > 1$. ($\sum t(t^2 - 1)$ could be written as $\sum_i(t_i^2 - 1)$ where i ranges over all observations and t_i is the number of observations tied with observation i, including itself.) The same result holds if zeros are present if they are omitted and n is reduced accordingly.

(b) If zeros are included for the ranking but given signed-rank zero, show that the null mean and variance are

$$E(T^+) = n(n + 1)/4 - v(v + 1)/4$$
$$\text{var}(T^+) = [n(n + 1)(2n + 1) - v(v + 1)(2v + 1) - \sum t(t^2 - 1)/2]/24$$

where v is the number of zeros and the sum is over all sets of nonzero ties.

(c) Show that the null distribution of T^+ is symmetric.

(d) Show that $T^+ + T^- = n(n + 1)/2 - v(v + 1)/2$.

53. List all possible ways of breaking ties for the data in (6.2) and verify Table 6.3.

54. Show that, under the null hypothesis of symmetry, the random method of breaking ties preserves the usual null distribution of the Wilcoxon statistic (even conditionally on the absolute values observed).

55. Show that 0 is an endpoint of the usual Wilcoxon confidence interval if and only if the ordinary Wilcoxon test would reject by one method of breaking the ties but not by another.

56. Verify that the $(k + 1)$th smallest Walsh average is 0 for $7 \le k \le 11$ for the data in (6.2).

57. Verify the results given after Equation (6.5) for the left tail of the null distribution of T^- for $n = 8$ by the average rank procedure, given a tie at ranks 1, 2, 3, and 4.

58. For the data 1, 1, 1, 2, 2, 2, 2, -3, show that
    (a) By the average rank procedure, the exact P-value is $25/2^8$ and the next P-value (the probability of a strictly smaller value of T^-) is $24/2^8$
    (b) However the ties are broken, the exact P-value is $25/2^8$ and the next P-value is $19/2^8$.
    (c) For one-tailed $\alpha = 24/2^8$, the sample is not significant by the average rank procedure, but after any tiebreaking it would be significant with probability $\frac{5}{6}$ by an exact randomized Wilcoxon test. (Thus this sample exemplifies the reverse of (6.3).)
    *(d) What are the corresponding confidence regions?

*59. Show that, if there are no nonzero ties, the signed-rank zero procedure satisfies the conditions (i)–(iii) of Sect. 6.5. (Hint: Show that $T_0^- \leq T^- \leq T_0^- + v(v + 1)/2$ where T_0^- is obtained by the signed-rank zero procedure and T^- by breaking ties randomly and v is the number of zeros. Use this to relate the critical value of T_0^- to the usual critical values.)

60. Show that the average rank procedure in conjunction with the signed-rank zero procedure satisfies the conditions (i)(b) and (ii) of Sect. 6.5. (Hint: Show that if all observations are increased equally, the signed-rank sum of the actual sample increases at least as much as that of the sample obtained by any reassignment of signs.)

61. For the data 0, -1, -2, -3, -4, 5, 6, 7, 8, 9, 10, 11, show that
    (a) By the reduced sample procedure, the null probability of a negative-rank sum *as small as or smaller than* that observed is $43/2^{11} = 0.0211$.
    (b) If the zero is retained and given the signed rank $+1$, the null probability of a *smaller* negative-rank sum is $87/2^{12} = 0.0212$, which neverthless exceeds the value in (a).
    *(c) For $43/2^{11} \leq \alpha \leq 87/2^{12}$, by the reduced sample procedure, the sample is significant in the positive direction, but becomes not significant when every observation is increased by the same small amount. The corresponding confidence region for μ is an interval with the interior point 0 removed.

62. Show that for the requirements to avoid anomalies in the presence of zeros and ties given in Sect. 6.5, condition (i)(b) holds if and only if condition (ii) holds.

*63. Show that, for the procedure of breaking the ties at random, either actually using standard tables or reporting the probability with which doing so would lead to rejection, conditions (i), (ii) and (iii) and also (i′) of Sect. 6.5 hold.

64. In order to show that, despite intuition, larger observations are not always greater evidence of positivity, consider the density

$$f(x) = \begin{cases} 0.4 & -1 < x < 0, \\ 0.8 - 0.4x & 0 \leq x < 1, \\ 0 & \text{otherwise.} \end{cases}$$

    (a) Show that this is a "positive" density, since $f(x) > f(-x)$ for $0 < x < 1$ and $f(x) \geq f(-x)$ for all $x > 0$.
    (b) Show that if a sample of size 2 is drawn from a population with this density, the signed ranks 1, -2 are more likely than the signed ranks -1, 2.

65. (a) Show that increasing one observation in a sample may decrease the ordinary t statistic.
    (b) More generally, show that t is a decreasing function of x_1 for $x_1 > \sum_2^n x_i^2 / \sum_2^n x_i$ if $\sum_2^n x_i > 0$, and that $t \to 1$ as $x_1 \to \infty$.

*66. Show that, if the Wilcoxon statistic is computed for each possible way of breaking the ties (as in Sect. 6.3), the simple average of the resulting values is numerically equal to the statistic obtained by the average rank and signed-rank zero procedures,

67. For the data $-1, -2, -3, -4, 5, 6, 7, 8, 9, 10, 11$,
    (a) Show that by the Wilcoxon signed-rank test the one-tailed P-value for this sample is $43/2^{11}$, so that the sample would be judged significantly positive at all levels $\alpha \geq 43/2^{11}$.
    (b) Suppose that an additional observation is obtained and its value is 0.5. Show that the signed-rank procedure for the new sample gives a next P-value of $87/2^{12}$. Hence this sample would lead to a conclusion of not significantly positive for $43/2^{11} = 86/2^{12} \leq \alpha \leq 87/2^{12}$. In other words, the addition of a positive observation to a significantly positive sample may make it not significant by the Wilcoxon signed-rank test.
    (c) Show that the ordinary t test, based on normal theory, gives a one-tailed P-value of 0.0165 for the original data, and that an additional positive observation, if small enough, decreases the value of t and hence increases the P-value. Thus the t test has the same property.

68. For the data $0, 0, -2, -3, -5, 6, 9, 11, 12, 15, 16$, and the negative-rank sum as test statistic,
    (a) Show that, by the signed-rank zero procedure, the exact P-value is $23/2^9$ and the next P-value is $19/2^9$.
    (b) Show that, by the reduced sample procedure, the exact P-value is $14/2^9$ and the next P-value is $10/2^9$.
    (c) If the zeros are given signed ranks $\pm 1, \pm 2$, what are the possible P-values?
    (d) Do the results in (c) agree or disagree with those in (a) and (b)?

*69. The modified Wilcoxon procedure of Sect. 5 agrees with the reduced sample procedure for the data given in (6.7) and in Problem 61. Why is the modified procedure not subject to the same objections?

*70. Construct examples showing that, for the modified Wilcoxon procedure of Sect. 5,
    (a) If ties are handled by the average rank procedure, neither condition (i)(a) nor (iii) of Sect. 6.5 need hold.
    (b) If zeros are handled by the reduced sample procedure, none of the conditions (i)–(iii) of Sect. 6.5 need hold.

71. Show that the Wilcoxon statistic, calculated using the average rank procedure and including the zeros in the ranking but giving them signed-rank zero, can be written as

$$T = \sum_{i \leq j} \sum \operatorname{sign}(X_i + X_j)$$

where sign $(0) = 0$.

*72. Show that the one-tailed Wilcoxon test in the appropriate direction using the average rank and signed-rank zero procedures for ties and zeros is consistent against alternatives for which the X_i are independent and $P(X_i + X_j > 0) - P(X_i + X_j < 0)$ is at least some fixed amount for all $i \neq j$, while the "conservative" procedure is not.

73. A manufacturer of suntan lotion is testing a new formula to see whether it provides more protection against sunburn than the old formula. Ten subjects are chosen. The two types of lotion are applied to the back of each subject, one on each side, randomly allocated. Each subject is then exposed to a controlled but intense amount of sun. Degree of sunburn was measured for each side of each subject, with the results shown below (higher numbers represent more severe sunburn).

Subject	Old Formula	New Formula
1	41	37
2	42	39
3	48	31
4	38	39
5	38	34
6	45	47
7	21	19
8	28	30
9	29	25
10	14	8

(a) Test the null hypothesis that the difference of degree of sunburn is symmetrically distributed about 0, against the one-sided alternative that the new formula is more effective than the old. Use a Wilcoxon test at level 0.05, handling ties by the average rank procedure and using Table D as an approximation.

(b) Compute the exact P-value by generating the appropriate tail of the distribution using average ranks.

(c) Find the range of P-values which results when the ties are broken.

(d) Do (b) and (c) of this problem always lead to the same decision when $\alpha = 0.05$? Find the range of α for which the decisions are the same.

(e) Find a 90% upper confidence bound for the median difference assuming that the distribution of differences is symmetric.

74. For the data given in Problem 73, use the sign test procedure of Chap. 2 to
(a) Find the P-value for testing the null hypothesis that the median difference is 0.
(b) Find an upper confidence bound at level 0.90 for the median difference.

75. The Johnson Rod Mill Company produces steel rods. When the process is operating properly, the rods have a median length of 10 meters. A sample of 10 rods, randomly selected from the production line, yielded the results listed below.

$$9.8, \ 10.0, \ 9.7, \ 9.9, \ 10.0, \ 10.0, \ 9.8, \ 9.7, \ 9.8, \ 9.9$$

Does the process seem to be operating properly? How would you recommend handling the ties?

76. The Brighton Steel Works orders a certain size casting in large quantities. Before the castings can be used, they must be machined to a specified tolerance. The machining is either done by the company or is subcontracted, according to the following decision rule:
"If average weight of casting exceeds 25 kilograms, subcontract the order for machining. If average weight of castings is 25 kilograms or less, do not subcontract."
The company developed this decision rule in an effort to reduce costs, because the weight of a casting is a good indication of the amount of machining that will be necessary while the cost of subcontracting the castings is a function of the number of castings to be machined rather than the amount of machining required by each casting.

    The following data are for a random sample taken from a lot of 100 castings.

Casting	1	2	3	4	5	6	7	8
Weight	24.3	25.8	25.4	24.8	25.2	25.1	25.5	24.6

    (a) What decision is suggested by the Wilcoxon signed-rank test at level 0.05?
    (b) What assumption of the Wilcoxon test is critical here?
    (c) What do you think of this method of making a decision?

77. The manufacturers of Fusion, "a new toothpaste for the post-atomic age," hired Anonymous Unlimited, an independent research organization, to test their product. Anonymous Unlimited induced children to go to the dentist and have their cavities counted and filled, and then to switch from their regular brand to Fusion. A year later they went to the dentist again. Advertisements blared the astounding news: 87.5% had fewer cavities.

    The actual data were as follows:

Child number	1	2	3	4	5	6	7	8
Cavities using regular brand	4	2	1	6	4	3	4	1
Cavities using Fusion	2	1	1	3	3	2	0	2

    Apply to these data the statistical methods you consider most applicable and comment on your choice of methods. What conclusions can be drawn from the experiment, under what assumptions, and with what reservations? How could the experiment have been improved (without changing its scope greatly)? Be brief.

78. A sail-maker wanted to know whether the sails he makes of dacron are better than the sails he makes of cotton. He made 5 suits of dacron sails and 5 suits of cotton sails, all for a certain type of boat. He obtained 10 boats of this type, labeled $A, B, \ldots,$ J, and had them sail in two races. He picked 5 of the 10 boats at random; these 5 (they were $A, C, E, G,$ and H) used dacron sails in the first race and cotton sails in the second race. The other five ($B, D, F, I,$ and J) used cotton sails in the first race and dacron sails in the second. The order of finish in the first race was $C, H, A, J,$ B, E, I, F, G, D; in the second race it was $A, B, H, J, I, C, D, F, E, G.$ Analyze these results to shed light on the sail-maker's question.

79. Generalize the relations in (3.1)–(3.3) among $T, T^{+},$ and T^{-} so that they apply to sums of all, positive, and negative signed constants c_k.

80. Represent the general test statistic based on the sum of signed constants as the linear combination

$$T' = \sum_k c_k S_k$$

where S_k denotes the sign of the observation with rank k in absolute value. Under the null hypothesis that the observations are independently, continuously distributed, symmetrically about 0, show that
  (a) $E(T') = 0$
  (b) $\mathrm{var}(T') = \sum_k c_k^2$
  (c) T' is symmetrically distributed about 0.

*81. Show that the signed ranks and the signs of the Walsh averages have the relationship stated in Theorem 7.1.

*82. In a sample of size 10, suppose that the Walsh averages $(X_i + X_j)/2$ are negative if both i and j are odd, and positive otherwise. What could be the signed ranks of $X_1, \ldots, X_{10}$?

83. Show that, if $c_{k+1} \geq c_k \geq 0$ for all k, the test based on the sum of the signed c_k's satisfies the hypothesis of Theorem 3.3 (increasing an observation never decreases the probability of rejection $\phi(X_1, \ldots, X_n)$).

*84. Consider n observations X_i such that there are no ties in the Walsh averages.
  (a) Show that, as μ increases, the signed ranks of the centered observations $X_i - \mu$ change only when μ equals some Walsh average $(X_i + X_j)/2$.
  (b) Show that, in the situation described in (a), the only changes are as follows. If $i = j$, the signed rank of X_i changes from 1 to -1. If $i \neq j$ and $X_i < X_j$, the signed rank of X_i changes from $-(k + 1)$ to $-(k + 2)$ and that of X_j from $(k + 2)$ to $(k + 1)$, where k is the number of observations between X_i and X_j.
  (c) Show that the sum of negative constants c_k increases by c_1 if $i = j$ and by $(c_{k+2} - c_{k+1})$ if $i \neq j$, while the sum of positive c_k's decreases by the same amount, and the sum of signed c_k's decreases by twice this amount.

85. Show that the confidence region for the population center of symmetry μ corresponding to a one- or two-sided test based on a sum of signed c_k's is an interval if $c_{k+1} \geq c_k \geq 0$ for all k.

86. Show that, if $c_k = 0$ for $k \leq n - m$ and $c_k = 1$ for $k > n - m$, then
  (a) The test of Sect. 7.1 is equivalent to carrying out a sign test on the m observations largest in absolute value.
  (b) The corresponding confidence limits are $(X_{(k+1)} + X_{(n-m+k+1)})/2$ and $(X_{(n-k)} + X_{(m-k)})/2$, where k and $m - k$ are the lower and upper critical values for the number of negative observations in a sample of m. (Noether [1973] suggests these confidence limits for easy calculation and studies their efficiency.)

87. Show that, if $c_k = 0$ for $k \leq n - m$ and $c_k = k + m - n$ for $k > n - m$, then
  (a) The test of Sect. 7.1 is equivalent to carrying out a Wilcoxon signed-rank test on the m observations largest in absolute value. (This fact could be exploited to reduce tabulation.)
  (b) The corresponding confidence bounds are the $(t + 1)$th smallest and largest among those Walsh averages $(X_{(i)} + X_{(j)})/2$ with $j - i \geq n - m$, where t is

the critical value of the negative rank sum in a sample of m. (Hint: Show that the rank of $|X_{(i)}|$ among the m largest in absolute value is the number of negative $X_{(i)} + X_{(j)}$ with $j - i \geq n - m$ if $X_{(i)} < 0$.)

(c) How could these confidence bounds be found graphically?

88. For order statistics $X_{(i)}$ of a sample of n from a distribution which is continuous and symmetric about 0, show that $P(X_{(i)} + X_{(j)} > 0$ and $X_{(k)} + X_{(l)} > 0) = P(S_1 \leq i - 1$ and $S_1 + S_2 \leq k - 1)$ for $i \leq k \leq l \leq j$, where S_1 and S_2 are independently binomial with parameters $\frac{1}{2}, n - j + i$ and $\frac{1}{2}, j - l + k - i$ respectively. (Hint: $X_{(i)} + X_{(j)} > 0$ if and only if there are less than i negative observations among the $n - j + i$ largest in absolute value. See also Problem 86.)

*89. (a) How many exact levels are available for a test based on a sum of signed c_k's if no two subsets of the c_k's have the same sum?

(b) How many distinct confidence limits corresponding to such a test can there be in any given sample? How can this number be so much less than the number in (a)?

*90. Is the test which corresponds to the confidence bound in (7.5) equivalent to a signed-rank test?

*91. Consider the test based on the sum of the signed c_k for $c_k = k + 2^{-k}$.

(a) Show that this test agrees with the Wilcoxon test except that it distinguishes rank orders with same signed-rank sum. Compare Mantel and Rahe [1980].

(b) Show that its P-value is always at least as small as the Wilcoxon P-value.

(c) Show that its P-value is always larger than the next Wilcoxon P-value. How near can it come?

(d) When is this test more powerful than the Wilcoxon test? Equivalent? Less powerful? What would be a fair way to compare the power of the two tests?

(e) Show that the expected one-tailed P-value (expected significance level) of this test under the null hypothesis is $0.5 (1 + 1/2^n)$. Show that it is $0.5 (1 + \sum p_i^2) \geq 0.5(1 + 1/N)$ for a test statistic having N possible values with respective probabilities p_i under the null hypothesis, with equality holding if and only if $p_i = 1/N$ for $i = 1, \ldots, N$.

(f) Corresponding to this test, what are the confidence bounds for an arbitrary center of symmetry μ? How do they compare with the Wilcoxon bounds?

(g) What changes occur if c_k is changed to $k - 2^{-k}$ for some values of k? What if $c_k = k \pm \varepsilon 2^{-k}, 0 < \varepsilon < 1$?

92. Show that multiplying all c_k's by the same positive constant has no effect on a test based on the sum of signed c_k's. What about a negative constant? What about adding a constant to all c_k's?

93. Given a procedure ϕ, let ψ consist of applying ϕ after permuting the observations randomly. Show that ψ is a permutation-invariant procedure.

*94. In a testing situation where permutation invariance is applicable (every permutation π carries null into null and alternative into alternative distributions), show that

(a) If ϕ is uniformly most powerful (at level α), then ϕ has the same power against F and F_π for all alternatives F and all permutations π, and there is a permutation-invariant test which is uniformly most powerful.

(b) The statement in (a) remains true when the words "most powerful" are replaced by "most powerful unbiased."

(c) The envelope power is the same at F and F_π for all F and π, and if there is a most stringent test (at level α), then there is a most stringent test which is permutation invariant, where we use the following definitions. Let θ index the alternative distributions, let $\alpha(\theta; \phi)$ be the power of ϕ against θ, let the "envelope power" $\alpha^*(\theta)$ be the maximum of $\alpha(\theta; \phi)$ over all ϕ at level α, and let $\delta(\phi)$ be the maximum of $\alpha(\theta; \phi) - \alpha^*(\theta)$ over all θ (the maximum shortfall of ϕ); ϕ^* is "most stringent" if it minimizes $\delta(\phi)$ among tests ϕ at level α.

(d) If there is a uniformly most powerful invariant test, then it is a most stringent test.

(e) What properties of the permutations as a class of transformations are significant for these results, and how?

*95. Let S be a set of transformations of the observations which are one-to-one and onto. Suppose that, in a testing problem, all transformations in S carry null into null and alternative into alternative distributions.

(a) Show that the same is true of all transformations in the group G generated by S (under composition).

(b) Let θ index the possible distributions of the observations. Show that S and G induce sets $\bar{S}$ and $\bar{G}$ of transformations of θ and that $\bar{G}$ is the group generated by $\bar{S}$.

(c) Show that $\bar{G}$ is homomorphic, but not necessarily isomorphic, to G.

*96. Show that the following classes of transformations are groups:

(a) All permutations of $X_1, \ldots, X_n$.

(b) All transformations of the form $g(X_1), \ldots, g(X_n)$ where g is a strictly increasing, odd function.

97. Let g be a strictly increasing, odd function. Show that $g(X)$ has a symmetric distribution if and only if X does.

98. Show that if $X_1, \ldots, X_n$ and $X'_1, \ldots, X'_n$ have the same signed ranks, then there exists a strictly increasing, odd function g such that $g(X_j) = X'_j, j = 1, \ldots, n$.

*99. Show that, given any set S of transformations of observations,

(a) There exists a statistic T (possibly multivariate) whose value is the same for two different samples if and only if one sample can be transformed into the other by some transformation in the group generated by S.

(b) A procedure is invariant under the transformations of S if and only if it depends only on T.

(c) Give a set of transformations for which T would be the vector of signs of the observations.

100. (a) Show that strictly increasing, odd transformations do not generally preserve the property of symmetry of a distribution about a value other than 0.

(b) Which transformations always do?

(c) Give a (multivariate) statistic T such that a one-sample procedure is invariant under all transformations in (b) and under permutations if and only if it depends only on T.

101. (a) Show that the most powerful signed-rank test against F is of the form (9.1) if all combinations of signed ranks are equally likely under the null hypothesis.

(b) Explicitly, how is k determined and what happens when $P_F(r_1, \ldots, r_n) = k$?

102. Verify formula (9.2) for the probability of signed ranks $r_1, \ldots, r_n$ in a sample from a density f_θ.

103. Show that $|X|_{(j)}$ in (9.4) is the jth order statistic in a sample of n from the cumulative distribution $2F_0 - 1$.

104. Show that a test of the form (9.6) is equivalent to one based on a sum of signed constants (9.7).

105. If h is defined by (9.3), show that

 (a) $h(x) = x/\sigma^2$ for the normal distribution with unknown mean θ and known variance σ^2.
 (b) $h(x) = 2F_0(x) - 1$ for the logistic distribution (9.11).

*106. Show that the locally most powerful signed-rank test against the Laplace family of alternatives

$$F_\theta(x) = \begin{cases} e^{(x-\theta)}/2 & x \le \theta \\ 1 - e^{-(x-\theta)}/2 & x > \theta \end{cases}$$

$$f_\theta(x) = e^{-|x-\theta|}/2$$

is equivalent to the sign test of Chap. 2.

*107. Let $c_j = E[\log(1 + U_j) - \log(1 - U_j)]$ where U_j has a beta distribution with parameters j and $n - j + 1$. Show that a test based on the sum of signed constants c_j is a locally most powerful signed-rank test of $\theta = 0$ for every Lehmann family of alternatives $F_\theta = F^{1+\theta}$ where F is the cdf of a continuous distribution symmetric about 0.

*108. Obtain the sign test by invoking the principle of invariance for a suitable class of transformations when the observations are not assumed identically distributed.

*109. If a sequence of (n variate) distributions F_ν converges in distribution to a distribution F as $\nu \to \infty$ and if A is a closed set, then $P_\nu(A) \ge \limsup P_\nu(A)$.
 (a) Use the fact stated above to show that, if a test whose "acceptance" region is closed has level α for all distributions of a family $\mathscr{F}$, then it has level α for all limits in distribution of sequences in $\mathscr{F}$.
 (b) Suppose that U is a continuous function of the observations and is an upper confidence bound for a parameter θ at level $1 - \alpha$ for all distributions of a family $\mathscr{F}$. Show that $P(U \ge \theta) \ge 1 - \alpha$ for all limits in distribution of sequences in the subset of $\mathscr{F}$ where the parameter value is θ.
 (c) Show that the (joint) distribution of a sample from a discontinuous, symmetric distribution is the limit in distribution of distributions of samples from continuous distributions symmetric about the same point. (Hint: Show it for sample size 1 and then use the fact that if $F_{i\nu}$ converges in distribution to F_i, then the joint distribution of independent observations from $F_{i\nu}$, $i = 1, \ldots, n$, converges to that for F_i, $i = 1, \ldots, n$ as $\nu \to \infty$.)
 (d) Show that if an upper confidence limit U for the center of symmetry μ of a population has level $1 - \alpha$ for all continuous symmetric distributions and is a continuous function of the observations, then $P(U \ge \mu) \ge 1 - \alpha$ for all symmetric distributions. (A similar result holds for lower confidence bounds

and for confidence intervals. Thus closed confidence regions are conservative for discrete distributions.)

(e) Show that the confidence bounds corresponding to the Wilcoxon signed-rank test satisfy the hypotheses of (d).

(f) Show that the statement in (e) also holds for tests based on sums of signed constants c_k with $0 \leq c_1 \leq c_2 \leq \cdots \leq c_n$. (See Problem 85.)

(g) Show that, in the hypothesis of (d), continuous distributions can be replaced by distributions having densities.

CHAPTER 4

One-Sample and Paired-Sample Inferences Based on the Method of Randomization

1 Introduction

The signed-rank tests of the previous chapter rely on the fact that all assignments of signs to the ranks of the absolute values of independent observations are equally likely under the null hypothesis of distributions symmetric about zero, or about some arbitrary point which is subtracted from each observation before ranking. The same fact is true also for the observations themselves, not merely for their ranks, and the idea underlying these tests can be applied to any function of the observations, not merely to a function of the signed ranks.

More specifically, consider any function of sets of sample observations, and the possible values of this function under all assignments of signs to a given set of observations, or equivalently, to their absolute values. Dividing the frequency of each possible value of the function by the total number of possible assignments of signs to the given observations generates a frequency distribution called the *randomization distribution*. In general, this distribution depends on the given set of observations through their absolute values. As its name indicates, the randomization distribution is derived merely from the conditions that, given the absolute values of the n observations, all 2^n possible randomizations of the signs, each either $+$ or $-$, attached to the n absolute values, are equally likely to occur. As discussed in the previous chapter, this condition holds under the null hypothesis of symmetry about zero. In the paired-sample case it follows from the physical act of randomization under the null hypothesis of no treatment effect whatever.

There are many interesting tests based on such a randomization distribution. Any statistic could be used as the test statistic, and any such test is called a *randomization test*. If the value of the test statistic is determined by the signed ranks, as for the sign test or the Wilcoxon signed-rank test, the test is a signed-rank test as defined in Chap. 3. A signed-rank test may also be called a *rank-randomization test*, for contrast with more general randomization tests. Similarly, an arbitrary randomization test may be called an *observation-randomization test* to indicate that the value of the test statistic is determined by the signed observations, as opposed to the signed ranks. (The terms randomization test, rank-randomization test and observation-randomization test generalize to the case of more than one sample, as we will see in Chap. 6.) The randomization distribution of a rank-randomization (signed-rank) test statistic does not depend on the particular set of observations obtained as long as their absolute values are all different. For an observation-randomization test, however, the randomization distribution of the test statistic does depend in general on the observations obtained, specifically on their absolute values. Since the randomization distribution treats the absolute values as given and depends only on them, observation-randomization tests are conditional tests, conditional on the absolute values observed.

The principle of randomization tests is usually attributed to R. A. Fisher; it is discussed in both of Fisher's first editions [1970, first edition 1925; and 1966, first edition 1935]. Many nonparametric tests are based on this principle, as it is easily applied to a wide variety of problems. The randomization may be a designation of sign, or sample label, or an actual rearrangement of symbols or numbers. The test criterion is frequently a classical or parametric test statistic applicable for the same situation, or some monotonic function thereof which is equivalent for the purpose but simplifies calculations. In all situations, the randomization distribution derives from the condition that all possible outcomes of the randomization are equally likely.

Since many randomization distributions are generated by permutations, randomization tests are frequently called *permutation tests*. This name will not be used here, however, because an interpretation of "permutation" which is broad enough to include designating signs is not natural, and we have already used this term in discussing "permutation invariant" tests (as in Sect. 8, Chap. 3). *Conditional tests* is another possible name. However, this is insufficiently specific since many statistical tests are conditional tests, but with different conditioning than here. Another term which appears in the literature is *Pitman tests*, since Pitman [1937a, 1937b, 1938] studied them extensively. The term randomization test could lead to confusion with randomized tests (Sect. 5, Chap. 1), which are entirely unrelated, but some nomenclature must be adopted and none is perfect.

In this chapter we will first discuss the one-sample randomization test based on the sample mean, and the corresponding confidence procedure. Then we will define the general class of randomization tests for the one-sample problem, study some properties of these tests, and obtain most powerful randomization tests. Two-sample observation-randomization tests will be covered in Chap. 6.

While the presentation here is limited to the one-sample case, all the procedures are equally applicable to treatment-control differences and in general to paired-sample observations when the differences of the pairs are used as a single sample of observations. The hypotheses and any assumptions then refer to these differences and their distributions, as do the properties of the statistical procedures based on them. (See Sect. 7, Chap. 2.) It is not necessary that a paired-sample randomization test be based on only these differences, but other tests are seldom needed and will not be discussed in this book.

2 Randomization Procedures Based on the Sample Mean and Equivalent Criteria

2.1 Tests

Given a sample of n independent, identically distributed observations $X_1, \ldots, X_n$, suppose we wish to test the null hypothesis that their common distribution is symmetric about zero. Rejection of this null hypothesis implies either that the distribution is not symmetric, or that it is symmetric about some point other than zero. If the symmetry part of the null hypothesis can reasonably be assumed as part of the model, this null hypothesis reduces to a location hypothesis, as $H_0: \mu = 0$. This is a natural assumption for the test we will study here. The alternative then also concerns the value of μ, and may be either one-sided or two-sided.

Consider the particular sample obtained as one member of the family of all possible samples having the same absolute values $|X_1|, \ldots, |X_n|$ but not necessarily the same algebraic signs. Under the null hypothesis, a priori, the sign of X_j is as likely to be positive as negative, and each of the possible sets of assignments of signs to all the absolute values is equally likely to arise. Since there are 2^n ways to assign plus and minus signs to the absolute values, there are 2^n members of the family of possible samples, which can be enumerated, and each has probability $1/2^n$.

Once some test criterion is selected, the value of the statistic can be calculated for each member of the family. The null distribution of this statistic, conditional on the absolute values of the observations, is then easily found. This is the randomization distribution of the test statistic. From this distribution, a P-value can be found and a test performed in the usual way.

A natural criterion to use in this one-sample problem is the sample mean. The calculations for generating the randomization distribution of the mean are lengthy unless n is small (or only an extreme tail or P-value is found). However, any statistic which is, for given absolute values, a monotonic function of the sample mean provides an equivalent randomization test because the values of the function occur in the same order. Some specific statistics which are somewhat easier to use than the sample mean $\overline{X}$ but are equivalent for the purpose of a randomization test (Problem 1) are the sum of all the sample observations $S = \sum_j X_j$, the sum S^+ of the positive observations, and the sum S^- of the negative observations with sign reversed (so that $S^- \geq 0$). Student's t statistic, calculated in the ordinary way, is also equivalent, as will be shown below.

The method of generating the randomization distribution and the equivalence of these test statistics is illustrated in Table 2.1. In practice, of

Table 2.1[a]

Sample Values: 0.3, −0.8, 0.4, 0.6, −0.2, 1.0, 0.9, 5.8, 2.1, 6.1

$\overline{X} = 1.62$ $S = 16.2$ $S^+ = 17.2$ $S^- = 1.0$

Sample Absolute Values

0.2	0.3	0.4	0.6	0.8	0.9	1.0	2.1	5.8	6.1	S	$\overline{X}$	S^+	S^-
+	+	+	+	+	+	+	+	+	+	18.2	1.82	18.2	0
−	+	+	+	+	+	+	+	+	+	17.8	1.78	18.0	0.2
+	−	+	+	+	+	+	+	+	+	17.6	1.76	17.9	0.3
+	+	−	+	+	+	+	+	+	+	17.4	1.74	17.8	0.4
−	−	+	+	+	+	+	+	+	+	17.2	1.72	17.7	0.5
−	+	−	+	+	+	+	+	+	+	17.0	1.70	17.6	0.6
+	+	+	−	+	+	+	+	+	+	17.0	1.70	17.6	0.6
+	−	−	+	+	+	+	+	+	+	16.8	1.68	17.5	0.7
−	+	+	−	+	+	+	+	+	+	16.6	1.66	17.4	0.8
+	+	+	+	−	+	+	+	+	+	16.6	1.66	17.4	0.8
−	−	−	+	+	+	+	+	+	+	16.4	1.64	17.3	0.9
+	−	+	−	+	+	+	+	+	+	16.4	1.64	17.3	0.9
+	+	+	+	+	−	+	+	+	+	16.4	1.64	17.3	0.9
−	+	+	+	−	+	+	+	+	+	16.2	1.62	17.2	1.0
+	+	−	−	+	+	+	+	+	+	16.2	1.62	17.2	1.0
+	+	+	+	+	+	−	+	+	+	16.2	1.62	17.2	1.0
−	+	+	+	+	−	+	+	+	+	16.0	1.60	17.1	1.1

[a] These data are from Manis, Melvin [1955], Social interaction and the self concept, *Journal of Abnormal and Social Psychology*, **51**, 362–370. The X_j are differences $X_j = X'_j − X''_j$, where X'_j is the decrease in the "distance" between a subject's self-concept and a friend's impression of that subject after a certain period of time, and X''_j is the corresponding decrease for a subject and a nonfriend; the subject-friend pair was matched with the subject-nonfriend pair according to the value of their "distance" at the beginning of the time period. Since the nonfriends were roommates assigned randomly to the subjects, the subjects were expected to have the same amount of contact with nonfriends as with friends during the time period. Manis' hypothesis II was that over a given period of time, there will be a greater increase in agreement between an individual's self-concept and his friend's impression of him than there will be between an individual's self-concept and his nonfriend's impression. This hypothesis is supported if the null hypothesis that the X_j are symmetric about 0 is rejected in favor of a positive alternative. Manis used the Wilcoxon signed-rank test, which gives a one-tailed P-value of 0.0137 (Problem 2).

course, the values of only one statistic would be calculated; we give details for several only to make their relationship more intuitive. The first step is to list the absolute values of the observations to which signs are to be attached. It is generally easier to predict which assignments lead to the extreme values if this listing is in order of absolute magnitude (increasing or decreasing). While there are $2^{10} = 1024$ different assignments of signs, Table 2.1 enumerates only those 17 cases which lead to the largest values of $\bar{X}$, S, or S^+, and the smallest values of S^-. $S^- = 1.0$ was observed for these data and Table 2.1 shows that only 16 of the 1024 cases give S^- that small; hence the one-tailed P-value from the randomization test is $\frac{16}{1024} = 0.0156$.

The calculation is most easily performed in terms of S^- when $\bar{X}$ is "large" as here, and in terms of S^+ when $\bar{X}$ is small. If the same value of S^+ or S^- occurs more than once, each occurrence must be counted separately. Although calculating the complete randomization distribution straightforwardly would require enumerating 2^n possibilities, this is never necessary for any randomization test. In order to find P (the P-value), the enumeration must include only those $2^n P$ assignments which lead to values as extreme as that observed in the appropriate direction (that is, less than or equal to an observed S^- or greater than or equal to an observed S^+, when $S^- \leq S^+$). Furthermore, if a nonrandomized test is desired at level α, a decision to reject ($P \leq \alpha$) can be reached by enumerating these $2^n P$ cases, and a decision to "accept" by identifying any $2^n \alpha$ cases as extreme as that observed. A test decision by direct counting therefore requires enumeration of only $2^n \alpha$ or $2^n P$ cases as extreme as that observed, whichever is smaller. To compute the P-value, of course, it is necessary to enumerate *every* point in the relevant tail, and it is difficult to select them in the correct order for enumeration except in the very extreme end of the relevant tail. Considerable care is required if the entire distribution is not calculated. A systematic approach is to enumerate according to the number of $+$ or $-$ signs (starting with 0). Clever tricks, such as the "branch and bound" method of mathematical programming, might reduce the work entailed. Even to calculate the entire distribution, it is sufficient to enumerate $2^n/2 = 2^{n-1}$ assignments, since the randomization distributions of S^+, S^-, $\bar{X}$, S and t are all symmetric (Problem 4). This also means that the natural two-tailed test is equal-tailed. Approximations for use when exact calculation is too difficult will be discussed in Sect. 2.5.

If the null hypothesis is generalized to state that the observations are symmetrically distributed about some point μ_0, or the symmetry is assumed and the null hypothesis is $H_0: \mu = \mu_0$, the same procedure can be used but applied to the observations $X_j - \mu_0$. That is, the randomization distribution is generated by assigning signs to the $|X_j - \mu_0|$.

Since, given the $|X_j|$, the statistics S, $\bar{X}$, S^+, and S^- are all linearly related, they provide equivalent randomization test criteria. Although the ordinary t statistic is not linearly related to these other statistics, it also provides an equivalent randomization test criterion, as we now show.

*Writing the sample variance as

$$\sum_j (X_j - \bar{X})^2/n = \left(\sum_j X_j^2/n\right) - \left(\sum_j X_j\right)^2/n^2 = \left(n \sum_j X_j^2 - S^2\right)/n^2,$$

we can write the ordinary t statistic as

$$t = \frac{(n-1)^{1/2}S}{(n \sum_j X_j^2 - S^2)^{1/2}} = \pm \left[\frac{n-1}{(n \sum_j X_j^2/S^2) - 1}\right]^{1/2}, \qquad (2.1)$$

where the sign is the same as the sign of S. For given absolute values $|X_1|$, $\ldots, |X_n|$, the sum $\sum_j X_j^2$ in the denominator is a constant, and hence t is a monotone increasing function of S and therefore equivalent as a test statistic.*

The fact that a randomization test based on the usual t statistic is valid for any symmetric null distribution does not imply either that t has Student's t distribution or that the usual t test is valid without the assumption of normality. The normal theory test compares the observed value of the t statistic with a critical value from Student's t distribution (which is fixed, depending only on the sample size), while the randomization test compares it with a critical value obtained from the randomization distribution (which varies from sample to sample even for a given sample size). Accordingly, even though the test statistics are equal, the test procedures are not equivalent because a P-value or critical value based on the randomization distribution does not in general agree with one based on Student's t distribution.

In addition, the large-sample theory of the randomization test does not point particularly to the Student's t test. To order $1/\sqrt{n}$, the accuracy of usual large-sample theory, Student's t and the normal distribution are indistinguishable. Any asymptotic calculation fine enough to distinguish them will show that the randomization distribution of the t statistic, Student's t distribution, and the normal distribution all differ by terms of the same order, namely $1/n$. Specifically, it is well-known (and easily seen by inspecting a table) that Student's t distribution differs from the normal distribution by terms of order $1/n$, while the approximations developed in Sect. 2.5 show that the randomization distribution of the t statistic differs from Student's t distribution by terms of the same order (smaller order if the population has normal kurtosis).

In short, the fact that we can perform a randomization test based on the t statistic does not imply that Student's t distribution provides either a valid test in small samples or asymptotically better validity than other distributions approaching normality at the rate $1/n$. For numerical results, see Sect. 2.5(c) and the references cited there.

2.2 Weakening the Assumptions

We have been assuming that the observations are independent and identically distributed with a symmetric common distribution under H_0. If the assumptions are relaxed so that the X_j are not identically distributed but are independent and symmetrically distributed about 0 (or about the hypothesized value μ_0), then the level of the test is not affected (Problem 5a). (The level of the corresponding confidence procedure to be discussed in Sect. 2.3 is preserved if the X_j are independent and symmetrically distributed about a common point μ.) The test is also valid for a null hypothesis of no treatment effect if the X_j are the treatment-control differences in a matched-pairs experiment with controls selected randomly from the pairs (Problem 6). If the X_j are continuously distributed, then the tests (and corresponding confidence bounds) have level $\alpha = k/2^n$ for some chosen integer k; if they are discontinuously distributed, then the procedures as described are conservative (Problems 5 and 8b; Problem 109 of Chap. 3).

These relaxations of the assumptions and others described in Sect. 3.5, Chap. 3 are possible here for the same reasons as given there.

2.3 Related Confidence Procedures

Under the symmetry assumption, the randomization test procedure can also be used to construct a confidence region for the value of μ, the center of symmetry. Unfortunately, the randomization distribution is different when different values of μ are subtracted from the observations. The confidence region is an interval, and its endpoints could be obtained by trial and error. (That is, successive values of μ are subtracted, larger or smaller than previous values as appropriate, until the value of S^+ or S^- equals the appropriate level α upper or lower critical value of a randomization test for center of symmetry 0.) The endpoints of the normal theory confidence interval for μ at the same level could be used as initial trial values.

However, as in the case of the Wilcoxon signed-rank sum procedure, a more systematic and convenient method is available for finding exact confidence limits; this was suggested by Tukey but developed in Hartigan [1969]. For a sample of n observations, there are $2^n - 1$ different subsamples (of all sizes, 1, 2, ..., n, but excluding the null subsample). Consider the subsample means in order of algebraic (not absolute) magnitude. It can be shown that the kth smallest and kth largest subsample means are the lower and upper confidence bounds respectively, each at level $1 - \alpha$, that correspond to the one-sample randomization test based on $\overline{X}$ (or any other equivalent randomization test criterion), where the one-tailed level is $\alpha = k/2^n$, so $k = 2^n\alpha$ (Problem 8). If it is infeasible or impractical to find the kth smallest or largest among all the subsample means, one can find the kth among a smaller number of subsample means, say m, selected either at random without replacement (Problem 9) or in a "balanced" manner determined by group theory (as explained by Hartigan [1969]); then $\alpha = k/(m + 1)$. Similar methods for randomization tests are discussed in Sect. 2.5 below.

As explained in Sect. 2.2, these confidence procedures are also valid under certain relaxations of the assumptions.

2.4 Properties of the Exact Randomization Distribution

As mentioned in Sect. 2.1, the statistics S, $\bar{X}$, S^+, S^- and t all provide equivalent randomization test criteria and all have symmetric distributions, although only S^+ and S^- are identically distributed.

The means and variances of these statistics under the randomization distribution are easily calculated. These moments are of course conditional on the absolute values of the observations. Since they are considered constants, we denote the values $|X_1|, \ldots, |X_n|$ by $a_1, \ldots, a_n$. Then for $S = \sum_j X_j$ (Problem 11), for instance, we have

$$E(S) = 0 \tag{2.2}$$

$$\operatorname{var}(S) = \sum_j a_j^2 = \sigma^2. \tag{2.3}$$

Thus the center of symmetry for S is zero, as it is for $\bar{X}$ and t, but not for S^+ or S^-. Note that σ^2 is defined by (2.3) and is the variance of the randomization distribution of S, not the population variance of the X_j, although it is an unbiased estimator of n times the latter under the null hypothesis.

2.5 Modifications and Approximations to the Exact Randomization Procedure

The exact randomization distribution can always be enumerated by computer. Complete enumeration is not necessarily required to find tails of the distribution, or individual tail probabilities. Mathematical programming methods, such as branch and bound, can be used. However, there may be situations where a randomization test is desired but it is infeasible or impractical to carry it out exactly. One might then use (a) simulation ("Monte Carlo" sampling) of the complete randomization set, (b) a deterministically restricted randomization set, or (c) the normal or some other approximation.

(a) *Simulation (Sampling the Randomization Set)*. One method of reducing the size of the computational problem is to use only some of the absolute values, selected either at random or on some systematic basis that does not depend on the observed signs. This appears unduly wasteful, and it is presumably more efficient to use a subset of the complete randomization set without discarding observations.

The most direct approach would be to sample from the randomization distribution, with or without replacement. For example, we could read n random digits from a table or generate them by computer and assign a + sign to $|X_j|$ if the jth random digit is even, a − sign otherwise; this would provide one observation from the randomization distribution. Having obtained the assignment of signs, it is a simple matter to compute the chosen test criterion, such as S here. The process of assigning signs at random and computing the test criterion can be repeated until a large number of values have been obtained. The relative frequencies of these values comprise the "simulated" randomization distribution.

The usual simulation method would be to sample as above with replacement and to estimate the P-value of a randomization test by the corresponding relative frequency in the simulated distribution. In the present situation, instead of considering this to be an estimate of P, one could redefine the test as one that rejects if the relative frequency of equally or more extreme values in the simulated randomization distribution is α or less. This relative frequency is the P-value of the redefined test, and the redefined test has level α. This holds whether the sampling is done with or without replacement, but to be precise the simulated samples must be augmented by the actual sample in computing the relative frequencies (Problem 9). Thus it is possible to obtain the desired significance level and a precise P-value without carrying out a large number of simulation trials, as would be required to estimate a small probability accurately. Of course, this procedure entails some loss of power; the fewer the trials carried out, the greater the loss. Dwass [1957] gives an indication of the size of the loss. Valid confidence bounds can be obtained similarly by sampling the subsample means (Problem 9).

More sophisticated methods of simulation, i.e., of sampling the randomization distribution, might produce much more accurate estimates with a given number of trials. Such methods would be expected to entail a smaller loss of power, and at least some of them would be expected to provide a precise level and P-value when the simulation process is regarded as part of the definition of the test, not merely as an approximation procedure. This will not be investigated further here, however.

(b) *A Deterministically Restricted Randomization Set.* Instead of considering all 2^n possible reassignments of signs to $X_1, \ldots, X_n$, one might restrict consideration to a subset. If the subset has the special property of being a group (a subgroup of the group of all reassignments of signs), then it can be used to define a valid test in the same way as the whole randomization set. Problem 12 illustrates the group property and Problem 13 illustrates the fact that the equally likely property may fail in the absence of the group property.

More specifically, suppose $J = (J_1, \ldots, J_n)$ stands for any vector of 1's and -1's. Let G be a set of such vectors with the property that, if J and J' belong to G, then $JJ' = (J_1 J'_1, \ldots, J_n J'_n)$ also belongs to G. (In particular, $J^2 = (1, \ldots, 1)$, the identity for this form of multiplication, belongs to G, and every J is its own inverse.) Consider the set G_X of values $(J_1 X_1, \ldots, J_n X_n)$ for all J in G (the "orbit" of $(X_1, \ldots, X_n)$). Under the randomization distribution, given the set G_X to which X belongs, all members of G_X are equally likely. Thus the value of the criterion at X can be compared with its values for all members of G_X just as before. The null hypothesis is rejected if the value at X is among the k smallest (or largest) of these values, and the one-tailed level is k divided by the number of members of G. The power, of course, depends on the choice of G.

The confidence bounds for the center of symmetry μ corresponding to the randomization test based on $\bar{X}$ with the randomization set restricted as described above can be obtained as follows (Problem 14). For each J in G, find the subsample mean of those X_j with $J_j = -1$. The kth smallest and kth largest of these subsample means are the confidence bounds.

(c) *Approximations to the Randomization Distribution.* We now discuss some approximations to the randomization distribution that are based on tabled probability distributions. We consider first the ordinary normal approximation based on the standardized form of the random variable S, which we denote here by Z for convenience. The mean and variance of S given in (2.2) and (2.3) lead to

$$Z = S/\sigma = S/\sqrt{\sum_j a_j^2}. \tag{2.4}$$

When the randomization distribution of Z in (2.4) is approximated by the standard normal distribution, the results of Zahl [1962] show that the error in the approximated tail probability is at most

$$\Psi_3 C(\Psi_3, \Psi_4, Z) e^{-Z^2/2} \tag{2.5}$$

where

$$\Psi_r = \sum_j a_j^r / \sigma^r \tag{2.6}$$

for $r = 3$ and $r = 4$ and C in (2.5) is a function which is generally between 0.5 and 2.0. The final factor $e^{-Z^2/2}$ in (2.5) is less than 0.2 for $|Z| \geq 1.8$ and less than 0.1 for $|Z| \geq 2.15$. Note that Ψ_3 is of order $n^{-1/2}$ (in fact, about $2.3n^{-1/2}$ for a typical normal sample), and Ψ_4 is of order n^{-1}. Moses [1952] gives the rule of thumb that this normal approximation should not be used unless $\max_j (a_j^2)/\sum_j a_j^2 \leq 5/2n$.

Although this normal approximation procedure appears to reflect properties specific to the randomization distribution while the ordinary t procedure does not, the following argument shows that this is an illusion and suggests that the latter procedure may actually be better than the former procedure as an approximation to the randomization procedure.

We showed in Sect. 2.1 that the randomization test based on the ordinary t statistic is equivalent to one based on S. If we write the ordinary t statistic in (2.1) in the form

$$t = Z \left(\frac{n-1}{n-Z^2} \right)^{1/2}, \tag{2.7}$$

it is clear that comparing Z with a percent point z_α of the standard normal distribution is equivalent to comparing t with

$$z_\alpha \left(\frac{n-1}{n-z_\alpha^2} \right)^{1/2}. \tag{2.8}$$

As a result, the normal approximation to the randomization test merely replaces the critical value t_α of the ordinary t test by (2.8), which is, like t_α, a constant that does not depend on the sample observations. (The two critical values, t_α and the quantity in (2.8), differ considerably in small samples except for α near 0.05, although both approach z_α as $n \to \infty$. See Problem 15.) If the randomization test is to be performed by comparing the t statistic (2.1) or (2.7) to some constant value, then under normality, t_α will be a better constant on the average than (2.8). This suggests that t_α will be better in general, at least for samples that are not too far from normal. Another bad feature of (2.8) is that it becomes infinite before α reaches zero, namely for $z_\alpha = \sqrt{n}$.

*So far it appears that we have not improved on the ordinary t test as an approximation to the randomization test. We now discuss some more refined approximations intended to give an improvement. The first approximation is motivated by the following reasoning. Recall that if t has Student's t distribution with $n - 1$ degrees of freedom, then t^2 has the F distribution with 1 and $n - 1$ degrees of freedom, and hence, equivalently, $t^2/(t^2 + n - 1) = Z^2/n$ has the beta distribution with parameters $\frac{1}{2}$ and $(n - 1)/2$. However, the moments of these F and beta distributions differ from the corresponding moments of the randomization distributions of t^2 and Z^2/n except in special cases (Problem 16). Hence, a better approximation should result if we adjust the parameters of the F or beta distribution so that its first two moments are equal to the first two moments of the corresponding randomization distribution. These approximations will match the first four moments of the randomization distributions of t and Z respectively, their first and third moments being equal to zero by symmetry.

We show below that the adjustment required to match these moments in the beta form (the F form is harder and not equivalent) is equivalent to multiplying the normal theory degrees of freedom for F, namely 1 and $n - 1$, by a constant correction factor d defined by

$$d = \frac{(n - 3)\sigma^4 + 2\sum_j a_j^4}{n(\sigma^4 - \sum_j a_j^4)} = \frac{n - 3 + 2\Psi_4}{n(1 - \Psi_4)} \tag{2.9}$$

where Ψ_4 is defined by (2.6) with $r = 4$. Hence this method of approximation is to treat t^2, the squared value of (2.7), as F distributed with fractional degrees of freedom d and $(n - 1)\,d$.

In order to get an idea of the size of d, the correction factor multiplying the degrees of freedom, we rewrite (2.9) in the form (Problem 17a)

$$d = 1 + \left(\frac{b_2 - 3}{n}\right)\left(1 - \frac{b_2}{n + 2}\right)^{-1} \tag{2.10}$$

where

$$b_2 = (n + 2)\sum_j a_j^4 \bigg/ \left(\sum_j a_j^2\right)^2 = (n + 2)\Psi_4. \tag{2.11}$$

Under the null hypothesis, b_2 is a consistent estimator of the kurtosis of the distribution of the X_j (Problem 17b); the kurtosis parameter equals 3 for normal distributions. Hence, from (2.10) we see that d will differ from 1 by order $1/n$, with the amount of correction depending on the departure of kurtosis from that of the normal distribution.

The small fractional number of degrees of freedom d in the numerator of the foregoing F distribution may sometimes preclude its use in applications. A further approximation would be to replace this F distribution by a scaled F distribution with 1 and k degrees of freedom, where the scale factor and k are chosen to give the same first two moments. This leads (see below and Problem 18d) to treating c times the t statistic as Student's t distributed with k degrees of freedom where

$$k = \frac{nd(4 - d) + d^2 + 4d - 8}{nd(1 - d) + d^2 + 4d - 2},\tag{2.12}$$

$$c = \left(\frac{k - 2}{k}\frac{d(n - 1)}{d(n - 1) - 2}\right)^{1/2}.\tag{2.13}$$

In order to get an idea of the size of the correction factor here, we write d in (2.10) in the form

$$d = 1 + \frac{\gamma}{n}\tag{2.14}$$

where γ is a measure of the deviation from normal kurtosis. Substituting (2.14) for d in (2.12) and (2.13) we obtain after some algebra

$$k = n - 1 + \frac{\gamma}{3 - \gamma}\left(n - 5 - \frac{3\gamma}{3 - \gamma}\right) + \text{terms of order } (1/n)$$

$$c = 1 + \frac{\gamma}{3n} + \text{terms of order } (1/n^2).$$

These results show clearly the order of magnitude of the corrections to the normal theory values $k = n - 1, c = 1$.

We illustrate both of these approximations using the Darwin data (where $n = 15$) introduced in Sect. 3.1, Chap. 3. Fisher [1966 and earlier editions] gives the one-tailed P-value in the randomization distribution as $863/2^{15} = 0.02634$. Using Student's t distribution with $n - 1 = 14$ degrees of freedom, Fisher obtains 0.02485 for the ordinary t test ($t = 2.148$), and 0.02529 with a continuity correction to allow for the discreteness of the measurements ($t = 2.139$). If we use the approximation that treats t^2 as F distributed with d and $(n - 1)d$ degrees of freedom, we first calculate d from (2.9) for these same data as $d = 0.937$; then the F distribution with degrees of freedom $d = 0.937$ and $(n - 1)d = 13.12$ gives the one-tailed P-value as 0.02643 without a continuity correction, and 0.02686 with one. If we use the method of approximation by scaled F (or t) for these same data, we calculate by (2.12) and (2.13) $k = 11.31$, $c = 0.9855$, $ct = 2.108$ (with a continuity correction). Treating ct as Student's t distributed with $k = 11.31$ degrees of freedom, we obtain a one-tailed P-value of 0.02912. In this example ct gives a less accurate approximation to the randomization distribution than the ordinary t, but presumably this is just a coincidence.*

*PROOF. Under normal theory, t^2 has an F distribution with 1 and $n-1$ degrees of freedom, or, equivalently, $W = t^2/(t^2 + n - 1)$ has a beta distribution with parameters $\frac{1}{2}$ and $(n-1)/2$. For the ordinary t statistic, (2.7) shows that

$$W = \frac{t^2}{t^2 + n - 1} = \frac{Z^2}{n}$$

so that under normal theory, the marginal distribution of Z^2/n is beta. This suggests approximating the randomization distribution of Z^2/n by a beta distribution with first two moments equal to the moments of the randomization distribution of Z^2/n, which are (Problem 18b)

$$E(Z^2/n) = 1/n \qquad E[(Z^2/n)^2] = (3 - 2\Psi_4)/n^2. \qquad (2.15)$$

The first and second moments of the beta distribution with parameters a and b are

$$E(W) = \frac{a}{a + b} \qquad E(W^2) = \frac{a(a + 1)}{(a + b)(a + b + 1)}. \qquad (2.16)$$

If we equate the moments in (2.15) to the corresponding moments in (2.16) and solve the resulting equations simultaneously for a and b, the solution is

$$a = \frac{n - 3 - 2\Psi_4}{2n(1 - \Psi_4)}, \qquad b = (n - 1)a. \qquad (2.17)$$

Thus we are led to approximate the randomization distribution of Z^2/n by a beta distribution with the parameters in (2.17). This procedure is equivalent (Problem 18c) to approximating the randomization distribution of t^2 by an F distribution with degrees of freedom $2a = d$ and $2b = (n - 1)d$ as at (2.9).*   □

Notice that we have given no formal statements or proofs about asymptotic distributions or asymptotic errors of approximation. Such formal results would be complicated by the fact that the randomization distribution is conditional upon the absolute values of the observations, which are themselves random. It is not enough to apply the central limit theorem and standard asymptotic methods and conclude that, for instance, the marginal distribution of Z is asymptotically standard normal. What we really want to know is that the conditional distribution of Z, given $|X_1|, \ldots, |X_n|$, "approaches" the standard normal distribution in some sense that we have not even defined. There are several natural definitions, but under any of them, the conditional result is much stronger than the unconditional one and does not follow from an ordinary central limit theorem (Problem 20).

3 The General Class of One-Sample Randomization Tests

3.1 Definition

The randomization test based on the sample mean, or any of the equivalent test criteria given in Sect. 2.1, relies for its level only on the assumption that, given the absolute values of the observations, all assignments of signs are equally likely. This randomization test and randomization distribution are therefore conditional on these absolute values. In particular, this means that the level of the randomization test is α if, under the null hypothesis, the conditional probability of rejection given $|X_1|, \ldots, |X_n|$ is at most α, and that the P-value is the corresponding conditional probability of a value of the test statistic equal to or more extreme than that observed.

Generalizing, we define a *one-sample randomization test* for center of symmetry equal to zero as a test which is conditional on the absolute values of the observations, having a null distribution that is the randomization distribution generated by assigning signs randomly to these absolute values. Tests which are members of this general class may depend on the actual values of the X_j. Signed-rank (or rank-randomization) tests are those which depend only on the signed ranks of the X_j. Any signed-rank test, including the Wilcoxon signed-rank test or any test based on a sum of signed constants (as defined in Sect. 7.1, Chap. 3) is a randomization test; however, not all randomization tests are signed-rank tests, as the test of Sect. 2.1 based on the sample mean or equivalent criteria is not.

The class of all randomization tests is even broader than it may seem. Two other specific examples of randomization tests are described below.

(1) Consider the composite test defined as choosing a constant c and applying a level α Wilcoxon signed-rank test whenever $\sum_j X_j^2 \geq c(\sum_j |X_j|)^2$, and a level α sign test otherwise. This is a randomization test because given the $|X_j|$, the conditional level of the composite test is α regardless of which test is used. Of course, any number of component tests may be used, but in order for such a composite test to be a randomization test, the rule for choosing which component test to use must depend only on the $|X_j|$. A rule of this kind aimed at achieving good power for a wide variety of distributions has come to be called "adaptive" [Hogg, 1974]. The possibility was recognized at least as early as 1956, when Stein [1956] showed that in this way, for large n, power can be attained arbitrarily close to that of the

most powerful test against every translation alternative regardless of the shape of the distribution, as long as its density is sufficiently regular and symmetric.

(2) Another possibility would be to use a level α Wilcoxon signed-rank test whenever $|X_1| > |X_2|$ and a level α sign test otherwise. This composite test is also a randomization test, but it is not permutation invariant, that is, the result of the test could be altered by a rearrangement of the X_j. Permutation invariance (see Sect. 8, Chap. 3) is not a requirement of randomization tests, although it is generally natural and advantageous, as we will see in the next subsection.

3.2 Properties

We now turn to some of the theoretical properties of members of the class of one-sample randomization tests for zero center of symmetry.

Critical Function and Level

Let $\phi(X_1, \ldots, X_n)$ denote the probability of rejection (critical function) of an arbitrary test. We write $X_j = a_j I_j$ where $a_j = |X_j|$ and

$$I_j = \begin{cases} -1 & \text{if } X_j < 0 \\ 1 & \text{if } X_j > 0. \end{cases}$$

Then we may write the critical function as

$$\phi(X_1, \ldots, X_n) = \phi(a_1 I_1, \ldots, a_n I_n). \tag{3.1}$$

The $a_j = |X_j|$ are to be treated as constants. Given the $|X_j| = a_j$, the conditional expected value of the critical function, under the null hypothesis H_0 of a distribution symmetric about zero, is simply the mean of its randomization distribution, or

$$E_0[\phi(X_1, \ldots, X_n)||X_1| = a_1, \ldots, |X_n| = a_n] = \Sigma\phi(\pm a_1, \ldots, \pm a_n)/2^n, \tag{3.2}$$

where the sum is over all the 2^n possible assignments of signs. This expected value is the conditional probability of a Type I error for the test. Accordingly, a test ϕ of H_0 has conditional level α if the quantity in (3.2) is less than or equal to α. If this holds for all $a_1, \ldots, a_n$, then the test is a randomization test. Any such test also has level α unconditionally (Problem 21) by the usual argument (see Sect. 6.3, Chap. 2).

In the notation just introduced, we may say that a signed-rank (or rank-randomization) test is one whose critical function depends on the $a_j = |X_j|$ only through their ranks, as well as the signs I_j.

Weakening the Assumptions

The statements made in Sect. 2.2 about weakening the assumptions apply without change to all members of the general class of one-sample randomization tests, as do the statements of Sect. 3.5, Chap. 3 provided the tests are monotonic where monotonicity is obviously called for.

Justifications of Randomization Tests

Section 8, Chap. 3 presented an argument based on concepts of invariance to justify restricting consideration to signed-rank tests. Since all signed-rank tests are randomization tests, this same argument justifies restricting consideration to randomization tests. This argument required a very strong assumption, however, namely invariance under increasing, odd transformations (Sect. 8.2, Chap. 3). We shall now discuss quite different assumptions leading to randomization tests (not necessarily rank tests).

Observations not identically distributed. It is elementary that, as remarked above, any test having conditional level α given the $|X_j|$ also has level α unconditionally. It can also be shown that, if the null hypothesis is sufficiently broad, then conversely, any test having (unconditional) level α must also have conditional level α given the $|X_j|$, and therefore must be a randomization test. Such a strong result is not available for independent, identically distributed observations, as will be discussed shortly. For independent, not identically distributed observations, however, either of the following null hypotheses is sufficiently broad to imply the conclusion (as are certain less broad hypotheses).

H_0: The X_j are independent, each having a distribution that is symmetric about zero.

H_0': H_0 holds and the distributions have densities.

Thus the statement is as follows.

Theorem 3.1. *If a test has level α for either H_0 or H_0', then the test is a randomization test.*

PROOF. In order to see this for H_0, suppose that the conditional level were greater than α, given some set of absolute values $|X_1| = a_1, \ldots, |X_n| = a_n$. Consider independent X_j with a distribution such that $P(X_j = a_j) = P(X_j = -a_j) = \frac{1}{2}$; then H_0 is satisfied (H_0' is not) but the null probability of rejection is greater than α. This contradiction shows that the supposition is impossible and any level α test of this null hypothesis has conditional level α given $|X_1|, \ldots, |X_n|$. The result for H_0' will not be proved here. See Lehmann [1959, Sect. 5.10] or Lehmann and Stein [1949].   □

Observations identically distributed. A similar result holds for observations which are independent and identically distributed, provided the null hypothesis is again sufficiently broad, and in addition the tests are unbiased against a sufficiently broad alternative hypothesis. The proof requires that the alternative include distributions which are arbitrarily close to each distribution under the null hypothesis, so that, by the usual argument, the test will have to have level exactly α under every null distribution. Furthermore the kind of randomization test obtained, which we will now define, is technically weaker, though practically the same.

The randomization test defined in Sect. 3.1 is based on the 2^n assignments of signs and does not consider permutations of the order of the absolute values of the observations. For identically distributed observations, all such permutations are equally likely. If we consider all possible orders of the absolute values as well as all assignments of signs, then we obtain a randomization distribution which has $(n!\,2^n)$ equally likely possibilities instead of just 2^n. We will call a test based on this latter randomization distribution an $(n!\,2^n)$-type randomization test and the type discussed earlier a 2^n-type. Under the null hypothesis that $X_1, \ldots, X_n$ are independent and identically distributed, symmetrically about zero, a conditional test given the order statistics of the absolute values is an $(n!\,2^n)$-type randomization test (and conversely), because all assignments of signs and all orderings of the absolute values are equally likely. To be more specific, if ϕ is a test function and $|X|_{(1)} \leq \cdots \leq |X|_{(n)}$ are the order statistics of the absolute values of the observations $|X_j|$, then the conditional level of ϕ is

$$E_0[\phi(X_1, \ldots, X_n) \mid |X|_{(1)}, \ldots, |X|_{(n)}] = \sum \phi(\pm |X|_{(i_1)}, \ldots, \pm |X|_{(i_n)})/2^n n!$$
(3.3)

where the sum is over all $2^n n!$ possible assignments of both the signs and the arrangements (permutations) $i_1, \ldots, i_n$ of the integers $1, \ldots, n$. Thus the conditional level of ϕ given the order statistics of the absolute values is its level under $(n!2^n)$-type randomization. A 2^n-type randomization test is conditional on the absolute values $|X_1|, \ldots, |X_n|$ in their observed order, and an $(n!2^n)$-type need not be. Any 2^n-type randomization test, being more conditional, is also an $(n!2^n)$-type randomization test, but not conversely (Problem 22a). For permutation invariant tests, however, the two types are equivalent, as will be discussed shortly.

We now give conditions which lead to $(n!2^n)$-type randomization tests. The first condition is that the test have level α for a sufficiently broad null hypothesis, such as

H_0'': The X_j are independent with a common distribution that is symmetric about zero.

H_0''': H_0'' holds and the common distribution has a density.

The second condition is that the test be unbiased against a sufficiently broad alternative hypothesis (so that every null distribution is a limit of alternative distributions). Alternatives which specify identical, symmetric distributions but are still sufficiently broad are

H_1'': The X_j are independent with a common distribution that is symmetric about some point $\mu \neq 0$.

H_1''': H_1'' holds and the common distribution has a density.

The complete statement is given as follows.

Theorem 3.2. *If a test has level α for H_0'' and is unbiased against H_1'', or has level α for H_0''' and is unbiased against H_1''', then it is an $(n!2^n)$-type randomization test.*

The proof of this property will be given shortly. As will be evident from this proof, other, less broad, null and alternative hypotheses would also imply the conclusion of this theorem.

The result in Theorem 3.2 "justifies" restricting consideration to $(n!2^n)$-type randomization tests when the observations are identically distributed, but we have not yet justified the further restriction to 2^n-type randomization tests. Recall, however, that in Sect. 8, Chap. 3, we defined a procedure as permutation invariant if it is unchanged by permutations of the observations. If a test is permutation invariant, then averaging over the permutations has no effect, and hence the two types of randomization are equivalent. As a result, a permutation-invariant test is a 2^n-type randomization test if and only if it is an $(n!2^n)$-type randomization test (Problem 22b). The reasons given in Sect. 8, Chap. 3 for using a test which does not depend on the order of the observations apply equally here. In particular, for independent, identically distributed observations, nothing whatever is gained by taking the order into account. Therefore, any test of H_0'' against H_1'', or H_0''' against H_1''', may as well be taken as permutation invariant, and if it is unbiased, it must be a randomization test (2^n-type, or equivalently, $(n!2^n)$-type).

PROOF OF THEOREM 3.2 We outline here a proof of Theorem 3.2 for continuous distributions only. A proof for the discrete case is requested in Problem 23. Suppose we have a test of H_0''' which has level α under every common continuous distribution that is symmetric about 0 and is unbiased against the alternative H_1''' of a common continuous distribution that is symmetric about some other point. In order to prove that this test must have conditional level α given $|X|_{(1)}, \ldots, |X|_{(n)}$, we need prove only that the $|X|_{(1)}, \ldots, |X|_{(n)}$ are sufficient and boundedly complete for the common boundary K of H_0''' and H_1''', which is H_0''' (by Theorem 6.1 of Chap. 2). The sufficiency part is trivial, and will be left as an exercise (Problem 25). To prove completeness, consider the family of symmetric densities given by

$$C(\theta_1, \ldots, \theta_n) \exp(\theta_1 |X| + \cdots + \theta_n |X|^n - X^{2n}). \tag{3.4}$$

Let $T_k = \sum_j |X_j|^k$ for $k = 1, \ldots, n$. Since the T_k determine the coefficients of the polynomial $g(x) = \prod_{j=1}^n (x - |X_j|)$, the order statistics $|X|_{(j)}$ are functions of the T_k and hence so is $\sum_j X_j^{2n}$. If

$$E[\phi(|X|_{(1)}, \ldots, |X|_{(n)})] = 0$$

under all distributions in (3.5), then

$$\int \cdots \int \phi(|X|_{(1)}, \ldots, |X|_{(n)}) J(t_1, \ldots, t_n) \exp\left(\sum_k \theta_k t_k\right) dt_1 \cdots dt_n = 0$$

where the integration is over the region of possible values of $T_1, \ldots, T_n$ and where $J(t_1, \ldots, t_n)$ is $\exp(-\sum_j X_j^{2n})$ times the Jacobian of the $|X|_{(j)}$ with respect to the T_k. It follows by the theory of Laplace transforms that the integrand is zero and hence that $\phi = 0$. For more details on this approach see Lehmann [1959, pp. 132–133]. Alternatively, see Fraser [1957b, pp. 28–31] or Lehmann [1959, pp. 152–153] for an approach that is derived from the discrete case. These sources deal with the usual order statistics $X_{(1)}, \ldots, X_{(n)}$ and arbitrary densities rather than $|X|_{(1)}, \ldots, |X|_{(n)}$ and symmetric densities; this difference affects the proof only slightly. $\square$

Matched Pairs

In a matched pairs experiment with, say, treatment and control observations on each pair, it is usually assumed that the test procedure will depend only on the treatment-control differences. Under this assumption, the foregoing results apply. Similar results can be given which do not require this assumption and are more in the spirit of matched pairs, referring to the structure of the situation. One such result is described below.

Let X_j' and X_j'' denote the observations on the jth pair, and consider the null hypothesis

H_0^+: The X_j' and X_j'' are independently normal with common variance $\sigma^2 > 0$ and arbitrary means μ_j' and μ_j'' respectively. The treatment is assigned randomly to one member of each pair and has no effect on the observed value X_j' or X_j'' of the treated unit.

It can be shown that if a test has level α for this null hypothesis, then it is a conditional test given the X_j' and X_j'', and relies only on the random assignment of the treatment within pairs. The proof is quite similar to that given above for H_0''; see the references cited there. Several remarks should be made, however.

First, the null hypothesis H_0^+ sets up a normal model with completely arbitrary unit effects and no treatment effect on any unit. The treatment-control difference X_j for the jth pair is equally likely to be $X_j' - X_j''$ or $X_j'' - X_j'$. Second, the distribution of this treatment-control difference for the jth pair is not normal but is a mixture of two normal distributions with means $\mu'_j - \mu''_j$ and $\mu''_j - \mu'_j$, and possibly extremely small variance. Thus, it may, in a sense, approximate a two-point distribution $X_j = \pm(\mu'_j - \mu''_j) = \pm v_j$, say. Third, the result holds for any null hypothesis that contains H_0^+, i.e., any null hypothesis including all the distributions in H_0^+ (Problem 26). Fourth, the result does not imply that the test is a randomization test based solely on the treatment-control differences. One might, for example, use a composite or "adaptive" test with components that depend only on the treatment-control differences but a rule for choosing the component test that depends on the sums within pairs (Problem 27).

4 Most Powerful Randomization Tests

4.1 General Case

Reasons for using randomization tests were given in Sect. 3.2. In this subsection we will see how to find that randomization test which is most powerful against any specified alternative distribution. The particular case of normal alternatives will be illustrated in the two subsections following.

We consider the usual (2^n-type) randomization distribution, under which, given the absolute values $|X_1|, \ldots, |X_n|$, there are 2^n possible assignments of signs, and hence 2^n possible samples, which are equally likely. The implication of using a randomization test is that this condition holds under the null hypothesis. Consider now an alternative distribution with joint density or discrete frequency function $f(x_1, \ldots, x_n)$. Under this alternative, given $|X_1|, \ldots, |X_n|$, the conditional probabilities of each of the 2^n possible samples $x_1, \ldots, x_n$ are proportional to $f(x_1, \ldots, x_n)$. By the Neyman–Pearson Lemma (Theorem 7.1 of Chap. 1), it follows (Problem 28) that among tests with conditional level α given $|X_1|, \ldots, |X_n|$, that is, among randomization tests, the conditional power against f is maximized by a test of the form

$$\text{reject if } f(X_1, \ldots, X_n) > k$$
$$\text{"accept" if } f(X_1, \ldots, X_n) < k. \tag{4.1}$$

Randomization may be necessary at k (that is, the test may be randomized). The choice of k and the randomization at k must be determined so that the test has conditional level exactly α.

In other words, the procedure for finding the conditionally most powerful randomization test is to consider the value of f at each $x_1, \ldots, x_n$ having the same absolute values as the observed $X_1, \ldots, X_n$. The possible samples $x_1, \ldots, x_n$ are placed in the rejection region in decreasing order of their corresponding values of f, starting with the largest value of f, until their null probabilities total α. The region will consist of the $\alpha 2^n$ possible samples $x_1, \ldots, x_n$ which produce the largest values of f if $\alpha 2^n$ is an integer and if the $(\alpha 2^n)$th and $(\alpha 2^n + 1)$th values of f are not tied. Ties may be broken arbitrarily. If $\alpha 2^n$ is not an integer, a randomized test will be necessary.

Since this test is the randomization test which maximizes the conditional power against f, it also maximizes the ordinary (unconditional) power against f (Problem 28). Thus we have shown how to find the (2^n-type) randomization test which is most powerful against any specified alternative f. The conditions given are necessary and sufficient. The method for ($n!\,2^n$)-type randomization tests is similar.

4.2 One-Sided Normal Alternatives

Consider now the alternative that $X_1, \ldots, X_n$ are a random sample from a normal distribution with mean μ and variance σ^2. Then the joint density is

$$f(x_1, \ldots, x_n) = \prod_{j=1}^{n} \frac{1}{\sqrt{2\pi}\sigma} e^{-(x_j - \mu)^2/2\sigma^2}$$

$$= (2\pi)^{-n/2} \sigma^{-n} e^{-(\Sigma x_j^2 + n\mu^2)/2\sigma^2} e^{\mu \Sigma x_j/\sigma^2}. \tag{4.2}$$

For given $|x_1|, \ldots, |x_n|$ (and μ and σ^2), this f is an increasing function of $\sum x_j$ if $\mu > 0$, and a decreasing function of $\sum x_j$ if $\mu < 0$. Therefore, rejecting if $f(X_1, \ldots, X_n)$ is one of the k largest of its possible values given $|X_1|, \ldots, |X_n|$ is equivalent to rejecting if $\sum X_j$ (or an equivalent statistic) is one of the k largest of its possible values given $|X_1|, \ldots, |X_n|$ when $\mu > 0$, and if it is one of the k smallest when $\mu < 0$. Thus the result of Sect. 4.1 shows that the upper-tailed observation-randomization test based on $S = \sum X_j$ (or $\bar{X}$) is the most powerful randomization test against any normal alternative with $\mu > 0$, and similarly for the lower-tailed test and $\mu < 0$. In short, the one-tailed randomization tests based on the sample mean are uniformly most powerful among randomization tests against one-sided normal alternatives.

4.3 Two-Sided Normal Alternatives

Consider now the two-sided alternative that $X_1, \ldots, X_n$ is a random sample from a normal distribution with mean $\mu \neq 0$ and variance σ^2. Since we found in the previous subsection that different randomization tests are uniformly most powerful against $\mu < 0$ and against $\mu > 0$, there is no uniformly most powerful randomization test against this two-sided alternative. However, the equal-tailed randomization test based on $\bar{X}$ is the uniformly most powerful test against $\mu \neq 0$, among unbiased randomization tests (Problem 29). Further, in the class of randomization tests which are invariant under the transformation carrying $X_1, \ldots, X_n$ into $-X_1, \ldots, -X_n$, we show below that the equal-tailed randomization test is again the uniformly most powerful test. This invariance means that the test is unaffected if the signs of all the observations are reversed. Notice that this transformation carries the alternative given by μ, σ^2 into that given by $-\mu, \sigma^2$, so the invariance rationale

(Sect. 8, Chap. 3) can be applied. In particular, any invariant test has the same power against μ, σ^2 as against $-\mu, \sigma^2$.

PROOF. From the last sentence it follows that any invariant test has the same power against μ, σ^2 as against the density h obtained by averaging the density for μ, σ^2 and the corresponding density for $-\mu, \sigma^2$, namely

$$h(x_1, \ldots, x_n) = \left[\prod_{j=1}^{n} \frac{1}{\sqrt{2\pi}\sigma} e^{-(x_j - \mu)^2/2\sigma^2} + \prod_{j=1}^{n} \frac{1}{\sqrt{2\pi}\sigma} e^{-(x_j + \mu)^2/2\sigma^2} \right] \Big/ 2.$$

(4.3)

Accordingly, if we show that the equal-tailed randomization test based on $\bar{X}$ is the most powerful randomization test against h, it will follow that this is the most powerful invariant randomization test against every μ, σ^2. By (4.2) we can write (4.3) as

$$h(x_1, \ldots, x_n) = (2\pi)^{-n/2} e^{-(\Sigma x_j^2 + n\mu^2)/2\sigma^2} (e^{\mu \Sigma x_j/\sigma^2} + e^{-\mu \Sigma x_j/\sigma^2})/2\sigma^n$$

$$= (2\pi)^{-n/2} \sigma^{-n} e^{-(\Sigma x_j^2 + n\mu^2)/2\sigma^2} \cosh(\mu \sum x_j/\sigma^2). \qquad (4.4)$$

For fixed $|x_1|, \ldots, |x_n|$, (4.4) is an increasing function of $|\mu \sum x_j/\sigma^2|$, and hence of $|\sum x_j|$ (Problem 31). It follows from (4.1) that the most powerful randomization test against h is that which rejects for large $|\sum x_j|$. Since this is simply the equal-tailed randomization test based on $\bar{X}$, the proof is complete. □

The density h given in (4.3) may be interpreted as the marginal density of a sample $X_1, \ldots, X_n$ from a normal population if the population mean is, a priori, equally likely to be μ or $-\mu$. The proof uses this prior distribution only as a device, but the invariance rationale itself breaks down if μ and $-\mu$ are not regarded in effect as equally likely a priori. Notice also that h is not one of the alternatives originally under discussion (but a mixture of two), and that h does not make $X_1, \ldots, X_n$ independent except conditionally. (In terms of the interpretation involving prior probabilities, $X_1, \ldots, X_n$ are conditionally independent given that the mean is μ, or $-\mu$, but they are not marginally independent.)

5 Observation-Randomization versus Rank-Randomization Tests

We have found that the one-sample observation-randomization test of Sect. 2 is valid with almost no distribution assumptions, and it is the most powerful randomization test against normal alternatives. One can show that the performance of this nonparametric test under normal alternatives is almost as good as that of the corresponding most powerful parametric test whose validity is dependent upon the normal distribution. Since the nonparametric test is completely valid under much less restrictive assumptions, one could argue that there may be much to gain and little to lose by using this nonparametric test. However, the Wilcoxon signed-rank test is much more popular in applications that this observation-randomization test (or any other one-sample nonparametric test), even though it is only locally most powerful among signed-rank tests against alternatives which are only close to normal.

In fact, observation-randomization tests are seldom used in practice, while rank-randomization tests often are. Rank-randomization tests are probably used primarily because they are so convenient in practical applications. The observation-randomization test criteria of Sect. 2 depend heavily on the actual values of the $|X_j|$. Accordingly, the general one-sample randomization distribution cannot be tabled for arbitrary samples of fixed size. However, the randomization distribution of a test criterion based on signed ranks is constant for all samples of fixed size for which there are no ties among the absolute values. Thus the null distribution can be tabled for any rank-randomization (signed-rank) test. Further, particularly for some of them, confidence interval procedures corresponding to rank-randomization tests are easier to use.

The convenience of rank-randomization tests is not so important with the availability of computers. A computer program could be used to generate or simulate the complete general randomization distribution for any set of data, and hence the distribution of an observation-randomization test statistic or confidence bound. The cost of a little extra effort in data analysis is small compared to the other costs of many experiments.

Besides convenience, however, rank-randomization tests frequently have good performance properties. They may be highly efficient, and more powerful than observation-randomization tests against many alternatives (see Sect. 8, Chap. 8). They are less sensitive to outliers. Further, as mentioned in Sect. 8, Chap. 3, in most situations little is lost by treating alike practically all samples which have the same signed ranks. These are perhaps sufficient reasons for using rank-randomization rather than observation-randomization procedures in most situations.

Observation-randomization tests are nevertheless of considerable theoretical and historical interest. They provide important examples of the basic spirit of general nonparametric statistics in that the exact probability distribution can always be found in the null case without making specific assumptions about the population and a strong type of optimality can be achieved under normality. Even more important historically is that observation-randomization tests provided the seed of the basic idea for rank-randomization tests and hence may have been instrumental in their development since 1935.

PROBLEMS

1. (a) Express $\bar{X}$, S, S^+, and S^- each in terms of each of the others and $\sum |X_j|$.
   (b) Show that the randomization tests based on these statistics are equivalent.

2. (a) Find the one-tailed P-value of the Wilcoxon signed-rank test for Manis' data given in Table 2.1.
   (b) For Manis' data, find lower and upper confidence bounds corresponding to the Wilcoxon signed-rank test at the one-sided level $\alpha = 10/1024$ for an assumed center of symmetry.
   (c) Do (b) for the randomization test based on $\bar{X}$.
   (d) How would you interpret the confidence bounds in (b) or (c)?
   (e) In Manis' data, some subjects had smaller initial "distance" than others, and hence less possibility of reducing the distance over time. How does this affect the various randomization tests of the null hypothesis of symmetry about 0? What alternative procedures might be considered, with what advantages and disadvantages?

3. For the data in Sect. 3.1, Chap. 3, Fisher [1966 and earlier editions] obtained the one-tailed P-value of 0.02634 for a randomization test based on the t statistic. How many values of t as small as or smaller than that observed must he have counted?

4. Show that the randomization distributions of $\bar{X}$, S, S^+, S^-, and t are all symmetric.

5. Consider the null hypothesis H_0 that the observations $X_1, \ldots, X_n$ are independently distributed, symetrically about 0, and the randomization test that rejects H_0 when $K \leq k$ for some chosen integer k, where K is the number of different assignments of signs for which $\sum_j \pm X_j \leq \sum_j X_j$ (including $\sum_j X_j$ itself). Show that
   (a) This test has level $\alpha \leq k/2^n$, irrespective of whether the observations are identically distributed. The P-value is $K/2^n$.
   (b) This test has level $\alpha = k/2^n$ if the observations are also continuously distributed.

6. Consider a matched-pairs experiment with the null hypothesis of no treatment effect on any observation. Show that randomization within pairs induces the randomization distribution of the treatment-control differences whatever the paired observations may be.

7. Show that the confidence bounds corresponding to the upper-tailed randomization test based on $\bar{X}$ are
   (a) $X_{(1)}$ at level $1/2^n$,
   (b) $(X_{(1)} + X_{(2)})/2$ at level $2/2^n$;
   (c) $\min\{X_{(3)}, (X_{(1)} + X_{(2)})/2\}$ at level $3/2^n$,
   where $X_{(i)}$ is the ith order statistic of a sample of size n. Compare Problem 36, Chap. 3.

*8. Given sample observations $X_1, \ldots, X_n$, let K be the number of different assignments of signs for which $\sum_j \pm X_j \leq \sum_j X_j$ (including $\sum_j X_j$ itself). Show that
   (a) The number of nonnegative subsample means is $K - 1$. (Hint: A subsample total equals $(\sum_j X_j - \sum_j \pm X_j)/2$ where a $-$ is assigned if X_j is in the subsample and a $+$ is assigned otherwise.)
   (b) The confidence bound for center of symmetry μ corresponding to the randomization test that rejects for $K \leq k$ is the kth smallest subsample mean.

(c) The test that rejects for $K \leq k$ has level $\alpha \leq k/2^n$ if the X_j are independent and symmetric about 0 (not necessarily identically distributed). The P-value is $K/2^n$. If the X_j are also continuously distributed, then $\alpha = k/2^n$.

(d) If the X_j are independent with a distribution that is continuous and symmetric about μ, then the $2^n - 1$ subsample means all differ with probability one, and they partition the real line into 2^n intervals that are equally likely to contain μ.

*9. (Randomization tests with randomly chosen subsets of assignments of signs.) Let $X_1, \ldots, X_n$ be independent observations with a distribution that is symmetric about 0 under the null hypothesis H_0. Define $Y_0 = \sum_j X_j$ and $Y_i = \sum_j \pm X_j = \sum_j \pm |X_j|$ for $i = 1, \ldots, m$, where the signs are drawn at random either (i) unrestrictedly, or (ii) without replacement from all 2^n possible assignments except that corresponding to Y_0. Let K denote the number of values of i for which $Y_i \leq Y_0$ for $0 \leq i \leq m$, let R_i denote the set of values of j for which the sign of X_j differs in the sums for Y_0 and Y_i, and let Z_i denote the mean of the X_j for which j is in R_i for R_i nonempty. Show that

(a) The two sums given above to define Y_i are equivalent.

(b) Y_0 is a random sample of size one from the order statistics of $Y_0, Y_1, \ldots, Y_m$ under H_0.

(c) The test that rejects H_0 if $K \leq k$ has level $\alpha \leq k/(m + 1)$ and P-value equal to $K/(m + 1)$.

(d) The $R_i, i = 1, \ldots, m$, are a random sample drawn with replacement from all 2^n possible subsets of the set $\{1, 2, \ldots, n\}$ in case (i) above, and without replacement from all $2^n - 1$ nonempty subsets in case (ii) above.

(e) If H is the number of nonnegative Z_i and E is the number of empty R_i for $1 \leq i \leq m$, then $K = H + 1 + E$. In case (ii) above, $E = 0$.

(f) The confidence limit for center of symmetry μ that corresponds to the test that rejects for $K \leq k$ is the $(k - E)$th smallest Z_i for $1 \leq i \leq m$ (which is the kth smallest in case (ii) above).

(g) If the X_j have a continuous distribution that is symmetric about μ, then in case (ii) the level of the test in (c) is $\alpha = k/(m + 1)$ and the Z_i partition the real line into $m + 1$ intervals that are equally likely to contain μ. (This result can also be proved from Problem 8d using Problem 10 below.)

*10. Let $W_1 < \cdots < W_N$ be continuously distributed with $P(W_i > 0) = i/(N + 1)$ for $1 \leq i \leq N$. Let $W'_1 < \cdots < W'_m$ be the order statistics of a sample of m drawn without replacement from $W_1, \ldots, W_N$. Show that $P(W'_i > 0) = i/(m + 1)$ for $1 \leq i \leq m$. (Hint: One method is to consider the case where the W_i are obtained by subtracting from the order statistics of a sample an additional observation from the same distribution).

11. (a) Find the center of symmetry and the variance of the randomization distributions of each of $\bar{X}, S, S^+, S^-$, and t.

(b) How are the moments of $\bar{X}, S, S^+$, and S^- related? Why do the moments of t not have a simple relationship to these moments?

12. Let G be the set of all n-dimensional vectors $J = (J_1, \ldots, J_n)$ of 1's and -1's such that an even number of elements are equal to -1. Let G_X be the set defined in procedure (b) of Sect. 2.4. Show that

(a) G has 2^{n-1} members.

(b) If no $X_j = 0$, then G_X also has 2^{n-1} members.

(c) If J and J' both belong to the set G, then the vector $(J_1 J'_1, \ldots, J_n J'_n)$ also belongs to G.

(d) Given that X belongs to a particular set G_Y, all members of G_Y are equally likely under the randomization distribution. What if X does not belong to G_Y?

*13. Let G be the set of all n-dimensional vectors of 1's and -1's such that the first two elements are not both equal to -1. Let G_X be the set defined in procedure (b) of Sect. 2.4. Show that

(a) G has $3(2^{n-2})$ members. (G does not satisfy Problem 12c.)

(b) $X = (X_1, \ldots, X_n)$ has the smallest or second smallest mean among members of G_X if all $X_j > 0$ except possibly X_1 or X_2 but not both.

(c) If no $X_j = 0$, then under the randomization distribution the probability is at least $3/2^n$ that X has the smallest or second smallest mean among members of G_X. (If X could be treated as a random member of G_X, the probability would be $(8/3)/2^n$, which is always smaller than $3/2^n$.)

*14. (a) Show that the confidence bound corresponding to a randomization test based on $\bar{X}$ with a randomization set restricted by means of a group G as described in procedure (b) of Sect. 2.4 is the kth smallest or largest subsample mean among subsamples corresponding to members of G, where k is the critical value of the randomization test and the subsample corresponding to a vector J in G consists of those X_j for which $J_j = -1$.

(b) What operation on the subsamples corresponds to the group multiplication in G?

15. (a) Make a small table to compare the values of $z_\alpha[(n-1)/(n-z_\alpha^2)]^{1/2}$, t_α, and z_α, where z_α is the upper α quantile of the standard normal distribution and t_α is that of the Student's t distribution with $n-1$ degrees of freedom. (A good picture can be obtained from the values $\alpha = 0.10, 0.05, 0.025, 0.01$; $n = 3, 6, 10, 20$.)

*(b) How do the values of t_α, $z_\alpha[(n-1)/(n-z_\alpha^2)]^{1/2}$, and z_α compare for large sample sizes?

(c) Find the ranges of the standardized randomization test statistic in (2.4) and the t statistic in (2.7), given the sample absolute values. Find the ranges unconditionally.

(d) What do the ranges found in (c) imply about the normal approximation (2.4) to the randomization test?

*(e) Show that the values of α for which $t_\alpha - z_\alpha[(n-1)/(n-z_\alpha^2)]^{1/2}$ is of smaller order than $1/n$ as $n \to \infty$ are the α values such that $z_\alpha = 0, \pm\sqrt{3}$.

16. (a) Find the moments of the randomization distribution of t^2 and compare them with the corresponding moments of the F distribution with 1 and $(n-1)$ degrees of freedom.

(b) Find the moments of the randomization distribution of $t^2/(t^2 + n - 1)$ and compare them with the corresponding moments of the beta distribution with parameters $1/2$ and $(n-1)/2$.

17. (a) Show that the expression for d in (2.10) is equivalent to the expression in (2.9).

(b) Show that b_2, as defined by (2.11), is a consistent estimator of the kurtosis of the distribution of the X_j under the null hypothesis of a distribution symmetric about zero.

*18. (a) Derive the first four moments of the randomization distribution of the statistic Z defined by (2.4).

   (b) Find the parameters of the beta distribution whose first two moments are the same as the corresponding moments of the randomization distribution of Z^2/n.

   (c) Show that approximating the randomization distribution of Z^2/n by the beta distribution in (b) is equivalent to approximating the randomization distribution of $t^2 = (n-1)Z^2/(n-Z^2)$ by an F distribution and find the degrees of freedom of this F distribution.

   (d) Let V have the F distribution with degrees of freedom 1 and k. Find the values of c and k such that c^2V has the same first two moments as the F distribution in (c).

19. Apply the approximations of (c) in Sect. 2.5 to the Darwin data in Sect. 3.1 of Chap. 3 to verify the results given in the text.

*20. (a) Give at least two possible definitions of convergence in distribution for conditional distributions.

   (b) Show that the definitions you gave in (a) imply the usual convergence in distribution of the marginal distributions.

   (c) Show that the converse of (b) does not hold.

   (d) Show that the ordinary central limit theorems do not apply to these definitions.

21. Show that a randomization test at level α has unconditional level α under the null hypothesis H_0 of a distribution symmetric about zero if the observations are independent and identically distributed.

22. Show explicitly, in terms of the expectation of the critical function ϕ under randomization, that

   (a) Any 2^n-type randomization test is an $(n!\,2^n)$-type randomization test, but not conversely.

   (b) A permutation-invariant test has the same level under both types of randomization.

*23. Show that for the null hypothesis H_0'': $X_1, \ldots, X_n$ are independent with a common distribution symmetric about zero, if a test is unbiased against the alternative H_1'': $X_1, \ldots, X_n$ are independent with a common distribution that is symmetric about some point $\mu \neq 0$ (or $\mu > 0$), then that test is an $(n!\,2^n)$-type randomization test. (Remember that the common distribution need not be continuous.)

*24. Show that, in Problem 23, if unbiasedness is required only against continuous alternatives, then the test need not be a randomization test for all discrete distributions. Why does this result not really contradict the results given in Sect. 3.2?

25. Show that the order statistics of the absolute values of the observations are sufficient statistics for a sample from an arbitrary distribution that is symmetric about zero.

*26. In a matched pairs experiment, show that if a test has level α for a null hypothesis containing H_0^+ as stated in Sect. 3.2, then this test is a conditional test given the observations X_j' and X_j''.

27. In a matched pairs experiment, give an example of a test which is conditional on the observations X_j' and X_j'' but does not depend only on the treatment-control differences. Why might such a test be desirable?

28. Show that, both conditionally and unconditionally, the most powerful randomization test against an alternative density or discrete frequency function $f(x_1, \ldots, x_n)$ is of the form given in (4.1).

*29. (a) Show that the equal-tailed randomization test based on $\bar{X}$ is a uniformly most powerful unbiased randomization test against the alternative that the observations are a random sample from a normal distribution with mean $\mu \neq 0$ and variance σ^2.

(b) Using the fact that the test in (a) is uniformly most powerful among tests which are invariant under reversal of signs, show that it is the most stringent randomization test against the same alternative.

*30. Show in general that under appropriate conditions
(a) If a uniformly most powerful invariant test and a uniformly most powerful unbiased test both exist, then they must be the same test.
(b) If a uniformly most powerful invariant test exists, then it is most stringent.

31. Show that, for fixed $|x_1|, \ldots, |x_n|$, the function $h(x_1, \ldots, x_n)$ given in (4.4) is an increasing function of $|\sum x_j|$.

32. Let $X_1, \ldots, X_n$ be independent and identically distributed with density $f(x) = (1/\sigma)\exp\{-\rho([x - \theta]/\sigma)\}$ where ρ is a known symmetric function and σ is a known scale factor. Let the null hypothesis be $H_0: \theta = 0$.
(a) Find the most powerful randomization test of H_0 against the alternative $H_1: \theta = \theta_1$ for θ_1 specified.
*(b) Under what circumstances does there exist a uniformly most powerful randomization test against the alternative $H_1': \theta > 0$.
(c) Show that the (locally) most powerful randomization test against the alternative $H_1': \theta > 0$ for small θ can be based on the statistic $\sum_j \Psi(x_j/\sigma)$ where $\Psi(x) = \rho'(x) = d\rho(x)/dx$. (See Sect. 9, Chap. 3.)
(d) Show that the tests in (a) and (c) are valid even if the σ and ρ assumed are incorrect.
(e) Show that the maximum likelihood estimate $\hat{\theta}$ of θ satisfies $\sum_j \Psi([x_j - \hat{\theta}]/\sigma) = 0$ where $\Psi(x) = \rho'(x)$. (Estimates of this form appear significantly in work of Edgeworth [1908–9] and Fisher [1935] and were named M-estimates in Huber [1964], which studies them extensively.)
(f) Show that the estimate that corresponds (in a suitable sense) to the test in (c) is the maximum likelihood estimate.

33. With reference to Problem 32,
(a) If σ is unknown, how could it be estimated without affecting the validity of the randomization tests in (a) and (c)?
(b) If in addition ρ is not fully known but has some unknown shape parameters, how could they be estimated without affecting the validity of the randomization tests in (a) and (c).
(c) What estimates would correspond to the tests in (a) and (b)?

CHAPTER 5

Two-Sample Rank Procedures for Location

1 Introduction

The previous chapters dealt with inference procedures applicable in one-sample (or paired-sample) problems. We now consider the situation where there are two mutually independent random samples, one from each of two populations. We discuss tests which apply to the null hypothesis that the two populations are identical and the confidence procedures related to these tests.

In choosing an appropriate test from among those available for the null hypothesis of identical populations, consideration should be given to the alternative hypothesis, since different tests are sensitive to different alternatives. The alternatives may be simply that the two populations differ in some unspecified way, but frequently some specific type of difference is of particular interest. A general alternative which is frequently important is that the observations from one population tend to be larger than the observations from the other ("stochastic dominance"). A particular case of this relationship occurs when the populations satisfy the shift assumption, which is explained explicitly in the next section. Frequently, the difference in "location" between the two populations is of primary interest. Under the shift assumption, this difference is the same whatever location parameter is chosen, and is the amount of shift required to make the two populations identical. Furthermore, we can develop confidence procedures (corresponding to the test procedures) which give confidence intervals for this difference in location, or shift.

The primary discussion of two-sample tests in this book is divided among three chapters. The median test, tests based on rank sums, and more general rank-randomization tests are discussed in this chapter, observation-randomization tests in Chap. 6, and Kolmogorov–Smirnov tests in Chap. 7. Chapter 8 discusses asymptotic relative efficiency generally, and specifically for both one and two-sample procedures.

Once the tests have been developed, it will be evident that they retain their significance levels and hence remain valid even when the basic assumptions are relaxed in certain ways. For instance, even independence of the observations is not required for tests of no treatment effect as long as the assignment to treatment or control groups is appropriately random. While these two-sample procedures do not require normal populations, they do not provide protection against inequality of variance, or other differences in shape, even in the null case.

2 The Shift Assumption

Suppose that $X_1, \ldots, X_m$ and $Y_1, \ldots, Y_n$ are mutually independent sets of observations drawn from two populations. We say that the Y population is the same as the X population except for a shift (the shift assumption) by the amount μ if

$$P(X \le x) = P(Y \le x + \mu) \quad \text{for all } x. \tag{2.1}$$

This condition is equivalent to saying that X and $Y - \mu$ have the same distribution, or that Y is distributed as $X + \mu$. In terms of c.d.f.'s, this relation is

$$F(x) = G(x + \mu) \quad \text{for all } x, \tag{2.2}$$

where F and G are the cumulative distribution functions of X and Y respectively. Under the shift assumption, the cumulative distribution function of the Y population is the same as that of the X population but shifted to the left if $\mu < 0$, and to the right if $\mu > 0$, as in Fig. 2.1 (a) and (b) respectively. If X and Y have densities f and g, then (2.1) or (2.2) is equivalent to

$$f(x) = g(x + \mu) \quad \text{for all } x. \tag{2.3}$$

Two arbitrary density functions which satisfy (2.3) are shown in Fig. 2.1 (c).

The shift assumption means that the two populations have the same shape, and in particular their variances must be equal if they exist (Problem 1). Two normal populations with the same variance satisfy the shift assumption, but two normal populations with different variances do not, nor do two Poisson or exponential populations with different parameters.

If the shift assumption holds, then μ, the amount of the shift, must equal the difference between the two population medians. It must also equal the

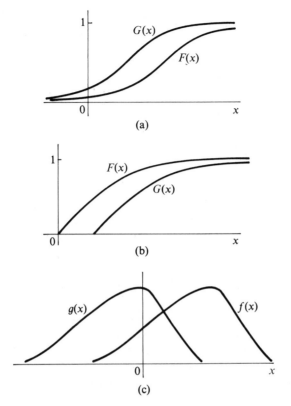

Figure 2.1 (a) $F(x) = G(x + \mu)$, F normal, $\mu < 0$. (b) $F(x) = G(x + \mu)$, F exponential, $\mu > 0$. (c) $f(x) = g(x + \mu)$, $\mu < 0$.

difference between the two population means, and indeed the difference between the two values of any other location parameter (if it exists), such as the mode, the midrange, or the average of the lower and upper quartiles. The mean need not equal the median (or any other location parameter) in either population, but the difference between the mean and the median must be the same for both populations. Since the populations are the same except for location, the difference in location is the same however it is measured, and it equals the shift μ (Problem 2). For this reason, the shift parameter is sometimes called the location parameter.

The confidence procedures in this chapter are developed under the shift assumption, and accordingly they provide confidence intervals for μ, the amount of the shift. If the shift assumption fails badly, the procedures will not perform as advertised since the confidence level will not ordinarily be valid.

On the other hand, the test procedures here can be developed and justified logically without assuming that the shift assumption (or any other relationship between the distributions) holds under the alternative hypothesis. The tests retain their level as long as $F = G$ under the null hypothesis. Thus, while "acceptance" of this null hypothesis may be interpreted as not rejecting the possibility that the shift assumption holds with $\mu = 0$, rejection of the null hypothesis implies no inference about whether the shift assumption holds for any other μ. Furthermore, the tests developed here appear as if they would be good against alternatives which include more than shifts, and certain mathematical properties to be discussed provide justification for this view. Of course similar statements apply to parametric tests of a difference in location for two otherwise identical populations, including the normal theory test for equal means, so these points are not new or special to nonparametric tests.

3 The Median Test, Other Two-Sample Sign Tests, and Related Confidence Procedures

The sign test discussed in Chap. 2 for the one-sample or paired-sample problem can be adapted to the two-sample problem in various ways to test the null hypothesis of identical distributions. Tests of this nature, which may be called two-sample sign tests, are covered in this section. The data relevant for a two-sample sign test can be presented in a 2 × 2 table whose entries are observed sample frequencies. The test is then carried out by *Fisher's exact test*, the general method appropriate for such data. One especially important two-sample sign test is usually called the *median test*; this test and its corresponding confidence procedures will be discussed in particular detail here.

3.1 Reduction of Data to a 2 × 2 Table

Suppose we have a set of m X observations and a set of n Y observations, and hence a total of $m + n = N$ observations in the two sets combined. The data can be summarized by a 2 × 2 table if all the observations are dichotomized. Consider the following two methods of dichotomizing the observations according to size:

(1) A particular number ξ is selected before looking at the data, and each observation is classified as being either "below ξ" or "above ξ." (For definiteness, we define "below" as "strictly less than" and above as "greater than or equal to." This ensures a unique and exhaustive dichotomy.)

(2) A particular integer t is selected and the t smallest observations in both sets combined are classified as "below," the rest as "above." (This definition is incomplete in certain cases of ties; see Sect. 3.3.)

In either case, the dichotomized sample data can be presented in a 2 × 2 table like that shown in Table 3.1. Note that the numbers A and B determine

Table 3.1

	X's	Y's	
Below	A	B	t
Above	$m - A$	$n - B$	$N - t$
	m	n	N

the entire table, since the column totals, m and n, are fixed by the sample sizes. In case (2), since t is also fixed by the procedure, A alone determines the entire table.

In case (1), A and B are determined by simply comparing each observation with the fixed ξ. In case (2), the most straightforward procedure is to combine both sets of observations into a single array but keep track of which observations are X's and which are Y's. Then A and B are the number of X's and Y's respectively among the t smallest in the single combined array.

For any given data set, the same 2×2 tables can be obtained by method (1) with various choices of ξ and by method (2) with various choices of t. Specifically, the t in a table obtained by method (1) gives the same table by method (2). Further, any method (2) table is also given by method (1) if ξ is selected in such a way that exactly t observations are "below ξ," that is, if ξ is any number greater than the (t)th smallest observation and smaller than or equal to the $(t + 1)$th smallest observation. (If no such ξ exists because of ties, then t is not a possible value by either method.) Note that method (2) does not require any consideration of ξ at all, however.

For example, consider the two samples[1] below where $m = 10$ and $n = 12$.

X_i: 13.8, 24.5, 20.7, 22.5, 26.5, 14.5, 6.4, 20.0, 17.1, 15.5

Y_j: 16.2, 23.9, 24.3, 17.8, 15.7, 14.9, 6.1, 11.1, 16.5, 17.9, 15.3, 14.3

In case (1), if we take $\xi = 18.0$, we find that 5 of the X's and 10 of the Y's are less than 18.0 so that $A = 5$ and $B = 10$; this gives Table 3.2. For case (2),

---

[1] These data are from Mosteller, F. and D. Wallace [1964, Sect. 4.8 at pp. 174–175], *Inference and Disputed Authorship: The Federalist*, Addison Wesley Publishing Co., Reading, Mass. The Y's are scores computed in a certain way for the 12 "Federalist Papers" whose authorship is in dispute between Hamilton and Madison. More specifically, Y_j is the natural logarithm of the odds provided by the data in favor of Madison's having written the jth disputed paper, under certain assumptions about the underlying model, except that it has been adjusted to allow for the length of the paper. The X's are scores computed in the same way for 10 segments of about 2,000 words each taken from material known to be by Madison. With the adjustment for length, the X's and Y's should come from approximately the same population if the model is reasonably good and if the disputed papers are by Madison. If the X's and Y's are not from the same population, this by no means indicates that the disputed papers are by Hamilton—the Y's are vastly different from the scores for Federalist Papers known to be by Hamilton. The indication would be rather that something remains to be explained, perhaps an inadequacy in the model. The adequacy of the model is explored extensively by Mosteller and Wallace.

Table 3.2					Table 3.3		
	X's	Y's			X's	Y's	
< 18.0	5	10	15	Below	3	5	8
≥ 18.0	5	2	7	Above	7	7	14
	10	12	22		10	12	22

if we take $t = 8$, we find that, of the eight smallest observations in both sets combined, 3 are X's and 5 are Y's; this gives Table 3.3. The same table would be obtained by method (1) for any ξ in the interval $15.3 < \xi \leq 15.5$.

If the X's and Y's have the same distribution, then for any number ξ, we have $P(X < \xi) = P(Y < \xi)$. This of course is equivalent to saying that for some fraction p, ξ is a quantile of the same order p in both populations. For method (1), where ξ is a preselected constant, it follows that A and B, the respective numbers of "successes" (observations less than ξ), are independently binomially distributed with the same parameter $p = P(X < \xi) = P(Y < \xi)$. The standard test for the equality of proportions then provides a test of identical distributions. When we apply the standard test for equality of proportions to a table obtained by method (1), we call it the *two-sample sign test with fixed* ξ. The standard test for equality of proportions (Fisher's exact test or an approximation) is conditional on the observed value of $A + B = t$. Under this conditioning, the marginal totals in Table 3.1, that is, m, n, t, and $N - t$, are all fixed. Hence once one of A, B, $C = m - A$, and $D = n - B$ is known, the others are determined, and the test may as well be based on the conditional distribution of say A given t.

In the two-sample sign test with fixed ξ, the dichotomizing point ξ is a constant chosen without knowledge of the observations. An unfortunate choice of ξ might produce a test with very low power. For example, if ξ is chosen too small, then most of the observations are likely to be above ξ; this makes t small and leaves A with a small range of possible values, which suggests that A will not be a powerful test statistic.

If we do not require that ξ be chosen in advance, a natural choice for ξ is some particular quantile of the combined sample of X's and Y's. As long as there are no tied observations, such a choice fixes t rather than ξ (as in method (2) above), since a sample quantile is an order statistic of fixed order. However, as noted previously, fixing ξ leads eventually to conditioning on t, albeit at a value depending on the data. If we are going to condition on t eventually anyway, might we not fix t initially as well?

A test based on A for preselected t (method (2)) is called a *quantile test*. When t is selected so that the observations are dichotomized at the combined sample median, the test is called the *median test*. Notice that the marginal totals in Table 3.1 are again fixed, so that any cell determines the entire table. Furthermore, the null distribution of A given t is the same as before. (Indeed, this conditional null distribution applies also when the choice of t (or ξ) is not made in advance but is based on the combined sample, as will be discussed.)

Specifically, if there are no ties at the combined sample median, the median test for N even is equivalent to always choosing the value $t = N/2$; for N odd it is equivalent to $t = (N - 1)/2$ if the combined sample median is counted as "above" rather than "below," as it is by our earlier, arbitrary convention. Ties will be discussed in Sect. 3.3.

As an example, we develop the 2×2 table that arises when the median test is applied to the Mosteller and Wallace data given earlier in this section. Since $N = 22$, we choose $t = 11$. The smallest $t = 11$ observations in the combined sample include four X's and seven Y's, as Table 3.4 shows. Fisher's exact test or an approximation may be applied to this 2×2 table. We need not consider the combined sample median explicitly, but it is any number between the eleventh and twelfth observations in the ordered pooled sample. These observations are 16.2 and 16.5 respectively. Dichotomizing at say 16.4 leads to Table 3.4, as would dichotomizing at any other ξ in the interval $16.2 < \xi \le 16.5$.

The two-sample sign test for fixed ξ is not a rank test because it does not depend only on the ranks of the two samples. This test would be appropriate to use when the measurement scale has only a small number of possible (or likely) values, since then it is natural to choose ξ equal to the central value expected.

The two-sample median and other quantile tests are particularly useful in analyzing data related to experiments involving life testing because they permit termination of the experiment before all units under test have expired. The information needed to perform the test is complete once t units have expired, and sometimes well before that (Problem 5). The control median test and the first-median test are variants of the two-sample quantile tests with particular forms of termination rules. These variants reach the same decisions as a two-sample quantile test (usually earlier) and hence coincide if sampling is terminated as soon as a decision is reached, except that the two tails may correspond to different two-sample quantile tests (Problems 6 and 7). Rosenbaum [1954] gives a test which is equivalent to a special case of a two-sample quantile test since it is based on the number of observations in one sample that are larger than the largest value in the other sample.

Table 3.4

	X's	Y's	
Below combined median	4	7	11
Above combined median	6	5	11
	10	12	22

3.2 Fisher's Exact Test for 2 × 2 Tables

The general test known as Fisher's exact test provides a method for analyzing 2 × 2 tables like Table 3.1 that arise in two-sample sign tests. In Fisher's test, the marginal totals m, n, t, and $N - t$ are all fixed, either initially or after conditioning, and the test is based on the conditional distribution of A given t.

Assume that the X's and Y's that produced Table 3.1 are independent. Under the null hypothesis that the X's and Y's have the same distribution, the data in Table 3.1 represent N independent observations from a single population. As long as either ξ or t is preselected, for any given t, m, and n all subsets of size t are equally likely to be the subset containing the t smallest observations, and hence any set of t observations out of the N is as likely as any other set of t to constitute the observations in the first row of Table 3.1. It follows that the conditional distribution of A given t is the hypergeometric distribution (Problem 8), with discrete frequency function given by

$$f(A \mid m, n, t) = \binom{m}{A} \binom{n}{t - A} \bigg/ \binom{N}{t}. \tag{3.1}$$

This is true, regardless of underlying distributions, if a given set of N units is separated into two samples by randomization (Problem 8).

The test based on this null distribution is referred to generally as *Fisher's exact test*. Other tests, such as the chi-square test for the equality of two proportions, are really approximations to the one based on (3.1). Fisher's exact test is appropriately applied in situations which may arise in the following three conceptually different ways.[2]

(1) (All marginal totals fixed initially). The margins may all be fixed by the rules leading to the table, as in a quantile test, such as the median test, where m and n are the sample sizes and the dichotomization is made so that t will have a certain value.

(2) (Equality of proportions). A and B are binomial random variables based on samples of size m and n respectively, and each has the same parameter p under the null hypothesis. Then conditioning on $A + B = t$ fixes all the marginal totals. This situation includes the two-sample sign test with fixed ξ, introduced in Sect. 3.1.

(3) (Double dichotomy). A single sample of N observations may be dichotomized in two distinct ways (for example, sex and employment status), the two dichotomies being independent under the null hypothesis. Then all the marginal totals are random variables, but conditioning on m and t fixes all the margins. This case does not arise in two-sample tests.

In our situation, as long as the choice of t in (1), or the dichotomizing point ξ in (2), depends only on the combined sample observations and not

---

[2] There are, however, 2 × 2 tables for which Fisher's exact test is not appropriate. (See Sect. 8, Chap. 2 for examples.)

on which observations are X's and which are Y's, the null distribution of A given t is given by (3.1).

The one-tailed P-value of Fisher's exact test is the cumulative probability in (3.1) of the observed A or less for the left tail, and the observed A or more for the right tail; these are tail probabilities in the hypergeometric distribution.

Tables of the hypergeometric distribution are available [for instance, Lieberman and Owen, 1961], but they are necessarily bulky. In order to perform the median test in the absence of ties, only one value of t is required for each combination of m and n, so that more convenient tables are possible. Table E is designed for use with the median test for $t = N/2$ if N is even and $t = (N \pm 1)/2$ if N is odd, when $m \leq n$. For example, it applies to Table 3.4 and gives a P-value for $A \leq 4$ of 0.335. If $m > n$ the designations X and Y can be interchanged so that A still represents the observed number "below" in the smaller sized sample. This is actually equivalent to basing the test on B instead of A, with large B corresponding to small A. If other values of t are required, as when there are ties at the combined sample median or naturally dichotomous observations (see Sect. 3.3), Table E cannot be used. Notice that A is symmetrically distributed for $t = N/2$.

In the absence of tables or outside the range of available tables, we must use a computer program or approximations to the null distribution. The most common approximation is based on Z^2, the chi-square statistic corrected for continuity, or on its signed square root, Z, which is approximately the normal deviate corresponding to the one-tailed P-value and can therefore be referred to Table A. An advantage of Z is that it reflects the direction of the sample difference; Z^2 masks this direction, and hence can only be used for two-tailed tests. Formulas for Z are

$$Z = \left(A \pm \frac{1}{2} - \frac{mt}{N}\right)\left[\frac{N^3}{mnt(N-t)}\right]^{1/2} \qquad (3.2a)$$

$$= \left[A(n-B) - (m-A)B \pm \frac{N}{2}\right]\left[\frac{N}{mnt(N-t)}\right]^{1/2}, \qquad (3.2b)$$

where the $\pm$ term represents the continuity correction and is to be taken as $+$ if A is included in the lower tail and $-$ if in the upper tail. For the data of Table 3.2, Equation (3.2b) gives

$$Z = [5(2) - 5(10) + 11]\left[\frac{22}{10(12)(15)(7)}\right]^{1/2} = -1.212,$$

corresponding to a normal distribution tail probability of 0.113 from Table A. The correct value is also 0.113, but such close agreement is unusual. Fisher and Yates [1963, Table VIII] give the critical values of Z for the exact test at $\alpha = 0.025$ and 0.005. These values indicate that the normal approximation is quite accurate in the case of the median test provided the smaller sample size is at least 12. If m/N and t/N are both far from $\frac{1}{2}$, however, the approximation is not very accurate for one-tailed probabilities.

The test based on chi-square is popularly known as "the" test for 2×2 tables, but it is really just an approximation to Fisher's exact test. The exact test seems to be less frequently used, probably because tables of the chi-square distribution are much more accessible and sample sizes are frequently large anyway.

*Several binomial approximations are also available, but they require the use of binomial tables which are themselves limited by having two more parameters than normal tables. They work best when the table is rearranged so that the two margins opposite A are the two smallest margins. That is, if necessary, interchange the columns to make $m \leq n$ and the rows to make $t \leq N - t$. The simplest binomial approximation is to treat A as binomial with parameters

$$n_1 = \min(m, t), \qquad p_1 = \frac{\max(m, t)}{N}. \tag{3.3}$$

This amounts to treating the largest margin as belonging to an infinite sample. It matches the actual mean and range of A, but gives a variance too large by omission of the finite population correction factor $(N - n_1)/(N - 1)$.

A second approximation [Sandiford, 1960] matches the mean and approximately the variance, but not the upper limit. The procedure is to treat A as binomial with parameters

$$n_2 = \text{an integer near } \frac{mt(N - 1)}{N(N - 1) - n(N - t)}, \qquad p_2 = \frac{mt}{n_2 N}. \tag{3.4}$$

An integer value of n_2 is used solely to facilitate entering binomial tables. Ord [1968] gives a correction factor that improves this approximation.

The probability of A or less may also be approximated by the corresponding binomial probability with parameters

$$n_3 = \min(m, t) = n_1, \qquad p_3 = \frac{\max(m, t) - A/2}{N - (n_3 - 1)/2}. \tag{3.5}$$

This is not equivalent to treating A as binomial, since p_3 depends on A. For this reason, it is not easy to see what mean and variance this approximation assigns to A, but it obviously gives the correct range. This is the first approximation given by Wise [1954], and is actually an upper bound on the probability when $A = 0$. It is based on approximating the sum of hypergeometric probabilities by the incomplete beta function plus a correction factor.

To apply these three approximation to data in Table 3.2, we first rearrange it in the form of Table 3.5.

Table 3.5

	X's	Y's	
≥ 18.0	5	2	7
< 18.0	5	10	15
	10	12	22

We then find from (3.3), (3.4) and (3.5):

$$n_1 = 7, \quad p_1 = \frac{10}{22} = 0.4546, \quad P\text{-value} = 0.159;$$

$$n_2 = 5, \quad p_2 = \frac{10(7)}{5(22)} = 0.6363, \quad P\text{-value} = 0.104;$$

$$n_3 = 7, \quad p_3 = \frac{10 - 2.5}{22 - 3} = 0.3947, \quad P\text{-value} = 0.091.$$

As mentioned earlier, the exact P-value is 0.113. Here the second approximation is better than the third. In other cases, the third approximation may be better than the second, and both are almost always better than the first.

These and other approximations based on the binomial, Poisson, normal and other distributions are discussed more fully in Lieberman and Owen [1961] and Johnson and Kotz [1969]. Peizer, extending Peizer and Pratt [1968] and Pratt [1968], developed an excellent normal approximation that is easily calculated. It has been refined and studied by Ling and Pratt [1981] and is given at the end of Table E. See also Molenaar [1970].

3.3 Ties

Ties present no problem in the two-sample sign test with fixed ξ because we defined "above" as meaning greater than or equal to and "below" as strictly below. Ties are also easily handled for a quantile test, but a brief discussion is needed here. Consider the median test, for example; then we intend to choose $t = N/2$ or $(N - 1)/2$. However, if ties occur at the median of the combined sample, then dichotomizing at the median will ordinarily lead to some other value of t, a smaller value when the observations equal to the median are counted as "above." The value of t could be preserved by breaking these ties at random, along the lines of Sect. 6 of Chap. 3. A more appealing procedure would be to dichotomize at a point slightly above or slightly below the median, whichever value makes t closer to $N/2$. This is equivalent to keeping the sample median as the dichotomizing point and redefining the terms "above" and "below" in order to make t as close as possible to $N/2$. In other words, the observations at the median are assigned to that category, "above" or "below," which contains fewer other observations. We are free to do this since t may be chosen as any function of the combined sample without changing the null distribution given in (3.1), as remarked earlier.

It may sometimes be preferable, especially if the observations are changes and the median occurs at "no change," to omit the observations at the median. Then "below" means strictly below and "above" means strictly above, and the sample sizes are the numbers of observations different from the median. The hypergeometric distribution continues to apply under the null hypothesis, however (Problem 9c).

In the special case where the observations are not only not all different, but also have only two possible values, the data are inherently in the form of a 2 × 2 table. Then there is no freedom of choice regarding the value of t and the situation is more like the case of having a fixed ξ. However, if one of the two possible values is called "below" and the other "above," this could be considered an extreme case of the situation with ties described earlier.

3.4 Corresponding Confidence Procedures

Suppose that the shift assumption, as stated in Eqs. (2.1), (2.2) or (2.3), holds so that X and $Y - \mu$ have the same distribution, but μ is unknown. In order to test a null hypothesis which specifies a particular value for μ using a two-sample sign test procedure, we could subtract μ from each Y_j and then apply a two-sample test for identical distributions to the observations $X_1, \ldots, X_m, Y_1 - \mu, \ldots, Y_n - \mu$. The confidence region corresponding to such a test consists of those values of μ which would be "accepted" when so tested. We could proceed by trial and error, testing various values of μ to see which ones lead to "acceptance." However, for a two-sample median or other quantile test, there is a very simple way to obtain these confidence limits explicitly, as we will now see.

Consider a two-sample quantile test at level α specifying t as the marginal total of the first row. Let a and a' be the lower and upper critical values of A, that is, $P(A \leq a) + P(A \geq a') \leq \alpha$ under the hypergeometric distribution with parameters m, n, and t. Then we would "accept" μ if there are at least $(a + 1)$ X's and at most $(a' - 1)$ X's among the t smallest of $X_1, \ldots, X_m$, $Y_1 - \mu, \ldots, Y_n - \mu$, and reject μ otherwise. This region of "acceptance" is the interval between two confidence limits which can be very simply stated in terms of the order statistics of the two samples as follows.

Let $X_{(1)}, \ldots, X_{(m)}$, be the X's rearranged in order of increasing (algebraic) value, so that $X_{(1)} \leq X_{(2)} \leq \cdots \leq X_{(m)}$. Define $Y_{(1)}, \ldots, Y_{(n)}$ similarly. Then the test procedure "accepts" μ if and only if (Problem 10)

$$Y_{(t-a'+1)} - X_{(a')} \leq \mu \leq Y_{(t-a)} - X_{(a+1)} \tag{3.6}$$

except perhaps at the endpoints, where the procedure has not been defined. Equation (3.6) then gives the confidence interval for the shift μ which corresponds to a two-sample quantile test with first row total t. The confidence level is $1 - \alpha$, where $\alpha = P(A \leq a) + P(A \geq a')$ according to the hypergeometric distribution for this m, n, and t. Of course, either the left-hand or right-hand side of (3.6) may be used separately as a one-sided confidence bound; then α is $P(A \geq a')$ or $P(A \leq a)$ respectively. If the distributions are not continuous, the confidence levels are conservative as long as the end points are included in the confidence regions (Problem 107).

The values of a and a' for given α, or of α for some selected a and a', can be found from Table E for $t = N/2$ if N is even, or $t = (N \pm 1)/2$ if N is odd, that is, the values of t corresponding to the median test. It may be desirable to use other values of t, especially since the choice of α for any one t is very limited. However, this requires a more extensive table, since computation of α must of course be based on the value of t actually used.

Suppose that $m = n = 10$ and we use $t = N/2 = 10$. Then Table E shows that choosing $a = 2$, $a' = 8$ gives $\alpha = 0.0115 + 0.0115 = 0.0230$. Thus at level $1 - \alpha = 0.9770$, (3.6) gives

$$Y_{(3)} - X_{(8)} \leq \mu \leq Y_{(8)} - X_{(3)}$$

as the confidence interval for the shift μ which corresponds to the median test.

Mood [1950, pp. 395–398] suggests confidence bounds of the form $Y_{(r)} - X_{(s)}$ without relating them to the median test and its counterpart with arbitrary t. (See also Mood and Graybill [1963, pp. 412–416], and Mood et al. [1974, pp. 521–522].) It is not obvious that the formula given there for α is equivalent to the hypergeometric formula (Problem 11). The confidence limits corresponding to the control median test, the first median test, and Rosenbaum's test are all of the form (3.6) except that the upper and lower limits employ different values of t (Problem 13).

3.5 Power

In this section we discuss methods of finding the power of two-sample sign tests against alternatives under which the X's and Y's are independent random samples from two different populations. Consider first the two-sample sign test with ξ fixed and selected in advance. Here A and B are independently binomially distributed with respective parameters $p_1 = P(X < \xi)$ and $p_2 = P(Y < \xi)$, that is, the proportions of the X and Y populations below ξ. Accordingly, the power of the two-sample sign test with fixed dichotomizing point ξ is simply the power of the test for equality of two proportions against the alternative p_1 and p_2. This is not a particularly nonparametric problem and will not be discussed further here.

Consider now the median test, or, more generally, any quantile test with the value of t fixed in advance. Suppose that the populations are continuous, so that we may ignore the possibility of ties and hence of not being able to attain the chosen t. Then, by (3.6), the two-tailed test rejects the null hypothesis of equal populations if and only if

$$Y_{(t-a'+1)} - X_{(a')} > 0 \quad \text{or} \quad Y_{(t-a)} - X_{(a+1)} < 0, \tag{3.7}$$

where a and a' are respectively the lower and upper critical values of A. The power of this test is then the probability of (3.7). Each inequality in (3.7) by itself gives a one-tailed test whose power is the probability of that inequality. Consider the second inequality. The order statistics $X_{(a+1)}$ and $Y_{(t-a)}$ are independent, and for any two specific populations, their densities can be simply expressed algebraically in terms of the population distributions (Problem 30, Chap. 2). The probability $P(Y_{(t-a)} - X_{(a+1)} < 0)$, and hence the power of a lower-tailed test, can therefore be obtained from these densities by a two-variable integration (Problem 14a). The power of an upper-tailed test can be obtained similarly. The sum of these probabilities gives the power of a two-tailed test since the two inequalities are mutually exclusive (Problem 14b). Hence the power of any quantile test with t fixed in advance is easily found using the expression (3.7) of the rejection region in terms of order statistics. Calculation of the power directly from the original definition of the test in terms of a 2×2 table appears much more difficult. The exact power of the median test in small samples was investigated in Gibbons [1964c] for various alternatives.

The approximate normality of the order statistic leads to an approximation for the probability of (3.7). Specifically, if m and n are large and t/N is not near 0 or 1, then a/m and $(t - a)/n$ will not be near 0 or 1 provided α is not too small (Problem 15). If X has a density which is continuous and nonzero at the quantile of order $(a + 1)/m$ for the X distribution, then it follows that $X_{(a+1)}$ is approximately normal with some mean $\mu(x, a + 1)$ and variance $\sigma^2(x, a + 1)$ (see Problem 31, Chap. 2). Similarly $Y_{(t-a)}$ is approximately normal with mean $\mu(y, t - a)$ and variance $\sigma^2(y, t - a)$, say. Since $X_{(a+1)}$ and $Y_{(t-a)}$ are independent, their difference is also approximately normal, and $P(Y_{(t-a)} - X_{(a+1)} < 0)$ can be approximated by the cumulative standard normal distribution $\Phi(z)$, with

$$z = \frac{\mu(x, a + 1) - \mu(y, t - a)}{[\sigma^2(x, a + 1) + \sigma^2(y, t - a)]^{1/2}}. \tag{3.8}$$

One approximation for the mean and variance is given by (Problem 16)

$$\mu(x, a + 1) = \text{the quantile of order } p \text{ of the } X \text{ distribution}, \tag{3.9}$$

$$\sigma^2(x, a + 1) = [p(1 - p)/m]\{f[\mu(x, a + 1)]\}^{-2}, \tag{3.10}$$

where $p = (a + 1)/m$ and f is the density of X; $\mu(y, t - a)$ and $\sigma^2(y, t - a)$ are similarly defined.

3.6 Consistency

The criterion for consistency in one-sample (or paired-sample) tests was given in Sect. 3.4 of Chap. 3. A test based on two samples is called *consistent* against a particular alternative if its power against that alternative approaches 1 as the sample sizes m and n approach infinity. Of course, the test must be defined for all sample sizes m and n. Thus consistency in the two-sample case is, strictly speaking, a property of a double sequence of tests, one test for each pair (m, n). Sometimes it is assumed that m and n are related in such a way that m/n approaches a positive, finite limit as m and n both approach infinity. We do not assume this, however, and permit m and n to vary independently.

The median test is consistent against any alternative such that X and Y have different medians. More precisely, for each m and n, consider a two-tailed median test, and suppose that the level in each tail is at least ε for every m and n, where ε is some positive constant which does not depend on m and n (that is, the level in each tail is bounded away from 0 as m and n approach ∞). Then the probability of rejection approaches 1 as m and n approach infinity if the X and Y populations have different medians. If the medians are not both unique, this must be interpreted as meaning that *no* median of X is also a median of Y. Figure 3.1 shows two sets of hypothetical c.d.f.'s where the medians are not both unique. In 3.1(a), the X and Y medians have no common value even though the c.d.f.'s are quite similar. In Fig. 3.1(b), the c.d.f's are quite disparate but they have a point in common which is a median of both populations. A necessary and sufficient condition for X and Y to have no common median (Problem 17), which applies whether the medians are unique or not, is that there exists a ξ, such that either

$$P(X \le \xi) < 0.5 < P(\xi \le Y) \quad \text{or} \quad P(Y \le \xi) < 0.5 < P(\xi \le X). \quad (3.11)$$

In the first case, ξ is smaller than any median of X but larger than any median of Y, while in the second case, ξ is larger than any median of X and smaller than any median of Y, as would be possible in Fig. 3.1(a).

Similarly, the one-tailed median test is consistent against alternatives with medians which differ in the appropriate direction.

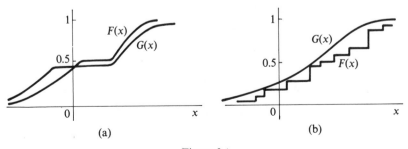

Figure 3.1

These facts can be proved (Problem 18a) using the fact that the left and right sides of (3.6) are both consistent estimators of the difference of the medians. Alternatively, (3.11) can be used along with the consistency of the two-sample sign test with fixed ξ (Problem 18b).

3.7 "Optimum" Properties

As developed earlier, the median test is to a considerable extent a two-sample analogue of the one-sample sign test. However, the optimum properties of the sign test (Sect. 3.2, Chap. 2) do not carry over to the median test, or to the other two-sample quantile tests (that is, when t is chosen in advance). A proof like that in Sect. 3.3 of Chap. 2 would require specifying families of distributions for which the entries in the 2×2 table with t fixed are sufficient statistics, and this seems impossible to do.

The usual test for 2×2 tables does have optimum properties like those of the ordinary binomial test (Sects. 7 and 8, Chap 1). While these properties are usually expressed in contexts where not all margins are fixed (equality of proportions and double dichotomy, in the terminology of Sect. 3.2), they can also be stated for all margins fixed, as in the case of the median test. However, this gives optimality only among tests based solely on the entries in the 2×2 table. Since the question of particular significance for non-parametric statistics is whether a test based on some other statistic might be better, we will not pursue the exact sense in which the median test is optimum among tests based on the entries in the 2×2 table.

The two-sample sign test with fixed ξ does have optimum properties analogous to those of the one-sample sign test. Specifically, consider Table 3.1 with "below" meaning "below ξ" where ξ is chosen in advance. Among tests of the null hypothesis that the probability below ξ is the same in both populations, Fisher's exact test applied to 2×2 tables like Table 3.1 has the following properties: (a) a one-tailed test is uniformly most powerful un-biased against the alternative that the probability below ξ is larger in the X than in the Y population, or smaller, whichever is appropriate; (b) any two-tailed test is admissible; and (c) a two-tailed, (conditionally) unbiased test is uniformly most powerful among unbiased tests against the alternative that the probability below ξ is not the same in both populations (Problem 19a). The restriction that the X's be identically distributed and the Y's be identically distributed can also be relaxed as in Sect. 3.2 of Chap. 2 (Problem 19b).

Thus the two-sample sign test with fixed ξ has some optimality properties that are truly nonparametric in nature, while the quantile tests apparently do not. This may seem to suggest that the sign test with fixed ξ is generally to be preferred to the median test. This is the case when one is interested in a single value of ξ, known in advance. Such situations are rare, however, and the advantages of optimality for fixed ξ are ordinarily outweighed by the advantages of the median test in being able to select the point of dichotomization sensibly in light of the combined sample.

3.8 Weakening the Assumptions

We have been assuming that all the observations are independent and identically distributed under the null hypothesis. The name and construction of the median test might suggest that its level would be retained as long as all observations are drawn from distributions with the same median. However, this is unfortunately not the case. If the X's are drawn from one population and the Y's from another population, where the medians are the same but the scale parameters differ, the level may be seriously affected even in large samples (see Pratt [1964] for further discussion and numerical, asymptotic results). The same point applies, a fortiori, to the corresponding confidence procedure. The two-sample sign test with fixed ξ is, of course, valid whenever the probability below ξ is the same for the two populations, but it would seldom happen that we know in advance a value of ξ for which this assumption holds under a null hypothesis that allows different scales in the two populations.

In a treatment-control comparison where the units are assigned at random to the two groups, the randomization itself guarantees the level of the median test for the null hypothesis that the treatment has no effect. To be more specific, suppose the X's refer to a control group and the Y's to a treatment group. Given the N units in the experiment, if any set of m of them is as likely as any other set of m to be the control group, and each unit would yield the same measurement whether treated or not, then the probability of rejection by the median test is the usual null probability for 2×2 tables obtained from the hypergeometric distribution. The randomization does not guarantee that the level is preserved in the corresponding confidence procedure, however, in the absence of some property such as no interaction between treatment and units, or the shift assumption. (See Sect. 2, Problem 21, and for further detail and discussion, Sect. 7 of Chap. 2, and Sect. 9 of Chap. 8.) The same statements apply to any other quantile test.

Another kind of weakening is possible for all one-tailed, two-sample sign tests, whether a quantile test or one with fixed ξ. Suppose the X_i and Y_j are independent, but not necessarily identically distributed, and consider any one-tailed, two-sample sign test with rejection region in the upper tail of A (too many X's are in the "below" category). This test rejects with probability at most α (the exact level when all observations are independently, identically distributed) if

$$P(X_i < z) \le P(Y_j < z) \quad \text{for all } z, i, \text{ and } j, \tag{3.12}$$

and with probability at least α if

$$P(X_i < z) \ge P(Y_j < z) \quad \text{for all } z, i, \text{ and } j. \tag{3.13}$$

(Compare Eqs. (3.21) and (3.22), Chap. 3.) Under (3.12), any X_i is less likely to be to the left of any specified point than any Y_j is, so that the distribution of every X_i is "to the right" of ("*stochastically larger*" than) the distribution of every Y_j. See Fig. 2.1(a) for a graphic illustration of this relationship. Similarly (3.13) means that all the X's are "*stochastically smaller*" than all the Y's. Since the probability of rejection is at most α when (3.12) holds, the null hypothesis could be broadened to include (3.12) without affecting the significance level. Similarly it is natural to broaden the alternative to include (3.13); since the probability of rejection is at least α when (3.13) holds, the test is by definition unbiased against (3.13).

On the other hand, against certain alternatives under which the X's are drawn from one population and the Y's from another population with a different median, the power of the median test is less than its level. The test is then biased against this alternative; this is true even for the one-tailed test in the indicated direction (Problem 22).

*The statements of the next-to-last paragraph are consequences (Problem 23) of the following fact, which is of interest in itself. Suppose a test ϕ is "increasing" in the Y direction in the sense that, if ϕ rejects for $X_1, \ldots, X_m$, $Y_1, \ldots, Y_n$ and any X_i is decreased or Y_j increased, ϕ still rejects. (The one-tailed two-sample sign tests rejecting if A is too large have this property, by Problem 23.) Then the probability that ϕ will reject increases (not necessarily strictly) when the distribution of X_j is moved to the left or the distribution of Y_j is moved to the right, that is, when the c.d.f. F_i of X_i is replaced by F_i^* where $F_i^*(x) \ge F_i(x)$ for all x, or when the c.d.f. G_j of Y_j is replaced by G_j^* where $G_j^*(y) \le G_j(y)$ for all y. Formally, for randomized tests ϕ, we have the following theorem.

Theorem 3.1 *If* $X_1, \ldots, X_m, Y_1, \ldots, Y_n$ *are independent with c.d.f.'s* F_i, G_j *and* $\phi(X_1, \ldots, X_m, Y_1, \ldots, Y_n)$ *is a randomized test function which is decreasing in each* X_i *and increasing in each* Y_j, *then the probability of rejection*

$$\alpha(F_1, \ldots, F_m, G_1, \ldots, G_n) = E_{F_1, \ldots, G_n}[\phi(X_1, \ldots, X_m, Y_1, \ldots, Y_n)] \tag{3.14}$$

is an increasing function of the F_i *and a decreasing function of the* G_j. *Specifically,*

$$\alpha(F_1, \ldots, F_m, G_1, \ldots, G_n) \le \alpha(F_1^*, \ldots, F_m^*, G_1^*, \ldots, G_n^*) \tag{3.15}$$

if $F_i^*(x) \ge F_i(x)$ *for all* x *and* i *and* $G_j^*(y) \le G_j(y)$ *for all* y *and* j.

The proof is similar to that of Theorem 3.3 of Chap. 3 and is left as Problem 42.*

4 Procedures Based on Sums of Ranks

In this section we consider again the situation of two mutually independent samples and the null hypothesis that the two populations sampled are identical. The median test of the previous section makes use of only the number of observations from each sample which are above or below the combined sample median. A test which takes into account more of the available relevant information might be expected to have greater power ordinarily. A simple but important and useful test of this type is the rank sum test, which will be discussed in this section. This test is frequently called the Wilcoxon test or Mann–Whitney test (or sometimes even the Mann–Whitney–Wilcoxon test). Kruskal [1957] has traced the history of rank sum tests as far back as Deuchler [1914]. However, the publications initiating the tests in modern terms are Wilcoxon [1945], Mann and Whitney [1947], and also Festinger [1946].

The properties of the rank sum will be developed here under the assumption that $X_1, \ldots, X_m, Y_1, \ldots, Y_n$ are independent observations drawn from two populations and for the null hypothesis that the populations are equivalent. The shift assumption is not necessary here and usually will not be made, although there is reason to think that the rank sum test has especially good performance for location alternatives. If we do make the shift assumption, the null hypothesis can be stated as $H_0: \mu = 0$ and further, the rank sum test procedure can easily be modified to provide a test of $H_0: \mu = \mu_0$. The corresponding confidence procedures for the shift parameter are explained in Sect. 4.3. In Sect. 4.6, we will see that the assumptions can be weakened in certain ways without affecting the level of the test or confidence procedure. We will usually assume that both populations are continuous so that we need not consider the possibility of ties either across or within samples. The problem of ties is considered explicitly in Sect. 4.7.

4.1 The Rank Sum Test Procedure

To carry out the rank sum test procedure, we first combine the m X's and n Y's into a single group of $m + n = N$ observations, which are all different because of the continuity assumption. We then arrange the pooled observations in order of magnitude, but keep track of which observations are from which sample. We assign the ranks $1, 2, \ldots, N$ to the combined ordered observations, with 1 for the smallest and N for the largest.

The data shown in Table 4.1 have been ranked by this procedure. (Often in practice the first row is omitted and the values which are from say the X sample are underlined or similarly indicated.)

The rank sum can be defined as the sum of the ranks in either sample; we use R_x to denote the sum of the ranks of the X observations and R_y for

Table 4.1[a]

Sample	Y	X	X	Y	Y	X	X	Y	X	Y
Value	1.25	1.75	3.25	4.25	5.25	6.25	6.75	7.25	9.00	10.00
Rank	1	2	3	4	5	6	7	8	9	10

[a] These data are from United States Senate [1953], *Hearings Before the Select Committee on Small Business*, Eighty-third Congress, First Session on Investigation of Battery Additive AD-X2 (March 31, June 22–26). The X's and Y's refer to untreated batteries and batteries treated with AD-X2 respectively. The values given here were obtained by averaging the ranks given on performance of the batteries by two representatives of the manufacturer. The assumptions of the beginning of this section are not satisfied, but the batteries for treatment were selected randomly from the 10 batteries, and this also validates the test, as discussed in Section 4.6.

the sum of the ranks of the Y observations. Since $R_x + R_y$ is the sum of all the ranks, $1 + 2 + \cdots + N = N(N + 1)/2$, we have

$$R_x + R_y = N(N + 1)/2 \tag{4.1}$$

and the tests based on R_x and R_y are therefore equivalent. In Table 4.1 we have

$$R_x = 2 + 3 + 6 + 7 + 9 = 27,$$

$$R_y = 1 + 4 + 5 + 8 + 10 = 28,$$

and

$$R_x + R_y = 55$$

which is in agreement with (4.1). Either statistic R_x or R_y is commonly called the Wilcoxon rank sum statistic.

The rank sum test can also be based on another equivalent statistic, usually called the Mann–Whitney statistic and denoted by U. U can be defined as the number of (X, Y) pairs for which $X > Y$, or informally as the number of X's greater than Y's (or Y's smaller than X's). Equivalently, U is the sum over $i = 1, \ldots, m$ of the number of Y's which are smaller than X_i, or the sum over $j = 1, \ldots, n$ of the number of X's larger than Y_j. In Table 4.1 there are 5 X's larger than the smallest Y, 3 X's larger than the second smallest Y, 3 larger than the third Y, 1 larger than the fourth Y, and 0 larger than the largest Y, giving a sum $U = 5 + 3 + 3 + 1 + 0 = 12$.

If we reverse the roles of X and Y, we can define an alternative Mann–Whitney statistic U' as the number of X's which are smaller than Y's (or Y's larger than X's). Then $U + U'$ is the total number of (X, Y) pairs, or

$$U + U' = mn. \tag{4.2}$$

The foregoing definitions lead fairly directly to a linear relation between the Mann-Whitney and Wilcoxon statistics; specifically (Problem 29),

$$U = R_x - m(m + 1)/2 = mn + n(n + 1)/2 - R_y, \qquad (4.3)$$

$$U' = R_y - n(n + 1)/2 = mn + m(m + 1)/2 - R_x. \qquad (4.4)$$

In some cases, it may be easier to find U or U' by calculating R_x or R_y and then using (4.3) or (4.4), rather than to find them directly.

Under the null hypothesis of identical populations, the ranks of the X's are equally likely to be any set of m from the integers $1, 2, \ldots, N$. In other words, the X ranks consistute a random sample of size m drawn without replacement from the set $\{1, 2, \ldots, N\}$. If, on the other hand, the Y population tends to have larger values than the X population, we would expect the X ranks to be smaller than under the null hypothesis. For one-sided alternatives in this direction, the appropriate regions for rejection are then small values of R_x or U, and large values of R_y or U'. The corresponding test against the alternative that the Y population tends to have larger values than the X population rejects for large values of R_x or U, and small values of R_y or U'. An equal-tailed test against the two-sided alternative of a difference in either direction rejects at level 2α if either of the foregoing one-tailed tests rejects at level α.

To test the null hypothesis that the Y population is the same as the X population except for a shift by the specified amount μ_0, the procedure above is followed without change except that μ_0 is subtracted from each Y before the combined samples are arranged in order of size. That is, the rank sum test is applied in exactly the same way as above, but to $X_1, \ldots, X_m, Y_1 - \mu_0, \ldots, Y_n - \mu_0$ instead of $X_1, \ldots, X_m, Y_1, \ldots, Y_n$. The corresponding confidence interval procedures will be discussed in Sect. 4.3.

For investigation of theoretical properties of the rank sum test, two representations of the test statistics are often convenient. For the Mann–Whitney statistic, it is natural to write

$$U = \sum_{i=1}^{m} \sum_{j=1}^{n} U_{ij} \qquad (4.5a)$$

where

$$U_{ij} = \begin{cases} 1 & \text{if } Y_j < X_i \\ 0 & \text{if } Y_j > X_i. \end{cases} \qquad (4.5b)$$

The Wilcoxon rank sum statistic is represented naturally as

$$R_x = \sum_{1}^{N} kI_k \qquad (4.6a)$$

where

$$I_k = \begin{cases} 1 & \text{if the observation with rank } k \text{ is an } X \\ 0 & \text{if the observation with rank } k \text{ is a } Y. \end{cases} \qquad (4.6b)$$

We can, of course, use whichever representation is more convenient for the purpose at hand, since the statistics are linearly related.

4.2 Null Distribution of the Rank Sum Statistics

The exact null distribution of any of these rank sum statistics is based on the fact stated before that, under the null hypothesis of identical distributions, the X ranks constitute a random sample of size m drawn without replacement from the first N integers, where $N = m + n$. Equivalently, all arrangements of the mX's and nY's in order of size are equally likely. With $\binom{N}{m}$ possible arrangements, each one occurs with probability $1/\binom{N}{m}$. This fact determines the null distribution of R_x, and by (4.1), (4.3) and (4.4), also that of R_y, U and U'.

The direct method of generating the null distribution of say R_x is to list all possible arrangements, calculate the value of R_x for each, and tally the results. Then

$$P(R_x = t) = v(t)/\binom{N}{m}$$

where $v(t)$ is the number of arrangements for which the sum of the X ranks equals t. For tabulation it is more efficient to use an easily developed recursive technique (Problem 30). Fix and Hodges [1955] present a more sophisticated approach, tabulating related quantities more compactly than is possible for the distribution itself (Problem 32).

The mean and variance of these rank sum statistics under the null hypothesis are most easily evaluated by using the fact that R_x is the sum of m observations drawn without replacement from the finite population consisting of $\{1, \ldots, N\}$. The mean and variance of this population (Problem 33) are $(N + 1)/2$ and $(N^2 - 1)/12$. The mean and variance (calculated using the finite-population correction factor) of the sample sum are therefore

$$E(R_x) = m(N + 1)/2 \tag{4.7}$$

$$\text{var}(R_x) = [m(N - m)/(N - 1)][(N^2 - 1)/12] = mn(N + 1)/12. \tag{4.8}$$

The means and variances of R_y, U and U' under the null hypothesis can then be found from the relationships given in (4.1), (4.3) and (4.4). The results (Problem 34) are

$$E(R_y) = n(N + 1)/2, \qquad \text{var}(R_y) = \text{var}(R_x) = mn(N + 1)/12, \tag{4.9}$$

$$E(U) = E(U') = mn/2, \qquad \text{var}(U) = \text{var}(U') = mn(N + 1)/12. \tag{4.10}$$

U and U' are identically distributed (Problem 35), and their possible values are all the integers between 0 and mn inclusive. The possible values of R_x are all the integers from $m(m + 1)/2$ to $m(2N - m + 1)/2$, and of R_y the integers from $n(n + 1)/2$ to $n(2N - n + 1)/2$ (Problem 36). The distributions of U, U', R_x and R_y are all symmetric about their respective means for any m and n (Problem 37).

Since all the rank sum statistics are equivalent, a table of the null distribution is needed for only one of these statistics. Table F at the back of the book gives the cumulative tail probabilities of R_x for $m \le n \le 10$. Only the smaller tail probability is given; each entry is both a lower tail probability for $R_x \le m(N + 1)/2$ and a symmetrically equivalent upper tail probability for $R_x \ge m(N + 1)/2$. In order to use this table, the sample with fewer observations should be labeled the X sample. More extensive tables are published in Harter and Owen [1970].

For m and n large, R_x, R_y, U and U' are all approximately normally distributed under the null hypothesis [Mann and Whitney, 1947], with the means and variances given above. Small tail probabilities are generally overestimated by the normal approximation. For sample sizes both smaller than 20, for example, it is better to omit the continuity correction of $\frac{1}{2}$ in such a way as to reduce the tail probability when the standardized normal variable is greater than 2 or, for comparison with critical values from normal tables, when the one-sided significance level is smaller than 0.025. See Jacobson [1963] for more detail.

4.3 Corresponding Confidence Procedures

We now turn to the problem of setting confidence bounds on μ when X_1, $\ldots$, X_m, and $Y_1, \ldots, Y_n$ are independent random samples from two continuous populations and the Y population is the same as the X population except for a shift by the amount μ. The confidence region for μ which corresponds to the rank sum test consists of all values of μ which would be "accepted" if the rank sum test procedure were applied to $X_1, \ldots, X_m$, $Y_1 - \mu, \ldots, Y_n - \mu$. That region could be found by trial and error, testing various values of μ to see which ones lead to "acceptance." A more systematic approach is possible, however, leading to a simpler method, as will now be described.

For the observations $X_1, \ldots, X_m, Y_1 - \mu, \ldots, Y_n - \mu$, the statistic U is the number of pairs $(X_i, Y_j - \mu)$ for which $X_i > Y_j - \mu$, or, equivalently $Y_j - X_i < \mu$. Accordingly, the rank sum test with rejection region $U \le k$ applied for a particular hypothesized value of μ would reject or accept according as μ is smaller or larger than the $(k + 1)$th of the mn differences $Y_j - X_i$ when these differences are arranged from smallest to largest. Thus, the $(k + 1)$th difference from the smallest is the lower confidence bound for μ corresponding to this test, with confidence level $1 - \alpha$, where α is equal to the null probability that $U \le k$. Similarly, the $(k + 1)$th difference from the largest is an upper confidence bound for μ at level $1 - \alpha$. A table of the null distribution of U is not given in this book, but the above procedure is easily carried out using R_x. If c is that number from Table F such that $P(R_x \le c) = \alpha$, then by (4.3) we have

$$k = c - m(m + 1)/2. \tag{4.11}$$

This is equivalent to saying that k is the number of possible values of R_x below the critical value at level α, that is, $k + 1$ is the rank of the lower tail critical value at level α of R_x among all possible values of R_x (Problem 39).

There is a convenient graphical procedure for finding the $(k + 1)$th from the smallest and/or largest of the differences $Y_j - X_i$ and hence for determining the confidence bounds. The method is illustrated in Fig. 4.1 for the data given in Table 4.1. Each of the N observations is plotted on a rectangular coordinate system, the X observations on the abscissa and Y

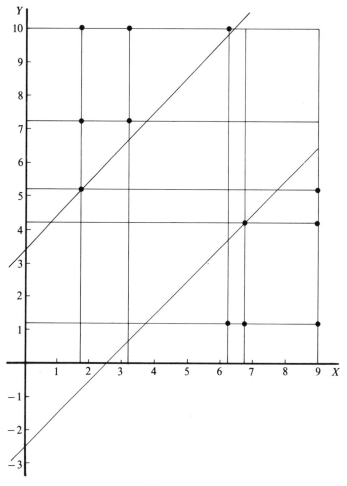

Figure 4.1

observations on the ordinate. A horizontal line is drawn through each Y_j and a vertical line through each X_i. Each of the mn intersections represents a pair of observations (X_i, Y_j). Note that an intersection with an axis does not count unless this axis is the line through an observation equal to 0. The 45° line $Y - X = \mu$, for any number μ as Y intercept, divides the (X, Y) pairs into two groups. Those on the left and above have $Y - X > \mu$, and those on the right and below have $Y - X < \mu$. Thus the upper confidence bound on μ is the Y-intercept of the 45° line which has k intersections to the upper left of it and passes through the $(k + 1)$th intersection. If greater accuracy is desired, the difference $Y_j - X_i$ can be computed for that (X_i, Y_j) pair which gives the $(k + 1)$th intersection. In close cases it may be necessary to compute more than one $Y_j - X_i$ to determine exactly which is the $(k + 1)$th.

We illustrate these calculations using the example of Table 4.1 even though the shift assumption is unreasonable in the situation producing these data. We have $m = n = 5$ and from Table F, $P(R_x \leq 20) = 0.041$ so that $c = 20$ for $\alpha = 0.041$. From (4.11), $k = 20 - 5(6)/2 = 5$. The line in Fig. 4.1 which has Y intercept 3.50 has five intersections above and to the left of it and passes through the sixth intersection (1.75, 5.25), so that $3.50(= 5.25 - 1.75)$ is an upper confidence bound for μ with confidence coefficient 0.959 $(= 1 - 0.041)$. Similarly, -2.50 is a lower bound at this same level, and $-2.50 < \mu < 3.50$ is a confidence interval with level $1 - 2(0.041) = 0.918$.

4.4 Approximate Power

The confidence procedure just developed, corresponding to the rank sum test procedure, required the shift assumption, but that assumption was not made to develop the test or its null distribution theory. This subsection and the next concern further properties of the test procedure, namely its power and consistency, under general alternatives which need not be shifts. We assume only that the X and Y observations are mutually independent, each set having the same continuous distribution.

While the exact power of the rank-sum tests can be evaluated for specific alternatives (see for example, Gibbons [1964c]), the present discussion is limited to asymptotic power. In Sect. 4.2, we stated that the null distributions of U, U', R_x and R_y are asymptotically normal. Since these statistics are also asymptotically normal under most alternative hypotheses, the power of the rank sum test against a specified alternative can be approximated by applying Table A to the appropriate standardized variable once the mean and variance under that alternative are evaluated. (The asymptotic normality and an appropriate standardization follow from results proved by Chernoff and Savage [1958] and Hoeffding [1948] for more general classes of statistics based on two sets of identically distributed observations. For the rank sum statistic, Capon [1961a] gives a generalization to nonidentically distributed sets of observations, along with general expressions for the mean and variance of the test statistic.)

In order to approximate the power of the rank sum test against specified alternative distribution, it will be convenient to introduce the probabilities:

$$p_1 = P(X_i > Y_j) \tag{4.12}$$

$$p_2 = P(X_i > Y_j \text{ and } X_k > Y_j) \tag{4.13}$$

$$p_3 = P(X_i > Y_j \text{ and } X_i > Y_l) \tag{4.14}$$

for all i, j, k, l with $i \neq k$ and $j \neq l$. Hence p_1 is the probability that an X variable exceeds a Y variable; p_2 is the probability that two different X variables both exceed a single Y variable; and p_3 is the probability that an X variable exceeds both of two different Y variables. Integral expressions for these three probabilities are given in Problem 41.

For any X and Y distributions, the moments of U needed for standardization can be expressed in terms of these probabilities as

$$E(U) = mnp_1 \tag{4.15}$$

$$\text{var}(U) = mn[(m - 1)(p_2 - p_1^2) + (n - 1)(p_3 - p_1^2) + p_1(1 - p_1)]. \tag{4.16}$$

We will now prove these results using the expression for U given in (4.5a), where U is the sum of mn indicator variables U_{ij}, defined by (4.5b) ($U_{ij} = 1$ if $X_i > Y_j$). These U_{ij} are Bernoulli random variables, identically distributed although not all independent. In terms of the probability p_1, their mean is

$$E(U_{ij}) = P(X_i > Y_j) = p_1 \quad \text{for all } i, j \tag{4.17}$$

so that

$$E(U) = \sum_{i=1}^{m} \sum_{j=1}^{n} E(U_{ij}) = mnp_1.$$

In terms of the probabilities p_1, p_2 and p_3, the second-order moments of the U_{ij} are

$$\text{var}(U_{ij}) = p_1(1 - p_1) \tag{4.18}$$

$$\text{cov}(U_{ij}, U_{kj}) = P(X_i > Y_j \text{ and } X_k > Y_j) - p_1^2$$

$$= p_2 - p_1^2 \tag{4.19}$$

$$\text{cov}(U_{ij}, U_{il}) = P(X_i > Y_j \text{ and } X_i > Y_l) - p_1^2$$

$$= p_3 - p_1^2 \tag{4.20}$$

$$\text{cov}(U_{ij}, U_{kl}) = 0, \tag{4.21}$$

for all i, j, k, l with $i \neq k$ and $j \neq l$. We now express the variance of U in terms of the moments of U_{ij} (Problem 43) as

$$\text{var}(U) = mn[\text{var}(U_{ij}) + (m-1)\,\text{cov}(U_{ij}, U_{kj}) + (n-1)\,\text{cov}(U_{ij}, U_{il})$$
$$+ (m-1)(n-1)\,\text{cov}(U_{ij}, U_{kl})] \tag{4.22}$$

for $i \neq k, j \neq l$. Substituting (4.18)–(4.21) into (4.22) gives (4.16) immediately.

4.5 Consistency

A two-sample test is called consistent against a particular alternative if the power of the test against that alternative approaches 1 as the sample sizes approach infinity. (For more detail, see Sect. 3.6.) The rank sum test is consistent against any alternative under which the X's and Y's are independent observations from two continuous populations with $p_1 \neq \frac{1}{2}$, where p_1 is the probability that an X exceeds a Y, as at Equation (4.12). More precisely, for each m and n, consider a two-tailed rank sum test with level at least ε in each tail, where ε is a positive constant. Then the probability of rejection approaches 1 as m and n approach infinity if the X and Y populations are such that $p_1 \neq \frac{1}{2}$. Similarly, a one-tailed rank sum test is consistent against alternatives with p_1 different from $\frac{1}{2}$ in the appropriate direction. These consistency properties are easily proved (Problem 46) using the fact that U/mn is a consistent estimator of p_1, as will be shown in Sect. 4.8.

It should be noted that there are situations in which the median test is consistent but the rank sum test is not, and vice versa. For example, if the X and Y distributions have different medians but $p_1 = \frac{1}{2}$, then the rank sum test is not consistent while the median test is. Similarly, if X and Y have equal medians but $p_1 \neq \frac{1}{2}$, then the rank sum test is consistent but the median test is not (Problem 47). If the X and Y populations differ and one stochastically dominates the other, then $p_1 \neq \frac{1}{2}$ and hence the rank sum test is consistent (Problem 51).

4.6 Weakening the Assumptions

The null distribution of the rank sum test statistic is derived under the assumption that the X ranks are equally likely to be any set of m out of the integers $1, 2, \ldots, N$. This assumption is satisfied if the X's and Y's are drawn from the same population. If the populations differ in any way, however, the level of the test is ordinarily affected. In particular, the assumption that $p_1 = \frac{1}{2}$, where $p_1 = P(X > Y)$, is not sufficient to guarantee the level. Even for populations which are symmetric about the same point and have the same shape, the level may be seriously affected if their variances differ. (See Pratt [1964] and also Problem 52.) The same observation applies, a fortiori, to the corresponding confidence procedure.

The assumptions can be weakened, however, in the same way as was possible for the median test (Sect. 3.8). In a treatment-control comparison, random assignment of the units to the two groups itself guarantees the level of the rank sum test for the null hypothesis that the treatment has no effect (but not the level of the corresponding confidence bounds for a shift without something like a shift assumption). In the ordinary case where the X's and Y's are all mutually independent, if they are not necessarily identically distributed, then the one-tailed test rejecting when U is too small (too few X's greater than Y's) rejects with probability at most α if the distribution of every X_i is "stochastically larger" than the distribution of every Y_j, as defined in Equation (3.12), and hence the null hypothesis could be broadened to include this possibility. The same test rejects with probability at least α if the distribution of every X_i is "stochastically smaller" than the distribution of every Y_j, as in Equation (3.13), so the test is "unbiased" against this alternative. (The proof, requested in Problem 54, is like that for the median test.) However, the test is not unbiased against the alternative that the X's are drawn from one population and the Y's from another and $p_1 < \frac{1}{2}$, as its power is less than α for some alternatives satisfying this condition (Problem 55).

4.7 Ties

Two or more observations which are equal in value are called tied. The ranks of tied observations have not yet been defined, and hence the rank sum test cannot be applied in the presence of ties without some further specification. We have avoided this difficulty so far by assuming continuous distributions and hence zero probability of a tie. In practice, we must have a method of dealing with ties because of discontinuous distributions or unrefined measurements. The discussion here will parallel but abridge the corresponding discussion of zeros and ties in Sect. 6 of Chap. 3; in particular, "zeros" have no counterpart here.

The confidence procedures (given in Sect. 4.3) for the amount of a shift depend only on the differences $Y_j - X_i$. Even when some of these differences are tied, we can determine the $(k + 1)$th from the smallest difference and this is still a lower confidence bound L for the shift μ with k defined as before. However, the exact confidence level now depends on whether L is included in the confidence interval or not. More precisely,

$$P(L \leq \mu) \geq 1 - \alpha \geq P(L < \mu), \tag{4.23}$$

where $1 - \alpha$ is the exact confidence level in the continuous case (Problem 56; see also Problems 62 and 107). A corresponding statement holds for an upper confidence bound, and for two-sided confidence limits. Thus the confidence procedures of Sect. 4.3 can still be used, but now it makes a theoretical difference whether or not the endpoints are included in the stated interval. Ordinarily this is of no practical consequence and the issue need not be resolved.

Since the confidence procedures are still applicable, they could be used to test the null hypothesis that the amount of the shift is $\mu = 0$, which is equivalent to the hypothesis of identical populations. If 0 is not an endpoint of the confidence interval, the corresponding test rejects or "accepts" the null hypothesis according as 0 is outside or inside the confidence interval. By Problem 57, this is equivalent to rejecting ("accepting") if the ordinary rank sum test rejects ("accepts") no matter how the ties are broken. If 0 is an endpoint of the confidence interval, it may be sufficient to state this fact and not actually carry the test procedure further. Another possibility is to be "conservative" and "accept" the null hypothesis in all borderline cases; this amounts to breaking the ties in the direction of "acceptance" and corresponds to including the endpoint in the confidence interval statement. When many ties are likely, however, both these possibilities may reduce the power considerably.

Two other basic methods of handling ties are the average rank method and breaking the ties, which we now discuss. Examples will be given shortly.

The *average rank* (or *midrank*) method assigns to each member of a group of tied observations the simple average of the ranks they would have if they were not tied. The rank sum statistic is then computed as before, but its null distribution is not the same as for observations without ties. The exact distribution conditional on the ties can be enumerated, or a normal approximation can be used (see below). The average rank procedure is equivalent to defining the Mann–Whitney U statistic as the number of (X_i, Y_j) pairs for which $X_i > Y_j$ plus one-half of the number for which $X_i = Y_j$, because U and R_x continue to be related by Equation (4.3) when U is defined in this way and R_x is computed from the average ranks (Problem 58).

Methods which *break the ties* assign distinct integer ranks to the tied observations. If the ties are boken randomly the usual null distribution of the rank sum statistic is preserved. Another possibility already mentioned is to break the ties in favor of acceptance.

Test procedures. To illustrate tests based on these methods of handling ties, consider the following samples, each arranged in order of magnitude.

$$X \text{ sample } (m = 4) \qquad 0, 1, 2, 3$$
$$Y \text{ sample } (n = 10) \qquad 1, 1, 2, 2, 3, 3, 3, 4, 4, 5 \qquad (4.24)$$

Suppose first that we are using the average rank method. There are, for example, three 1's tied at positions 2, 3, and 4. Accordingly, each of these 1's is given the average rank of 3, and similarly for the other sets of tied ranks. The average rank results are given in Table 4.2.

Under the null hypothesis that the X and Y populations are the same, the distribution of the X rank sum R_x obtained by the average rank method

Table 4.2

Observation	0	1	1	1	2	2	2	3	3	3	3	4	4	5
Sample	X	X	Y	Y	X	Y	Y	X	Y	Y	Y	Y	Y	Y
Average Rank	1	3	3	3	6	6	6	9.5	9.5	9.5	9.5	12.5	12.5	14

can again be determined from the fact that each possible set of m ranks is equally likely to be that belonging to the X observations, but this distribution is now conditional on the positions of the ties in the combined sample, or equivalently, on the average ranks present. For the data in (4.24), we have $m = 4$ and $n = 10$ so that there are $\binom{N}{m} = \binom{14}{4} = 1001$ ways to select a set of $m = 4$ ranks out of the $N = 14$ ranks. Table 4.2 shows that there are only 6 different average ranks present, in a pattern consisting of one 1, three 3's, three 6's, four 9.5's, two 12.5's, and one 14. Of the 1001 possible selections of $m = 4$ average ranks given this pattern, one selection (three 3's and one 1) gives $R_x = 10$, nine selections give $R_x = 13$, three give $R_x = 15$, etc. A portion of the lower tail of the distribution of R_x, given this pattern of ties, is shown in Table 4.3. Note that the distribution is very uneven and lumpy, as is frequently the case when many ties occur.

For the data in (4.24), the X rank sum can be found from Table 4.2 as $R_x = 19.5$. From the distribution in Table 4.3, we see that under the null hypothesis, given the ties observed, the exact probability of an X rank sum as small as or smaller than that observed is $P(R_x \le 19.5) = 84/1001 = 0.084$.

Enumeration of the exact distribution of R_x based on average ranks in the presence of ties can be lengthy, but it is not difficult to carry out by computer. In favorable circumstances, the distribution of R_x given in Table F could be used, but it applies exactly only when no ties are present; in general, it should not be used when the average rank method is applied in examples with many ties. For the data in (4.24), Table F gives the P-values $P(R_x \le 19) = 0.071$ and $P(R_x \le 20) = 0.094$; these results, although close to the true P-value 0.084 found in the paragraph above, are not correct, and in other examples the discrepancy may be greater. Another possibility is to use the normal approximation once the relevant mean and variance conditional on the ties observed are obtained (Problem 59). This procedure may also be very inaccurate in the presence of many ties because the exact distribution is generally lumpy, as noted for Table 4.3. Simulation could also be used. Lehman [1961] performed an interesting but limited comparison between the exact and approximate distributions with ties for the case $m = n = 5$.

Table 4.3

r	10	13	15	16	16.5	18	18.5	19	19.5	21	...
$1001\, P(R_x = r)$	1	9	3	9	12	9	4	1	36	3	...

We next consider applying some tiebreaking methods to the data in
(4.24). The three observations which are ones could be assigned the ranks
2, 3, 4 in any order, so that any one of 2, 3, or 4 could be an X rank and the
other two Y ranks, and similarly for the remaining groups of ties. Thus there
are $\binom{3}{1}\binom{3}{1}\binom{4}{1} = 36$ different possible ways of breaking the ties.

If each group of ties is broken separately by some random procedure, the
method is called *breaking the ties at random*. This procedure, though un-
appealing, preserves the usual null distribution of the rank sum statistic;
hence the P-value or a critical value can be found from Table F.

The two extreme methods of breaking the ties giving (i) the smallest, and
(ii) the largest, rank sum are shown in Table 4.4 for the data in (4.24); they
give $R_x = 16$ and $R_x = 23$ respectively. As a result, we know that any method
of breaking the ties gives $16 \leq R_x \leq 23$, and the corresponding range of
P-values from Table F is $0.027 \leq P \leq 0.187$. In particular, breaking the
ties at random would give some P-value in this interval.

The "*conservative*" procedure is to break all ties in favor of acceptance.
In this example, where we are rejecting when R_x is small, the "conservative"
value is $R_x = 23$ and the corresponding "conservative" P-value from Table
F is $P(R_x \leq 23) = 0.187$.

The other extreme would be to break all ties in favor of rejection and
therefore obain $R_x = 16$ and a corresponding P-value of 0.027 from Table F.
At the 0.05 level then, or any other level between 0.027 and 0.187, the rank
sum test would reject by one of the two extreme resolutions of ties but not
by the other. Thus looking at the range of possibilities under tiebreaking
would not provide an unambiguous decision for these data.

In some situations, of course, the two extreme cases of tiebreaking may
lead to the same decision. One might therefore hope to have the best of both
worlds by using tiebreaking when it is unambiguous and average ranks when
tiebreaking is indeterminate. This is not a valid shortcut, however, because
the level of the average rank test is calculated on the assumption that average
ranks will always be used, and average ranks can lead to opposite decisions
from tiebreaking even if tiebreaking is unambiguous. The best-of-both-
worlds test amounts to looking at more than "ancillary" data before deciding
on a test procedure, and hence its level would be very difficult to determine.

To illustrate the difficulty, consider the data in Table 4.5 where the ties
are in the same positions as for the data in (4.24).

Table 4.4

Observation	0	1 1 1	2 2 2	3 3 3 3	4 4	5
Sample	X	X Y Y	X Y Y	X Y Y Y	Y Y	Y
Ranks (i)	1	2 3 4	5 6 7	8 9 10 11	12 13	14
Ranks (ii)	1	4 2 3	7 5 6	11 8 9 10	12 13	14

Table 4.5

Observation	0	1	1	1	2	2	2	3	3	3	3	4	4	5
Sample	X	Y	Y	Y	X	X	X	Y	Y	Y	Y	Y	Y	Y

Notice in Table 4.5 that all the ties are within samples. In such a case, any method of breaking the ties gives the same X ranks, namely 1, 5, 6, 7, and the same rank sum $R_x = 19$. If we ignore the ties and use Table F, the probability of an even smaller rank sum is $P(R_x < 19) = 0.053$, and of one as small or smaller is $P(R_x \leq 19) = 0.071$. At the one-sided level 0.05, the populations would not be judged significantly different when the ties are broken, no matter how they are broken; tiebreaking leads to no ambiguity.

On the other hand, if the average rank method is used on the data in Table 4.5, the rank sum is again $R_x = 19$, but the null distribution of the average rank test statistic given the ties observed is that in Table 4.3, not Table F. Table 4.3 gives $P(R_x \leq 19) = 0.048$, so that the average rank test judges this sample as significant at the 0.05 level, the opposite conclusion from tiebreaking.

The null distribution in Table 4.3 is correct only if the average rank test is used on all samples with ties in the same positions. If the average rank method would be used in some cases, such as the data in (4.24), and if the null distribution would be calculated conditional on the pattern of ties observed, then the average rank method must be used in all cases, including those where the tiebreaking is unambiguous, such as the data in Table 4.5. This example shows that, unfortunately, trying to obtain the best of both worlds affects the level of the average rank procedure.

Choice of Procedure. Which test procedure should we use? The following requirements (analogous to those stated in Sect. 6.5 of Chap. 3 for the one-sample case) seem intuitively desirable.

(i) A sample which is significant in the direction $X < Y$ shall not become insignificant nor an insignificant sample significant in the direction $X > Y$ when (a) some Y's are increased or X's decreased, or (b) all Y's are increased or all X's decreased by an equal amount.

(ii) Those values of an assumed shift μ that would be "accepted" if tested shall form an interval. (This says that the confidence region corresponding to the test shall be an interval, and is equivalent to (i)(b) (Problem 60).)

(iii) A sample shall be judged significant in either direction if it is significant in that direction no matter how the ties are broken; similarly for not significant.

The average rank test procedure satisfies (i) (b) and (ii) but not (i) (a) and (iii) (Problem 61). This procedure presumably gives better power, at least in any ordinary situation, than breaking the ties either conservatively or randomly. The regular tables do not apply, however, and the null distribution must be generated for the set of average ranks observed.

The "conservative" procedure, that is, breaking the ties and choosing the value of the test statistic that is least favorable to rejection, satisfies all of these requirements. However, the true significance level is unknown and may sometimes be much smaller than the nominal level, resulting in a considerable reduction in power over the average rank procedure.

Breaking the ties at random permits use of the ordinary tables and satisfies all of the requirements above (Problem 62). However, the introduction of extraneous randomness in an artificial way is objectionable in itself, and presumably reduces the power.

The confidence bounds for an assumed shift μ corresponding to any method of breaking ties are those obtained in Sect. 4.3. Whether or not the confidence bounds are included in the confidence interval depends on how the ties are broken. The confidence regions corresponding to the average rank procedure may be different, although they are also intervals (Problem 63).

4.8 Point and Confidence Interval Estimation of $P(X > Y)$

There are situations in which one is interested in the probability that a randomly selected member of the X population will exceed an independent, randomly selected member of the Y population. This probability is the parameter $p_1 = P(X > Y)$ defined earlier by (4.12). Suppose, for example, that X is the strength of a manufactured item and Y is the maximum stress to which it will be subjected when installed in an assembly (Birnbaum, 1956]. If $X > Y$, the component will not fail in use. In such a case, p_1 is a parameter of clear economic importance. It might also be of interest in noneconomic contexts. In a comparison of two populations, it is frequently desirable to say something about how much they differ, in addition to, or instead of, performing a test of the hypothesis that they are the same. The difference between the population means or medians, and the amount of the shift μ if the shift assumption is made, are defined only if the difference between two items can be measured on some numerical scale. A point estimate or confidence interval for these quantities has meaning only to the extent that the scale has meaning. However, in the absence of such a meaningful scale, as long as the items can be ranked, p_1 is still meaningful. Accordingly, we will now discuss point and interval estimation of p_1, but again under the assumption that ties occur with probability zero.

Since p_1 is the probability that an X exceeds a Y, a natural estimator is the proportion of (X_i, Y_j) pairs for which $X_i > Y_j$, that is, U/mn. This estimator has expected value p_1 by (4.15), and the variance can be found from (4.16). Hence U/mn is unbiased for p_1, and it is consistent (Problem 65), that is, for every $\varepsilon > 0$,

$$P\left(\left| \frac{U}{mn} - p_1 \right| > \varepsilon \right) \to 0 \quad \text{as } m \text{ and } n \to \infty. \tag{4.24}$$

If the class of possible distributions is broad enough, for instance, if the X's and Y's may be drawn from any two continuous distributions, then no other unbiased estimator has variance as small (Problem 66). In a sufficiently restricted class of distributions, unbiased estimators with smaller variance may exist, but, as in Sect. 3.2 of Chap. 2, the greater the apparent advantages of such estimators, the greater risk accompanies their use. If the two populations are normal with means μ_x, μ_y and common variance σ^2, for example, the value of p_1 is $p_1' = \Phi[(\mu_x - \mu_y)/\sigma\sqrt{2}]$. A natural estimator of p_1', and hence of p_1, is $\hat{p}_1 = \Phi[(\overline{X} - \overline{Y})/s\sqrt{2}]$, where $\overline{X}$ and $\overline{Y}$ are the sample means, s is the usual estimate of the standard deviation, and Φ is the standard normal c.d.f. (This estimator is not quite unbiased for p_1' under normality, but could be made so (Problem 67) by an adjustment similar to that given by Ellison [1964] for the one-sample case.) Under the normality assumption, and in many other circumstances, $\hat{p}_1$ has smaller variance than U/mn, in fact, substantially smaller when p_1 is near 0 or 1 (Problem 68). However, its values tend to cluster around p_1', which equals p_1 for normal distributions but not in general. A slight departure from normality can easily lead to an important difference between p_1 and p_1'. (Typical goodness-of-fit tests of normality shed almost no light on this particular question.) In fact, such information as the observations provide for estimating p_1 beyond the value of U/mn is relatively little and difficult to extract. All this is especially true when the advantage of the estimator $\hat{p}_1$ over p_1 would be greatest if the assumption of normality were literally correct, namely when p_1 is near 0 or 1. Of course, one might focus interest on the parameter p_1' or even on $(\mu_x - \mu_y)/\sigma\sqrt{2}$ instead of on p_1. However, if p_1 is really the relevant parameter, this does not solve the real problem and might encourage misunderstandings. When p_1 is near $\frac{1}{2}$, U/mn may have smaller variance than $\hat{p}_1$ and is not likely to have much larger variance, so there is little to be gained from using some other estimator (see also Sects. 3 and 10, Chap. 8).

Based on the normal approximation to the distribution of U, we can obtain an approximate confidence interval for p_1 from the inequality

$$|U - mnp_1| \le z\sqrt{\mathrm{var}(U)} \tag{4.25}$$

where z is an appropriate standard normal deviate. Var(U) could be estimated by (4.16) with p_2 and p_3 replaced by estimates obtained from the data. A natural estimator of p_2 is the proportion of triples (X_i, X_k, Y_j), $i \ne k$, for which $X_i > Y_j$ and $X_k > Y_j$, and a natural estimator of p_3 is the proportion of triples (X_i, Y_j, Y_l), $j \ne l$, with $X_i > Y_j$ and $X_i > Y_l$.

Alternatively, var(U) might be replaced by an upper bound. One upper bound is given by the inequality (proof below)

$$\mathrm{var}(U) \le mnp_1(1 - p_1)\max(m, n). \tag{4.26}$$

This bound on the variance can be made sharper if the class of distributions is restricted. For example, if the X population is stochastically smaller than the Y, that is, $F(t) \geq G(t)$ for all t, then the variance satisfies (Birnbaum and Klose [1957]; Rustagi [1962])

$$\text{var}(U) \leq mn[(1 - 2p_1)^{3/2}(2m - n - 1) + (n - 2m + 1)$$
$$+ 3p_1(2m - 1) - 3p_1^2(m + n - 1)]/3 \quad (4.27)$$

for $m \leq n$, and similarly for $m > n$ with m and n interchanged in (4.27). If $\text{var}(U)$ is replaced by either of these upper bounds, the right-hand side of (4.25) still depends on p_1, so that an interval for p_1 is not immediately obtained. The inequalities resulting in (4.25) could be solved for p_1 (Problem 69), or the estimate U/mn could be substituted for p_1 on the right-hand side of (4.25) to produce endpoints which do not involve p_1.

PROOF. The inequality in (4.26) follows from (4.22) and the inequalities below. For $i \neq k, j \neq l$,

$$\text{cov}(U_{ij}, U_{kj}) + \text{cov}(U_{ij}, U_{il}) = \text{cov}(U_{ij}, U_{kj}) + \text{cov}(1 - U_{ij}, 1 - U_{il})$$
$$= P(X_i > Y_j \text{ and } X_k > Y_j) + P(X_i < Y_j \text{ and } X_i < Y_l) - p_1^2 - (1 - p_1)^2$$
$$\leq 1 - P(X_k < Y_j \text{ and } X_i > Y_l) - p_1^2 - (1 - p_1)^2$$
$$= 1 - (1 - p_1)p_1 - p_1^2 - (1 - p_1)^2 = p_1(1 - p_1) \quad (4.28)$$

$$\text{cov}(U_{ij}, U_{kj}) \leq [\text{var}(U_{ij}) \, \text{var}(U_{kj})]^{1/2} = p_1(1 - p_1) \quad (4.29)$$

$$\text{cov}(U_{ij}, U_{il}) \leq [\text{var}(U_{ij}) \, \text{var}(U_{il})]^{1/2} = p_1(1 - p_1). \quad (4.30)$$

In obtaining (4.28) we used the fact that the three events ($X_i > Y_j$ and $X_k > Y_j$), ($X_i < Y_j$ and $X_i < Y_l$) and ($X_k < Y_j$ and $X_i > Y_l$) are mutually exclusive, and hence the sum of their probabilities is at most 1. For $m \leq n$, say, we write Equation (4.22) as

$$\text{var}(U) = mn\{\text{var}(U_{ij}) + (m - 1)[\text{cov}(U_{ij}, U_{kj}) + \text{cov}(U_{ij}, U_{il})]$$
$$+ (n - m) \, \text{cov}(U_{ij}, U_{il}) + (m - 1)(n - 1) \, \text{cov}(U_{ij}, U_{kl})\}$$

and substitute (4.18), (4.21), (4.28) and (4.30) to obtain the desired result, (4.26). □

5 Procedures Based on Sums of Scores

Suppose, as at the beginning of Sect. 4, that $X_1, \ldots, X_m$ and $Y_1, \ldots, Y_n$ are mutually independent random samples from continuous populations (so that "ties" need not be considered). In order to test the null hypothesis of identical populations, the rank sum test could be generalized to use arbitrary constants which are not necessarily ranks. Once the combined sample observations are arranged in order of magnitude, any set of numbers c_1, $c_2, \ldots, c_N$ (positive or negative), which we call *scores*, can be assigned to the observations. (The constant c_k is associated with X_i if X_i has rank k in the combined sample.) Then the sum of scores of observations from say the X sample provides a test statistic much like R_x.

For the data of Table 4.1, for any constants $c_1, \ldots, c_N$ we have the following assignment of scores.

Sample	Y	X	X	Y	Y	Y	X	Y	X	Y
Value	1.25	1.75	3.25	4.25	5.25	6.25	6.75	7.25	9.00	10.00
Rank	1	2	3	4	5	6	7	8	9	10
Score	c_1	c_2	c_3	c_4	c_5	c_6	c_7	c_8	c_9	c_{10}

The null distribution of the sum of scores test statistic can be determined by enumeration in a manner analogous to that described for R_x in Sect. 4.2, since under the null hypothesis the X scores constitute a random sample of m scores drawn without replacement from the N available. A table of P-values could then be constructed for any particular set of constants c_k. Alternatively, a normal approximation could be used, by standardizing with the mean and variance given in Problem 77a.

For a test of the null hypothesis that the Y population is the same as the X population except for a shift by the amount μ, the foregoing test is applied to $X_1, \ldots, X_m, Y_1 - \mu, \ldots, Y_n - \mu$. The set of all values of μ which would be "accepted" when so tested forms a confidence region for the amount of the shift, under the model of the shift assumption. The confidence region will be an interval if the c_k form a monotone sequence ($c_{k+1} - c_k$ has the same sign for all k), and each confidence bound will be one of the mn differences $Y_j - X_i$ (see Sect. 6 and Problem 76).

The general sum of scores statistic can be written in a form analogous to (4.6) as

$$\sum_{1}^{N} c_k I_k \tag{5.1a}$$

where

$$I_k = \begin{cases} 1 & \text{if the observation with rank } k \text{ is an } X \\ 0 & \text{if the observation with rank } k \text{ is a } Y. \end{cases} \tag{5.1b}$$

Similar statistics analogous to R_y are also easily defined. For any particular set of scores c_k, the sum of scores statistics for the two samples are again linearly related and hence equivalent as test statistics.

Many different two-sample rank tests are of this general type, including all the ones we have studied so far in this chapter. If $c_k = k$ for $k = 1, \ldots, N$, the sum of scores test is simply the rank sum test. If $c_k = 1$ for $k \leq N/2$ and

$c_k = 0$ otherwise, it is the median test. If $c_k = 1$ for $k \leq t$ and $c_k = 0$ otherwise, it is a two-sample quantile test with the observations divided into "below" and "above" at the (t)th order statistic of the combined sample. The two-sample sign test for fixed ξ, however, is not a sum of scores test, because t would have to be chosen such that t observations are less than ξ and hence the scores are not fixed in advance; as remarked earlier, it is not a rank test either.

Fisher and Yates [1963 and earlier editions starting in 1938] suggested using as c_k the expected value of the kth from the smallest in a random sample of N from the standard normal distribution; this choice arises naturally in Sect. 7. Tables of these c_k, called normal scores, are given by Fisher and Yates [1963] to two decimal places, and with more precision by Teichroew [1956] and Harter [1961]. Terry [1952] gives tables of the null distribution of the resulting test statistic and discusses approximations. Even the most straightforward normal approximation involves the sum $\sum_1^N c_k^2$ (Problem 77), which is tabulated by Fisher and Yates [1963] for $N \leq 50$. This sum can be found from the individual c_k^2 values given in Teichroew [1956] for $N \leq 20$. The small sample power of this test (and other two-sample sum of scores tests) was investigated in Gibbons [1964c] for various alternatives. An optimality property of this test and the rank sum test will be discussed in Sect. 7.

Van der Waerden [1952, 1953] suggested setting c_k equal to the quantile of order $k/(N + 1)$ for the standard normal distribution; this test is nearly the same as the Fisher–Yates–Terry test without requiring special tables of c_k. The constants here are especially conveniently obtained from tables of rational quantiles of the normal distribution, for instance, Fisher and Yates [1963] and van der Waerden and Nievergelt [1956]. The latter reference also gives the null distribution of the test statistic, and van der Waerden [1956] discusses approximations to this distribution.

The two foregoing normal scores tests actually differ little from the Wilcoxon rank sum test at typical levels, at least in small samples. They do offer a much wider choice of natural significance levels however. That is, in the tail of the null distribution, the rank sum test statistic treats as alike possible sample outcomes which are distinguished by the normal scores test statistic. The comments of the last two paragraphs of Sect. 7.1 of Chap. 3 apply here also, with appropriate modification.

In order to emphasize the fact that the various sums of scores tests do distinguish between arrangements of the X and Y variables in different, but related ways, the values of the extreme small values of these test statistics and the corresponding cumulative (left-tail) frequencies are shown in Table 5.1 for the sample sizes $m = 6$, $n = 6$. These frequencies are divided by $\binom{12}{6} = 924$ to obtain probabilities. The fact that the rank sum statistic differentiates rank orders in a much less refined way is obvious from this listing. The Fisher–Yates–Terry (T_2) and van der Waerden (T_3) test statistics give almost identical orderings for that part of the tail which is listed (49/924 or 5.3%); the orderings are identical for the first 4.2% (39/924) of the tail

Table 5.1 Cumulative Left-Tail Frequencies $\sum f$ for Rank Sum Statistic R_x, Fisher–Yates–Terry Statistics T_2 and van der Waerden Statistic T_3

R_x	$\sum f$	$-T_2$	$\sum f$	$-T_3$	$\sum f$
21	1	4.49	1	4.07	1
22	2	4.28	2	3.88	2
23	4	4.07	4	3.68	4
24	7	3.86	5	3.49	5
25	12	3.85	7	3.48	7
26	19	3.66	8	3.29	8
27	30	3.64	10	3.28	10
28	43	3.59	12	3.24	12
29	61	3.44	14	3.09	14
30	83	3.42	15	3.07	15
31	111	3.38	17	3.05	17
		3.27	19	2.96	19
		3.23	21	2.89	21
		3.21	22	2.88	22
		3.18	24	2.85	24
		3.16	26	2.84	26
		3.06	28	2.76	28
		3.00	30	2.68	30
		2.97	32	2.66	32
		2.95	34	2.64	34
		2.90	35	2.60	35
		2.86	37	2.57	37
		2.84	39	2.55[a]	41
		2.79	40	2.48	42
		2.76	42	2.45	48
		2.74	48	2.41	49
		2.70	49		

[a] The four values listed for 2.55 are actually only two pairs of true ties. To four decimal places, two values are equal to 2.5524 and two equal 2.5522.

and between 4.7% (43/924) and 5.3% inclusive. Futhermore, all three statistics are almost monotonic functions of one another, although frequently one stays constant while another changes. In other words, in the portion of the tail listed for all three statistics (a little over 5%), the possible sets of X ranks can be put in an order such that they enter the critical region in almost this order for all three tests, the only difference being how many enter at one time. The one exception is where $-T_2 = 2.79$ and $-T_3 = 2.55$. For larger sample sizes there would be more differences, although those between T_2 and T_3 are always minor.

6 Two-Sample Rank Tests and the $Y - X$ Differences

Tests depending only on the ranks of the X_i and Y_j in the combined sample may be called *two-sample rank tests*. They include those based on sums of scores, but could also be of other forms.

We have seen in detail that the rank sum statistic is the number of positive differences $Y_j - X_i$, and that the corresponding confidence bound for shift is an order statistic of these differences. More generally, the confidence bound corresponding to any two-sample rank test is necessarily one of these differences, as mentioned in the last section. This follows from the fact that, as μ varies, the ranks obtained from $X_1, \ldots, X_m, Y_1 - \mu, \ldots,$ $Y_n - \mu$ change only at the values $\mu = Y_j - X_i$.

As the values of the $Y_j - X_i$ relate to confidence intervals for shift, so their signs relate to tests of zero shift (identical populations). The two-sample rank tests are exactly those tests depending only on the signs of these differences, as long as order within samples is ignored, which in practice it would be. In other words, two-sample permutation-invariant tests (see Sect. 7) are rank tests if and only if they depend only on the signs of the $Y_{(j)} - X_{(i)}$, the differences of the order statistics of the two samples separately.

We shall not go into further detail about these relationships here, because the one-sample situation is very similar and it was discussed at length in Sect. 7 of Chap. 3. We do note, however, that the confidence bound corresponding to a one-tailed two-sample quantile test was identified in Sect. 3.4 as a particular difference $Y_{(j)} - X_{(i)}$, and of course the test itself depends only on the sign of the same difference. Thus a quantile test is the simplest case from this point of view, as from others.

7 Invariance and Two-Sample Rank Procedures

The properties of permutation invariance and invariance under other transformations were defined for one-sample procedures in Sect. 8 of Chap. 3. We give here analogous definitions of these properties in the two-sample case, and also some arguments for restricting consideration to these types of invariant procedures. Since the arguments are also analogous to those discussed extensively in Sect. 8 of Chap. 3, they will be repeated here only briefly. They lead to a definition of two-sample, permutation-invariant rank tests as those having both of these invariance properties.

A procedure $\phi(X_1, \ldots, X_m, Y_1, \ldots, Y_n)$ will be called *permutation invariant* if it is unchanged by permutations of the X_i and by permutations of the Y_j. In other words, ϕ is not changed if the order of the X's or the order of the Y's is changed. In mathematical notation, ϕ is permutation invariant if

$$\phi(X_{i_1}, \ldots, X_{i_m}, Y_{j_1}, \ldots, Y_{j_n}) = \phi(X_1, \ldots, X_m, Y_1, \ldots, Y_n) \qquad (7.1)$$

for all permutations $i_1, \ldots, i_m$ of $1, \ldots, m$ and $j_1, \ldots, j_n$ of $1, \ldots, n$.

A procedure is permutation invariant if and only if it is a function solely of the order statistics of the two samples separately, that is, of $X_{(1)}, \ldots, X_{(m)}$ and $Y_{(1)}, \ldots, Y_{(n)}$ alone. Here $X_{(i)}$ denotes the ith from the smallest among the X's and $Y_{(j)}$ is the jth from the smallest among the Y's (not among the X's and Y's combined).

In the kind of situation we have been discussing, it is natural to use a permutation invariant procedure, and all of the procedures discussed in this chapter have been of this type. There are also two concrete arguments against using procedures that are not permutation invariant but depend on the order of the X's or the order of the Y's separately. First, in some models the order statistics of the two samples form a sufficient statistic so that nothing can be gained by looking beyond them. Second, even when this condition fails, there is a direct invariance argument that may apply.

The order statistics of the two samples, $X_{(1)}, \ldots, X_{(m)}, Y_{(1)}, \ldots, Y_{(n)}$, form a sufficient statistic if the X_i are identically distributed and the Y_j are identically distributed and all are independent. (We have often made this assumption. For sufficiency, it must hold under every contemplated distribution. In testing, it must hold under the alternative as well as the null hypothesis. Notice that it is not necessary, however, that the X's have the same distribution as the Y's.) By the properties of sufficiency, it follows that, under the stated conditions, given any procedure there exists an equivalent, permutation invariant procedure. This procedure (possibly randomized) is based on the order statistics alone and has exactly the same operating characteristics as any given procedure based on the observations and their order within each sample separately.

If the X's, or the Y's, are not necessarily identically distributed, then the order statistics of the two samples no longer form a sufficient statistic. In this case, the foregoing, very strong, justification of permutation invariance is no longer applicable. However, it may still seem unreasonable to permit the procedure to depend on the order of the X's or of the Y's separately. If the observations provide intuitively the same evidence on the matter in hand however the X's or Y's are rearranged separately, then a "reasonable" procedure ϕ would be unaffected by rearrangement of the X's or of the Y's, that is, would be permutation invariant. The discussion of the "principle of invariance" in Sect. 8 of Chap. 3 applies here also with appropriate changes (Problem 88), and hence will not be further elaborated.

Any procedure which is a function of the order statistics in the two samples separately is permutation invariant. It need not be a function of the ranks. Rank procedures, however, are the only ones which are also invariant under another, very large and very different class of transformations, specifically, those produced by strictly increasing functions. Invariance under these transformations might also be desirable, and we now consider them.

Suppose for convenience that the X's and the Y's are independent random samples from two populations and we are testing the null hypothesis that the populations are the same against the alternative that they are not. In other words, the X's and Y's are all independent, the X's are identically distributed, the Y's are identically distributed, the null hypothesis is that the X's and Y's have the same distribution and the alternative is that they do not.

Let g be any strictly increasing function. If $X_1, \ldots, X_m, Y_1, \ldots, Y_n$ satisfy the null hypothesis, then so also do $g(X_1), \ldots, g(X_m), g(Y_1), \ldots, g(Y_n)$, and the same applies to the alternative hypothesis. Accordingly, we can "invoke the principle of invariance" and require that a test treat $X_1, \ldots, Y_n$ in the same way as $g(X_1), \ldots, g(Y_n)$. If this is required for all strictly increasing functions g, then any two sets of observations with the same X ranks and Y ranks must be treated alike, because any set of observations can be carried into any other set with the same ranks by such a g (Problem 89). In short, tests based on the ranks of the observations are the only tests which are invariant under all strictly increasing transformations g.

The same argument applies to other null and alternative hypotheses of the sort we have been considering, provided only that all strictly increasing transformations g carry null distributions into null distributions and alternatives into alternatives. This holds, for instance, if the earlier alternative hypothesis is tightened to require that the X's be stochastically larger than the Y's (Sects. 3.8 and 4.6), or relaxed to permit the X's or Y's or both to be not necessarily identically distributed (Problem 90). It does not hold under the shift assumption, however (Problem 90d).

Arguments were given earlier for restricting consideration to permutation invariant procedures, that is, for excluding procedures which depend on the order of the X's or of the Y's separately. Applying the argument for permutation invariance, along with the argument for procedures which are invariant under transformations by any strictly increasing function, that is, rank tests, leads to restricting consideration to *permutation invariant rank tests*. These tests depend only on the ranks of the X's and Y's in the combined sample, without regard to their order within the separate samples. The null distributions of their test statistics can be generated by taking as equally likely the $\binom{N}{m}$ separations of the ranks $1, \ldots, N$ into m X ranks and n Y ranks.

These procedures can also be defined in terms of the following indicator variables:

$$I_j = \begin{cases} 1 & \text{if the } j\text{th smallest observation among } X_1, \ldots, Y_n \text{ is an } X \\ 0 & \text{otherwise} \end{cases}$$

for $j = 1, \ldots, N$. A two-sample test is a permutation invariant rank test if and only if its critical function $\phi(X_1, \ldots, Y_n)$ is a function only of $I_1, \ldots, I_N$ (Problem 91).

As discussed for the one-sample case in Sect. 8 of Chap 3, the argument for invariance under strictly increasing functions is far less compelling than the argument for permutation invariance based on sufficiency. Indeed one might not want to treat $X_1, \ldots, Y_n$ and $g(X_1), \ldots, g(Y_n)$ alike in some extreme instances. However, if one is content to treat alike practically all sets of observations with the same ranks, then restricting consideration to two-sample tests is justifiable. In such a case, little is lost by using a rank test. The choice of which rank test remains, of course.

8 Locally Most Powerful Rank Tests

There are many two-sample rank tests. In this section we consider the question of which one has the best power function. As in the one-sample case, there is no two-sample rank test which is uniformly most powerful against the classes of alternatives usually considered. We must therefore be content with more limited objectives. In Sect. 8.1 we will first find that two-sample rank test which is most powerful against any particular alternative pair of distributions for X and Y. We will then consider a one-parameter family of alternatives approaching the null hypothesis of equal distributions and find the "locally most powerful" two-sample rank test against that family, in general, and in some specific cases. This test is always based on a sum of scores. In Sect. 8.2 we will investigate the class of all locally most powerful rank tests, finding that all sets of scores can arise in this way. We assume throughout that the observations are independent, are identically distributed in each sample, and have continuous distributions so that ties within or across samples have probability zero.

8.1 Most Powerful and Locally Most Powerful Rank Tests Against Given Alternatives

Let $r_1, \ldots, r_m, r'_1, \ldots, r'_n$ be the respective ranks corresponding to the observations $X_1, \ldots, X_m, Y_1, \ldots, Y_n$ after they are pooled and arranged from smallest to largest. Thus r_i is the rank of X_i and r'_j is the rank of Y_j in the combined sample. If we distinguish different orders of the X's and Y's within samples, then there are $N!$ possible arrangements of these ranks ($N = m + n$). We could argue that, by sufficiency, it is not necessary to distinguish order within samples, but omit this step because our derivations will reach this conclusion automatically. We will derive the most powerful tests among *all* rank tests, not merely among permutation-invariant rank tests. We will see that the resulting test *is* permutation invariant, but proving this first would not facilitate the derivation.

As usual, consider the null hypothesis of identical X and Y populations. Under this hypothesis, the $N!$ possible arrangements of the ranks are all equally likely. By the Neyman–Pearson Lemma (Theorem 7.1 of Chap. 1), it follows that, among rank tests at level α, the most powerful test against a simple (completely specified) alternative K rejects if the probability under K of the observed rank arrangement is greater than a constant k, and "accepts" if it is less than k, where k and the probability of rejection at k are chosen to make the level exactly α (Problem 92). This may be expressed as

$$\text{reject if } P_K(r_1, \ldots, r_m, r'_1, \ldots, r'_n) > k$$
$$\text{"accept" if } P_K(r_1, \ldots, r_m, r'_1, \ldots, r'_n) < k, \qquad (8.1)$$

where the argument of P_K denotes the observed rank arrangement. If two or more rank arrangements have probability exactly k under the alternative K, the probabilities of rejection when these arrangements occur need not be the same; they need only bring the level of the test to exactly α.

The most powerful rank test against the simple alternative K depends, of course, on K. Even if we restrict consideration to normal alternatives with $\mu_2 > \mu_1$ and common variance σ^2, the most powerful rank test depends on the values of the parameters. (This contrasts with Sect. 4.2 of Chap. 6, where we will see that the most powerful observation-randomization test depends only on whether μ_2 is larger or smaller than μ_1.) If we consider the situation when μ_2 is close to μ_1, however, we will find there is a "locally most powerful" rank test against one-sided normal alternatives, namely the one-tailed Fisher–Yates–Terry test. (This is a sum of scores test where the score c_j is the expected value of the jth from the smallest in a standard normal sample of size N, as explained in Sect. 5.)

Specifically, any non-randomized, Fisher–Yates–Terry test which is one-tailed in the appropriate direction is the unique most powerful rank test at its level against every normal alternative with common variance and with $0 < (\mu_2 - \mu_1)/\sigma < \varepsilon$, for some sufficiently small, positive ε. The same test uniquely maximizes the derivative of the power function at the null hypothesis $\theta = 0$ when the power is regarded as a function of $\theta = (\mu_2 - \mu_1)/\sigma$, rather than of the parameters separately. (The same is true of the partial derivative with respect to μ_2, with μ_1 and σ fixed, etc., but for any rank test, the power is a function of θ alone (Problem 93) and it is natural to regard it as such.) These statements refer only to levels achievable by a nonrandomized, one-tailed Fisher–Yates–Terry test. At other levels, the derivative of the power at $\theta = 0$ is maximized among rank tests by a (randomized) Fisher–Yates–Terry test, though not always uniquely, and maximizing the power in a neighborhood of $\theta = 0$ may require treating differently those different rank arrangements for which the sum of scores equals its critical value k (Problem 93). Similar statements about alternatives in the opposite direction also hold, of course.

We will also prove corresponding results for the rank sum test when the alternative distributions are logistic. The logistic distribution with mean μ and scale parameter σ (and variance $\sigma^2\pi^2/3$) has cumulative distribution function and density

$$F_{\mu,\sigma}(x) = \frac{1}{1 + e^{-(x-\mu)/\sigma}}, \qquad f_{\mu,\sigma}(x) = \frac{e^{-(x-\mu)/\sigma}}{\sigma[1 + e^{-(x-\mu)/\sigma}]^2}. \qquad (8.2)$$

(See Fig. 9.1, Chap. 3.) Specifically, consider the alternative that the X's and Y's are drawn independently from logistic distributions with means μ_1 and μ_2 and common scale parameter σ. We will show that any non-randomized, rank sum test which is one-tailed in the appropriate direction is the unique most powerful rank test at its level against every such alternative with $0 < (\mu_2 - \mu_1)/\sigma < \varepsilon$, for some sufficiently small, positive ε. Among rank tests it also uniquely maximizes the derivative of the power with respect to $\theta = (\mu_2 - \mu_1)/\sigma$ at $\theta = 0$. At other levels, the derivative of the power at $\theta = 0$ is maximized by a (randomized) rank sum test, though not always uniquely, and maximizing the power near $\theta = 0$ may require differential treatment of different borderline rank arrangements (Problem 95).

These results will be derived by a general method so that tests with similar properties can be derived for any one-parameter, one-sided alternative, and for any alternative reducible to this form (by a strictly monotonic transformation of the data that is allowed to depend on nuisance parameters).

Although we have already used the term "locally most powerful," it has not actually been defined here. One definition, consistent with that given somewhat informally for the one-sample case in Sect. 9 of Chap. 3, is that a test is *locally most powerful* among tests in some designated class against some designated alternative if it has maximum power among all such tests at all alternative distributions which are sufficiently close in some specified sense to the null hypothesis. If we deal with an alternative which can be indexed by a parameter θ that equals 0 when the null hypothesis is true and is positive under the alternative, and if we define "close" in the obvious way, then this definition will require that the test be uniformly most powerful in some interval $(0, \varepsilon)$ with $\varepsilon > 0$. Such a test will also maximize the slope of the power function at $\theta = 0$. The converse is not always true, however, and is not automatic even when true; this difficulty is easily overlooked.

PROOFS. Consider a family of distributions F_θ for the X population and a family G_θ for the Y population, both depending on a one-dimensional parameter θ. Suppose that $\theta = 0$ satisfies the null hypothesis of identical distributions, that is, $F_0 = G_0$, and consider the alternative $\theta > 0$.

Let $\underline{r} = (r_1, \ldots, r_m, r'_1, \ldots, r'_n)$ denote the rank arrangement and $P_\theta(\underline{r})$ its probability under the alternative θ. Since all rank arrangements are equally likely under the null hypothesis, it follows (Problem 96) that a rank test maximizes the derivative of the power at $\theta = 0$ if and only if it is of the form

$$\text{reject if } \frac{d}{d\theta} P_\theta(\underline{r}) > k \text{ at } \theta = 0$$

$$\text{"accept" if } \frac{d}{d\theta} P_\theta(\underline{r}) < k \text{ at } \theta = 0. \tag{8.3}$$

The value of k and the probability of rejection for $\underline{r}$ on the boundary given by k need only be chosen so that the test has level exactly α. More circumspect behavior at k may be required to maximize power for small θ, however. Since

$$P_\theta(\underline{r}) = \frac{1}{N!} + \theta \frac{d}{d\theta} P_\theta(\underline{r}) + \text{terms of smaller order}$$

as $\theta \to 0$, it follows from (8.1) that the most powerful rank test against θ is again of the form (8.3) for sufficiently small θ; however, if two or more rank arrangements $\underline{r}$ lie on the boundary, those with larger values of the remainder terms must be favored for the rejection region.

At certain levels there is no room for randomization or other choice at the boundary, and the situation is simple. Specifically, a test of the form

$$\text{reject if } \frac{d}{d\theta} P_\theta(\underline{r}) \geq k \text{ at } \theta = 0$$

$$\text{"accept" otherwise,} \tag{8.4}$$

among rank tests at its level, *uniquely* maximizes both the derivative of the power at $\theta = 0$ and the power against θ for all θ in some interval $0 < \theta < \varepsilon$ (Problem 96). $\qquad\square$

We now seek more convenient expressions for the derivative needed in these tests. If the X and Y populations have densities f_θ and g_θ under the alternative θ, we may write $P_\theta(\underline{r})$ as

$$P_\theta(\underline{r}) = \int \cdots \int_R \prod_{i=1}^m f_\theta(x_i) \prod_{j=1}^n g_\theta(y_j)\, dx_1 \cdots dx_m dy_1 \cdots dy_n \tag{8.5}$$

where R is the region in $(X_1, \ldots, X_m, Y_1, \ldots, Y_n)$-space where the rank arrangement is $\underline{r}$. Assume it is legitimate to differentiate (8.5) under the integral sign, and let

$$h_1(x) = \frac{\partial}{\partial\theta} \log f_\theta(x)\Big|_{\theta=0} = \frac{1}{f_0(x)} \frac{\partial}{\partial\theta} f_\theta(x)\Big|_{\theta=0}$$

$$\tag{8.6}$$

$$h_2(y) = \frac{\partial}{\partial\theta} \log g_\theta(y)\Big|_{\theta=0} = \frac{1}{g_0(y)} \frac{\partial}{\partial\theta} g_\theta(y)\Big|_{\theta=0}.$$

Then

$$\frac{d}{d\theta} P_\theta(\underline{r})\Big|_{\theta=0} = \int \cdots \int_R \left[\sum_i h_1(x_i) + \sum_j h_2(y_j)\right] \prod_i f_0(x_i)\, dx_i \prod_j g_0(y_j)\, dy_j$$

$$= \left\{\sum_i E_0[h_1(Z_{(r_i)})] + \sum_j E_0[h_2(Z_{(r_j)})]\right\}/N! \tag{8.7}$$

where, on the right-hand side, $Z_{(1)} < \cdots < Z_{(N)}$ are an ordered sample of N from the distribution with density $f_0 = g_0$. Therefore the test (8.4), for example, is equivalent to

$$\text{reject if } \sum_i E_0[h_1(Z_{(r_i)})] + \sum_j E_0[h_2(Z_{(r_j')})] \geq k$$

"accept" otherwise

(8.8)

where the constant k may differ from formula to formula. This result may also be written in the form

$$\text{reject if } \sum_1^N c_j I_j \geq k$$

"accept" otherwise

(8.9)

where

$$I_j = \begin{cases} 1 & \text{if } Z_{(j)} \text{ is an } X \\ 0 & \text{if } Z_{(j)} \text{ is a } Y \end{cases}$$

and

$$c_j = \lambda E_0[h_1(Z_{(j)}) - h_2(Z_{(j)})] + \gamma, \qquad j = 1, \dots, N, \qquad (8.10)$$

for arbitrary constants γ and λ, λ positive. Since the test in (8.9) is equivalent to that in (8.4), (8.9) also has the property that, among rank tests at its level, it uniquely maximizes the derivative of the power at $\theta = 0$ and uniquely maximizes the power against θ for all θ in some interval $0 < \theta < \varepsilon$. Notice that the test is therefore based on a sum of scores in the sense of Sect. 5.

Similarly, (8.3) is equivalent to

$$\text{reject if } \sum_1^N c_j I_j > k$$

"accept" if $\sum_1^N c_j I_j < k$,

(8.11)

and at any level α, a rank test maximizes the derivative of the power at $\theta = 0$ if and only if it is of the form (8.11), where the constant k and the probability of rejection when $\sum_1^N c_j I_j = k$ are such that the test has level exactly α.

Similar statements hold for $\theta < 0$, with rejection when $\sum_1^N c_j I_j$ is too small.

In the case of normal shift, say F_θ is $N(\theta, 1)$ and G_θ is $N(0, 1)$, (8.6) becomes

$$h_1(x) = x, \qquad h_2(y) = 0,$$

and the $Z_{(j)}$ are an ordered sample from $N(0, 1)$. The c_j given by (8.10) with $\lambda = 1, \gamma = 0$ are therefore the expectations of the normal order statistics.

Thus we obtain the Fisher–Yates–Terry test as the locally most powerful rank test, with the properties stated earlier, for normal shift alternatives.

In the logistic case, let X have the distribution (8.2) with parameters $\mu = \theta$ and $\sigma = 1$ while Y has parameters $\mu = 0$ and $\sigma = 1$. Since the logistic distribution with $\sigma = 1$ has $f = F(1 - F)$, we have in (8.6)

$$h_1(x) = \frac{\partial}{\partial \theta} \log F_{0,1}(x - \theta)[1 - F_{0,1}(x - \theta)] \Big|_{\theta = 0} = 2F_{0,1}(x) - 1 \quad (8.12)$$

and $h_2(y) = 0$. The $Z_{(j)}$ are now order statistics from the logistic distribution $F_{0,1}$, so

$$h_1(Z_{(j)}) = 2U_j - 1 \quad (8.13)$$

where $U_j = F_{0,1}(Z_{(j)})$ is an order statistic from the uniform distribution on $(0, 1)$. Therefore (8.10) becomes

$$c_j = \lambda E(2U_j - 1) + \gamma = 2\lambda j/(N + 1) - \lambda + \gamma. \quad (8.14)$$

A test based on these c_j is equivalent to one with $c_j = j$, and thus the rank sum test is the locally most powerful rank test against logistic shift alternatives.

The locally most powerful rank test against Lehmann-type alternatives is investigated in Gibbons [1964a].

8.2 The Class of Locally Most Powerful Rank Tests

In the preceding subsection we showed that for any particular one-parameter family of alternatives, the sum of scores test with scores given by (8.10) is locally most powerful among rank tests. We will now investigate the family of all locally most powerful rank tests that we could obtain by varying the alternative. As a preliminary, we note that, although the exact properties stated after (8.10) and (8.11) imply that all locally most powerful rank tests must be of the form (8.9) or (8.11), this form is unfortunately more inclusive than expected or intended. The difficulty is that (8.9), though restrictive, applies only at certain levels, while (8.11) as it stands leaves so much open that it is completely unrestrictive. Every rank test can be expressed in the form (8.11) by taking the c_j equal for all j. Constant c_j could be excluded, but there are also similar, less degenerate ways of using the form (8.11) to express many tests that we would not want to call sum-of-scores tests. Thus we may say that all locally most powerful rank tests are sum-of-scores tests, but we lack a good definition of the latter for purposes of this statement.

We shall not pursue this definitional problem, but rather turn to the more interesting converse question—does every sum-of-scores test arise in this way for some one-parameter family of alternatives? We prove below that the answer is Yes. Specifically, given any arbitrary scores c_j, there is a one-parameter family of alternatives given by densities f_θ, g_θ such that the c_j satisfy (8.10) for some $\lambda > 0$ and some γ. Given any c_j, therefore, there exists a one-parameter alternative such that, among tests at the same level, any test of the form (8.9) uniquely maximizes the derivative of the power at $\theta = 0$ and uniquely maximizes the power against θ for all θ in some interval $0 < \theta < \varepsilon$, and any test of the form (8.11) maximizes the derivative of the power at $\theta = 0$.

Roughly speaking then, the class of locally most powerful rank tests is identical with the class of sum-of-scores tests. Intuitively, it may seem unreasonable that the c_j should be utterly arbitrary. The intuition not reflected in the theoretical result is that, while any particular set of scores c_j is locally most powerful against some alternative, this may not be at all like the kind of alternative against which good power is desired. For example, if good power is desired against alternatives which are one-tailed in the direction of X larger than Y and natural in other respects, then (presumably) increasing one of the X's should not decrease the test statistic. This implies that the c_j should be monotonically increasing in j. In general, if the class of alternatives is sufficiently restricted, the locally most powerful rank tests will not yield all sets of scores. The sets of scores which arise from restricted classes of alternatives are complicated, however, and will not be discussed here. We note, though, that stochastic dominance alone does not imply monotonic c_j (Problem 103). The reader is referred to Uzawa [1960] for a complete presentation of the conditions on the c_j which result when certain restrictions are placed on the family of alternatives.

PROOF. We will show that every set of scores $c_1, \ldots, c_N$ satisfies (8.10) for some positive λ, some γ, and some one-parameter family of the following kind. Let the Y distribution be uniform on the interval $(0, 1)$ and let

$$f_\theta(x) = 1 + \theta h(x), \qquad 0 \le x \le 1 \tag{8.15}$$

where h is a bounded function on the interval $(0, 1)$ with

$$\int_0^1 h(x)\, dx = 0. \tag{8.16}$$

Then f_0 is a density for sufficiently small θ, and (8.10) becomes

$$c_j = \lambda E[h(U_j)] + \gamma, \qquad j = 1, \ldots, N, \tag{8.17}$$

where $U_1 < \cdots < U_N$ are the order statistics in a sample of N from the uniform distribution on $(0, 1)$. Given a set of scores c_j, our task is to find a γ, a positive λ, and a bounded function h satisfying (8.16) and (8.17). It suffices to find a bounded function $q(u)$ for $0 \le u \le 1$ such that

$$c_j = E[q(U_j)], \qquad j = 1, \ldots, N, \tag{8.18}$$

as then $h(u) = q(u) + b$ will be bounded, will satisfy (8.16) for some b, and will satisfy (8.17) with $\lambda = 1$, $\gamma = -b$. We will use a polynomial as the bounded function $q(u)$. Now it is true (Problem 94d) that

$$E(U_j^k) = \frac{j(j+1)\cdots(j+k-1)}{(N+1)\cdots(N+k)}, \qquad j = 1, \ldots, N.$$

Therefore, for $q(u) = \sum_k a_k (N+1)\cdots(N+k)u^k$, Equation (8.18) becomes

$$c_j = \sum_k a_k j(j+1)\cdots(j+k-1), \qquad j = 1, \ldots, N, \qquad (8.19)$$

and it remains to find a_k which satisfy (8.19). The right-hand side of (8.19) can be considered a polynomial in j. There is certainly a polynomial $\sum b_k j^k$ such that

$$c_j = \sum_k b_k j^k, \qquad j = 1, \ldots, N. \qquad (8.20)$$

Therefore (8.19) will be satisfied if the two polynomials are identical, that is, if a_k can be chosen so that, as polynomials in j,

$$\sum_k a_k j(j+1)\cdots(j+k-1) = \sum_k b_k j^k. \qquad (8.21)$$

It is easy to prove by induction on k (Problem 104) that j^k is a linear combination of the terms $j(j+1)\cdots(j+l-1)$ with $l \le k$. It follows that there are a_k satisfying (8.21) and the proof is complete. $\qquad \square$

PROBLEMS

1. Let F and G be any two c.d.f.'s that satisfy the shift assumption $F(x) = G(x + \mu)$. Show that they have the same central moments for all k, that is $E_F\{[X - E_F(X)]^k\} = E_G\{[X - E_G(X)]^k\}$ for all k; in particular, their variances are equal.

2. Let X and Y be any two random variables with c.d.f.'s F and G respectively that satisfy the shift assumption $F(x) = G(x + \mu)$. Show that μ is the difference in their locations, no matter how location is measured; that is, μ is the difference between means, modes, medians, p-points, etc., whenever these quantities exist and are unique. What happens if the p-points are not unique?

3. For the Mosteller and Wallace data given in Section 3.1 find the one-tailed P-value according to
   (a) A two-sample sign test with fixed $\xi = 17.0$.
   (b) A two-sample quantile test with $t = 13$.
   (c) All two-sample sign tests with fixed ξ for $14 \le \xi \le 20$.
   (d) All two-sample quantile tests for $6 \le t \le 16$.

4. Given a sample of m X's and n Y's with no ties, construct a path of N steps in the plane, starting at the origin, such that the kth step is one unit to the right if the kth smallest observation is an X and one unit upward if it is a Y.
   (a) Show that there is a one-to-one correspondence between the possible paths and the possible rank arrangements of the X's and Y's.
   (b) Describe the acceptance and rejection regions of a two-sample quantile test in terms of these paths.

*5. (Early decision). In a situation where the observations are obtained in order, as in life testing, the outcome of a test or other procedure may be known early so that sampling could be curtailed before making all observations. This is especially true for some tests based on ranks and some confidence procedures based on order statistics. Answer the following questions about early decision rules, assuming there is no chance of ties. (Hint: Problem 4 may be helpful here.)

(a) How early can a decision be reached for (i) a one-tailed (two-tailed), two-sample quantile test, (ii) a one-tailed (two-tailed), two-sample sign test with fixed ξ, (iii) a lower (upper) confidence bound corresponding to a two-sample quantile test, (iv) a two-sided confidence interval corresponding to a two-sample quantile test.

(b) Show that a one-tailed quantile test based on A for preselected t that rejects for $A \leq a$ (in the notation of Table 3.1) reaches a decision of reject at the wth observation if and only if $t - a \leq w \leq t$ and the wth observation is $Y_{(t-a)}$.

(c) Show that the probability of rejection in (b) under the null hypothesis is $\binom{w-1}{t-a-1}\binom{N-w}{n-t+a}/\binom{N}{n}$. (This result can be obtained conditionally, or equivalently, by considering a 3×2 table with row totals $w - 1$, 1, and $N - w$. Other probabilistic arguments lead to alternative formulas such as $[(t - a)/w]\binom{m}{w-t+a}\binom{n}{t-a}/\binom{N}{w}$.)

(d) Show that the test in (b) reaches a decision of "accept" at the wth observation if and only if $a + 1 \leq w \leq t$ and the wth observation is $X_{(a+1)}$, and that this has probability $\binom{w-1}{a}\binom{N-w}{m-a-1}/\binom{N}{m}$ under the null hypothesis.

(e) Find the null frequency function of the number of observations required for a decision by a one-tailed quantile test.

(f) Show that a two-tailed quantile test that rejects for $A \leq a$ or $A \geq a'$ reaches a decision of "accept" at the wth observation if and only if $t - a' + a + 1 < w \leq t$ and the wth observation is either $Y_{(t-a'+1)}$ or $X_{(a+1)}$.

(g) Show that, under the null hypothesis, the probability that a two-tailed quantile test requires w observations for a decision is

$$P = \left[\binom{w-1}{t-a-1}\binom{N-w}{n-t+a} + \binom{w-1}{a'-1}\binom{N-w}{m-a'}\right]/\binom{N}{m}$$

$$\text{if } w \leq t - a' + a + 1,$$

and, with the same formula for P, is

$$P + \left[\binom{w-1}{t-a'}\binom{N-w}{n-t+a'-1} + \binom{w-1}{a}\binom{N-w}{m-a-1}\right]/\binom{N}{m}$$

$$\text{if } t - a' + a + 2 \leq w \leq t.$$

(h) What can be said about the time required for a decision in parts (d) and (g)?

(i) How could P-values be defined when a curtailed sampling procedure is used?

*6. Given two mutually independent random samples of m X's and n Y's from populations with continuous c.d.f.'s F and G respectively, let U be the number of Y's that are smaller than the median of the X sample. If the X sample is regarded as the control group and the Y sample as the treatment group, the control median test proposed by Kimball et al. [1957] (see also Gastwirth [1968]) is to reject the null hypothesis $F = G$ if U is small. Generalizing to an arbitrary quantile, let U be the number of Y's which are smaller than the kth smallest of the m X observations. (Hint: Problem 4 may be helpful.)

(a) Show that the null frequency function of U is

$$P(U = u) = \binom{k + u - 1}{u}\binom{N - k - u}{n - u}\bigg/\binom{N}{n} \quad \text{for } u = 0, 1, \ldots, n.$$

(Compare Problem 5c and d. $U = u$ if and only if the $(k + u)$th smallest observation is $X_{(k)}$.)

(b) Show (preferably without algebra or reference to the statement in (a)) that if $U = u$ is observed, then the P-value of a test rejecting the null hypothesis when U is small is a hypergeometric tail probability associated with the following 2×2 table. (This relates the "hypergeometric waiting-time" distribution to the ordinary hypergeometric distribution.)

	X's	Y's	
Below	k	u	$k + u$
Above	$m - k$	$n - u$	$N - k - u$
	m	n	N

(c) Show that a one-tailed test based on U, which might be called a control quantile test, always reaches the same decision as a suitably chosen, one-tailed, two-sample quantile test, and vice versa, but this is not true for two-tailed tests. (The lower tail is of primary interest.)

(d) What is the confidence bound corresponding to a one-tailed test based on U?

(e) In the situation of Problem 5, show that a decision of reject cannot be reached early by a lower-tailed control quantile test, and a decision of "accept" cannot be reached early by an upper-tailed control quantile test. When can the other decisions be reached?

(f) Show that, if sampling is curtailed as soon as a decision can be reached, the one-tailed control quantile and ordinary quantile tests coincide in all respects (stopping time and decision).

*7. In the situation described in Problem 6, let V be the number of X's that are smaller than the median of the Y sample. The first-median test, proposed by Gastwirth [1968], is based on U if the median of the X sample is smaller than the median of the Y sample, and on V otherwise. Hence it permits an earlier decision than the control median test in some circumstances, especially in two-tailed tests. Generalizing to an arbitrary quantile, let $X_{(k)}$ be the kth smallest among the m X's and $Y_{(l)}$ the (l)th smallest among the n Y's and let U be the number of Y's smaller than $X_{(k)}$ and V the number of X's smaller than $Y_{(l)}$. Note that U has the distribution given in Problem 6a. (Hint: Problem 4 may also be helpful in answering the questions below.)

(a) Show that V has the same null distribution as U but with different parameters. What are the parameters?

(b) Let the test statistic be U if $X_{(k)} < Y_{(l)}$ and V otherwise. Show that U is the test statistic if and only if $U \le l - 1$ and V is the test statistic if and only if $V \le k - 1$.

(c) Let the critical value be u if $X_{(k)} < Y_{(l)}$ and v otherwise, where $u \leq l - 1$ and $v \leq k - 1$; express the level of this test as a sum of two hypergeometric probabilities. Such a test might be called a first-quantile test. (Gastwirth [1968] considers the case $u = v$, $m = 2k - 1$, $n = 2l - 1$.)

(d) Show that each tail of a first-quantile test always reaches the same decision as a suitably chosen, one-tailed, two-sample quantile test, and vice versa, but the two-tailed, two-sample quantile tests reach the same decision as only those first-quantile tests having $k + u = l + v$. (They reach "accept" decisions at the same time, but the first-quantile tests reach reject decisions sooner. See, however, parts (j) and (k) of this problem.)

(e) Find a convenient expression for the null conditional distribution of U given that U is the test statistic.

(f) If $k = l$, show that the test statistic is min (U, V).

(g) Show that if $m = 2k - 1$ and $n = 2l - 1$ the test statistic is U with probability $\frac{1}{2}$ under the null distribution.

(h) Show that if $k = l$ and $m = n$ then U and V are identically distributed and the test statistic is U with probability $\frac{1}{2}$ under the null distribution.

(i) In (h), the two tails of the test are alike, so a two-tailed P-value can be defined naturally as twice the one-tailed P-value. Discuss the problems of defining the P-value of first-quantile tests in other situations.

(j) In the situation of Problem 5, show that a decision of reject cannot be reached early by a first-quantile test. When can a decision of "accept" be reached?

(k) If sampling is curtailed as soon as a decision can be reached, show that the two-tailed, two-sample quantile tests coincide in all respects with the first-quantile test having $k + u = l + v$.

(l) Show that, under the null hypothesis, the probability that a first quantile test requires w observations for a decision is

$$P = \left[\binom{w-1}{l-1}\binom{N-w}{n-1} + \binom{w-1}{k-1}\binom{N-w}{m-k} \right] \Big/ \binom{N}{m} \quad \text{if } w \leq u + v + 1,$$

and, with the same formula for P, is

$$P + \left[\binom{w-1}{u}\binom{N-w}{n-u-1} + \binom{w-1}{v}\binom{N-w}{m-v-1} \right] \Big/ \binom{N}{m}$$

$$\text{if } u + v + 2 \leq w < t.$$

8. Prove that, in each of the three situations described in Sect. 3.2, the conditional distribution of A given t, or the distribution of A for fixed t, is the hypergeometric distribution given in (3.1),

(a) If the N observations are drawn from the same population.

(b) If N given units are assigned to the two columns by randomization.

9. Given two samples, suppose that ties occur at the median of the combined sample and all those values at the median are counted as "above."

(a) Under what circumstances will the margin t used in the two-sample median test be unchanged?

(b) Show that ties reduce t otherwise.

(c) Show that, if the values at the median are omitted, the hypergeometric distribution still applies to the resulting 2×2 table under the null hypothesis.

10. Prove that the median test accepts (rejects) the hypothesis $\mu = \mu_0$ under the shift assumption if the value μ_0 is interior (exterior) to the random interval (3.6).

11. (a) Verify the following results given by Mood [1950, p. 396]. Under the shift assumption when $s' \geq s$ and $r \geq r'$

$$P(Y_{(r')} - X_{(s')} < \mu < Y_{(r)} - X_{(s)})$$
$$= P(X_{(s)} < Y_{(r)} - \mu \text{ and } X_{(s')} > Y_{(r')} - \mu)$$
$$= 1 - P(X_{(s)} > Y_{(r)} - \mu) - P(X_{(s')} < Y_{(r')} - \mu)$$

where

$$P(X_{(s)} > Y_{(r)} - \mu) = \sum_{i=0}^{s-1} \binom{r+i-1}{r-1}\binom{N-r-i}{N-r} \Big/ \binom{N}{m}$$

$$P(X_{(s')} < Y_{(r')} - \mu) = \sum_{i=s'}^{m} \binom{r'+i-1}{r'-1}\binom{N-r'-i}{n-r'} \Big/ \binom{N}{m}.$$

(b) Relate these "hypergeometric waiting-time" tail probabilities to ordinary hypergeometric tail probabilities.

*12. To test the null hypothesis that two samples come from the same population, Rosenbaum [1954] suggests a test based on the number of observations in the X sample which lie outside an extreme value of the Y sample. Define the test statistic S as the number of X's larger than the largest Y.
   (a) Show that the null frequency function of S is $f(s) = n\binom{m}{s}/N\binom{N-1}{s}$. Against alternatives in the direction $X > Y$, a test based on S would reject in the upper tail. Rosenbaum gives tables of upper critical values of S for tests at conservative levels 0.01 and 0.05.
   (b) For m and n both large and approximately equal, show that $f(s)$ is approximately equal to $2^{-(s+1)}$, so that the upper-tail critical values of S are 5 and 7 for approximate $\alpha = 0.01$ and 0.05, respectively. (Part (e) strengthens this result.)
   (c) What two-sample quantile test is equivalent to rejecting for $S \leq s$?
   (d) Express the P-value of Rosenbaum's test as a hypergeometric tail probability. (Note that the P-values for different s correspond to different two-sample quantile tests.)
   (e) Show that $f(s) \leq n(m/N)^s/(N-s)$ and $P(S \geq s) \leq N(m/N)^s/(N-s)$ under the null hypothesis.

*13. Use differences of X and Y order statistics to express the confidence limits for a shift parameter that correspond to
   (a) The control median test (see Problem 6).
   (b) The first-median test (see Problem 7).
   (c) The control quantile test (see Problem 6).
   (d) The first-quantile test (see Problem 7).
   (e) Rosenbaum's test (see Problem 12).

14. (a) For the one-tailed, two-sample quantile test against an alternative of two different population densities, express the power as a two-variable integral involving the two population cumulative and density functions.
   (b) Show that the power of the two-tailed test is the sum of two such integrals.

15. With the notation of Table 3.1 for the two-sample quantile test, show that a/m and $(t - a)/n$ are bounded away from 0 and 1 under suitable conditions on m, n, t, and α.

*16. (a) Argue that if X has density f and c.d.f. F, then the order statistic $X_{(i)}$ is approximately normal with mean $F^{-1}(i/m)$ and variance $i(m - i)/m^3\{f[F^{-1}(i/m)]\}^2$ under suitable conditions on i, m, and F. (One such argument uses the normal approximation to the binomial distribution of the empirical cumulative distribution.)

   (b) What kind of precise limit statements along these lines would you expect to hold, under how broad conditions?

   (c) Sketch a proof, perhaps under more restrictive conditions.

17. Show that (3.11) is a necessary and sufficient condition for two random variables X and Y to have different medians, whatever median is chosen for each if the median is not unique.

18. Show that the median test is consistent against alternatives with different population medians

   (a) Using the idea of consistent estimation and its relation to consistent tests (see Sect. 3.4, Problems 26 and 28 of Chap. 3 and Problem 1 of Chap. 1).

   (b) Using the consistency of the sign test with ξ fixed as in (3.11).

*19. (a) Show that the two-sample sign test with fixed ξ has the optimum properties stated in Sect. 3.7 when the X's come from one population and the Y's from another.

   (b) Show that these properties also hold for suitable alternatives under which neither the X's nor the Y's need be identically distributed.

*20. Show that the hypergeometric probability of a or less in Table 3.1 is less than the binomial probability of a or less for the binomial parameters m and $p = t/N$ if $a \le (mt/N) - 1$, and greater than if $a \ge mt/N$ [Johnson and Kotz, 1969; Uhlmann, 1966].

*21. (a) Argue or show by example that in a treatment-control experiment, if the treatment effect is not an identical shift in every unit, the level of the confidence procedure corresponding to the median test may be less than nominal despite random assignment of units to groups, even when both population distributions are symmetric and the treatment effect is defined as the difference between the two centers of symmetry. (One type of example has one group much less dispersed than the other and uses Problem 20).

   (b) In (a), show that if the populations have the same shape but different scales, then asymptotically the confidence level is always less than nominal. Assume the population has nonzero density at the median. (Hint: Use Problem 16.)

*22. Show that the lower-tailed median test is biased against the alternative that the median of the X population is larger than the median of the Y population. (An example of power less than α can be constructed as in Problem 21a if the difference in medians is very small.)

23. Use Theorem 3.1 to show that a suitable median test rejects with probability at most α under (3.12) and at least α under (3.13). Show that, more generally, the test functions of one-tailed, two-sample sign tests rejecting for A large are increasing in the Y direction.

24. Prove Theorem 3.1, that a test is monotonic in the distributions of the observations if the critical function is monotonic in the observations.

25. (a) Use (3.8) to (3.10) to give an easily evaluated expression for the approximate power of the one-tailed median test against the alternative that the X and Y populations are both normal but with possibly different means and variances.
    (b) Evaluate this power expression for $m = 6$, $n = 9$, $\alpha = 0.10$, population means differing by 1, and population variances in the ratios 0.25, 1, and 4.

26. A sample of size t is drawn without replacement from a finite dichotomous population of size N which contains exactly m elements with the value 1 and $n = N - m$ with the value 0. For the sample observations $X_1, \ldots, X_t$, the number of 1's in the sample is $\sum_1^t X_i = t\overline{X}$.
    (a) Show that

$$E(t\overline{X}) = mt/N, \qquad \mathrm{var}(t\overline{X}) = mnt(N - t)/N^2(N - 1).$$

    (b) What is the probability distribution of $t\overline{X}$?
    (c) Relate this situation to the two-sample quantile test statistic of Sect. 3.
    (d) Relate this situation to the corresponding situation with a sample of size m and a population containing t 1's and $N - t$ 0's.

27. Let $\hat{\xi}$ be the median of a combined sample of m X and n Y random variables. Let $F_m(\hat{\xi})$ and $G_n(\hat{\xi})$ denote the respective sample proportions of X's and Y's which are smaller than $\hat{\xi}$ so that the median test statistic can be written as $A = mF_m(\hat{\xi})$.
    (a) Show that the one-tailed median test that rejects for small values of A is equivalent to a test rejecting for small values of $F_m(\hat{\xi}) - G_n(\hat{\xi})$.
    (b) If $n \to \infty$ while m remains fixed, show that $\hat{\xi}$ converges to the median of the Y population, find the limiting value of the two-sample median test statistic, and show that the median test approaches the one-sample sign test (Moses [1964]).

*28. What happens to the two-sample quantile test if $N \to \infty$ with t fixed?

29. Verify the linear relationships between the Mann-Whitney and Wilcoxon statistics given in (4.3) and (4.4).

30. (a) If $r_{m,n}(u)$ denotes the number of arrangements of m X and n Y variables such that the value of U, the number of (X, Y) pairs for which $X > Y$, equals u, show that

$$r_{m,n}(u) = r_{m,n-1}(u) + r_{m-1,n}(u - n)$$

for all $u = 0, 1, \ldots, mn$ and all positive integer-valued m and n, with the following initial and boundary conditions for all $m \geq 0$ and $n \geq 0$:

$$r_{m,n}(u) = 0 \quad \text{for all } u < 0 \text{ and all } u > mn$$

$$r_{m,0}(0) = r_{0,n}(0) = 1.$$

This provides a simple recursive relation for generating the frequencies of values of U and hence for the null probability function $p_{m,n}(u) = P(U = u)$ for sample sizes m and n, using

$$(m + n)p_{m,n}(u) = np_{m,n-1}(u) + mp_{m-1,n}(u - n).$$

(b) What change is required in order to generate directly the null cumulative probabilities $F_{m,n}(u) = P(U \leq u)$?

(c) What change is required in order to generate the null probability function of the X rank sum R_x? The null cumulative function of R_x?

31. Use the recursive method developed in Problem 30 to generate the complete null distribution of U or R_x for all $m + n \leq 6$. Check your results using Table F.

*32. Derive results for the two-sample case that are analogs of the one-sample results given in Problem 13 of Chap. 3. (Fix and Hodges [1955] give these two-sample results and a table. Their work is inspired Problem 13 of Chap. 3.)

33. Let R be uniformly distributed on the integers $1, 2, \ldots, N$ and let S be independently uniformly distributed on $(0, 1)$.

(a) Show that S has mean $\frac{1}{2}$ and variance $\frac{1}{12}$.

(b) Show that $R + S$ is uniformly distributed on $(1, N + 1)$ and hence has the same distribution as $NS + 1$.

(c) Use (a) and (b) to show that $R + S$ has mean $N/2 + 1$ and variance $N^2/12$.

(d) Use (c) and (a) to show that R has mean $(N + 1)/2$ and variance $(N^2 - 1)/12$.

34. Use the null mean and variance of R_x given in (4.7) and (4.8) to verify the corresponding moments of R_y and U given in (4.9) and (4.10).

35. Show that the null distributions of U and U' are identical.

36. Show that the possible values of U, U', R_x, and R_y are as stated in the paragraph following (4.10).

37. Show that U and U' are symmetrically distributed about $mn/2$, R_x is symmetric about $m(N + 1)/2$, and R_y is symmetric about $n(N + 1)/2$.

38. (a) Show that the continuity correction for the approximate normal deviate of the rank sum test statistic is $\sqrt{3/mn(N + 1)}$.

(b) Show that for $0.1 < m/N < 0.9$, this continuity correction is less than $1/\sqrt{0.03N^3}$ and hence less than 0.02 if $N \geq 44$, less than 0.01 if $N \geq 70$.

(c) Show that for $0.1 < m/N < 0.9$, the continuity correction in the normal approximation (3.2) for the median test is less than $\sqrt{N/0.3}\sqrt{N^2 - 1}$ and hence less than 0.02 if $N \geq 27778$, less than 0.01 if $N \geq 111112$.

39. Show that the value of k given in (4.11) and needed for the confidence bound is one less than the rank (among all possible R_x) of the lower tail critical value at level α and hence $k + 1$ is the rank.

40. Represent two samples of sizes m and n by a path as explained in Problem 4. This path separates the rectangle with corners $(0, 0)$, $(0, n)$, $(m, 0)$, and (m, n) into two parts. Show that the upper left part has area U and the lower right part has area U', where U and U' are the Mann-Whitney statistics.

41. Given continuous c.d.f.'s F and G, show that p_1, p_2, and p_3 as given in Equations (4.12)–(4.14) can be written as follows, where $dF(x)$ can be replaced by $f(x)\,dx$ if F has density f, and similarly for $dG(y)$.

(a) $p_1 = \int G(x)\,dF(x) = 1 - \int F(y)\,dG(y)$.

(b) $p_2 = \int [1 - F(y)]^2\,dG(y) = 2\int [1 - F(x)]G(x)\,dF(x)$.

(c) $p_3 = \int G^2(x)\,dF(x) = 1 - 2\int F(y)G(y)\,dG(y)$.

42. Calculate p_1, p_2, and p_3 defined in Equations (4.12)–(4.14) if X and Y are drawn from identical populations, and substitute these results in (4.15) and (4.16) to verify the results given in (4.10).

43. Derive the expression for var(U) in Equation (4.22) by using (4.18)–(4.21).

44. Natural estimators of p_1, p_2, and p_3, as defined in Equations (4.12)–(4.14), are the corresponding proportions of sample comparisons (X_i, X_j) which satisfy the respective inequalities.
    (a) Show that these estimators for p_1 and p_2 can be expressed as

    $$\hat{p}_1 = \sum_i (R_i - i)/mn = U/mn$$

    $$\hat{p}_2 = 2 \sum_i (R_i - i)(m - i)/mn(m - 1),$$

    where R_i is the rank of $X_{(i)}$ in the combined sample.
    (b) Give a similar expression for $\hat{p}_3$.
    (c) Find the expected value of these estimators under the null hypothesis of identical distributions.

45. Let $X_1, \ldots, X_m$ and $Y_1, \ldots, Y_n$ be independent random samples from the continuous distributions F and G respectively.
    (a) Show that the Mann-Whitney statistic U can be written as $n \sum_{i=1}^{m} G_n(X_i)$, where $G_n(y)$ is the proportion of Y's that are smaller than y.
    (b) If $n \to \infty$ while m remains fixed, $U/n \to \sum_{i=1}^{m} G(X_i) = W$, say (in probability, and with probability one). Characterize the distribution of W under the null hypothesis.
    (c) What is the limit of the rank sum test as $n \to \infty$ while m remains fixed? It amounts to a one-sample test (see Moses [1964]). Is this one-sample test a rank test? A nonparametric test?
    (d) Show that $E(W/m) = P(X > Y) = p_1$ as defined in (4.12) (see Moses [1964]).

46. Show that the rank sum test is consistent against the alternatives given in Sect. 4.5.

47. Show that, for some alternatives, the median test is consistent and the rank sum test is not, and vice versa.

48. Suppose that X and Y are independent but not identically distributed, and that X is stochastically larger than Y.
    (a) Show that $P(X > Y) > \frac{1}{2}$.
    (b) Show that $E(X) > E(Y)$ if both expectations exist and are not both infinite.
    (c) Show that the median of X is at least as large as the median of Y, provided at least one is unique, no matter how the other is chosen.
    (d) What happens in (c) if neither X nor Y has a unique median? (Of course, (c) and (d) apply equally to quantiles of any order.)

49. Show that if $F(x) = [G(x)]^k$ for all x (Lehmann alternative), then X is stochastically larger than Y for $k > 1$, smaller for $k < 1$.

50. Show that a nonzero shift implies stochastic dominance.

51. Show that the rank sum test is consistent against stochastic dominance, and hence by Problem 50, is consistent against shifts.

52. Suppose the X and Y populations have the same median but, with high probability, X is close to the median and Y is not. Show that
   (a) In the notation of Sect. 4.4, $p_1 = \frac{1}{2}$, $p_2 = \frac{1}{2}$, and $p_3 = \frac{1}{4}$, approximately.
   (b) $E(U) = mn/2$ and $\mathrm{var}(U) = m^2 n/4$, approximately.
   (c) As approximated, the mean is the same as under the null hypothesis, but the variance is larger for $n > 2m - 1$, smaller for $n < 2m - 1$.
   (d) The probability of rejection by the rank sum test must be greater than α if $n > 2m - 1$, less if $n < 2m - 1$, for some levels α and some populations of the type described.

*53. (a) What questions about the differences $Y_j - X_i$ correspond to questions about the Walsh averages in Problem 37 of Chap. 3?
   (b) Investigate some of the questions raised in (a).

54. Prove that the rank sum test is unbiased against the alternative that the distribution of every X_i is stochastically smaller than the distribution of every Y_j.

55. Use Theorem 3.1 to show that the rank sum test rejects with probability at most α under (3.12) and at least α under (3.13).

56. For a sample of X's and a sample of Y's, possibly with ties, let L be the $(k + 1)$th smallest difference $Y_j - X_i$. Show that if the Y population differs from the X population only by a shift μ, then $P(L \le \mu) \ge 1 - \alpha \ge P(L < \mu)$ where $1 - \alpha$ is the exact probability in the continuous case. (Hint: Consider breaking ties randomly. See also Problem 107.)

57. Consider the confidence interval for shift corresponding to the rank sum test.
   (a) Show that if 0 is not an endpoint, then all methods of breaking any ties lead to the same decision, namely, to "accept" if 0 is outside and reject if 0 is inside.
   (b) Show that if 0 is an endpoint, then there are tied observations and breaking ties one way leads to rejection, another way to acceptance.

58. Show that the value of R_x computed using the average rank procedure is equal to $U + m(m + 1)/2$, where U is equal to the number of (X_i, Y_j) pairs for which $X_i > Y_j$ plus one-half of the number for which $X_i = Y_j$.

59. Show that when ties are handled by the average rank procedure, under H_0,
   (a) The means of R_x and U are not affected by ties but the variance is reduced to

$$\mathrm{var}(R_x) = \mathrm{var}(U) = mn[N(N^2 - 1) - \sum t(t^2 - 1)]/12N(N - 1)$$

   where t is the number of observations tied at a given rank and the sum is over all sets of tied observations (as in Problem 52 of Chap. 3).
   (b) The distributions of R_x and U may not be symmetrical.

60. Show that conditions (i)(b) and (ii) in Section 4.7 are equivalent.

61. Show that the average rank test procedure satisfies (i)(b) and (ii) but not (i)(a) and (iii) in Section 4.7.

62. Show that breaking ties randomly
   (a) Preserves the null distribution of U, even conditionally on the pattern of ties present.
   (b) Satisfies all the conditions (i)–(iii) in Section 4.7.

*63. Show that the confidence regions for shift corresponding to the average rank procedure are always intervals.

64. Show that the multiplicities of the average ranks and hence the exact pattern of ties can be determined from the ranks without multiplicities. (For instance, the average ranks 1, 3, 6, 9.5, 12.5, 14 can arise only from a combined sample with the pattern of ties displayed in Table 4.2.)

65. Use the result in Problem 1 of Chap. 1 to show that U/mn is a consistent estimator of $p_1 = P(X_i > Y_j)$.

66. Show that U/mn is a minimum variance unbiased estimator of p_1 if the only restriction is that the X and Y distributions are continuous.

*67. Under the assumption that the X and Y populations are normal with arbitrary means, express the minimum variance unbiased estimator of p_1 as conveniently as possible for the case of variances
(a) Known.
(b) Common but arbitrary (unknown).
(c) Arbitrary, possibly not all the same.

*68. (a) Show that if the X and Y populations are normal with common variance, then for large m and n, the variance of $\Phi[(\bar{X} - \bar{Y})/s\sqrt{2}]$ is approximately $\frac{1}{2}\phi^2(\xi)(1/m + 1/n + \xi^2/N)$ where $\xi = (\mu_x - \mu_y)/\sigma\sqrt{2}$.
(b) Compare the result in (a) to the variance of U/mn.

69. How might one find the values of p_1 for which $(U - mnp_1)^2/z^2$ equals
(a) The right hand side of (4.26)?
(b) The right hand side of (4.27)?

*70. Show that the bound on var(U) given by (4.27) is never greater than that given by (4.26). (Hint: The inequality to be proved is equivalent to $[(1 - 2p)^{3/2} - 1 + 3p](2m - n - 1) \leq 3p^2(m - 1)$ for $0 < p < \frac{1}{2}$. You may wish to prove and use $1 - 3p \leq (1 - 2p)^{3/2} \leq 1 - 3p + 2p^2$ for $0 < p < \frac{1}{2}$.)

71. Mr. Greenthumb has come to consult you about the following problem. To test the effect of a certain type of fertilizer on the growth of spinach, he divided his spinach field into 20 plots, picked ten plots at random and fertilized them, and left the other ten unfertilized. Upon harvesting he obtained the following yields, in bushels.

Unfertilized plots 6.1, 10.2, 8.7, 6.4, 7.3, 10.9, 7.7, 8.4, 9.0, 9.8

Fertilized plots   10.1, 11.2, 12.3, 9.2, 12.0, 11.9, 9.6, 10.8, 10.3, 12.7

When you question him persistently, he admits to thinking that, in the absence of fertilizer, the yields should be independently distributed, approximately normally with the same mean and variance for all plots. He thinks that the effect of the fertilizer is to increase the yield by some amount; he is sure it does no harm.

The people paying for his research do not like distribution assumptions, however, and he has consented to analyze the data without such assumptions.
(a) What methods of analyzing his data would you suggest he consider? What would you tell him about these methods? Be as precise as you can, but remember that Mr. Greenthumb, though highly intelligent, is nevertheless not a statistician.

(b) Mr. Greenthumb would like to make a preliminary report right away. For this purpose, analyze the data in some quick but reasonable way, even if it is not the way you consider optimum.

72. Twenty mice are placed randomly in individual cages and the cages are divided randomly into two groups, each of size 10. All the mice are infected with tuberculosis; then the mice in the second group are each given a certain drug (B) while those in the first group are given a placebo (A). Since the drug (B) is known to be nontoxic, those mice in the treatment group would not be expected to die sooner than the control group. The number of days to death after infection are shown below.

Control (A): 5, 6, 7, 7, 8, 8, 8, 9, 10, 12
Drug   (B): 7, 8, 8, 8, 9, 9, 12, 13, 14, 17

(a) Test the hypothesis that the drug is without effect, using the rank sum test and the average rank procedure to handle ties.
(b) Find the smallest and largest P-values when the ties are broken.
(c) Find a lower confidence bound for the effect of the drug, using a level of 0.90.

73. A professor decided that since it was necessary to give tests in overcrowded classrooms, the temptation for eyes to wander should be minimized. He decided to give two sets of tests, with the only difference being the order in which the questions appeared. The tests were distributed in such a way that no student could gain information if his eyes wandered. The test results are given below. Determine whether there is a significant difference between the average grades for these sets, and find a confidence interval for the difference.

Set A: 78, 68, 78, 90, 66, 75, 50, 42, 80, 74
Set B: 82, 81, 83, 95, 91

74. Suppose X and Y samples each have possible values $1, 2, \ldots, r$ and the value i is observed m_i times in the X sample, n_i times in the Y sample, $\sum_i m_i = m$ and $\sum_i n_i = n$. The observed frequencies can be arranged in an $r \times 2$ table with ith row entries m_i, n_i and row total $N_i = m_i + n_i$. Consider using the rank sum test with ties handled by the average rank procedure.
(a) Express the test statistic in terms of the observed frequencies.
(b) How does this test relate to the ordinary chi-square test for an $r \times 2$ table when $r = 2$? When $r > 2$?
(c) If the possible values are some arbitrary numbers $a_1, a_2, \ldots, a_r$ instead of $1, 2, \ldots, r$, what effect would this have?

*75. (Early decision in rank sum tests). Suppose X and Y observations are obtained in order of magnitude and a one-tailed rank sum test is to be used based on sample sizes m and n and rejecting for $R_x \leq t$. Assume no ties. The results below are based on Alling [1963]. The similar problem for censored samples is discussed in Halperin and Ware [1974].
(a) Given the first $N' = m' + n'$ observations, derive expressions for the minimum and maximum possible values of R_x.
(b) When can a decision to reject first be reached? An "accept" decision?
(c) Show that a decision to reject can be first reached only after observing an X, an "accept" decision after observing a Y.
(d) Show that a decision can always be reached early and that this is true of all rank tests.

76. Show that the confidence region for a shift parameter corresponding to a one-tailed or two-tailed test based on a sum of scores is an interval if the c_k form a monotone sequence.

77. (a) Show that the null distribution of the sum of the X scores in a sample of m X's and n Y's has mean $m \sum_1^N c_k/N$ and variance $mn[\sum_1^N c_k^2/N - (\sum_1^N c_k/N)^2]/(N - 1)$, in the notation of Sect. 5.
    (b) Show without further algebra that the sum of the Y scores has the same variance as the sum of the X scores (under all circumstances).
    (c) Use the result in (a) to verify (4.7) and (4.8) for the rank sum test.
    (d) Use the result in (a) to obtain the null mean and variance of the median test statistic A.

78. (a) Argue that if c_k is approximately $J[(k - 0.5)/N]$ or $J[k/(N + 1)]$ for some function J, then $\sum_1^N c_k/N$ is approximately $\int_0^1 J(u)\,du$ and $\sum_1^N c_k^2/N$ is approximately $\int_0^1 J^2(u)\,du$.
    (b) What kinds of conditions would be needed to make the approximations in (a) good?
    (c) What function J corresponds to the rank sum test?
    (d) Compare the approximate and exact values for the rank sum test.
    (e) What function J corresponds to the Fisher–Yates–Terry test? The van der Waerden test?
    (f) For these test statistics, what approximate values arise by applying the result in (a) to the mean and variance in Problem 77(a)?

*79. Consider a test based on the sum of scores c_k for $c_k = k + 2^{-k}$. Formulate the two-sample counterpart of Problem 91 of Chap. 3 and answer the questions posed.

80. Use Theorem 3.1 to show that a sum of scores test rejects with probability at most α under (3.12) and at least α under (3.13) if the scores form a monotonic sequence.

81. Do (a)–(f) below using the Fisher–Yates–Terry procedure.
    (a) Suppose the following are two independent random samples, the second drawn from the same distribution as the first except for translation by an amount θ (to the right if $\theta > 0$). Test at a level near 0.01 the null hypothesis $\theta = 2$ against the alternative $\theta < 2$.

    First sample: $-0.2,\ 0.4,\ -0.8,\ -1,\ 0$
    Second sample: $0.5,\ 1.1,\ -0.3,\ 0.6,\ 2,\ 0,\ 0.8,\ 1.5$

    (b) Give the exact level of the test used in (a).
    *(c) Give the confidence bound for θ which corresponds to the test used in (a).
    (d) Suppose the following are two independent random samples, the second drawn from the same distribution as the first except for translation by an amount θ. Test at level near 0.10 the null hypothesis $\theta = 0$ against the alternative $\theta \neq 0$.

    First sample: $79,\ 13,\ 138,\ 129,\ 59,\ 76,\ 75,\ 53$
    Second sample: $96,\ 141,\ 133,\ 107,\ 102,\ 129,\ 110,\ 104$

    (e) Give the exact level of the test used in (d).
    *(f) Give the confidence interval corresponding to the test used in (d).

82. Do (a)–(f) of Problem 81 using the median test and related confidence procedure.

83. Do (a)–(f) of Problem 81 using the rank sum test and related confidence procedure.

84. Suppose that m X observations are independent and follow the uniform distribution on $(0, 1)$, and that n Y observations are independent and have the density

$$g(y) = (1 + \alpha)y^\alpha \quad \text{if } 0 \le y \le 1 \quad \text{for } \alpha > -1.$$

If the X and Y observations are mutually independent, find the probabilities of all $\binom{N}{m}$ possible rank arrangements if
(a) $m = n = 2$.
(b) $m = 2, n = 3$.
(c) $m = n = 3$.

85. What is the relation between the densities in Problem 84 and Lehmann alternatives $F(x) = [G(x)]^k$ for all x?

86. Suppose that X_1, X_2 have density $f(x) = 1$ for $0 \le x \le 1$; Y_1, Y_2 have density $g(y) = 1$ for $-\varepsilon \le y \le 0$ and $\varepsilon \le y \le 1$; and all are independent. Find the probability of each possible set of combined ranks. (Note that this probability is not increasing in the X ranks even though $F(x) \le G(x)$ for all x.) Find a function Q such that $F(x) = Q[G(x)]$.

87. Given an arbitrary two-sample procedure, how can a permutation-invariant procedure be defined which has identical statistical properties whenever the two samples are drawn independently from two populations?

88. Suppose that $X_1, \ldots, X_m, Y_1, \ldots, Y_n$ are independent but neither the X's nor the Y's are necessarily identically distributed. How might one argue for using a permutation-invariant procedure? What are some circumstances under which permutation invariance would clearly be inappropriate?

89. Given arbitrary X_i, Y_j, X_i', Y_j' for $i = 1, \ldots, m$, $j = 1, \ldots, n$, show that the X_i and Y_j have the same ranks in the combined sample of all X_i, Y_j as do the X_i', Y_j' in the combined sample of all X_i', Y_j', if and only if there exists a strictly increasing function g such that $g(X_i) = X_i'$ and $g(Y_j) = Y_j'$ for all i, j.

90. (a) Show that all strictly increasing transformations of the observations leave the following hypothesis invariant: $X_1, \ldots, X_m$ and $Y_1, \ldots, Y_n$ are independent and the X population is stochastically larger than the Y population.
    (b) Show the same for the hypothesis that $X_1, \ldots, X_m, Y_1, \ldots, Y_n$ are independent (but not necessarily identically distributed).
    (c) Show the same for the hypothesis that $X_1, \ldots, X_m, Y_1, \ldots, Y_n$ are independent and the X's are identically distributed.
    (d) Show that this invariance does not apply to the shift hypothesis.

91. Show that a two-sample test is a permutation-invariant rank test if and only if its critical function depends only on the indicator variables I_j defined in Sect. 6.

92. Show that a rank test is the most powerful rank test at level α against a simple alternative K if and only if it has the form (8.1) and level exactly α.

*93. Suppose the X population is $N(\theta, 1)$ and the Y population is $N(0, 1)$. Use Problem 97 to show that

(a) $P_\theta(\underline{r}) = E\left\{\exp\left[\theta \sum_{i=1}^{m} Z_{(r_i)} - m\theta^2/2\right]\right\}/N!$

$$= \frac{1}{N!}\left\{1 + \theta E\left(\sum_{i=1}^{m} Z_{(r_i)}\right) + \frac{1}{2}\theta^2\left[E\left(\sum_{i=1}^{m} Z_{(r_i)}\right)^2 - m\right]\right\} + 0(\theta^3)$$

where the $Z_{(j)}$ are order statistics from $N(0, 1)$ and $0(\theta^3)$ indicates terms of order θ^3.

(b) For sufficiently small positive θ, the most powerful rank test at level $3/\binom{N}{m}$ rejects if the X ranks are any permutation of $(n + 1, n + 2, \ldots, N), (n, n + 2, n + 3, \ldots, N)$, or $(n + 1 - k, n + k, n + 3, n + 4, \ldots, N)$, where $k = 1$ or 2, and it remains to be determined whether $k = 1$ or $k = 2$.

(c) If $m = n$, both choices for k give the same coefficient of θ and hence the most powerful test is whichever choice gives the larger value of $\text{var}(Z_{(n+1-k)} + Z_{(n+k)} + Z_{(n+3)} + \cdots + Z_{(N)})$. With appropriate tables of the variances and covariances of normal order statistics, e.g., Teichroew [1956], it can be verified that the two choices are not equally powerful in general.

(d) There are many sets of scores for which the sum of scores would give the most powerful test, including the choice of k. The Fisher–Yates–Terry scores, however, would not distinguish between $k = 1$ and $k = 2$ when $m = n$.

(e) What rank tests maximize the derivative of the power at $\theta = 0$?

(f) Reconcile (c) and (d) with the statement that the Fisher-Yates-Terry test is the locally most powerful rank test against normal alternatives.

94. If $U_1 < U_2 < \cdots < U_N$ are order statistics of a sample of N from the uniform distribution on $(0, 1)$, show that

(a) $E(U_i) = i/(N + 1)$.

(b) $\text{cov}(U_i, U_j) = i(N + 1 - j)/(N + 1)^2(N + 2)$ for $i < j$.

(c) $\text{var}(U_i) = i(N + 1 - i)/(N + 1)^2(N + 2)$.

(d) $E(U_i^r) = i(i + 1)\cdots(i + r - 1)/(N + 1)\cdots(N + r)$.

*95. Suppose the X population is logistic $(\theta, 1)$ and the Y population is logistic $(0, 1)$. Use Problem 97 to show that

(a) $P_\theta(\underline{r}) = E\left\{\prod_{i=1}^{m} e^{-\theta}[1 + (e^{-\theta} - 1)U_{r_i}]^{-2}\right\}/N!$

$$= \frac{e^{-m\theta}}{N!}\left\{1 - 2\lambda E\left(\sum_{i=1}^{m} U_{r_i}\right) + \lambda^2 E\left[2\left(\sum_{i=1}^{m} U_{r_i}\right)^2 + \sum_{i=1}^{m} U_{r_i}^2\right]\right\} + 0(\theta^3)$$

$$= \frac{1}{N!}\left[1 + \theta E\left[\sum_{i=1}^{m}(2U_{r_i} - 1)\right] + \theta^2 E\left\{\frac{1}{2}\left[\sum_{i=1}^{m}(2U_{r_i} - 1)\right]^2\right.\right.$$

$$\left.\left. - \sum_{i=1}^{m} U_{r_i}(1 - U_{r_i})\right\}\right] + 0(\theta^3)$$

where the U_k are order statistics from the uniform distribution on $(0, 1)$, $\lambda = e^{-\theta} - 1$, and $0(\theta^3)$ indicates terms of order θ^3.

(b) The statement of Problem 93b holds here also.

(c) Both choices for k give the same coefficients of θ and θ^2 and hence determination of the most powerful choice requires either further terms or a different approach. (Hint: Use Problem 94. Note that $E(U_{n+1-k} + U_{n+k})$, $\text{cov}(U_{n+1-k} + U_{n+k}, U_j)$, and $E[(U_{n+1-k} + U_{n+k})U_j]$ do not depend on k for $j > n + k$.

(d) The derivative of the power at $\theta = 0$ is maximized by either choice of k (and thus by a nonrandomized test).

(e) A randomized rank sum test also maximizes the derivative of the power at $\theta = 0$. What is its critical function and how does this test relate to the foregoing nonrandomized test?

96. In the two-sample problem, consider a one-parameter family of alternatives indexed by θ, where $\theta = 0$ gives the null hypothesis. Let $\alpha(\theta)$ be the power of an arbitrary rank test and $\alpha'(0)$ the derivative of the power at $\theta = 0$. Show that

(a) A rank test maximizes $\alpha'(0)$ among rank tests at level α if and only if it has the form (8.3) and level exactly α.

(b) A rank test of the form (8.4) uniquely maximizes both $\alpha'(0)$ and $\alpha(\theta)$ for all θ in some interval $0 < \theta < \varepsilon$.

97. Suppose that two populations have c.d.f.'s F and G and densities f and g, and that f vanishes whenever g vanishes. Show that the rank arrangement $\underline{r}$ has probability

$$P_\theta(\underline{r}) = E\left[\prod_{i=1}^{m} f(Z_{(r_i)})/g(Z_{(r_i)})\right]\bigg/N!$$

where $Z_{(1)} < \cdots < Z_{(N)}$ are the order statistics of a sample of N from the distribution with density g.

*98. Show that the locally most powerful rank test against continuous Lehmann alternatives $F_\theta = G^{1+\theta}$ is based on the scores $c_j = E(\log U_j)$ where U_j has the beta distribution with parameters j and $N - j + 1$.

*99. (a) Show that the locally most powerful rank test against an alternative with monotone likelihood ratio $f_\theta(x)/g_\theta(x)$ which is, say, nondecreasing in x for every θ, is based on monotonic scores c_j.

(b) Show by example that the result in (a) need not hold if the property of monotone likelihood ratios is replaced by stochastic dominance (which is weaker).

100. Show that, when the observations are obtained in order, a two-tailed quantile test that rejects for $A \le a$ and $A \ge a'$ first reaches a decision at the second smallest of $X_{(a+1)}$, $X_{(a')}$, $Y_{(t-a'+1)}$, $Y_{(t-a)}$.

101. Suppose that two populations have c.d.f.'s F and G and densities f and g, and that f vanishes whenever g vanishes. Show that

(a) $H(u)$ can be defined so that $F(x) = H[G(x)]$ for all x.

(b) $H'(u) = h(u) = f[G^{-1}(u)]/g[G^{-1}(u)]$.

(c) $P(\underline{r}) = E[\prod_{i=1}^{m} h(U_{(r_i)})]/N!$ where $U_{(1)} < \cdots < U_{(N)}$ are uniform order statistics on $(0, 1)$.

(d) For the parametric alternative $F = F_\theta$, $G = F_0$,

$$\frac{d}{d\theta}P_\theta(\underline{r})\bigg|_{\theta=0} = \frac{1}{N!}\sum_{i=1}^{m} E\left\{\frac{\partial}{\partial\theta}\log f_\theta[F_0^{-1}(U_{(r_i)})]\bigg|_{\theta=0}\right\}$$

and this agrees with (8.7).

*102. Derive an expression for the scores that give the locally most powerful rank test for alternatives under which the X and Y distributions have densities in
(a) An exponential family $p_\theta(x) = C(\theta)h(x)e^{Q(\theta)T(x)}$.
(b) A shift parameter family $p_\theta(x) = p(x - \theta)$.

*103. (a) Show that the most powerful rank test against a simple alternative with densities f, g and likelihood ratio $f(x)/g(x)$ which is nondecreasing in x rejects when the X ranks are $r_1, \ldots, r_m$ if it rejects when they are $r'_1, \ldots, r'_m$ and $r'_i \leq r_i$ for all i. More generally, show that the critical function of this test as a function of the X ranks is nondecreasing in each. (Hint: Use Problem 97.)
(b) Show by example that the most powerful rank test against a simple alternative with X stochastically larger than Y need not have this property.

104. Show that for any k, x^k is a linear combination of terms of the form $x(x + 1) \cdots (x + l - 1)$ with $l \leq k$. (For instance, $x^2 = -x + x(x + 1)$.)

*105. In the *one*-sample problem, consider a family of densities $f_\theta(x) = 0.5 + 0.5\theta g(x)$, $-1 \leq x \leq 1$, with g bounded on $[-1, 1]$ and $\int_{-1}^{1} g(x)\, dx = 0$. Show that
(a) The locally most powerful signed-rank test of $\theta = 0$ against $\theta > 0$ is based on the sum of signed constants c_j with $c_j = E[g(U_j) - g(-U_j)]$ where U_j has the beta distribution with parameters j and $n - j + 1$.
(b) If $g(x) = \sum_k a_k(n + 1) \cdots (n + k)x|x|^{k-1}$ where k runs from 1 to a finite upper limit, then g satisfies the conditions given above and leads to $c_j = \sum_k a_k j(j + 1) \cdots (j + k - 1)$.
(c) Any set of constants c_j arises in this way from some family f_θ of the type specified and hence is locally most powerful against some alternative of this type.

106. Consider a crossover design in which m units receive the active treatment before the control and n units receive the control before the treatment. If the $m + n$ response differences (later − earlier) are taken as the basic data, they may be analyzed as in a two-sample problem.
(a) What conditions here imply the usual "nonparametric" null hypotheses for the two-sample problem? Are they "nonparametric?"
(b) Under what conditions would an estimate of the population difference in the two-sample problem be an estimate of the treatment effect?
(c) Show that, under suitable conditions, a test of the null hypothesis of no treatment effect can be made conditionally on the number of units with a response change, using a 2×2 table with one row for increases and one for decreases.

*107. Do Problem 109 of Chap. 3 with two independent samples in place of one sample and with the shift assumption in place of the symmetry assumption.

CHAPTER 6

Two-Sample Inferences Based on the Method of Randomization

1 Introduction

In Chap. 4, the principle, method, and rationale of randomization tests applicable to the one-sample (or paired-sample) case and the null hypothesis of symmetry about a specified point were illustrated and some properties of these tests were discussed. In this chapter we employ the same ideas in the case of two independent sets of observations and the null hypothesis of identical populations.

As in Chap. 5, we may have either two random samples independently drawn from two populations or $N = m + n$ units, m of which are chosen at random for some treatment. Under the null hypothesis that the two populations are identical, or that the treatment has no effect, given the N observations, all $\binom{N}{m}$ distinguishable separations of them into two groups, m labeled X and n labeled Y, are equally likely. This fact was used to generate the null distributions of all the test statistics considered in the last chapter. The distribution of a two-sample statistic generated by taking all possible separations of the observations into two groups, X and Y, as equally likely is called its *randomization distribution*. Hence this distribution applies under the null hypotheses mentioned above. A *two-sample randomization test* then is one whose level or P-value is determined by this randomization distribution.

The two-sample tests discussed in Chap. 5 are called rank tests, since the test statistic employed in each case is a function of the ranks of the X's and Y's in the combined group. However, they are also members of the class of two-sample randomization tests and hence might be called *rank-randomization tests*. On the other hand, randomization tests may use a test statistic which is not a function of the ranks alone; rather, the statistic may be any function of the actual values of the observations. To emphasize this, these randomization tests might be called *observation-randomization tests*. In general, the randomization distribution of an observation-randomization test statistic must be generated anew for each different set of observations. Since the randomization distribution applies under the null hypothesis conditionally on any given set of N observed values, these randomization tests are conditional tests, conditional on the N values observed.

In this chapter we will first discuss the two-sample observation-randomization test based on the difference between the two sample means (or any equivalent criterion), and the corresponding confidence procedure. Then we will introduce the general class of two-sample randomization tests and study most powerful randomization tests.

2 Randomization Procedures Based on the Difference Between Sample Means and Equivalent Criteria

2.1 Tests

Given two sets of mutually independent observations, $X_1, \ldots, X_m$ and $Y_1, \ldots, Y_n$, suppose we wish to test the null hypothesis that their distributions are identical. Rejection of this hypothesis implies a difference, but not any particular type of difference. If the shift model of Chap. 5 is assumed, or if we are particularly interested in detecting a difference in location, this situation can be considered a location problem. Then we have the null hypothesis $H_0 : \mu = 0$, where μ is the shift parameter, and a corresponding one-sided or two-sided alternative. The test we will study here is particularly appropriate under the shift model.

A natural choice for a two-sample observation-randomization test statistic is the difference between the two sample means $\overline{Y} - \overline{X}$. This is the statistic underlying the procedures used for normal populations in parametric statistics, and we would expect it to be particularly sensitive to differences in location if the population shape is actually nearly normal. A randomization test could be carried out by computing $\overline{Y} - \overline{X}$ for each of the $\binom{N}{m}$ separations of the actual observations into m X's and n Y's, which are equally likely under the null hypothesis. We can then see how far out the particular observed $\overline{Y} - \overline{X}$ is the tail of the randomization distribution thus generated, and carry out a test as usual.

Such calculations are lengthy. They can be simplified, unfortunately only a little, by using some statistic more convenient than $\overline{Y} - \overline{X}$ but equivalent for the purpose. Some specific statistics which are equivalent since they are monotonic functions of $\overline{Y} - \overline{X}$ given the combined set of observations

(Problem 1) are the ordinary two-sample t statistic (less convenient!), the sum $\sum_1^m X_i$ of the X's, and $S^* = \sum_{r_i > m} X_i - \sum_{r'_j \leq m} Y_j$ where r_i is the rank of X_i and r'_j is the rank of Y_j in the combined ordered sample.

The calculations are most easily performed using S^* when $\bar{Y} - \bar{X}$ is "large," that is in the upper tail, and using a corresponding statistic defined in Problem 1 when $\bar{Y} - \bar{X}$ is in the lower tail. If the same value of S^* occurs more than once, each occurrence must be counted separately. (Theoretically this has probability zero if continuous distributions are assumed, but in practice it may occur.) One will ordinarily try to avoid enumerating all $\binom{N}{m}$ possible separations. In order to find a P-value, only those $\binom{N}{m}P$ separations which lead to values of S^* equal to or more extreme (in the appropriate direction) than that observed must be enumerated. If a nonrandomized test is desired at level α, a decision to reject can be reached by enumerating these same $\binom{N}{m}P$ separations, with $P < \alpha$, and a decision to "accept" requires identifying any $\binom{N}{m}\alpha$ cases as extreme as that observed. For rejection, or a P-value, every point in the relevant tail must be included; since it is difficult to select the cases in the correct order, considerable care is required.

The procedure for generating the entire randomization distribution is illustrated in Table 2.1. For $m = 3$, $n = 4$, there are $\binom{7}{3} = 35$ distinguishable separations into X's and Y's. Each separation is listed in the table and the value of S^* calculated for each. (The $\sum X$ column is included only to illustrate the evaluation of a different test statistic and to make the test more intuitive; the $\bar{Y} - \bar{X}$ column is included for use in Sect. 2.4.) The observed value of S^* is 1.5 and only three of the enumerated separations produce an S^* that small. Hence the one-tailed P-value by the randomization test is $3/35 = 0.0857$. Since the randomization distribution is far from symmetric here, different ways of relating the two tails would give appreciably different two-tailed P-values. (See also Sect. 2.4 below, Problem 5, and Sect. 4.5, Chap. 1.)

If the null hypothesis is that the Y population is the same as the X population except for a shift by a specified amount μ_0, then the foregoing randomization test may be applied to $X_1, \ldots, X_m, Y_1 - \mu_0, \ldots, Y_n - \mu_0$. The corresponding confidence procedure will be discussed in Sect. 2.3.

2.2 Weakening the Assumptions

In Sect. 2.1, the validity of the randomization distribution as the null distribution for the randomization test was based on the assumption that the X and Y samples come from identical populations under the null hypothesis. As noted in Sect. 1, the randomization distribution is also valid if N given units are randomly separated into two groups, one of which is treated, and the null hypothesis is that the treatment has no effect on any unit.

The probability of rejection is ordinarily affected however, and sometimes increased, if the samples are drawn from two populations which differ in

Table 2.1^a

X sample   −0.2, 0.9, 2.0

Y sample   0.5, 6.5, 11.5, 14.3

$\overline{X} = 0.9$ $\overline{Y} = 8.2$ $m = 3$ $n = 4$ $S^* = 1.5$

| | | Sample Separations | | | | | | | |
−0.2	0.5	0.9	2.0	6.5	11.5	14.3	$\sum X$	S^*	$\overline{Y} - \overline{X}$
X	X	X	Y	Y	Y	Y	1.2	0	8.16
X	X	Y	X	Y	Y	Y	2.3	1.1	7.53
X	Y	X	X	Y	Y	Y	2.7	1.5	7.30
Y	X	X	X	Y	Y	Y	3.4	2.2	6.89
X	X	Y	Y	X	Y	Y	6.8	5.6	4.91
X	Y	X	Y	X	Y	Y	7.2	6.0	4.68
Y	X	X	Y	X	Y	Y	7.9	6.7	4.27
X	Y	Y	X	X	Y	Y	8.3	7.1	4.03
Y	X	Y	X	X	Y	Y	9.0	7.8	3.62
Y	Y	X	X	X	Y	Y	9.4	8.2	3.39
X	X	Y	Y	Y	X	Y	11.8	10.6	1.99
X	Y	X	Y	Y	X	Y	12.2	11.0	1.76
Y	X	X	Y	Y	X	Y	12.9	11.7	1.35
X	Y	Y	X	Y	X	Y	13.3	12.1	1.12
Y	X	Y	X	Y	X	Y	14.0	12.8	0.71
Y	Y	X	X	Y	X	Y	14.4	13.2	0.48
X	X	Y	Y	Y	Y	X	14.6	13.4	0.36
X	Y	X	Y	Y	Y	X	15.0	13.8	0.12
Y	X	X	Y	Y	Y	X	15.7	14.5	−0.28
X	Y	Y	X	Y	Y	X	16.1	14.9	−0.52
Y	X	Y	X	Y	Y	X	16.8	15.6	−0.92
Y	Y	X	X	Y	Y	X	17.2	16.0	−1.16
X	Y	Y	Y	X	X	Y	17.8	16.6	−1.51
Y	X	Y	Y	X	X	Y	18.5	17.3	−1.92
Y	Y	X	Y	X	X	Y	18.9	17.7	−2.15
Y	Y	Y	X	X	X	Y	20.0	18.8	−2.79
X	Y	Y	Y	X	Y	X	20.6	19.4	−3.14
Y	X	Y	Y	X	Y	X	21.3	20.3	−3.55
Y	Y	X	Y	X	Y	X	21.7	20.5	−3.78
Y	Y	Y	X	X	Y	X	22.8	21.6	−4.42
X	Y	Y	Y	Y	X	X	25.6	24.4	−6.06
Y	X	Y	Y	Y	X	X	26.3	25.1	−6.47
Y	Y	X	Y	Y	X	X	26.7	25.5	−6.70
Y	Y	Y	X	Y	X	X	27.8	26.6	−7.34
Y	Y	Y	Y	X	X	X	32.3	31.1	−10.22

^{a} These data are percent change in retail sales of Alabama drug stores from April 1971 to April 1972. The X values are for three counties selected at random from those Alabama counties classified as SMSA's (Standard Metropolitan Statistical Areas), and the Y values are for randomly selected other counties. Small samples were selected so that generation of the entire randomization distribution could be illustrated. A null hypothesis of practical importance here is that the average percent change in retail sales during this period for metropolitan areas in Alabama is not smaller than the corresponding change for less urban areas.

any way, and in particular if the population variances differ. Nevertheless, it is true that when the X's and Y's are all mutually independent with possibly different distributions, the one-tailed randomization test that rejects when $\bar{Y} - \bar{X}$ is too large retains its level for the null hypothesis that every X_i is "stochastically larger" than every Y_j. The same statements hold with large and larger replaced by small and smaller.

Related remarks apply, of course, to the related confidence procedures. For further discussion and detail, see Sects. 3.8 and 4.6 of Chap. 5 and Problem 17.

2.3 Related Confidence Procedures

Under the shift assumption, the randomization test procedure can also be used to construct a confidence region for the amount of the shift μ. Unfortunately, the randomization distribution is different when different values of μ are subtracted from the Y's. The confidence region is nevertheless an interval, and its endpoints could be obtained by trial and error by subtracting successive values of μ, larger or smaller than previous values as appropriate, until the value of the test statistic equals the appropriate upper or lower critical value of a randomization test at level α for shift 0. The endpoints of the normal theory confidence interval for the difference of means at the same level could be used as initial trial values of μ.

However, as in the corresponding one-sample problem, there is a more systematic and convenient method which can be used to find the confidence limits for μ exactly. Consider all pairs of equal-sized subsamples of X's and Y's. Specifically, for the sample of m X's, consider all possible subsamples of size r, and for the sample of n Y's, consider all possible subsamples of the same size r. Take all possible pairs of these equal-sized subsamples of X's and Y's for all r, $1 \leq r \leq \min(m, n)$. The total number of different pairs is

$$\sum \binom{m}{r}\binom{n}{r} = \sum \binom{m}{m-r}\binom{n}{r} = \binom{m+n}{m} - 1 = \binom{N}{m} - 1.$$

For each pair, take the difference of the subsample means, say the Y subsample mean minus the X subsample mean. Consider the $\binom{N}{m} - 1$ differences in order of algebraic (not absolute) value. It can be shown that the kth smallest and kth largest of these differences of equal-sized subsample means are the lower and upper confidence bounds respectively, each at level $1 - \alpha$, that correspond to the two-sample randomization test based on $\bar{Y} - \bar{X}$ (or any other equivalent test criterion), where $\alpha = k/\binom{N}{m}$ and hence $k = \binom{N}{m}\alpha$ (Problem 2).

To save labor, instead of using all $\binom{N}{m} - 1$ differences one could use a smaller number, say M, if they are selected either at random without replacement (Problem 3) or in a "balanced" manner determined by group theory; then $\alpha = k/(M + 1)$. Both methods are discussed briefly in Sect. 2.5;

more detail for a single sample from a symmetric population was given in Sect. 2.5 of Chap. 4.

2.4 Properties of the Exact Randomization Distribution

Unfortunately, the randomization distribution of $\bar{Y} - \bar{X}$ (or any equivalent randomization test statistic) is symmetric in general only for $m = n$; for $m \neq n$, the two tails are not mirror images except by a remarkable coincidence (Problem 5). This lack of symmetry makes generation of the null distribution more tedious and approximation based on moments more difficult. Furthermore, it has the methodological consequence that there is more than one possible definition of the two-tailed procedure. For example, the randomization test based on the absolute value of the difference of means $|\bar{Y} - \bar{X}|$ is not equal-tailed; a test which rejects if $|\bar{Y} - \bar{X}|$ is in the upper α tail of its randomization distribution (using the absolute values of the differences) is not the same as a test which rejects if $\bar{Y} - \bar{X}$ is in the upper or lower $\alpha/2$ tail of its randomization distribution (Problem 6). Presumably, however, the discrepancy is ordinarily small. In the example of Table 2.1, the upper tailed P-value for $|\bar{Y} - \bar{X}| = 7.30$ is 5/35, as compared to 3/35 for $\bar{Y} - \bar{X} = 7.30$.

We will now derive the exact mean and variance of the randomization distribution of $\bar{X}$ and of $\bar{Y} - \bar{X}$. For this purpose it is convenient to write the latter statistic as a function of only one of the sample means, say $\bar{X}$. If we denote the $m + n = N$ observations in the combined sample by $a_1, a_2, \ldots, a_N$, we can write

$$\bar{Y} - \bar{X} = \frac{\sum_1^N a_j - \sum_1^m X_j}{n} - \bar{X}$$

$$= \left(\sum_1^N a_j - N\bar{X} \right) \Big/ n. \qquad (2.1)$$

For the randomization distribution we can interpet $\bar{X}$ as the mean of a sample of m drawn without replacement from the population consisting of $a_1, \ldots, a_N$. The mean and variance of this population are

$$\bar{a} = \sum_1^N a_j / N = \left(\sum_1^m X_j + \sum_1^n Y_j \right) \Big/ N$$

and

$$\sum_1^N (a_j - \bar{a})^2 / N = S_2 / N$$

where

$$S_r = \sum_1^N (a_j - \bar{a})^r = \sum_1^m (X_j - \bar{a})^r + \sum_1^n (Y_j - \bar{a})^r. \qquad (2.2)$$

Therefore the randomization distribution of $\bar{X}$ has mean

$$E(\bar{X}) = \bar{a} \tag{2.3}$$

and variance (including the correction factor for finite populations)

$$\text{var}(\bar{X}) = \frac{N - m}{N - 1} (S_2/N)/m = nS_2/mN(N - 1). \tag{2.4}$$

Furthermore, by (2.1), (2.3) and (2.4) the mean and variance σ^2 of the randomization distribution of $\bar{Y} - \bar{X}$ are given by

$$E(\bar{Y} - \bar{X}) = 0 \tag{2.5}$$

$$\text{var}(\bar{Y} - \bar{X}) = \sigma^2 = NS_2/mn(N - 1). \tag{2.6}$$

The corresponding moments of several other equivalent test criteria are easily derived (Problem 4).

2.5 Approximations to the Exact Randomization Distribution

If a randomization test is desired but cannot be carried out exactly, whether by direct enumeration of the randomization distribution or otherwise, one can use (a) simulation, (b) a restricted randomization set, or (c) the normal or some other approximation. These procedures are described briefly here; see Sect. 2.5 of Chap. 4 for more details.

(a) *Simulation.* Each sample observation is associated with a number from 1 to N and some method is used to draw m different numbers between 1 and N at random. The observations associated with the m numbers generated are then labeled X, the rest are labeled Y, and the test criterion is calculated. The whole process is repeated many times and the relative frequency distribution obtained for the test criterion is the simulated randomization distribution. This provides an estimate of the P-value of the randomization test, or the exact P-value of a redefined test, as explained in Sect. 2.5 of Chap. 4.

(b) *Restricted Randomization Set.* Instead of generating separations into X's and Y's randomly, one can generate them systematically by means of a group of transformations. Conceptually, the simplest method would be to consider a subgroup G of the group of all permutations of the integers $1, \ldots, N$, apply each permutation in G to the given observations arranged in the order $X_1, \ldots, X_m, Y_1, \ldots, Y_n$, take the first m of the permuted values as X's and the remaining n as Y's, and calculate the test statistic. The proportion of the calculated values of the test statistic which are at least as extreme as the observed value is the P-value of the test based on this restricted randomization set. The corresponding confidence bound for a shift parameter μ can be obtained as follows. Assume that the identity is the only permutation in G that does not change the separation. (Nothing is gained by considering other permutations that do not change the separation.) For each permutation in G except the identity, take the mean of all Y's that are permuted to become X's and subtract the mean of all X's that are permuted to become Y's. The kth smallest (or kth largest) of these differences of subsample means is the lower (or upper) confidence bound corresponding to the test having k points in the critical region, and the one-tailed α is k divided by the size of the subgroup.

(c) *Approximations.* As in the one-sample case in Chap 4, several approximations that are based on tabled probability distributions are possible. Four will be given here, but none of them reflects the asymmetry of the randomization distribution of $\bar{Y} - \bar{X}$.

A natural approximation is the standard normal distribution for the standardized randomization test statistic $(\bar{Y} - \bar{X})/\sigma$ where σ is given by (2.6). The following reasoning suggests, however, that a better approximation may be obtained by calculating the ordinary two-sample t statistic and treating it as Student's t distributed with $N - 2$ degrees of freedom.

The ordinary two-sample t statistic for equality of means assuming equal variances can be written here as (Problem 10)

$$t = \left(\frac{\bar{Y} - \bar{X}}{\sigma}\right)\left[\frac{N - 2}{N - 1 - (\bar{Y} - \bar{X})^2/\sigma^2}\right]^{1/2}. \tag{2.7}$$

Comparing $(\bar{Y} - \bar{X})/\sigma$ with a percent point z_α of the standard normal distribution is then equivalent to comparing t with the constant

$$z_\alpha\left(\frac{N - 2}{N - 1 - z_\alpha^2}\right)^{1/2}, \tag{2.8}$$

rather than with the corresponding percent point t_α of Student's t distribution with $N - 2$ degrees of freedom. (The quantity in (2.8) and t_α differ appreciably in small samples except for α near 0.05, although both approach z_α as $N \to \infty$. See Problem 15 of Chap. 4.) If the randomization test is to be performed by comparing the t statistic to a constant value, then t_α will be a better constant on the average than (2.8) under normality. This suggests that it will be better in general, at least for combined samples which are not highly non-normal. Moses [1952] suggests the rule of thumb that this approximation can be used when $\frac{1}{4} \leq m/n \leq 4$ and the kurtosis for the pooled sample is close to 3.

  *By matching the moments of $t^2/(t^2 + N - 2)$ under the randomization distribution with the corresponding moments of the beta distribution, we obtain another approximation. As derived below, it is equivalent to treating t^2 as F distributed with fractional degrees of freedom d and $(N - 2)d$, where

$$d = \frac{2(1 - D)}{(N - 1)D - 1} \tag{2.9}$$

$$D = \frac{N}{mn}\left[\frac{3(m - 1)(n - 1)}{(N - 2)(N - 3)}\left(1 - 2\frac{S_4}{S_2^2}\right) + \frac{S_4}{S_2^2}\right], \tag{2.10}$$

and S_r is given in (2.2). Recall, however, that an upper-tailed F probability corresponds to a two-tailed t probability.

An idea of the size of the correction factor d is obtained by rewriting (2.9) in the form (Problem 11)

$$d = 1 + \left(\frac{N+1}{N-1}\right)c_2 \left[\frac{2mn(N-2)}{6mn - N^2 - N} - c_2\right]^{-1} \tag{2.11}$$

where

$$c_2 = \frac{N(N+1)S_4/S_2^2 - 3(N-1)}{(N-2)(N-3)/(N-1)}. \tag{2.12}$$

Under the null hypothesis, c_2 is a consistent estimator (Problem 12) of the population kurtosis minus 3, which is 0 for normal distributions.

A further approximation, which avoids the small fractional number of degrees of freedom d in the numerator of F, is to treat c times the t statistic as Student's t distributed with k degrees of freedom where k and c are given in (2.11) and (2.12) of Chap. 4 respectively with n replaced by $N - 1$. This step is exactly the same as in Sect. 2.4 of Chap. 4 (Problem 13).*

*PROOF. Under normal theory, t^2 has an F distribution with 1 and $N - 2$ degrees of freedom, or equivalently $t^2/(t^2 + N - 2)$ has a beta distribution with parameters $\frac{1}{2}$ and $(N - 2)/2$. We also have, by (2.7),

$$\frac{t^2}{t^2 + N - 2} = \frac{(\bar{Y} - \bar{X})^2}{(N-1)\sigma^2}. \tag{2.13}$$

This suggests approximating the randomization distribution of $(\bar{Y} - \bar{X})^2/(N-1)\sigma^2$ by a beta distribution with the same first two moments. This randomization distribution has first two moments (Problem 14a)

$$E[(\bar{Y} - \bar{X})^2/(N-1)\sigma^2] = 1/(N-1) \tag{2.14}$$

$$E[(\bar{Y} - \bar{X})^4/(N-1)^2\sigma^4 = \frac{N[3(mn - N + 1) + (N^2 + N - 6mn)S_4/S_2^2]}{mn(N-1)(N-2)(N-3)}$$

$$= D/(N-1) \tag{2.15}$$

where D is given in (2.10). Equating the moments in (2.14) and (2.15) to the corresponding moments of the beta distribution with parameters a and b as given in (2.16) of Chap. 4 gives the relations (Problem 14b)

$$a = \frac{1-D}{(N-1)D-1}, \qquad b = (N-2)a. \tag{2.16}$$

Thus we are led to approximate the randomization distribution of $(\bar{Y} - \bar{X})^2/$ $(N - 1)\sigma^2$ by a beta distribution with the parameters in (2.16). This is equivalent (Problem 14c) to approximating the randomization distribution of t^2 by an F distribution with degrees of freedom $2a = d$ and $2b = (N - 2)d$, as at (2.9).*   □

*Matching the first two moments of the randomization distribution of $(\bar{Y} - \bar{X})^2/(N - 1)\sigma^2$ with those of the beta distribution is like (but not equivalent to) matching the first two moments of the randomization distribution of t^2. The latter is like matching the second and fourth moments of the randomization distribution of t. Treating the randomization distribution of t as symmetric matches also its first moment, but not ordinarily its third (Problem 15). Hence the approximation at (2.9) and (2.10) has done something like matching the first, second and fourth moments of t under the randomization distribution.

We have not discussed the asymptotic normality of the randomization distribution of $(\bar{Y} - \bar{X})/\sigma$. It is complicated because we really want to know whether the conditional distribution of $(\bar{Y} - \bar{X})/\sigma$, given the N observations actually obtained (but not which are X's and which are Y's), "approaches" the standard normal distribution in some sense, not merely whether the marginal distribution of $(\bar{Y} - \bar{X})/\sigma$ is asymptotically standard normal. A similar comment applies to observation-randomization tests generally, but not to rank-randomization tests. See also the last paragraph of Sect. 2.5 of Chap. 4.*

3 The Class of Two-Sample Randomization Tests

3.1 Definition

The level of the randomization test based on the difference between the means of two independent samples, or on any other equivalent test criterion, relies only on the assumption that, given the observations, say $a_1, \ldots, a_N$, all possible separations into $X_1, \ldots, X_m, Y_1, \ldots, Y_n$ are equally likely, as they are under the null hypothesis of identical populations, or the null hypothesis that the treatment has no effect when the treatment group is selected randomly from the whole set. Thus it is a conditional test, conditional on $a_1, \ldots, a_N$. In particular, it has level α if, under the null hypothesis, the conditional probability of rejection given $a_1, \ldots, a_N$ is at most α, and its P-value is the corresponding conditional probability of a value of the test statistic equal to or more extreme than that observed.

More generally, as stated in Sect. 1, a *two-sample randomization test* is a test which is conditional on the observations $a_1, \ldots, a_N$, its null distribution being the randomization distribution generated by randomly separating the combined observations into two groups of sizes m and n. Such tests may depend on the actual values of $a_1, \ldots, a_N$, and may be called *observation-randomization* tests to emphasize this possibility. Rank-randomization tests are those which use only functions of the ranks of the two samples within the combined set of N. All of the tests presented in Chap. 5 were rank-randomization tests. The class of all randomization tests is broader than the examples given so far might suggest; it includes analogues of the one-sample examples given in Sect. 3.1 of Chap. 4 (Problem 19).

The procedures we have considered have all been permutation-invariant, that is, not affected by rearrangements of the observations within the X sample or within the Y sample (Sect. 7 of Chap. 5). For such procedures, two arrangements of $a_1, \ldots, a_N$ into $X_1, \ldots, X_m, Y_1, \ldots, Y_n$ may as well be regarded as identical if they are the same except for the order of the X's or the order of the Y's or both, and we need distinguish only $\binom{N}{m}$ distinct possibilities, which we have previously called "separations." Furthermore, in some situations the indices (labels) within samples lack reality or meaning and it is then impossible or meaningless to distinguish more possibilities, which would require distinguishing order within samples. (Sometimes, for instance, only the order statistics of each sample are available.) There are situations, however, in which it is possible and even sensible to take into account the order within one or both samples, and there are procedures that do so (Problem 20). It then becomes necessary to distinguish two types of randomization distribution, according as permutations within samples are or are not included in the randomization set.

An $N!$-type randomization distribution is one which includes permutations—the equally likely possibilities are all $N!$ arrangements of the N given values $a_1, \ldots, a_N$ into $X_1 \ldots, X_m, Y_1, \ldots, Y_n$; reorderings within samples are allowed and counted separately. An $\binom{N}{m}$-type randomization distribution excludes permutations—the equally likely possibilities are only the $\binom{N}{m}$ separations of the N given values $a_1, \ldots, a_N$ into an X sample and a Y sample without reordering; that is, a set of m a's is drawn without replacement and, with the a's in their original order, X_i is the value of the ith a value in the set drawn and Y_j is the value of the jth a value not in this set. An $\binom{N}{m}$-type randomization test is necessarily an $N!$-type randomization test (since it is more conditional), but the converse is not true (see below). To conclude from some given conditions that a test must be an $\binom{N}{m}$-type randomization test is stronger than to conclude that it must be an $N!$ type. Whether all randomization tests of the $N!$ type are valid, or only those of the $\binom{N}{m}$ type (or not even all of those), depends on the null hypothesis (see Problem 21). For a permutation-invariant procedure, the two types of randomization distribution are equivalent but the $\binom{N}{m}$ type is simpler; the $N!$ type merely counts repeatedly, $m!n!$ times altogether, each possibility counted by the $\binom{N}{m}$ type. (Related points arose in Sects. 7 and 8.1 of Chap 5, and the corresponding aspects of the one-sample problem were discussed in Sect. 3.2 of Chap. 4.)

3.2 Properties

We now turn to some of the theoretical properties of members of the class of two-sample randomization tests.

Critical Function and Level

Denote the critical function, that is, the probability of rejection, of an arbitrary test by $\phi(X_1, \ldots, X_m; Y_1, \ldots, Y_n)$. Consider first a null hypothesis H_0 under which, given the observations $a_1, \ldots, a_N$, all $N!$ arrangements of them into two samples $X_1, \ldots, X_m; Y_1, \ldots, Y_n$ are equally likely. Then under H_0, the conditional expected value given the a_j of $\phi(X_1, \ldots, X_m; Y_1, \ldots, Y_n)$ is simply the mean of its $N!$-type randomization distribution, or

$$E_0[\phi(X_1, \ldots, X_m; Y_1, \ldots, Y_n)|a_1, \ldots, a_N]$$
$$= \sum \phi(a_{\pi_1}, \ldots, a_{\pi_m}; a_{\pi_{m+1}}, \ldots, a_{\pi_N})/N! \tag{3.1}$$

where the sum is over the $N!$ permutations $\pi_1, \ldots, \pi_N$ of the integers $1, \ldots, N$. Alternatively, consider a null hypothesis H_0 under which, given the observations $a_1, \ldots, a_N$, all $\binom{N}{m}$ separations into an X sample of size m and a Y sample of size n are equally likely. Then under H_0, the conditional expected value given the a_j of $\phi(X_1, \ldots, X_m; Y_1, \ldots, Y_n)$ is simply the mean of the $\binom{N}{m}$-type randomization distribution, or

$$E_0[\phi(X_1, \ldots, X_m; Y_1, \ldots, Y_n)|a_1, \ldots, a_N]$$
$$= \sum \phi(a_{\pi_1}, \ldots, a_{\pi_m}; a_{\pi_{m+1}}, \ldots, a_{\pi_N})\bigg/\binom{N}{m} \tag{3.2}$$

where the sum is over the $\binom{N}{m}$ separations of the integers into two sets $\{\pi_1, \ldots, \pi_m\}$ and $\{\pi_{m+1}, \ldots, \pi_N\}$ with $\pi_1 < \cdots < \pi_m$ and $\pi_{m+1} < \cdots < \pi_N$.

The expected value (3.1) or (3.2), whichever applies, is the conditional probability of a Type I error for the test. Accordingly, a test ϕ has conditional level α, given the a_j, if the quantity (3.1) or (3.2) is less than or equal to α. If this holds for all $a_1, \ldots, a_N$, then the test is a randomization test, of the $N!$ type or the $\binom{N}{m}$ type respectively. Any such test also has level α unconditionally by the usual argument. We shall see that, conversely, a test having unconditional level α must, under certain circumstances, have conditional level α given the observations $a_1, \ldots, a_N$, that is, must be an $N!$-type or $\binom{N}{m}$-type randomization test at level α.

The statements in Sect. 2.2 about weakening the assumptions apply here also as long as, for those statements referring to stochastic dominance, the critical function is suitably monotonic in the X_i and Y_j.

Justifications of Randomization Tests

Section 7 of Chap. 5 presented an argument based on concepts of invariance to justify restricting consideration to rank tests, which are a particular case of randomization tests. This argument required, however, the very strong assumption of invariance under all strictly increasing transformations of the observations. We shall now discuss quite different assumptions that lead to randomization tests (not necessarily rank tests).

Observations Not Identically Distributed Under the Null Hypothesis. We have already noted that, under suitable hypotheses, a randomization test has conditional level α given the observations and hence unconditional level α. It can also be shown that, if the null hypothesis is sufficiently broad, then every test at level α is a randomization test. Such a strong result is not available for independent, identically distributed observations, as will be discussed shortly. If identical distributions are not required, however, and the null hypothesis is sufficiently broad, all level α tests are $\binom{N}{m}$-type randomization tests. This holds under either of the following null hypotheses:

H_0: The observations $Z_1, \ldots, Z_N$ are independent with arbitrary distributions, and $X_1, \ldots, X_m$; $Y_1, \ldots, Y_n$ are a random separation of $Z_1, \ldots, Z_N$ into an X sample and a Y sample.

H_0': H_0 holds and the $X_1, \ldots, Y_n$ have densities.

The same conclusion also holds under less broad hypotheses, including

H_0'': H_0 holds and the Z_j are normally distributed with arbitrary means μ_j and common variance σ^2.

Note that H_0, for instance, does *not* say that $X_1 \ldots, X_m$; $Y_1, \ldots, Y_n$ are independently distributed with arbitrary distributions. This would place no restriction whatever on the relation between the X's and the Y's and hence could not serve usefully as a null hypothesis. By a random separation we mean that, given $Z_1, \ldots, Z_N$, the X's are a random sample of the Z's without replacement but in their original order, while the Y's are the remaining Z's in their original order.

We have in mind the kind of situation in which, for example, there are N available experimental units, of which m are to be chosen at random to receive a treatment and the rest to serve as controls. If the null hypothesis is that the treatment has no effect whatever on any unit, then the random selection of the units to be treated guarantees the validity of the level of any randomization test. Are there any other tests at a specified level α? Suppose that, if all experimental units were untreated (or if the treatment had no effect whatever), one would be willing to assume no more than H_0, that is, that the N observations on the N experimental units are independently distributed with arbitrary distributions, possibly all different. The fact stated above is that the only tests having level α under such a weak assumption are the randomization tests at level α. Furthermore, the null hypothesis can be made considerably more restrictive without upsetting this conclusion. For example, the conclusion holds for a normal model with common variance but arbitrary unit effects, as in H_0''. It therefore holds for any null hypothesis which permits this normal model (Problem 24b). It also holds if the unit effects are arbitrary constants (Problem 24a).

These properties are summarized in Theorem 3.1.

Theorem 3.1. *If a test has level α for H_0, H_0', or H_0'', then it is an $\binom{N}{m}$-type randomization test. Conversely, a randomization test has level α for H_0 and hence for H_0' and H_0''.*

The proof of Theorem 3.1 for H_0 is requested in Problem 24c. The proofs for H_0' and H_0'' involve measure-theoretic considerations and will not be considered here. See Lehmann [1959, Sect. 5.10] or Lehmann and Stein [1949].

Observations Identically Distributed Under the Null Hypothesis. In some cases it may be reasonable to assume that all the X's and Y's are independent and identically distributed under the null hypothesis. Consider, for instance, a situation in which a random sample of m units $X_1, \ldots, X_m$ is drawn from one population and a random sample of n units $Y_1, \ldots, Y_n$ is drawn from another. If in fact the populations are identical, then all N X's and Y's are simply a random sample from a single population; that is, they are independent and identically distributed. An example of this occurs if m units are drawn at random from a large population and treated in some way and n units are independently drawn at random from the same population to serve as controls. If the treatment has no effect on the population distribution of the characteristic measured, even though it may affect the individual units, then $X_1, \ldots, Y_n$ are again independent and identically distributed.

This imposes conditions on the null hypothesis that were not imposed earlier. Nevertheless, we can prove that a test having level α for a sufficiently broad null hypothesis of this kind must be an $N!$-type randomization test if we also impose the additional condition on the test that it be unbiased against a sufficiently broad class of alternatives. Specifically, suppose we wish to test one of the following null hypotheses:

H_0: The variables $X_1, \ldots, Y_n$ are independently, identically distributed.

H_0': The variables $X_1, \ldots, Y_n$ are independently, identically distributed and their common distribution has a density.

Suppose also that the test ϕ, at level α, is unbiased against an alternative hypothesis which includes distributions arbitrarily close to each distribution of the null hypothesis. Then, by the usual argument (Sect. 6.3, Chap 2), ϕ

must have level exactly α under every null distribution included in the null hypothesis. This in turn implies that ϕ has conditional level α given the combined sample observations but not their assignment to $X_1, \ldots, X_m$; $Y_1, \ldots, Y_n$. (The proof, Problem 25b, is like that in Sect. 3.2 of Chap. 4.) Since the conditional null distribution is the N!-type randomization distribution, it follows that if ϕ is an unbiased test of H_0 or H_0' against a sufficiently broad alternative, then it is an N!-type randomization test. Alternatives which are sufficiently broad are (Problem 25a)

H_1: $X_1, \ldots, X_m$; $Y_1, \ldots, Y_n$ are drawn independently from two populations which are the same except for a shift.

H_1': H_1 holds and the populations have densities.

These properties are summarized in Theorem 3.2.

Theorem 3.2. *If a test has level α for H_0 or H_0' and is unbiased against one-sided or two-sided shifts, then it is an N!-type randomization test.*

The result in Theorem 3.2 "justifies" restricting consideration to N!-type randomization tests when the X's and Y's are independent and identically distributed within samples. Under these circumstances, the reasons given in Sect. 7 of Chap. 5 for using a permutation-invariant procedure also apply. As mentioned earlier, for permutation-invariant tests, the distinction between N!-type and $\binom{N}{m}$-type randomization is of no consequence. Thus we have "justified" restriction to permutation-invariant randomization tests. Of course, the weaknesses in the justification are the requirements for level strictly α or less under a very broad null hypothesis and power strictly α or more for a very wide class of alternative distributions.

4 Most Powerful Randomization Tests

4.1 General Case

Reasons for using randomization tests were given in Sect. 3.2. In this subsection we will see how to find that randomization test which is most powerful against any specific alternative distribution. The particular case of normal shift alternatives will be illustrated in the two subsections following.

Suppose, for definiteness, that we are considering all of the N!-type randomization tests. Then under the randomization distribution, given the observations $a_1, \ldots, a_N$, all N! possible arrangements into $X_1, \ldots, X_m$, $Y_1, \ldots, Y_n$ are equally likely. A randomization test is valid as long as this condition is satisfied under the null hypothesis. Consider now an alternative with joint density or discrete frequency function $f(x_1, \ldots, x_m, y_1, \ldots, y_n)$. Under this alternative, given $a_1, \ldots, a_N$, the conditional probabilities of each of the $N!$ possible arrangements $X_1, \ldots, X_m, Y_1 \ldots Y_n$ are proportional to $f(X_1, \ldots, X_m, Y_1, \ldots, Y_n)$. By the Neyman–Pearson Lemma (Theorem 7.1 of Chap. 1), it follows (Problem 26) that among randomization tests, the conditional power against f is maximized by a test of the form

$$\text{reject if } f(X_1, \ldots, X_m, Y_1, \ldots, Y_n) > k$$
$$\text{``accept'' if } f(X_1, \ldots, X_m, Y_1, \ldots, Y_n) < k. \qquad (4.1)$$

Randomization may be necessary at k. The choice of k and the randomization at k must be determined so that the test has conditional level exactly α. That is, we consider the value of f at each arrangement $x_1, \ldots, x_m, y_1, \ldots, y_n$ which arises by rearrangement of the observed $X_1, \ldots, X_m, Y_1, \ldots, Y_n$. The arrangements $x_1, \ldots, x_m, y_1, \ldots, y_n$ are placed in the rejection region in decreasing order of their corresponding values of f, starting with the largest value of f, until their null probabilities total α. The region will consist of the $\alpha N!$ possible arrangements which produce the largest values of f if $\alpha N!$ is an integer and if the $\alpha N!$th and $(\alpha N! + 1)$th values of f are not tied. Ties may be broken arbitrarily. If $\alpha N!$ is not an integer, a randomized test will be necessary.

Since this test is the randomization test which maximizes the conditional power against f, it also maximizes the ordinary (unconditional) power against f. (Problem 28 of Chap. 4 covers this for the one-sample case.) Thus we have shown how to find the most powerful $N!$-type randomization test against any specified alternative f.

The conditions given are necessary and sufficient. The method for $\binom{N}{m}$-type randomization tests is similar. Of course, if a most powerful $N!$-type randomization test is an $\binom{N}{m}$-type randomization test, then it is also most powerful in the smaller class of $\binom{N}{m}$-type randomization tests.

4.2 One-Sided Normal Alternatives

Consider now the alternative that the X's are normal with mean μ_1 and variance σ^2, the Y's are normal with mean μ_2 and the same variance σ^2, and all are independent. It follows from (4.1) (Problem 27) that the upper-tailed randomization test based on $\bar{Y} - \bar{X}$ (or an equivalent statistic) is the most powerful randomization test (of either the $N!$-type or the $\binom{N}{m}$-type) against any such alternative with $\mu_2 > \mu_1$; that is, it is the uniformly most powerful randomization test against $\mu_2 > \mu_1$. Similarly, the lower-tailed randomization test based on $\bar{Y} - \bar{X}$ is the uniformly most powerful randomization test against $\mu_2 < \mu_1$. Note that the one-tailed tests here do not depend on the values of the parameters, in contrast to the most powerful rank tests of Sect. 8.1 of Chap. 5.

4.3 Two-Sided Normal Alternatives

Suppose that the X's and Y's are normal with common variance, as above, and consider the alternative $\mu_1 \neq \mu_2$. There is no uniformly most powerful randomization test against this alternative, different randomization tests being most powerful against $\mu_1 < \mu_2$ and $\mu_1 > \mu_2$. It is apparently unknown whether there is a uniformly most powerful unbiased randomization test. We shall prove, however, that the randomization test rejecting for large $|\overline{Y} - \overline{X}|$ has two other properties. This test may be thought of as a two-tailed randomization test based on $\overline{Y} - \overline{X}$, but as mentioned earlier, unless $m = n$, it is not ordinarily the equal-tailed randomization test based on $\overline{Y} - \overline{X}$ because the randomization distribution of $\overline{Y} - \overline{X}$ is not ordinarily symmetric unless $m = n$ (Problem 5).

One property of the randomization test rejecting for large $|\overline{Y} - \overline{X}|$ is that it is uniformly most powerful against $\mu_1 \neq \mu_2$ among randomization tests which are invariant under transformations carrying $X_1, \ldots, X_m$, $Y_1, \ldots, Y_n$ into $c - X_1, \ldots, c - X_m, c - Y_1, \ldots, c - Y_n$, where c is an arbitrary constant. Notice that such a transformation carries the alternative given by μ_1, μ_2, σ into that given by $c - \mu_1, c - \mu_2, \sigma$, so the invariance rationale (Sect. 8, Chap. 3) can be applied. In particular, any invariant test has the same power against all alternatives with the same $\mu_1 - \mu_2$ and the same σ (Problem 28). The statement is that no randomization test which is invariant under all such transformations is more powerful against even one alternative with $\mu_1 \neq \mu_2$ than the randomization test rejecting when $|\overline{Y} - \overline{X}|$ is too large.

*This randomization test is also the "most stringent" randomization test in the situation under discussion. This property is defined as follows. Let $\alpha^*(\mu_1, \mu_2, \sigma)$, be the power of the most powerful randomization test against the alternative given by (μ_1, μ_2, σ). Then the power $\alpha(\mu_1, \mu_2, \sigma)$ of any other randomization test ϕ is at most $\alpha^*(\mu_1, \mu_2, \sigma)$, and the difference measures how far short of optimum the test ϕ is. Accordingly, we define

$$\Delta = \max_{\mu_1, \mu_2, \sigma} \left[\alpha^*(\mu_1, \mu_2, \sigma) - \alpha(\mu_1, \mu_2, \sigma) \right]$$

as the maximum amount by which ϕ falls short of optimum. The randomization test which rejects for large $|\overline{Y} - \overline{X}|$ has the property that it minimizes Δ among randomization tests and hence is most stringent. Specifically, the maximum amount by which the power of ϕ is less than that of the best randomization test against each alternative separately is minimized among randomization tests by the randomization test rejecting for large $|\overline{Y} - \overline{X}|$. For no $\varepsilon > 0$ is there a randomization test which comes within $\Delta - \varepsilon$ of the optimum at every alternative, and the randomization test rejecting for large $|\overline{Y} - \overline{X}|$ comes within Δ everywhere. Notice that this property does not in itself exclude the possibility that another randomization test is much better against most alternatives but slightly worse against some (Problem 31).*

*PROOF. The same basic device can be used to obtain both of the above properties. (This is not surprising in light of the fact that a uniformly most powerful invariant test is most stringent under quite general conditions [Lehmann, 1959, p. 340].)

Consider the alternatives given by (μ_1, μ_2, σ) and $(c - \mu_1, c - \mu_2, \sigma)$. The average power of any test against these two alternatives is the same as its power against

$$
h(x_1, \ldots, x_m, y_1, \ldots, y_n) = \left[\prod_{i=1}^{m} g(x_i; \mu_1, \sigma) \prod_{j=1}^{n} g(y_j; \mu_2, \sigma) \right.
$$
$$
\left. + \prod_{i=1}^{m} g(x_i; c - \mu_1, \sigma) \prod_{j=1}^{n} g(y_j; c - \mu_2, \sigma) \right] \bigg/ 2 \quad (4.2)
$$

where $g(z; \mu, \sigma)$ is the normal density with mean μ and variance σ^2 (Problem 32a). We shall show that, for any (μ_1, μ_2, σ), there is a c such that the randomization test rejecting for large $|\bar{Y} - \bar{X}|$ is the most powerful randomization test against h. From this and some further arguments, the desired conclusions follow (Problem 32c).

By straightforward calculation, the first term on the right-hand side of (4.2) is a multiple of

$$
\exp\left\{ -\frac{1}{2\sigma^2} \left[\sum_i (x_i - \bar{\mu})^2 + \sum_j (y_j - \bar{\mu})^2 \right. \right.
$$
$$
\left. \left. + \frac{2mn}{N} (\mu_2 - \mu_1)(\bar{x} - \bar{y}) + \frac{mn}{N} (\mu_2 - \mu_1)^2 \right] \right\} \quad (4.3)
$$

where $\bar{x} = \sum_i x_i/m$, $\bar{y} = \sum_j y_j/n$, and

$$
\bar{\mu} = (m\mu_1 + n\mu_2)/N. \quad (4.4)
$$

The second term on the right-hand side of (4.2) is the same as (4.3) but with $c - \mu_1$ in place of μ_1 and $c - \mu_2$ in place of μ_2, and therefore with

$$
c - \bar{\mu} = [m(c - \mu_1) + n(c - \mu_2)]/N \quad (4.5)
$$

in place of $\bar{\mu}$. If $c = 2\bar{\mu}$, so that the two quantities in (4.4) and (4.5) are equal, then the density in (4.2) becomes

$$
\frac{1}{(2\pi\sigma^2)^{N/2}} \left[\exp\left\{ -\frac{1}{2\sigma^2} \left[\sum_1^m (x_i - \bar{\mu})^2 + \sum_1^n (y_j - \bar{\mu})^2 + \frac{mn}{N} (\mu_2 - \mu_1)^2 \right] \right\} \right]
$$
$$
\cosh\left[\frac{mn(\mu_2 - \mu_1)}{N\sigma^2} (\bar{x} - \bar{y}) \right] \quad (4.6)
$$

where $\cosh(t) = (e^t + e^{-t})/2$, which is an increasing function of $|t|$. The most powerful randomization test against this density is that rejecting for large $|\bar{Y} - \bar{X}|$ (Problem 32b), as was to be proved.*   □

PROBLEMS

1. Show that the randomization tests based on $\bar{Y} - \bar{X}$, $\bar{Y}$, $\sum_1^m X_i$, $\sum_1^n Y_j$, S^*, S^{**}, the ordinary two-sample t statistic, and r, are all equivalent; here $S^* = \sum_{r_i > m} X_i - \sum_{r'_j \leq m} Y_j$ and $S^{**} = \sum_{r'_j > n} Y_j - \sum_{r_i \leq n} X_i$ where r_i is the rank of X_i and r'_j is the rank of Y_j in the combined ordered sample, and r is the ordinary product-moment correlation coefficient between the N observations and the N indicator variables defined by

$$I_k = \begin{cases} 1 & \text{if the observation with rank } k \text{ is an } X \\ 0 & \text{if the observation with rank } k \text{ is a } Y. \end{cases}$$

*2. Show that the confidence bounds for shift corresponding to the one-tailed, two-sample randomization tests based on $\bar{Y} - \bar{X}$ at level $\alpha = k/\binom{N}{m}$ are the kth smallest and largest of the differences of equal-sized subsample means.

*3. Consider the $\binom{N}{m} - 1$ differences of equal-sized subsample means in the two-sample problem. Show that under the shift assumption
   (a) The $\binom{N}{m}$ intervals into which these differences divide the real line are equally likely to contain the true shift μ.
   (b) If M of the differences are selected at random without replacement, then the kth smallest is a lower confidence bound for μ at level $\alpha = k/(M + 1)$.

4. Find the mean and variance of the randomization distribution of the statistics $\bar{Y}$, S^*, and S^{**} defined in Problem 1.

5. (a) Show that the randomization distribution of $\bar{Y} - \bar{X}$ is symmetric about 0 if (i) $m = n$, or (ii) the combined sample is symmetric.
   *(b) Can you construct an example in which the randomization distribution of $\bar{Y} - \bar{X}$ is symmetric about 0 but neither (i) nor (ii) of (a) holds? (The authors have not done so.)

6. Consider the samples X: 0.5, 0.9, 2.0 and Y: -0.2, 6.5, 11.5, 14.3. Show that
   (a) The randomization distribution of $\bar{Y} - \bar{X}$ is that given in Table 2.1.
   (b) The randomization test based on $|\bar{Y} - \bar{X}|$ has upper-tailed P-value 6/35 and hence rejects the null hypothesis at level $\alpha = 6/35$.
   (c) The equal-tailed randomization test based on $\bar{Y} - \bar{X}$ at level $\alpha = 6/35$ and the lower-tailed test at level 3/35 both "accept" the null hypothesis. Find the P-values.

7. Suppose the following are two independent random samples, the second drawn from the same distribution as the first except for a shift of the amount μ.

$$X: 0.2, 0.6, 1.2 \quad \text{and} \quad Y: 1.0, 1.8, 2.3, 2.4, 4.1$$

   (a) Use the randomization test based on the difference of sample means (or an equivalent randomization test statistic) to test the null hypothesis $\mu = 2$ against the alternative $\mu < 2$, at a level near 0.01.
   (b) Give the exact P-value of the test in (a).
   (c) Give the confidence bound for μ which corresponds to the test used in (a).
   (d) Find the approximate P-value based on (i) the standard normal distribution and (ii) Student's t distribution with $N - 2 = 6$ degrees of freedom.

8. Show that the randomization test based on $\bar{Y} - \bar{X}$ and the two-sample rank sum test are equivalent for one-tailed $\alpha \leq 2/\binom{N}{m}$ but not for $\alpha = 3/\binom{N}{m}$ or $4/\binom{N}{m}$ (assume $m \geq 4$, $n \geq 4$).

*9. Express the confidence bounds for shift corresponding to the randomization test based on $\bar{Y} - \bar{X}$ in terms of the order statistics of the two samples separately for $\alpha = k/\binom{N}{m}$, $k = 1, 2, 3, 4$.

10. Verify the expression for t given in (2.7).

11. Verify that (2.9) and (2.11) are equivalent expressions for the degrees-of-freedom correction factor d.

12. Show that c_2 as given by (2.12) is a consistent estimator of the population kurtosis minus 3 if the X's and Y's are independent and identically distributed with a suitable number of finite moments.

13. Show that the step relating F with fractional degrees of freedom to a scaled t distribution in Sect. 2.5 is the same as in Sect. 2.5 of Chap. 4 with n replaced by $N - 1$ (see Problem 18 of Chap. 4).

14. (a) Derive formulas (2.14) and (2.15) for the first two moments of the randomization distribution of $(\bar{Y} - \bar{X})^2/(N - 1)\sigma^2$.
    (b) Show that the beta distribution with the same first two moments has parameters given by (2.16).
    (c) Show that approximating the randomization distribution of $(\bar{Y} - \bar{X})^2/(N-1)\sigma^2$ by this beta distribution is equivalent to approximating the randomization distribution of t^2 by an F distribution with d and $(N - 2)d$ degrees of freedom, where d is given by (2.9).

15. Show that the randomization distribution of t has mean 0 but need not have third moment 0.

*16. Show that a group of transformations can be used to restrict the randomization set as described in (b) of Sect. 2.5.

*17. Let $T(x_1, \ldots, x_m, y_1, \ldots, y_n)$ be nondecreasing in each x_i and nonincreasing in each y_j and consider randomization tests based on T. Show that
    (a) The corresponding confidence regions are intervals.
    (b) The level of the lower-tailed test remains valid when the observations are mutually independent and every X_i is "stochastically larger" than every Y_j.
    (c) The lower-tailed test is unbiased against alternatives under which the observations are mutually independent and every X_i is "stochastically smaller" than every Y_j.
    Note: Remember that the critical value of a randomization test depends on the observations and hence is not constant.

18. (a) Show that the randomization distribution of $\sum X_i$ is the same as the null distribution of the sum of the X scores for scores a_k (Sect. 5, Chap. 5).
    (b) Relate formulas (2.3) and (2.4) for the mean and variance of the randomization distribution of $\bar{X}$ to the corresponding results given in Problem 77a of Chap. 5 for sum of scores tests.
    (c) Why, despite (a), is the randomization test not a sum of scores test?

19. Invent an "adaptive" two-sample randomization test and discuss the rationale for it.

20. In each of the following situations, identify the real world counterparts of $X_1, \ldots,$ X_m, $Y_1, \ldots,$ Y_n and $a_1, \ldots, a_N$. Would it be possible to distinguish order within samples? Meaningful? Desirable? If a randomization test is to be used, should it be $\binom{N}{m}$ type? $N!$-type? What null hypothesis would be appropriate? Why? The situations are sketchily described; give further details as you need or desire.
    (a) A library ceiling has 50 light bulbs. Two types of bulbs are used in an experiment and the lifetime of each bulb is recorded.
    (b) In a library and a less well ventilated hallway with the same type of bulbs, bulb lifetimes are recorded to see if they are affected by ventilation.
    (c) A set of patients who are deemed appropriate and have given consent receives the usual medication for some type of high fever. A randomly selected subset of this set is also given a standard dose of an additional new drug. The temperature change in a four-hour period is recorded for all patients.
    (d) Same as (c) except that the dose of the new drug is varied from 20% below to 20% above the standard level.

21. In one of the situations of Problem 20 or some other situation, describe a randomization test which is not permutation invariant and why it might be desirable to use it.

22. (a) Show that a randomization test is $N!$ type if it is $\binom{N}{m}$ type.
    (b) In what sense is an $\binom{N}{m}$-type randomization test more conditional than an $N!$ type?

23. Given k and n, in how many ways can one select integers $j_1, j_2, \ldots, j_n$ which are all different and satisfy $1 \le j_1 < j_2 < \cdots < j_k \le n$ and $1 \le j_{k+1} < j_{k+2} < \cdots < j_n \le n$?

24. (a) Show that all level α tests are $\binom{N}{m}$-type randomization tests if the null hypothesis is that $X_1, \ldots, X_m; Y_1, \ldots, Y_n$ are a random separation of $a_1, \ldots, a_N$ into an X sample and a Y sample, where $a_1, \ldots, a_N$ are arbitrary constants.
    (b) Suppose it is known that all tests having level α for some null hypothesis H_0^* are $\binom{N}{m}$-type randomization tests. Show that the same is true for every H_0^{**} which contains H_0^*.
    (c) Show that all level α tests are $\binom{N}{m}$-type randomization tests for H_0: The observations $Z_1, \ldots, Z_N$ are independent with arbitrary distributions, and $X_1, \ldots, X_m;$ $Y_1, \ldots, Y_n$ are a random separation of $Z_1, \ldots, Z_N$ into an X sample and a Y sample.

25. (a) Show that if a test has level α for the H_0 or H_0' given after Theorem 3.1 and is unbiased against one-sided (or two-sided) shift alternatives, then it has level exactly α under every null distribution.
    (b) Show that the result in (a) in turn implies that the test has conditional level α given the combined sample observations.

26. (a) Show that a test is the most powerful $N!$-type randomization test at level α against a simple alternative if and only if it has level exactly α and is of the form (4.1) where k is a function of the order statistics of the combined sample.
    *(b) What change occurs in the statement in (a) for $\binom{N}{m}$-type randomization tests?

27. Show that a one-tailed randomization test based on $\bar{Y} - \bar{X}$ (or any equivalent statistic) is uniformly most powerful against one-sided normal alternatives with common variance.

28. (a) Show that if a test is invariant under transformations carrying X into $c - X$ and Y into $c - Y$ for all c, then its power against the alternative that X is $N(\mu_1, \sigma^2)$ and Y is $N(\mu_2, \sigma^2)$ depends only on $\mu_1 - \mu_2$ and σ.

   (b) Show that if the test in (a) is also invariant under changes of scale (carrying X into bX and Y into bY) then its power depends only on $(\mu_1 - \mu_2)/\sigma$.

  *(c) Why were changes of scale not considered in Sect. 4.3?

*29. Show that the randomization test based on $|\overline{Y} - \overline{X}|$ has the "most stringent" property of Sect. 4.3

   (a) If the alternative is restricted to the region $|\mu_1 - \mu_2| > b\sigma$ where b is some given constant.

   (b) If $\alpha^*(\mu_1, \mu_2, \sigma)$ is redefined as the maximum power achievable by any test.

*30. Show that uniformly most powerful invariant tests are "generally" most stringent under suitable conditions. What condition is most important?

31. In the situation of Sect. 4.3, draw hypothetical graphs of $\alpha^*(\mu_1, \mu_2, \sigma)$ and of $\alpha(\mu_1, \mu_2, \sigma)$ for the level α randomization test based on $|\overline{Y} - \overline{X}|$ as functions of $(\mu_2 - \mu_1)/\sigma$. Indicate how to find Δ from these graphs. What do the properties of $\alpha(\mu_1, \mu_2, \sigma)$ as "uniformly most powerful invariant" and "most stringent" imply about the graphs of the power of other randomization tests? (Assume the other randomization tests are invariant under increasing linear transformations so that their power depends only on $(\mu_1 - \mu_2)/\sigma$, but do not assume that they are invariant under changes of sign.) What can you say about the power of a one-tailed randomization test based on $\overline{Y} - \overline{X}$ at level α?

32. (a) Show that the power of a test against the density h given by (4.2) is the average of its power against the two normal alternatives given by

$$(\mu_1, \mu_2, \sigma) \quad \text{and} \quad (c - \mu_1, c - \mu_2, \sigma).$$

   (b) Show that the most powerful randomization test against h is that rejecting for large $|\overline{Y} - \overline{X}|$.

  *(c) From this, show that the randomization test rejecting for large $|\overline{Y} - \overline{X}|$ is uniformly most powerful invariant and most stringent against normal alternatives as stated in Sect. 4.3.

CHAPTER 7

Kolmogorov–Smirnov
Two-Sample Tests

1 Introduction

We have not previously discussed the use of criteria suggested by direct comparison of empirical (sample) cumulative distribution functions with one another or with hypothetical c.d.f.'s ("goodness of fit"). This important approach leads to a wide variety of procedures which stand apart from the procedures of earlier chapters in several respects. They are expressed in a different form. The relevant statistics are not approximately or asymptotically normally distributed. The theory of their asymptotic behavior is fascinating and raises different kinds of problems requiring different kinds of tools. The mathematical interest of these and other problems has played a larger role than statistical questions in motivating the extensive literature about them, although there is also some excellent work on statistically important questions.

The one-sample test procedures require that a completely specified distribution be hypothesized. In this sense they are not "nonparametric." Although the test statistics are "distribution-free" as that term is usually defined, they relate to null hypotheses that are entirely different from those in Chaps. 2–4, which require only symmetry. The two-sample procedures, however, relate to the "nonparametric" null hypothesis of identical distributions used in Chaps. 5 and 6.

The Kolmogorov–Smirnov criterion of maximum difference, defined below, has received the most attention. Another natural criterion is the Cramér–von Mises integrated squared difference. These and other variations of them involving "weights" are the only specific criteria developed in this tradition which have been broadly investigated for statistical purposes.

Pearson's chi-square goodness-of-fit criterion, which compares cell frequencies rather than cumulative frequencies, is even more popular. It could be viewed as a comparison of changes in c.d.f.'s, rather than a comparison of c.d.f.'s themselves.

We will discuss only the Kolmogorov–Smirnov criterion here because it is the only well-developed goodness-of-fit criterion that is at all competitive with the other procedures considered in this book against shift and similar alternatives. Also, simple confidence procedures can be based on it, as well as tests. The scope of this book does not extend to testing goodness of fit in the usual sense, especially not to preliminary tests to be followed by further analysis. A brief comment on preliminary testing is in order, however, because preliminary testing followed by a parametric analysis if the null hypothesis is "accepted" may be considered an alternative to nonparametric methods. The deviations from parametric assumptions that are most likely to occur are typically large enough to have a significant and perhaps even disastrous effect on a parametric analysis, and yet are not large in a goodness-of-fit sense. Typical goodness-of-fit tests, or others that could be devised, have low power against these alternatives for usual sample sizes, and hence provide little and inadequate protection. The accept/reject approach is not really appropriate to the purpose. Type II errors are serious while a Type I error is not (devising an unnecessary alternative analysis), and power is low. All this suggests that an appropriate significance level will be so large that one might as well assume rejection without even testing. The questions are whether the parametric analysis may be seriously invalidated by deviations from assumptions, and if so what to do about it. The answer to the first question usually depends far more on the particular form of analysis and the background of the data than on anything detectable by a goodness-of-fit statistic. The answer to the second question lies in sensitivity analyses and robust procedures, including nonparametric methods.

For reasons evident from these preliminary remarks, we give in this chapter a separate introduction to Kolmogorov–Smirnov procedures, emphasizing the two-sample case, and include more references than usual because we cannot go deeply into the theory.

2 Empirical Distribution Function

Given any sample from an unspecified population, a natural estimate of the unknown cumulative distribution function of the population is the *empirical* (or *sample*) *distribution function* of the sample, defined, at any real number t, as the proportion of sample observations which do not exceed t. For a sample of size m, the empirical distribution function will be denoted by F_m and may be defined in terms of the order statistics $X_{(1)} \le X_{(2)} \le \cdots \le X_{(m)}$ by

$$F_m(t) = \begin{cases} 0 & \text{if } t < X_{(1)} \\ j/m & \text{if } X_{(j)} \le t < X_{(j+1)}, \ 1 \le j < m \\ 1 & \text{if } t \ge X_{(m)}. \end{cases} \tag{2.1}$$

The following properties of $F_m(t)$ are easily proved (Problems 1 and 2) for observations which are independently and identically distributed with c.d.f. F.

(a) The random variable $mF_m(t)$ follows the binomial distribution with parameters m and $F(t)$.
(b) The mean and variance of $F_m(t)$ are

$$E[F_m(t)] = F(t), \qquad \mathrm{var}[F_m(t)] = F(t)[1 - F(t)]/m.$$

(c) $F_m(t)$ is a consistent estimator of $F(t)$ for fixed t.
(d) $F_m(t)$ converges uniformly to $F(t)$ in probability, that is,

$$P[\,|F_m(t) - F(t)| < \varepsilon \text{ for all } t] \to 1 \text{ as } m \to \infty, \text{ for all } \varepsilon > 0.$$

(e) $F_m(t)$ converges uniformly to $F(t)$ with probability one (Glivenko–Cantelli Theorem).
(f) $F_m(t)$ is asymptotically normal with mean and variance given in (b).
(g) The empirical distribution is the mean of the indicator random variables defined by

$$\delta_i(t) = \begin{cases} 1 & \text{if } X_i \le t \\ 0 & \text{otherwise,} \end{cases}$$

that is,

$$F_m(t) = \sum_{i=1}^{m} \delta_i(t)/m.$$

In particular, $F_1(t) = \delta_1(t)$. The covariance between values of $\delta_i(t)$ for the same i but different t is

$$\mathrm{cov}[\delta_i(t_1), \delta_i(t_2)] = \begin{cases} F(t_1)[1 - F(t_2)] & \text{if } t_1 \le t_2 \\ F(t_2)[1 - F(t_1)] & \text{if } t_2 \le t_1 \end{cases}$$

$$= \sigma(t_1, t_2), \text{ say.}$$

(h) $\mathrm{cov}[F_m(t_1), F_m(t_2)] = \sigma(t_1, t_2)/m$.
(i) For any fixed $t_1, \ldots, t_k$, the random variables $F_m(t_1), \ldots, F_m(t_k)$ are asymptotically multivariate normal with mean vector $[F(t_1), \ldots, F(t_k)]$ and covariance matrix with (i, j) element $\sigma(t_i, t_j)/m$. This can be proved by applying the multivariate Central Limit Theorem for identically distributed random variables to the vectors $\delta_i = [\delta_i(t_1), \ldots, \delta_i(t_k)]$ or by way of the multinomial distribution of the increments $F_m(t_1), F_m(t_2) - F_m(t_1), \ldots, F_m(t_k) - F_m(t_{k-1}), 1 - F_m(t_k)$.

3 Two-Sample Kolmogorov–Smirnov Statistics

In the previous two chapters we discussed a variety of tests for the situation where two independent samples, $X_1, \ldots, X_m$ and $Y_1, \ldots, Y_n$, are drawn from populations with unspecified c.d.f.'s F and G respectively, and the null hypothesis is $F = G$. Now the empirical distribution functions of the X and Y samples, denoted by $F_m(t)$ and $G_n(t)$ respectively, estimate their respective population c.d.f.'s. If the null hypothesis is true, there should be close agreement between $F_m(t)$ and $G_n(t)$ for all values of t. Some overall measure of the agreement or disagreement between the functions F_m and G_n is then a natural test statistic. The Kolmogorov–Smirnov two-sample test (sometimes called simply the Smirnov test) is based on the maximum difference; of course, $F_m(t)$ and $G_n(t)$ are in close agreement at all values of t if the maximum difference is small. Specifically, the test statistics are given by

$$D_{mn} = \max_t |G_n(t) - F_m(t)| \qquad (3.1)$$

$$D_{mn}^+ = \max_t [G_n(t) - F_m(t)] \qquad (3.2)$$

$$D_{mn}^- = \max_t [F_m(t) - G_n(t)], \qquad (3.3)$$

where (3.1) is called the two-sided statistic since the absolute value measures differences in both directions, and (3.2) and (3.3) are called the one-sided statistics. Appropriate critical regions are to reject $F = G$ if D_{mn} is "too large" for a two-sided test, if D_{mn}^+ is "too large" for a one-sided alternative $G \geq F$ and if D_{mn}^- is "too large" for a one-sided alternative $G \leq F$. (Assume each of these alternatives excludes $F(t) = G(t)$ for all t.)

Tests based on these statistics would appear to be sensitive to all types of departures from the null hypothesis $F = G$, and hence not especially sensitive to a particular type of difference between F and G. However, even for location alternatives, against which most of the two-sample tests presented in this book are designed to perform well, the Kolmogorov–Smirnov statistics are sometimes quite powerful. They are primarily useful, however, when any type of difference is of interest.

Alternative expressions of the Kolmogorov–Smirnov statistics are more easily evaluated, and they also show that the maxima are achieved. In (3.2) for instance, note that reducing t to the next smaller Y_j does not change $G_n(t)$ and hence can only increase the maximand. Therefore, we can write

$$D_{mn}^+ = \max_j [G_n(Y_j) - F_m(Y_j)]$$

$$= \max_j [(j/n) - F_m(Y_{(j)})]$$

$$= \max_j [(j/n) - (M_j/m)] \qquad (3.4)$$

where $Y_{(j)}$ is the jth smallest Y and M_j is the number of X's less than or equal to $Y_{(j)}$. Similarly,

$$D_{mn}^- = \max_i [F_m(X_i) - G_n(X_i)]$$

$$= \max_i [(i/m) - (N_i/n)] \qquad (3.5)$$

where N_i is the number of Y's less than or equal to $X_{(i)}$, the ith smallest X. The two-sided statistic is simply

$$D_{mn} = \max(D_{mn}^+, D_{mn}^-). \tag{3.6}$$

These representations (and others, Problem 3), also make it evident that the Kolmogorov–Smirnov statistics depend only on the ranks of the X's and Y's in the combined sample. Thus the tests based on them are two-sample rank tests. In particular, their distributions under the null hypothesis that the X's and Y's are independent and identically distributed with an un-specified common distribution do not depend on what that common distribution is as long as it has a continuous c.d.f. (This result is also evident otherwise; Problem 3.) The same null distributions hold also in the situation of say a treatment-control experiment where m units are selected at random from N to receive treatment, if the null hypothesis is no treatment effect and the distribution of the characteristic being measured is continuous so that ties have probability 0. (Ties are discussed in Sect. 5.)

4 Null Distribution Theory

In order to apply the Kolmogorov–Smirnov one-sided or two-sided tests, we need to find the null distributions of the statistics so that critical values or P-values can be determined. An obvious and straightforward method of enumerating the possibilities in order to obtain the null distribution of the Kolmogorov–Smirnov statistics (and many others) will be described in Sect. 4.1. Using this method, Kim and Jennrich prepared extensive tables of the exact null probability distribution of D_{mn}; these tables appear in Harter and Owen [1970]. A few values are given in Table G of this book. Section 4.2 contains brief comments on the relation between the one-tailed and two-tailed procedures. In Sect. 4.3 we derive simple expressions in closed form for the null distributions of both the one-sided and two-sided statistics for the case $m = n$. These are useful in themselves and asymptotic distributions are easily derived from them, as in Sect. 4.4, which also gives a heuristic argument that the same asymptotic distributions apply for $m \neq n$. Much of the mathematical interest lies in making the asymptotic distribution theory rigorous. It is also needed as an approximation for finite m and n beyond the range of tables. Hodges [1957], however, reports a brief numerical investigation of accuracy, and shows that this approximation can be quite poor and the accuracy fluctuates wildly as m, n change. In particular, it is much better for $m = n$ than for $m = n + 1$, same n.

The null distributions are usually attributed to Gnedenko and the Russian school (see, for example, Gnedenko and Korolyuk [1951] or Gnedenko [1954]), but many others have worked on finding convenient methods of computation and expression. See, for example, Massey [1951a], Drion [1952], Tsao [1954]. Korolyuk [1954] and [1955], Blackman [1956], Carvalho [1959], and Depaix [1962]. Hodges [1957] includes a useful review of algorithmic methods. (See also Steck [1969] for results that also make use of the first place where the maximum occurs.) Hájek and Šidák [1967] give a good summary of the results known about the Kolmogorov–Smirnov statistics. Darling [1957] also gives a valuable exposition on these and related statistics and a rather complete guide to the literature through 1956. Barton and Mallows [1965] give an Appendix with references on subsequent developments.

Throughout this section we assume (as does most of the literature most of the time) that the common distribution of the independent random variables X and Y is continuous, so that, with probability one, no two observations are equal. Then we can ignore the possibility of ties either within or between samples. Ties will be discussed in Sect. 5.

4.1 An Algorithm for the Exact Null Distribution

We now describe a simple counting or addition procedure for finding the probability $P[D_{mn} < c] = P[\max_t |G_n(t) - F_m(t)| < c]$ under the null hypothesis.

Represent the combined arrangement of X's and Y's by a path in the plane (as in Problem 4, Chap. 5) that starts at $(0, 0)$ and moves one unit up or to the right according as the smallest observation is a Y or an X, from there moves one more unit up or to the right according as the second smallest is a Y or an X, etc. Thus, the kth step is one unit up if the kth smallest observation is a Y, one unit to the right if it is an X, and the path terminates at (m, n). (For example, Fig. 4.1 represents the combined sample arrangement $X\ Y\ Y\ X\ X\ Y\ Y$ with $m = 3$, $n = 4$.) In this manner, all possible arrangements of observations from the two samples are placed in one-to-one correspondence with all paths of this sort starting at $(0, 0)$ and ending at (m, n). Under the null hypothesis, the paths are equally likely.

For any t, let u and v be the number of X's and Y's, respectively, that do not exceed t. Then $F_m(t) = u/m$ and $G_n(t) = v/n$. Furthermore, the point (u, v) lies on the path representing the sample, and as t varies, the point

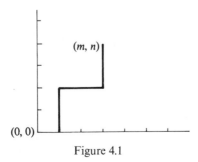

Figure 4.1

(u, v) reaches every lattice point (point with integer coordinates) on the path. Thus the event $D_{mn} < c$ occurs if and only if each such lattice point (u, v) on the path satisfies

$$|(u/m) - (v/n)| < c. \tag{4.1}$$

Consider this event geometrically. The expression $v/n = u/m$ is the equation of the diagonal line which connects the origin $(0, 0)$ and the terminal point of the path (m, n); the vertical distance from any point (u, v) on the path to this line is $|v - (nu/m)|$. Accordingly, $D_{mn} < c$ occurs if and only if the path stays always within a vertical distance of cn from the diagonal connecting $(0, 0)$ to (m, n) (or equivalently, a horizontal distance of cm).

Let $A(u', v')$ be the number of paths from $(0, 0)$ to (u', v') which stay within this distance, i.e., which satisfy (4.1) for $u \leq u'$, $v \leq v'$. (This number depends also on m, n, and c, but since they are fixed throughout we do not display this dependence notationally.) Since every path from $(0, 0)$ to (m, n) has equal probability under the null hypothesis, the probability we seek is

$$P(D_{mn} < c) = A(m, n) \bigg/ \binom{m + n}{m} = A(m, n) \bigg/ \binom{N}{m}. \tag{4.2}$$

It is obvious that $A(u, v)$ satisfiies the recursive relation

$$A(u, v) = \begin{cases} A(u - 1, v) + A(u, v - 1) & \text{if } (u, v) \text{ satisfies (4.1)} \\ 0 & \text{otherwise} \end{cases} \tag{4.3}$$

with boundary conditions

$$A(0, 0) = 1, \qquad A(u, v) = 0 \quad \text{if } u < 0 \text{ or } v < 0.$$

By carrying out this recursion as far as $u = m$, $v = n$, the quantity $A(m, n)$ needed in (4.2) is obtained. For small m and n, this can be done by hand by drawing the two boundary lines $(u/m) - (v/n) = \pm c$ in the rectangle with corners $(0, 0)$, $(m, 0)$, $(0, n)$ and (m, n) and entering the values of $A(u, v)$ at the points (u, v) successively which are inside (not touching) these boundary lines (Problem 4). Unfortunately, the whole recursion must be carried out anew for each m, n, and c. For larger but not too large sample sizes, it is easily done by computer. Some properties of $A(u, v)$ are given in Problem 6.

An analogous procedure can be used to find the probability of staying within any given boundary (Problem 7). In particular, it can be used for

$$P(D_{mn}^{+} < c). \tag{4.4}$$

We simply define $A(u', v')$ as the number of paths from $(0, 0)$ to (u', v') along which $(u/m) - (v/n) < c$, that is, which stay below the upper boundary line at a distance cn above the diagonal. An alternative procedure may be preferable however (Korolyuk [1955]; see also Hodges [1957] and Problem 23). It is clear by symmetry (Problem 9) that

$$P(D_{mn}^{+} < c) = P(D_{nm}^{+} < c) = P(D_{mn}^{-} < c) = P(D_{nm}^{-} < c) \tag{4.5}$$

for all m, n, and c.

4.2 Relation Between One-Tailed and Two-Tailed Procedures

The critical regions of the Kolmogorov–Smirnov tests are the complements of the events just considered; the level or P-value is the probability that the sample path reaches (or crosses) the relevant boundary. Note that the upper and lower one-tailed procedures are not based on the same statistic, and their critical regions are not in general disjoint. The events $D_{mn}^+ \geq c$ and $D_{mn}^- \geq c$ both occur for the same sample if the path reaches both boundaries. Therefore, although the two-tailed critical region is the union of two one-tailed critical regions with equal critical values, the two-tailed level may be less than twice the one-tailed level for that critical value, and the two-tailed P-value may be less than twice the one-tailed P-value (but not more). For $c > 0.5$, rejection by both one-tailed tests at once is impossible. Thus

$$P(D_{mn} \geq c) = P(D_{mn}^- \geq c) + P(D_{mn}^+ \geq c) = 2P(D_{mn}^+ \geq c) \text{ for } c > 0.5. \quad (4.6)$$

If m and n are both odd, this also holds for $c > 0.5 - 0.5/\max(m, n)$ (Problem 10d). Although this is not true for all smaller values of c (Problem 10e), the discrepancy is very small in the tails, even in relative terms. (For m and n large or equal, it is less than twice the fourth power of the one-tailed probability and hence the relative error is less than 1% for one-tailed probabilities of 0.2 or less (Problem 11). This may be true for all m and n. For numerical results and references, see Miller [1956].) In short, for practical purposes, if the P-value is small enough to be of interest, the two-tailed P-value may be taken to be twice the one-tailed P-value. Thus one-tailed and two-tailed critical values are related in the usual way in most cases, although complete statements about exact critical values are complicated because of the discreteness of the distribution.

In calculating power, the possibility of a path reaching both boundaries may have a more serious effect.

4.3 Exact Formulas for Equal Sample Sizes

In the special case where $m = n$, expressions can be given in closed form for the null c.d.f.'s of both the one-sided and two-sided Kolmogorov–Smirnov statistics. Specifically, we will show that, for $k = 1, 2, \ldots, n$,

$$P(D_{nn}^+ \geq k/n) = \binom{2n}{n-k} \bigg/ \binom{2n}{n} = (n!)^2/(n+k)!(n-k)! \quad (4.7)$$

$$P(D_{nn} \geq k/n) = 2\left[\binom{2n}{n-k} - \binom{2n}{n-2k} + \binom{2n}{n-3k} - \cdots\right]\bigg/\binom{2n}{n}$$

$$= 2\sum_{i=1}^{[n/k]}(-1)^{i+1}\binom{2n}{n-ik}\bigg/\binom{2n}{n} \quad (4.8)$$

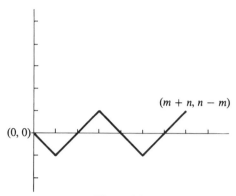

Figure 4.2

where $[n/k]$ denotes the largest integer not exceeding n/k. This gives the entire distribution in each case, since the statistics can take on values only of the form k/n when $m = n$. These formulas are easily evaluated, but they apply only to the case $m = n$.

We now derive these results (following the methods of Drion [1952] and Carvalho [1959], although they are much older). We again represent the combined sample arrangement by a path starting at the origin, but the argument will be easier to follow if we make the steps diagonal. Specifically, we move one unit to the right and one unit up for each Y, one unit to the right and one unit down for each X. This gives the same path as before except for a rotation clockwise by $45°$ (and a change of scale). Figure 4.2 depicts the path constructed by this rule for the same sample as Fig. 4.1, $X \, Y \, Y \, X \, X \, Y \, Y$.

The path now ends at $(m + n, \, n - m)$, which is $(2n, 0)$ when $m = n$. Furthermore, by analysis like that at (4.1), the event $D_{nn}^+ \geq k/n$ occurs if and only if the path reaches a height of at least k units above the horizontal axis before it terminates at $(2n, 0)$. We shall prove shortly that the number of such paths is $\binom{2n}{n-k}$. Since all paths are equally likely under the null hypothesis, and since the total number is $\binom{2n}{n}$, it follows that the probability $P(D_{nn}^+ \geq k/n)$ is $\binom{2n}{n-k}/\binom{2n}{n}$ as given in (4.7). The number of paths reaching height k is given by setting $l = 0$ in the following lemma, which will be needed later for both negative and positive l and k.

Lemma 4.1. *Let $N(k, l)$ be the number of paths going from $(0, 0)$ to $(2n, 2l)$ and reaching a height of k units. Suppose k is not between 0 and $2l$. (If it is, all paths terminating at height $2l$ obviously reach height k.) Then*

$$N(k, l) = \binom{2n}{n - k + l} = \binom{2n}{n + k - l}. \tag{4.9}$$

PROOF. Paths of this sort can be put into one-to-one correspondence with paths terminating at height $2k - 2l$ by reflecting that portion of each path to the right of the point where it last reaches height k, as in Fig. 4.3. The number

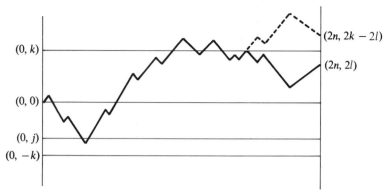

Figure 4.3

of these reflected paths is simply the total number of paths corresponding to samples having $n - k + l$ X's and $n + k - l$ Y's, namely $\binom{2n}{n-k+l}$. □

For the two-sided statistic, consider the event $D_{nn} \geq k/n$. It occurs if and only if the path reaches at least k units above or below the horizontal axis. By symmetry, the number of paths reaching height $-k$ is the same as the number reaching height k, which was found above. The difficulty is that some paths may reach both k and $-k$. We will count paths according to the boundary they reach first. It is convenient to extend the notation of Lemma 4.1, letting

$N(j, k, l) =$ the number of paths going from $(0, 0)$ to $(2n, 2l)$ reaching heights j and k, j first;

$N(\text{not } j, k, l) =$ the number of paths going from $(0, 0)$ to $(2n, 2l)$ reaching height k without having reached height j.

(Note that, in either case, heights j and k may subsequently be reached any number of times.) In this notation the number of paths satisfying $D_{nn} \geq k/n$ is the number reaching the upper boundary first plus the number reaching the lower boundary first, which is

$$N(\text{not } -k, k, 0) + N(\text{not } k, -k, 0) = 2N(\text{not } -k, k, 0) \quad (4.10)$$

by symmetry.

It follows immediately from the definitions that

$$N(\text{not } j, k, l) = N(k, l) - N(j, k, l). \quad (4.11)$$

Furthermore, the reflection used in the proof of Lemma 4.1 now shows (see Fig. 4.3; this is the crucial step) that

$$N(j, k, l) = N(j, k, k - l)$$
$$= N(\text{not } k, j, k - l) \quad \text{if } j \leq k \leq 2k - 2l \text{ or } j \geq k \geq 2k - 2l.$$

$$(4.12)$$

By applying (4.11) and (4.12) alternately, we can now evaluate (4.10). This amounts to repeated application of the reflection principle. Each application turns out to move the terminal height further from 0, until eventually no path can reach this height (the last term of (4.11) vanishes) and the process stops. We thus obtain, using (4.9) also,

$$N(\text{not } -k, k, 0) = N(k, 0) - N(-k, k, 0)$$

$$= \binom{2n}{n-k} - N(\text{not } k, -k, k),$$

$$N(\text{not } k, -k, k) = N(-k, k) - N(k, -k, k)$$

$$= \binom{2n}{n-2k} - N(\text{not } -k, k, -2k),$$

$$N(\text{not } -k, k, -2k) = N(k, -2k) - N(-k, k, -2k)$$

$$= \binom{2n}{n-3k} - N(\text{not } k, -k, 3k),$$

etc. Continuing and substituting back each time, we obtain

$$N(\text{not } -k, k, 0) = \binom{2n}{n-k} - \binom{2n}{n-2k} + \binom{2n}{n-3k} - \cdots . \quad (4.13)$$

If we define $\binom{2n}{n-jk} = 0$ for $n < jk$, this series terminates exactly when the number of paths corresponding to the next term is 0. Multiplying (4.13) by 2 gives the number of paths satisfying $D_{nn} \geq k/n$, and dividing this result by the total number of paths, $\binom{2n}{n}$ as before, then gives (4.8).

4.4 Asymptotic Null Distributions

For $m = n$, the exact formulas (4.7) and (4.8) for the null tail probabilities lead directly to the asymptotic null distributions of the one-sided and two-sided Kolmogorov–Smirnov statistics. We first investigate the behavior of (4.7) for large k and n. The right-hand side can be written as

$$\binom{2n}{n-k} \bigg/ \binom{2n}{n} = \frac{n(n-1)\cdots(n-k+1)}{(n+k)(n+k-1)\cdots(n+1)}$$

$$= \left(1 - \frac{k}{n+k}\right)\left(1 - \frac{k}{n+k-1}\right)\cdots\left(1 - \frac{k}{n+1}\right)$$

$$= \left[1 - \frac{k}{n} + o\left(\frac{k}{n}\right)\right]^k \quad \text{if } k/n \to 0 \quad (4.14)$$

where $o(x)$ denotes terms of order smaller than x. If we now substitute

$$1 + x + o(x) = e^{x + o(x)}$$

with $x = -k/n$ in (4.14), we obtain

$$\binom{2n}{n-k} \bigg/ \binom{2n}{n} = e^{-k^2/n + o(k^2/n)}. \tag{4.15}$$

To obtain limits other than 0 or 1, k must be of order $\sqrt{n}$. Normalizing by $\sqrt{n/2}$, for reasons evident later, we have by (4.7)

$$P(\sqrt{n/2}D_{nn}^+ \geq \lambda) = P(D_{nn}^+ \geq \lambda\sqrt{2n}/n)$$

$$= \binom{2n}{n-k} \bigg/ \binom{2n}{n}$$

where k now and hereafter is the largest integer not exceeding $\lambda\sqrt{2n}$. Thus, for λ fixed, $k^2/n \to 2\lambda^2$ and we find by (4.15) that

$$\lim_{n \to \infty} P(\sqrt{n/2}D_{nn}^+ \geq \lambda) = e^{-2\lambda^2}. \tag{4.16}$$

This is a very simple, easily calculated expression for the asymptotic probabilities. We note also that $n(D_{nn}^+)^2$ is asymptotically exponentially distributed and $2n(D_{nn}^+)^2$ is asymptotically chi-square distributed with 2 degrees of freedom (Problem 28), so that exponential or chi-square tables could be used.

The limiting distribution of the two-sided statistic is found the same way but using (4.8) as follows.

$$P(\sqrt{n/2}D_{nn} \geq \lambda) = P(D_{nn} \geq \lambda\sqrt{2n}/n)$$

$$= 2\sum_{i=1}^{[n/k]}(-1)^{i+1}\binom{2n}{n-ik} \bigg/ \binom{2n}{n} \tag{4.17}$$

where k is the largest integer not exceeding $\lambda\sqrt{2n}$, as before. Substituting ik for k in (4.15) we find

$$\lim_{n \to \infty}\binom{2n}{n-ik} \bigg/ \binom{2n}{n} = \lim_{n \to \infty} e^{-i^2 k^2/n} = e^{-2i^2\lambda^2}. \tag{4.18}$$

If we now substitute (4.18) in (4.17) we obtain the result

$$\lim_{n \to \infty} P(\sqrt{n/2}D_{nn} \geq \lambda) = 2\sum_{i=1}^{\infty}(-1)^{i+1}e^{-2i^2\lambda^2}. \tag{4.19}$$

Taking the limit as $n \to \infty$ term by term in (4.17) can be justified by the fact that the sum is, for each n, and in the limit, an alternating series whose terms decrease monotonically in absolute value to 0 as $i \to \infty$, and therefore dominate the sum of all later terms.

Now we will show heuristically that if D_{mn}^+ and D_{mn} are suitably standardized, their limiting distributions do not depend on how m and n approach ∞, and hence are the same for all m and n as for $m = n$, so that the expressions on the right-hand sides of (4.16) and (4.19) apply also to unequal sample sizes.

For two independent samples, with arbitrary but fixed $t_1, \ldots, t_k$, the random vectors

$$[F_m(t_1), \ldots, F_m(t_k)] \tag{4.20a}$$

and

$$[G_n(t_1), \ldots, G_n(t_k)] \tag{4.20b}$$

are independent. By the property (i) in Sect. 2, if the observations are identically distributed with c.d.f. F, these random vectors are both asymptotically multivariate normal with the same mean vector and with covariances $\sigma(t_i, t_j)/m$ and $\sigma(t_i, t_j)/n$ respectively. It follows that the vector

$$\sqrt{mn/(m+n)}[G_n(t_1) - F_m(t_1)], \ldots, \sqrt{mn/(m+n)}[G_n(t_k) - F_m(t_k)]$$

is asymptotically multivariate normal with mean vector 0 and the covariances $\sigma(t_i, t_j)$. Hence for each $t_1, \ldots, t_k$, the quantities $\sqrt{mn/(m+n)}[G_n(t_i) - F_m(t_i)]$, $i = 1, \ldots, k$, have the same limiting joint distribution however m and n approach ∞. This suggests that the maximum, or the maximum absolute value, will exhibit this same property, that is, that (4.16) and (4.19) generalize to

$$P(\sqrt{mn/(m+n)}D_{mn}^+ \geq \lambda) \to e^{-2\lambda^2} \tag{4.21}$$

$$P(\sqrt{mn/(m+n)}D_{mn} \geq \lambda) \to 2\sum_{i=1}^{\infty}(-1)^{i+1}e^{-2i^2\lambda^2} \tag{4.22}$$

as m and n approach ∞ in any manner whatever. It is far from a proof, however, because the joint limiting distributions of the values of a function on arbitrary finite sets do not suffice to determine the limiting distribution of its maximum.

The asymptotic distribution of the two-sided, one-sample statistic (see Sect. 7) was derived by Kolmogorov [1933]. Smirnov [1939] simplified this proof and proved corresponding results for the one-sided and two-sample cases as well. Feller [1948] used generating functions in connection with recursive relations. Doob [1949] made a transformation to the Wiener process (Problem 55) which shows heurtistically how these and other asymptotic distributions relate to first passage probabilites for that process and he derived the results in this way; see also Kac [1949]. Donsker [1952] justified this approach rigorously. A simple proof by van der Waerden [1963] and later work by Steck [1969] should also be mentioned. For further references, see the literature cited in the introduction to this section.

5 Ties

The finite sample and asymptotic null distributions of the Kolmogorov–Smirnov statistics discussed in Sect. 4 hold exactly only for a common distribution F which is continuous, and they do not depend on F as long as it is continuous. On the other hand, if F is discontinuous, the distributions do depend on F and hence the statistics are no longer distribution free. However, the tests are conservative; that is, the critical values at level α for F continuous are conservative if F is not continuous. Equivalently, $P(D_{mn} \geq c)$ and $P(D_{mn}^{+} \geq c)$ are smaller for F discontinuous than for F continuous.

One way to see this is to observe that the presence of ties reduces the range of possibilities over which the maximum is taken and hence can only make the statistics smaller. Specifically, consider D_{mn}^{+}, as defined by (3.4) for instance, and let $D_{mn}^{+}*$ be the value of D_{mn}^{+} which would be obtained by breaking the ties at random. Define D_{mn}^{*} similarly. The null distributions of $D_{mn}^{+}*$ and D_{mn}^{*} are the same as in the continuous case, because all sample arrangements of X's and Y's remain equally likely if ties are broken at random.

Furthermore, $D_{mn}^{+} \leq D_{mn}^{+}*$ because breaking ties can decrease but cannot increase the number of X's less than or equal to $Y_{(j)}$; similarly, $D_{mn} \leq D_{mn}^{*}$. Therefore the upper tail probabilities of D_{mn}^{+} and D_{mn} are no larger for discontinuous F than for continuous F; actually they are smaller (Problem 29). The general method of Problem 106 of Chap. 5 does not apply directly here because the Kolmogorov–Smirnov statistics are not continuous functions of the observations (Problem 30a). It can, however, be extended to "lower-semicontinuous" functions, which these statistics are (Problem 30b).

6 Performance

Suppose that the X's and Y's are independent random samples from populations with c.d.f.'s F and G respectively. We have already seen that the sample c.d.f. $F_m(t)$ for the X sample is a consistent estimator of the population c.d.f. $F(t)$, not merely at each t, but in the stronger sense of (d) in Sect. 2; we call this property "strong consistency" temporarily. An equivalent statement (Problem 2b) is that

$$D_m = \sup_t |F_m(t) - F(t)| \to 0 \text{ in probability.}$$

Here D_m is the one-sample Kolmogorov–Smirnov statistic for the hypothesized distribution F, discussed later in Sect. 7. Because there may be no maximizing t, D_m is defined as a supremum (sup), which means least upper bound. (Similarly, the infimum (inf) is the greatest lower bound. For any set of numbers, the supremum and infimum always exist, but either or both may be infinite.)

Actually, stronger properties than strong consistency have already been given. The Glivenko–Cantelli Theorem says that $D_m \to 0$ with probability one, and the asymptotic results discussed in Sect. 4.4 imply that D_m is of order $1/\sqrt{m}$ in probability. All we need here, however, is that F_m is a strongly consistent estimator of F. Similarly, G_n is a strongly consistent estimator of G. If m and n both approach ∞, it follows that $G_n - F_m$ is a strongly consistent estimator of $G - F$ and hence that D_{mn}, D_{mn}^+, and D_{mn}^- are consistent estimators of the corresponding population quantities, namely, the maxima (or suprema) of $|G(t) - F(t)|$, $G(t) - F(t)$, and $F(t) - G(t)$ respectively (Problem 31).

Consistency properties of the Kolmogorov–Smirnov tests follow in turn. Under the null hypothesis $F = G$, all three population quantities are 0, and the statistics are distribution free if the common population is continuous and stochastically smaller otherwise. Therefore, at any fixed level, the critical values of all three statistics approach 0 as m and n approach ∞. On the other hand, the statistics converge in probability to positive values, and consequently each test is consistent, whenever the corresponding population quantity is positive. Specifically, the two-sided test which rejects for large values of D_{mn} is consistent against all alternatives F, G with $F \neq G$, that is, with $F(t) \neq G(t)$ for some t. (Details of the proof are left as Problem 32.) The one-sided test which rejects for large values of D_{mn}^+ is consistent against all alternatives F, G with $F(t) < G(t)$ for some t, and similarly for D_{mn}^- and alternatives with $F(t) > G(t)$ for some t. Note that these one-sided alternatives include stochastic dominance and the shift model of Sect. 2 of Chap. 5 as special cases.

We now derive quite simply a lower bound on the power of the one-sided test and the behavior of this bound as m, $n \to \infty$ [Massey, 1950b]. The bound and its asymptotic behavior provide useful insight and will be relevant in Chap. 8. Let $c_{mn, \alpha}$ denote the right-tailed critical value of D_{mn}^+ for a test at level α. Define $\Delta = \sup_t [G(t) - F(t)]$ and suppose, for convenience, that the maximum is actually achieved so that $\Delta = G(t_0) - F(t_0)$ for some t_0. We know that $\Delta > 0$ if $F(t) < G(t)$ for some t. Since D_{mn}^+ is certainly never less than $G_n(t_0) - F_m(t_0)$, the power of the test satisfies the inequality

$$P(D_{mn}^+ \geq c_{mn, \alpha}) \geq P[G_n(t_0) - F_m(t_0) \geq c_{mn, \alpha}]. \tag{6.1}$$

Since $mF_m(t_0)$ and $nG_n(t_0)$ are independent and binomially distributed (see (a) of Sect. 2), the right-hand side of (6.1) can be evaluated without great difficulty for any specified F and G. Furthermore, for m and n large $c_{mn, \alpha}$ in (6.1) can be approximated by means of (4.21) and the binomial distribution can be approximated by the normal distribution (see (f) of Sect. 2). The result is (Problem 33)

$$P(D_{mn}^+ \geq c_{mn, \alpha}) \geq 1 - \Phi[(c_\alpha \sqrt{N/mn} - \Delta)/\sigma] \text{ approximately,} \tag{6.2}$$

where

$$c_\alpha = \sqrt{-\log \alpha/2} \tag{6.3}$$

and

$$\sigma = \sqrt{\{F(t_0)[1 - F(t_0)]/m\} + \{G(t_0)[1 - G(t_0)]/n\}}. \tag{6.4}$$

Using the further inequality $\sigma \le \sqrt{N/4mn}$, we obtain

$$P(D_{mn}^+ \ge c_{mn,\alpha}) \ge 1 - \Phi[2(c_\alpha - \Delta\sqrt{mn/N})] \text{ approximately.} \quad (6.5)$$

If both m and n approach infinity, the right-hand side of (6.5) approaches 1 since $\sqrt{mn/N} \to \infty$, and the power, which is larger, must also approach 1. Thus (6.5) implies in particular the consistency of the test based on D_{mn}^+ against the alternative $F(t) < G(t)$ for some t, as stated earlier. It is a stronger result than consistency, however, since it also gives a lower bound on the rate at which the power approaches 1. Since the power of optimum parametric tests is typically of the same order of magnitude, as we will show in Chap. 8, this lower bound also shows that the Kolmogorov–Smirnov tests are not infinitely less efficient than other tests, as one might have feared. Rather, their power is of the same order of magnitude, whether greater or less. Rigorous statements along these lines unfortunately require analysis of the order of magnitude of the neglected terms, which is difficult; for example, the normal approximation to the binomial distribution has relative error approaching infinity in the smaller tail when the approximate normal deviate becomes infinite at the rate $\sqrt{n}$, as it does here for fixed Δ.

Equation (6.1) gives a lower bound for the power function of the one-sided Kolmogorov–Smirnov test and an analogous bound is easily found for the two-sided test, as is an upper bound for the one-sided test (Problem 34). These bounds can always be evaluated for simple alternatives which specify both F and G, but of course how sharp they are is not known in general.

Some specific studies of power have been reported in the literature. For example, Dixon [1954], Epstein [1955] and van der Waerden [1953] made some power comparisons for various tests against alternatives of normal distributions differing only in location. These results show that the Kolmogorov–Smirnov tests are less powerful than the Wilcoxon, van der Waerden, Terry or t tests, but more powerful than the median test, for normal alternatives. While such results are indeed of interest, they cannot really provide any general conclusions for other alternatives. The calculation of power functions under more general, nonparametric alternatives would be valuable. Lehmann [1953] considered this problem, and Steck [1969] found formulas for the distribution of the one-sided statistic under Lehmann alternatives (one c.d.f. is a power of the other). Unfortunately, it appears that power is difficult to compute in general except by simulation.

Under most alternatives even the limiting distributions of these statistics are unknown so that it is not possible to compute asymptotic power either. However, bounds on the power can be calculated for large m and n using (6.5) and similar inequalities mentioned earlier. Capon [1965] used this approach to find a lower bound for the asymptotic (Pitman) efficiency of these tests, which can be applied to any specified distributions. Klotz [1967] found similar results for asymptotic efficiency as defined by Bahadur.

Pitman efficiency will be discussed extensively in Chap. 8 of this book and the Kolmogorov–Smirnov procedures are considered in Sect. 11 of that chapter.

7 One-Sample Kolmogorov–Smirnov Statistics

So far in this chapter we have been concerned only with inferences based on two independent samples using the Kolmogorov–Smirnov two-sample statistics. Analogous one-sample statistics can be used for certain kinds of inferences concerning the distribution of a single random sample.

For example, given n independent, identically distributed random variables $X_1, \ldots, X_n$ from a c.d.f. F, suppose we wish to test the null hypothesis $H_0: F = F_0$. Let $F_n(t)$ denote the empirical distribution function of this sample, and consider the following three statistics which are called the Kolmogorov–Smirnov one-sample statistics:

$$D_n = \sup_t |F_n(t) - F_0(t)| \tag{7.1}$$

$$D_n^+ = \sup_t [F_n(t) - F_0(t)] \tag{7.2}$$

$$D_n^- = \sup_t [F_0(t) - F_n(t)]. \tag{7.3}$$

Against the two-sided alternative $F(t) \neq F_0(t)$ for some t, a test rejecting the null hypothesis for large values of D_n is natural, and is consistent (Problem 35). Similarly, a test based on D_n^+ is consistent against all alternatives F with $F(t) > F_0(t)$ for some t, and one based on D_n^- is consistent against $F(t) < F_0(t)$ for some t. Curiously enough, however, each of these tests is biased (slightly) against the corresponding alternative (Massey [1952a] (Problem 36)). When F_0 is not continuous, tests using the null distribution or critical values for the continuous case are conservative (Problem 37).

Of course, the c.d.f. F_0 must be fully specified in order to calculate the value of any of these one-sample test statistics. In this sense, the tests are "parametric," and accordingly are discussed only briefly here. The statistics are however "nonparametric," or at least distribution-free, in the sense that when $F = F_0$, their distributions do not depend on F_0 as long as F_0 is continuous (Problem 38). Because of this, the problem of deriving their exact null distribution in the continuous case need be solved for only one continuous F_0, for instance, the uniform distribution on $(0, 1)$. These null distributions are continuous. Further, the one-sided statistics, D_n^+ and D_n^-, have identical distributions by symmetry. Massey [1950a] (see also Kolmogorov [1933]) derived a recursive relation for the null probability $P(D_n \leq k/n)$ for integer values of k; his method applies also to D_n^+. Birnbaum and Tingey [1951] found an expression in closed form for the entire cumulative null distribution of the one-sided statistics (Problem 50); Smirnov [1944] and Dempster [1955] showed that the asymptotic distribution can be derived from this expression (Problem 52). References for the asymptotic null distribution of $\sqrt{n}D_n$ and $\sqrt{n}D_n^+$ were given at the end of Sect. 4.

Tables of the exact c.d.f. of D_n are given in Birnbaum [1952]; these are useful to find P-values. Massey [1951b] gives critical values of D_n for $n = 1$, $2, \ldots, 20, 25, 30, 35$, $\alpha = 0.01, 0.05, 0.10, 0.15, 0.20$, and Miller [1956] gives critical values of D_n^+ for $n = 1, 2, \ldots, 100$, $\alpha = 0.005, 0.01, 0.025, 0.05, 0.10$. The latter tables may also be used for D_n at level 2α for $2\alpha \leq 0.10$ (and at level $2\alpha = 0.20$ with probably small error).

Each one-sample Kolmogorov–Smirnov statistic is the limit of the corresponding two-sample statistic as one of the sample sizes becomes infinite, in a strong sense of convergence in probability like the strong consistency discussed in Sect. 6 (Problem 39); for instance, $F_n - G_m$ converges to $F_n - G$ as $m \to \infty$. Presumably, then, the large sample properties of the one-sample procedures could be deduced from those of the two-sample procedures. In fact, however, it may be be more straightforward to deduce the asymptotic behavior of the two-sample statistics from that of the individual empirical distribution functions of each of the two samples separately.

In addition to hypothesis testing, the one-sample statistics can also be used to find confidence regions for the continuous population c.d.f. F; these regions might well be called nonparametric. Let $D_{n,\alpha}^+$ be the critical value of D_n^+ at level α. Then with confidence coefficient $1 - \alpha$, $F(x)$ everywhere lies on or above the lower confidence bound given by (Problem 40)

$$L_n(x) = \max[F_n(x) - D_{n,\alpha}^+, 0]. \tag{7.4a}$$

Similarly, an upper confidence bound is

$$U_n(x) = \min[F_n(x) + D_{n,\alpha}^+, 1]. \tag{7.4b}$$

A two-sided confidence band with confidence coefficient α is the region between $L_n(x)$ and $U_n(x)$ with $D_{n,\alpha}^+$ in (7.4a) and (7.4b) replaced by the level α critical value of D_n; of course, $D_{n,\alpha/2}^+$ could also be used, but it is ordinarily slightly larger (Problems 41 and 42). The simplest procedure in application is to graph the observed empirical distribution function F_n as a step function and plot parallel step functions at a distance $D_{n,\alpha}^+$ (or $D_{n,\alpha}$) above or (and) below F_n as appropriate but truncate them at 0 and 1. A c.d.f. would be "accepted" as a null hypothesis when tested by a Kolmogorov–Smirnov test procedure if and only if it lies entirely within the corresponding band.

This confidence property is quite special. The set of all "acceptable" c.d.f.'s is so simply describable only for tests based on statistics of the Kolmogorov–Smirnov type (including certain generalizations discussed by Anderson and Darling [1952]). No such simple description is possible for test statistics based on other measures of distance between c.d.f.'s, such as the Cramér–von Mises statistics (see Problem 57). In the framework of simultaneous inference, one might start with confidence regions for $F(x)$ based on $F_n(x)$ at each x and seek an adjustment to make them valid with overall (simultaneous) confidence $1 - \alpha$. If we start with the confidence regions in the form $|F(x) - F_n(x)| \leq c$, then the two-sided Kolmogorov–Smirnov confidence band results. If we start with $|F(x) - F_n(x)| \leq c_\alpha W[F(x)]$ for an arbitrary function W, such as $W[F(x)] = \sqrt{F(x)[1 - F(x)]}$, then a different band results, corresponding to the statistic $\sup_x |F_n(x) - F(x)|/W[F(x)]$. W determines the relative emphasis on short confidence intervals in different portions of the distribution. See Anderson and Darling [1952] for further discussion.

The D_n or D_n^+ statistic also provides procedures for determining the minimum sample size required to state with a predetermined probability $1 - \alpha$ that if $F(x)$ is estimated by $F_n(x)$, the error in the estimate will nowhere exceed a fixed value ε. For instance, we can use tables of the null distribution of D_n to find the smallest integer n such that $P(D_n < \varepsilon) \geq 1 - \alpha$ (or $P(D_n^+ < \varepsilon)$ $\geq 1 - \alpha$). This is the minimum n for which the two-sided (or one-sided) confidence bands described earlier lie within ε of the empirical distribution function.

PROBLEMS

1. Let F_m be the empirical distribution function of a random sample $X_1, \ldots, X_m$ from a population with c.d.f. F. Show that
   (a) $mF_m(t)$ is binomial with parameters m and $F(t)$.
   (b) $F_m(t)$ has mean $F(t)$ and variance $F(t)[1 - F(t)]/m$.
   (c) $F_m(t)$ is a consistent estimator of $F(t)$ for each t.
   (d) $F_m(t)$ is asymptotically normal with the mean and variance in (b).
   (e) $F_m(t_1)$ and $F_m(t_2)$ have covariance $\sigma(t_1, t_2) = \{\min[F(t_1), F(t_2)] - F(t_1)F(t_2)\}/m$.
   (f) $F_m(t_1), \ldots, F_m(t_k)$ are asymptotically multivariate normal with the means, variances, and covariances given in (b) and (e), for each $t_1, \ldots, t_k$.

*2. (a) In the situation of Problem 1, show that $F_m(t)$ converges uniformly to $F(t)$ in probability, as defined in (d) of Sect. 2, or (c) below. (A proof can be based on consistency and monotonicity alone.)
   (b) Show that the result in (a) is equivalent to $\sup_t |F_m(t) - F(t)| \to 0$ in probability.
   (c) Let "$Z_m(t) \to 0$ uniformly in t, in probability" mean that for all $\varepsilon > 0$, $P[|Z_m(t)| < \varepsilon$ for all $t] \to 1$ as $m \to \infty$. Let "$Z_m(t) \to 0$ in probability, uniformly in t" mean that for all $\varepsilon > 0$, $P[|Z_m(t)| < \varepsilon] \to 1$ uniformly in t as $m \to \infty$. Show by example that the second condition is weaker than the first.

3. Use the probability integral transformation to show that the Kolmogorov–Smirnov two-sample statistics are distribution-free for samples which are independently, continuously, and identically distributed.

4. Use the algorithm of Sect. 4.1 to show that the P-value of D_{mn} is 0.3429 for the sample arrangement $X\ Y\ Y\ X\ X\ Y\ Y$.

5. Show that the value of D_{mn} is always an integer divided by the least common multiple of m and n.

6. In the notation of Section 4.1, show that for $0 \leq u \leq m, 0 \leq v \leq n, 0 \leq k \leq m + n$,

$$A(m, n) = \sum_{u=0}^{m} A(u, k - u)A(m - u, n - k + u).$$

The range of u can be restricted to $\max(0, k - n) \leq u \leq \min(k, m)$. Using this expression for $A(m, n)$ with $k = (m + n)/2$ or $(m + n + 1)/2$ permits the recursion to be terminated at $u + v = k$. (See Hodges [1957].)

7. What change is required in the definition of $A(u, v)$ in Section 4.1 to obtain the exact null distribution of D_{mn}^+?

8. Use the result of Problem 7 to find the P-value of D_{mn}^+ for the sample arrangement $X\ Y\ Y\ X\ X\ Y\ Y$ (the same arrangement as Problem 4).

9. Show the symmetry relations (4.5) for the null distributions of the one-sided Kolmogorov–Smirnov statistics.

10. Assume that all possible arrangements of two samples have positive probability. Show that
   (a) $P(D_{mn} \geq c) = P(D_{mn}^+ \geq c) + P(D_{mn}^- \geq c) - P[\min(D_{mn}^+, D_{mn}^-) \geq c]$.
   (b) D_{mn}^+ and D_{mn}^- cannot both exceed 0.5.
   (c) If m or n is even, samples exist with $D_{mn}^+ = D_{mn}^- = 0.5$.
   (d) If m and n are both odd, the largest c for which $\min(D_{mn}^+, D_{mn}^-) \geq c$ is possible and hence the largest c for which $P(D_{mn} \geq c) < P(D_{mn}^+ \geq c) + P(D_{mn}^- \geq c)$ is $c' = 0.5 - [0.5/\max(m, n)]$.
   (e) For $m = 5, n = 7$, 16/35 is a possible value of D_{mn}^+ but 17/35 is not. Since $c' = 6/14 = 15/35$ here, this illustrates that values of D_{mn}^+ between c' and 0.5 of the form k/M, where k is an integer and M is the least common multiple of m and n, are sometimes possible and sometimes impossible.
   (f) If c is the critical value of D_{mn}^+ at exact level α and if $c > 0.5$ or m and n are both odd and $c > 0.5 - [0.5/\max(m, n)]$, then c is the critical value of D_{mn} at exact level 2α. This is not always true if the word exact is omitted.

11. Let P_1 be the larger of the two one-tailed Kolmogorov–Smirnov P-values and P_2 be the two-tailed P-value. Show that $0 \leq 2P_1 - P_2 \leq 2P_1^4$
   (a) Asymptotically, that is, when the right-hand sides of (4.21) and (4.22) are used for P_1 and P_2 respectively.
   *(b) For $m = n$. (Hint: Show that $2P_1 - P_2 \leq 2\binom{2n}{n-2k}/\binom{2n}{n} \leq 2P_1^4 = 2[\binom{2n}{n-k}/\binom{2n}{n}]^4$, where P_1 and P_2 are given by (4.7) and (4.8), by showing that

$$\left[1 - \frac{2k}{m + k}\right]\left[1 - \frac{2k}{m}\right] \leq \left[1 - \frac{k}{m}\right]^4$$

for $1 \leq k \leq m$. See (4.14).)

12. Define "k is between c and d" as meaning $c \leq k \leq d$ or $d \leq k \leq c$. In the notation of Section 4.2, show that
   (a) $N(j, k, l) = N(-j, -k, -l)$.
   (b) $N(j, k, l) = N(j, l) - N(k, j, l)$ if k is between j and $2l$.
   (c) $N(\text{not } j, k, l) = N(k, j, l)$ if j is between k and $2l$.
   (d) $N(j, k, l) = N(j, k - l) - N(k, j, k - l)$ if k is between j and $2k - 2l$.
   (e) $N(\text{not } j, k, l) = N(k, l) - N(\text{not } k, j, k - l)$ if k is between j and $2k - 2l$.
   (f) $N(\text{not } j, k, l) = N(\text{not } j, k, k - l)$.
   (g) $N(\text{not } j, k, l) = N(j, k - l) - N(\text{not } k, j, k - l)$ if j is between k and $2k - 2l$.
   (h) k is between j and $2k - 2l$ if and only if k is not between j and $2l$.

13. Derive the formula for the number of paths which satisfy $D_{nn} \geq k/n$ using
    (a) Part (c) of Problem 12.
    (b) Part (d) of Problem 12.

14. At (4.12), if the portion of the path to the left of the point where the path reaches height k is reflected, instead of the portion to the right, what result is obtained and how does it relate to (4.12)?

15. Let $N(i, j, k, l)$ and $N(i, \text{not } j, k, l)$ be the numbers of paths starting at height i, reaching height k after, and without, respectively, having first reached height j, and terminating at height l, where the steps are diagonal as in Sect. 4.3. Give properties of N like those given in Sect. 4.3 and Problem 12. Assume that the sum $(i + l + \text{the number of steps})$ is even. (What happens if this sum is odd?)

16. The exact null probability distribution of any of the Kolmogorov–Smirnov two-sample statistics can always be determined by enumerating all possible arrangements of the X and Y observations and computing the value of the statistic for each. Enumerate the $\binom{6}{3}$ arrangements for $m = n = 3$. Determine the exact null distributions of D_{mn} and D_{mn}^+, both from the enumeration and from (4.7) and (4.8).

17. Show that the Mann–Whitney statistics can be expressed in terms of the empirical distribution functions as follows:

$$U = n \sum_{i=1}^{m} G_n(X_i), \qquad U' = m \sum_{j=1}^{n} F_m(Y_j).$$

18. Define the indicator variables $I_k = 1$ if the kth smallest observation in the combined ordered sample of m X's and n Y's is an X, and $I_k = 0$ otherwise, $k = 1, 2, \ldots, m + n = N$. Show that the two-sample Kolmogorov–Smirnov statistics can be expressed as

$$D_{mn}^+ = \frac{N}{mn} \max_{1 \leq j \leq N} \left[\frac{jm}{N} - \sum_{k=1}^{j} I_k \right]$$

$$D_{mn} = \frac{N}{mn} \max_{1 \leq j \leq N} \left| \frac{jm}{N} - \sum_{k=1}^{j} I_k \right|$$

and if $m = n$,

$$D_{nn}^+ = \frac{1}{n} \max_{1 \leq j \leq 2n} \left[j - 2 \sum_{k=1}^{j} I_k \right]$$

$$D_{nn} = \frac{1}{n} \max_{1 \leq j \leq 2n} \left| j - 2 \sum_{k=1}^{j} I_k \right|.$$

19. Show that the one-sided, two-sample Kolmogorov–Smirnov statistics can be expressed [e.g., Steck, 1969] in terms of the X ranks r_i and the Y ranks r'_j, where r_i is the rank of the ith smallest X in the combined sample and r'_j is that of the jth smallest Y, as

$$mnD_{mn}^+ = \max_{1 \leq i \leq m} (mr_i - Ni + n) = \max_{1 \leq j \leq n} (Nj - nr'_j)$$

$$mnD_{mn}^- = \max_{1 \leq i \leq m} (Ni - mr_i) = \max_{1 \leq j \leq n} (nr'_j - Nj + m).$$

20. (a) Express the one-sided, two-sample Kolmogorov–Smirnov statistics in terms of $a_1, \ldots, a_{m+1}$ where a_j is the number of Y's between $X_{(j-1)}$ and $X_{(j)}$ for $j = 2, \ldots, m$ and a_1 and a_{m+1} are the number of Y's below $X_{(1)}$ and above $X_{(m)}$ respectively.
    (b) What is the distribution of $(a_1, \ldots, a_{m+1})$ under the null hypothesis?

*21. Show that for $m = n$ the mean of the two-sample, one-sided Kolmogorov–Smirnov statistic D_{mn}^+ under the null hypothesis is

$$E(D_{nn}^+) = \left[2^{2n} \bigg/ \binom{2n}{n} - 1 \right] \bigg/ 2n.$$

*22. Show by elementary considerations that $P(D_{nn} \geq 1/n) = 1$ always, and show that (4.8) gives the same answer under the null hypothesis. (Hint: Manipulations are like Problem 21.)

*23. Show that the null distribution of D_{mn}^+ can be obtained by the following recursive procedure. Let v_i be the smallest integer greater than or equal to $(ni/m) + nc$ for $0 \leq i \leq I$ where $I = [m(1 - c)]$ is the greatest integer in $m(1 - c)$, and $N_i = i + v_i$. Let $M_0 = 1$ and

$$M_i = \binom{N_i}{i} - \sum_{j=0}^{i-1} M_j \binom{N_i - N_j}{i - j}, \qquad i = 1, \ldots, I.$$

Then

$$P(D_{mn}^+ \geq c) = \sum_{i=0}^{I} M_i \binom{N - N_i}{m - i} \bigg/ \binom{N}{m}.$$

(Hint: In the representation of Sect. 4.1, the N_ith step is the first to reach the boundary and (i, v_i) is the point reached. Altogether there are

$$\binom{N_i}{i}$$

paths to this point, of which M_i reach the boundary first at the last (N_ith) step and

$$M_j \binom{N_i - N_j}{i - j}$$

at the N_ith step. There are $\binom{N}{m}$ paths to (m, n), of which

$$M_i \binom{N - N_i}{m - i}$$

reach the boundary first at the N_ith step. For calculation, choosing $m \leq n$ minimizes the number of terms, and replacing N_i by $N_i - 1$ (without changing N_j) in the recursive formula for M_i reduces their size somewhat and can be justified by observing that M_i is the number of paths to $(i, v_i - 1)$ not reaching the boundary.) See Hodges [1957], Korolyuk [1955], and Steck [1969].

24. Show that the one-sided, two-sample Kolmogorov–Smirnov statistics can be expressed as

$$D_{mn}^+ = \max_{i,j} \left\{ \frac{j}{n} - \frac{i}{m} : X_{(i+1)} > Y_{(j)} \right\}, \qquad D_{mn}^- = \max_{i,j} \left\{ \frac{i}{m} - \frac{j}{n} : Y_{(j+1)} > X_{(i)} \right\}.$$

25. Show that the confidence bounds for a shift parameter corresponding to the two-sample Kolmogorov–Smirnov tests with critical value c (rejecting for $D_{mn}^+ \geq c$, $D_{mn}^- \geq c$, or $D_{mn} \geq c$) are as follows. The upper bound is

$$\min_{i,j} \left\{ Y_{(j)} - X_{(i)} : \frac{j}{n} - \frac{i-1}{m} \geq c \right\} = \min_j \{ Y_{(j)} - X_{(i_j)} \}$$
$$= \min_i \{ Y_{(j_i)} - X_{(i)} \}$$

where i_j is the smallest integer exceeding $(mj/n) - mc$ and j_i is the smallest integer not less than $[n(i-1)/m] + nc$. The lower bound is

$$\max_{i,j} \left\{ Y_{(j)} - X_{(i)} : \frac{i}{m} - \frac{j-1}{n} \geq c \right\} = \max_j \{ Y_{(j)} - X_{(k_j)} \}$$
$$= \max_i \{ Y_{(l_i)} - X_{(i)} \}$$

where k_j is the smallest integer not less than $[m(j-1)/n] + mc$ and l_i is the smallest integer exceeding $(ni/m) - nc$.

26. Compare the confidence bounds for a shift parameter corresponding to the two-sample Kolmogorov–Smirnov tests (Problem 25) with critical value $c \geq 1 - [1/\min(m, n)]$ to those corresponding to the rank sum test.

27. Show that the largest possible values of the one-sided, two-sample Kolmogorov–Smirnov statistic D_{mn}^+ with $m \leq n$ and the associated tail probabilities under the null hypothesis can be expressed as follows where k is the largest integer in n/m:
    (a) $P(D_{mn}^+ = 1) = 1/\binom{N}{m}$ for all $m \leq n$.
    (b) $P(D_{mn}^+ \geq 1 - 1/n) = (m + 1)/\binom{N}{m}$ for $m < n$.
    (c) $P(D_{mn}^+ \geq 1 - 1/m) = N/\binom{N}{m}$ for $m \leq n < 2m$.
    (d) $P(D_{mn}^+ \geq 1 - i/n) = \binom{m+i}{m}/\binom{N}{m}$ for $0 < i < n/m$.
    (e) $P(D_{mn}^+ \geq 1 - 1/m) = [\binom{m+k}{m} + n - k]/\binom{N}{m}$.
    (f) $P(D_{mn}^+ \geq 1 - i/n) = [\binom{m+i}{m} + \binom{m+i-k-1}{m-1}(n - i)]/\binom{N}{m}$ for $n/m < i < 2n/m$.
    (g) $P(D_{mn}^+ \geq 1 - 1/m - i/n) = [\binom{m+k+i}{m} + \binom{m+i-1}{m-1}(n - k - i)]/\binom{N}{m}$ for $0 < i < n/m$.

28. Show that the asymptotic null distribution of $(2mn/N)(D_{mn}^+)^2$ is exponential and that of $(4mn/N)(D_{mn}^+)^2$ is chi-square with 2 degrees of freedom.

*29. Show that $P(D_{mn} \geq c)$ and $P(D_{mn}^+ \geq c)$ are strictly smaller for discontinuous than for continuous F for all c, $0 < c \leq 1$, when all observations are independent with c.d.f. F. (Hint: Consider the possibility that all observations are tied.)

30. Show that
    (a) The Kolmogorov–Smirnov statistics are discontinuous functions of the observations.
    (b) The Kolmogorov–Smirnov statistics are lower-semicontinuous, where $h(x)$ is defined as lower-semicontinuous at x_0 if, for every $\varepsilon > 0$, there is a neighborhood of x_0 where $f(x) \geq f(x_0) - \varepsilon$.
    *(c) The level of any test for discontinuously distributed observations is no greater than its level for continuously distributed observations if the acceptance region of the test is of the form $T \leq c$ where T is a lower-semicontinuous function of the observations.

31. In the situation and terminology of Sect. 6, show that $G_n - F_m$ is a strongly consistent estimator of $G - F$ and use this to show that the three two-sample Kolmogorov–Smirnov statistics are consistent estimators of the corresponding population quantities.

32. Show that the one-sided and two-sided Kolmogorov–Smirnov two-sample tests are consistent against the alternatives stated in Sect. 6.

33. Derive the asymptotic lower bound (6.2) on the power of the one-sided two-sample Kolmogorov–Smirnov test.

34. In the situation of Sect. 6, show that
    (a) The power of the one-sided Kolmogorov–Smirnov test is at most the null probability that $D_{mn}^+ \geq c_{mn,\alpha} - \Delta$.
    (b) For large m and n, the probability in (a) is approximately
    $$\exp\{-2[\max(c_\alpha - \Delta\sqrt{mn/N}, 0)]^2\}.$$
    (c) The two-sided Kolmogorov–Smirnov test with critical value $c_{mn,\alpha}$ has power at least
    $$P[G_n(x_0) - F_m(x_0) \geq c_{mn,\alpha}] + P[G_n(x_0) - F_m(x_0) \leq -c_{mn,\alpha}].$$
    (d) For large m and n the quantity in (c) is approximately
    $$\Phi[(\Delta - c_\alpha\sqrt{N/mn})/\sigma] + \Phi[(-\Delta - c_\alpha\sqrt{N/mn})/\sigma]$$
    $$\geq \Phi[2(\Delta\sqrt{mn/N} - c_\alpha)] + \Phi[2(-\Delta\sqrt{mn/N} - c_\alpha)],$$
    where c_α is the value of λ for which (4.18) equals α.
    *(e) Parts (c) and (d) and (6.1)–(6.4) are valid for any x_0, with $\Delta = G(x_0) - F(x_0)$. What choice of x_0 gives the tightest lower bound in (6.3), (6.4), and part (d)?

35. Show that the Kolmogorov–Smirnov one-sample tests are consistent as stated in Sect 7.

36. Show that the Kolmogorov–Smirnov one-sample tests are biased. (Hint: Let Z be a function of X such that $X < Z < a$ for $X < a$, $X > Z > b$ for $X > b$, and $Z = X$ otherwise, where $F_0(a) = 1 - F_0(b) =$ critical value of the test statistic. An X sample rejects whenever the corresponding Z sample does, but not conversely.) [Massey, 1950b].

*37. Show that the Kolmogorov–Smirnov one-sample statistics are stochastically smaller under the null hypothesis when F_0 is discontinuous than when it is continuous, and hence that the P-values and critical values that are exact for the continuous case are conservative in the discontinuous case.

38. Show that the one-sample Kolmogorov–Smirnov statistics are distribution-free for a sample from a population with a continuous c.d.f. $F_0(x)$. (Hint: Let $U = F_0(X)$.)

39. Show that the two-sample Kolmogorov–Smirnov statistics approach the corresponding one-sample statistics as one sample size becomes infinite. Define the type of convergence you use.

40. Show that the lower, upper, and two-sided confidence bands defined by the critical values of the Kolmogorov–Smirnov one-sample statistics each have probability at least $1 - \alpha$ of completely covering the true c.d.f. sampled, whatever it may be.

41. Show that a one-sample Kolmogorov–Smirnov test would "accept" the c.d.f. F_0 if and only if F_0 lies entirely within the corresponding confidence band. Assume the same critical value is used for F_0 discontinuous as for F_0 continuous.

42. For F_0 continuous, show that
   (a) The P-value of D_n is twice the smaller of the two corresponding one-sided P-values if $D_n \geq 0.5$, and less than twice if $0 < D_n < 0.5$.
   (b) The critical value $D_{n,\alpha} = D^+_{n,\alpha/2}$ if $D^+_{n,\alpha/2} \geq 0.5$. Otherwise $D_{n,\alpha} < D^+_{n,\alpha/2}$.

43. Use a symmetry argument to show that the null distribution of D^-_n is identical to that of D^+_n.

44. Show that, in the definitions (7.1)–(7.3) of the Kolmogorov–Smirnov one-sample statistics, the supremum is always achieved for D^+_n but may not be for D^-_n or D_n.

45. Show that for F_0 continuous the Kolmogorov–Smirnov one-sample statistics defined in (7.1)–(7.3) can be written as

$$D^+_n = \max_{1 \leq i \leq n} \, [(i/n) - F_0(X_{(i)})]$$

$$D^-_n = \max_{1 \leq i \leq n} \, [F_0(X_{(i)}) - (i - 1)/n]$$

$$D_n = \max_{1 \leq i \leq n} \, \{\max[(i/n) - F_0(X_{(i)}), F_0(X_{(i)}) - (i - 1)/n]\}.$$

46. Show that under the null hypothesis $F = F_0$ for F_0 continuous, the null distribution of the one-sample Kolmogorov–Smirnov statistics can be expressed as follows, where $U_1 < U_2 < \cdots < U_n$ are the order statistics of a sample of size n from the uniform distribution on $(0, 1)$.
   (a) $P(D^+_n \leq c) = P[U_i \geq (i/n) - c \text{ for } i = 1, \ldots, n]$.
   (b) $P(D_n \leq c) = P\{(i/n) - c \leq U_i \leq [(i - 1)/n] + c \text{ for } i = 1, \ldots, n\}$.
   (c) $P(D^+_n \leq c) = n! \displaystyle\int_{a_n} \int_{a_{n-1}} \cdots \int_{a_2} \int_{a_1}^{u_n} \, du_1 \cdots du_n$ where $a_i = (i/n) - c$ for $i > nc$,

   $a_i = 0$ otherwise, and $0 \leq c \leq 1$.

   (d) $P(D_n \leq c) = \displaystyle\int_{1-c}^{1-(1/n)+c} \int_{1-(1/n)-c}^{1-(2/n)+c} \cdots \int_{(2/n)-c}^{(1/n)+c} \int_{(1/n)-c}^{c} f(u_1, \ldots, u_n) du_1 \cdots du_n$

   where $f(u_1, \ldots, u_n) = n!$ for $0 < u_1 < \cdots < u_n < 1$ and 0 otherwise, and $1/2n \leq c \leq 1$.

   (e) $P(D_n \leq c) = n! \displaystyle\int_{a_n}^{1} \int_{a_{n-1}}^{b_{n-1}} \cdots \int_{a_1}^{b_1} \, du_1 \cdots du_n$ where a_i is defined in (c), $b_i =$

   $\min\{[(i - 1)/n] + c, u_{i+1}\}$, and $1/2n \leq c \leq 1$.

47. Illustrate the use of the integral in Problem 46(c) by evaluating $P(D^+_n \leq c)$ for all c when $n = 2$. Use this result to show that the upper-tail critical value of D^+_n is 0.776 for $n = 2$, $\alpha = 0.05$.

48. Illustrate the use of the integral in Problem 46(d) or (e) by evaluating $P(D_n \leq c)$ for all c when $n = 2$. Use this result to show that the critical value of D_n is 0.8419 for $n = 2$, $\alpha = 0.05$.

*49. Let F_m be the empirical distribution of a sample of size m from a population with continuous c.d.f. F. Let b be a constant, $b \geq 1$.

(a) Verify that $P[F_m(t) \leq bF(t)$ for all $t] = 1 - (1/b)$ for $m = 1$ and $m = 2$.

(b) Prove the result in (a) for all m. (Hint: Take F uniform on $(0, 1)$. Let Z be the sample maximum. Given Z, the remainder of the sample is distributed uniformly on $(0, Z)$. The result for $m - 1$ implies that $P[F_m(t) \leq bt$ for $0 < t < Z|Z] = 1 - [(m - 1)/mbZ]$. The remaining requirement for the event to occur is $Z \geq 1/b$.) This result is due to Daniels [1945], and a special case is given in Dempster [1959].

*50. Show that the null distribution of D_n^+ can be expressed in the following form due to Birnbaum and Tingey [1951].

$$P(D_n^+ \geq c) = (1 - c)^n + c \sum_{j=1}^{[n(1-c)]} \binom{n}{j}\left(1 - c - \frac{j}{n}\right)^{n-j}\left(c + \frac{j}{n}\right)^{j-1}.$$

(Hint: Referring to Problem 46(a), let the last i for which $U_i < (i/n) - c$ be $n - j$. Then exactly $n - j$ of the U's are smaller than $1 - c - (j/n)$, and this has probability $\binom{n}{j}[1 - c - (j/n)]^{n-j}[c + (j/n)]^j$. Furthermore, the remaining U's are conditionally uniformly distributed on the interval $[1 - c - (j/n), 1]$ and at most k of them are in $[1 - c - (j/n), 1 - c + \{(k - j)/n\}]$. By Problem 49, this has conditional probability $1 - \{(j/n)/[c + (j/n)]\} = c/[c + (j/n)]$. Multiplying these two probabilities gives the jth term of the sum, which is the probability that the $(n - j)$th order statistic is the last at which the empirical c.d.f. exceeds the upper bound. See also Chapman [1958], Dempster [1959], Dwass [1959], Pyke [1959], and Problem 52.)

51. Verify directly, from both Problem 46(c) and Problem 50, that under the null hypothesis, $P(D_n^+ > 0.447) = 0.10$ for $n = 5$.

*52. Generalize the Birnbaum–Tingey formula in Problem 50 to a formula for the null probability that $\sup_t [F_m(t) - bF(t)] \geq c$ for arbitrary constants $c > 0$ and $b > 1 - c$. See the references cited in Problem 50.

*53. Derive the asymptotic null distribution of D_n^+ from the Birnbaum–Tingey formula in Problem 50 (Dempster [1955]).

54. Under the null hypothesis show that

(a) D_n^+ is uniformly distributed on $(0, 1)$ for $n = 1$.

(b) The density of D_n^+ for $n = 2$ is

$$h(x) = \begin{cases} 1 + 2x & 0 \leq x \leq \frac{1}{2} \\ 2(1 - x) & \frac{1}{2} \leq x \leq 1 \\ 0 & \text{otherwise.} \end{cases}$$

55. Let F_m be the empirical distribution function of a random sample of size m from the uniform distribution on $(0, 1)$. Define

$$X_m(t) = \sqrt{m}[F_m(t) - t]$$

$$Z_m(t) = (t + 1)X_m[t/(t + 1)]$$

for all $0 \leq t \leq 1$.

(a) Find $E[X_m(t)]$, $E[Z_m(t)]$, $\text{var}[X_m(t)]$, $\text{var}[Z_m(t)]$, $\text{cov}[X_m(t), X_m(u)]$, $\text{cov}[Z_m(t), Z_m(u)]$, for all $0 \leq t \leq u \leq 1$ and all m.

(b) What regions for Z_m correspond to the regions $D_m^+ < c$ and $D_m < c$ for the Kolmogorov–Smirnov one-sample statistics?

56. (a) Under the null hypothesis that a sample comes from a normal population, consider the Kolmogorov–Smirnov one-sample statistics with parameters estimated by the sample mean $\overline{X}$ and standard deviation s, namely, $\hat{D}_n^+ = \sup_t \{F_n(t) - \Phi[(t - \overline{X})/s]\}$ and $\hat{D}_n = \sup_t |F_n(t) - \Phi[(t - \overline{X})/s]|$, where Φ is the standard normal c.d.f. Show that their null distributions do not depend on the mean and variance of the normal population.

(b) Give an analogous result for the null hypothesis of an exponential population.

(c) Does an analogous result hold for all parametric null hypotheses?

57. Measures of the distance between F_m and G_n other than their maximum difference can also be used as distribution-free tests of the null hypothesis $F = G$ that two continuous distributions are identical. The one called the Cramér–von Mises statistic is defined as

$$\omega_{mn}^2 = \frac{mn}{N} \int_{-\infty}^{\infty} [F_m(t) - G_n(t)]^2 \, d\left[\frac{mF_m(t) + nG_n(t)}{N}\right].$$

(a) Prove that ω_{mn}^2 is distribution-free.

(b) What is the appropriate critical region for the alternative $F \neq G$.

(c) Show that ω_{mn}^2 can also be expressed as $\sum_{j=1}^{N} [(jm/N) - \sum_{k=1}^{j} I_k]^2/mn$ where $I_k = 1$ if the kth smallest observation in the combined ordered sample is an X, and $I_k = 0$ otherwise.

(d) Express ω_{mn}^2 in terms of the ranks of the X's and Y's in the combined sample. See Cramér [1928], von Mises [1931], Anderson and Darling [1952], and Darling [1957] for further properties of ω_{mn}^2 and other related measures of the distance between F_m and G_n.

CHAPTER 8

Asymptotic Relative Efficiency

1 Introduction

In any given inference situation, many statistical procedures may be available, both parametric and nonparametric. Some measure of their relative merits is needed, especially as regards their performance or operating characteristics. For instance, a comparison of the power functions of various tests of the same (or essentially the same) hypotheses would be of interest. It is frequently more convenient, and also more suggestive, to use a measure of relative merit called the relative efficiency.

The *relative efficiency* of a procedure Ω_1 with respect to a procedure Ω_2 is defined as the ratio of sample sizes needed to achieve the same performance, namely n_2/n_1, where n_2 is the number of observations required to achieve the same performance using Ω_2 as can be achieved using Ω_1 with n_1 observations. Thus, in particular, the relative efficiency of Ω_1 with respect to Ω_2 is less than or greater than 1 according as Ω_2 requires fewer or more observations than Ω_1 to achieve the same performance.

The use of this definition poses certain problems. For one thing, the comparison procedure Ω_2, at least, must be defined for each possible sample size n_2. Thus it should really be regarded as a sequence of procedures, one for each sample size. The fact that n_2 is not a continuous variable poses another slight problem, because typically there will be no integer n_2 for which the performance of Ω_2, in some specified respect, exactly matches the performance of Ω_1 with n_1 observations.

The main problem, however, is that there are many ways to specify what it means to "achieve the same performance." Each specification will produce its own value of n_2 and hence of the relative efficiency. This multiplicity of values can confuse comparisons.

345

In testing, for example, one could ask that Ω_2 achieve the same power as Ω_1 for any specific alternative, or the same average power for some weighted average over a set of alternatives. The relative efficiency, sometimes called the *power efficiency* in this case, is thus a function of all those variables which determine power, including n_1 and α as well as the alternative distribution. Usually, however, it varies much less than the power as a function of these variables. Consequently, the relative performance of two tests can usually be described much more concisely in terms of relative efficiency than in terms of power functions directly. In some cases, conveniently, the relative efficiency is approximately constant, so that the entire comparison reduces to a single number.

Sometimes the relative efficiency has a lower bound close to one everywhere in the range of importance. If, for example, the relative efficiency of a nonparametric test with respect to a parametric test is never much below 1, the nonparametric test is wasting at most some small fraction of the observations. One might then feel that the advantages, like simplicity and broader applicability, of the nonparametric test outweigh this small waste. Such a conclusion should not be reached casually, however, without serious assessment of the real value of such advantages and of increasing power.

For point estimation, relative efficiency is usually defined as the ratio of sample sizes needed to achieve the same variances or mean squared errors of the estimators. Other functions of the error (other "loss" functions) could be used instead of the square for matching. Matching one function does not ordinarily match others or the error distributions exactly. The relative efficiency of two estimators also depends on the assumed distribution and n_1, just as for tests it depends on n_1 and the alternative where the power is matched.

For confidence intervals, relative efficiency might be defined by matching expected lengths. This would not entirely match the distributions of the endpoints, of course. The probability of covering some specified false value could be matched instead, and the result would then be a function of the false value used. In any case, the relative efficiency of two confidence procedures will depend on the confidence level, as well as the true distribution and n_1.

Since relative efficiency generally depends on so many factors, its implications may be difficult to assess and interpret. This problem often disappears conveniently when limits are taken. The *asymptotic relative efficiency* of a procedure Ω_1 with respect to a procedure Ω_2 is defined roughly as the limit of the relative efficiency as $n_1 \to \infty$. Here, of course, both Ω_1 and Ω_2 must be sequences of procedures, defined for arbitrarily large sample sizes n_1 and n_2. It would seem that the limit required in this definition might well fail to exist, and that when it does exist it might depend on essentially the same variables as the relative efficiencies for finite sample sizes. We shall see, however, that things become much simpler as the sample sizes approach infinity, and a single number will describe a great many features of the asymptotic relative behavior of two procedures. Because of this very great convenience, asymptotic relative efficiency is widely used for comparisons of procedures, even though it is only a large-sample property.

This chapter is devoted to a study of asymptotic relative efficiency. We first investigate heuristically what typically happens as the sample sizes approach infinity in the case of tests, point estimates and confidence bounds (Sects. 2–4 respectively, followed by an example in Sect. 5). We will then be in a position to list the specific properties of asymptotic relative efficiency (Sect. 6) and explain how it is calculated (Sect. 7). The theory will then be illustrated (Sect. 8) by applying the results to the one-sample procedures which were discussed in Chaps. 2–4. Specifically, we will derive numerical values of the asymptotic relative efficiency for various pairs of procedures for some selected families of distributions. Thereafter we will discuss matched pairs (Sect. 9), the two-sample shift problem (Sect. 10), and finally the Kolmogorov–Smirnov procedures (Sect. 11), which behave differently from the others and pose different technical problems.

2 Asymptotic Behavior of Tests: Heuristic Discussion

The object of this section is to explore, without worrying about rigor, how typical tests behave in large samples. We consider the power of an individual test as well as the relative performance of two tests. The development will suggest various natural definitions of asymptotic efficiency of one test relative to another, and will show that we can expect a certain formula to apply to these definitions, and also to the corresponding point estimators and confidence bounds. The calculations throughout will be approximate, but the approximate statements should become exact in the limit as the sample sizes approach infinity. The discussion will be heuristic, with no formulations or proofs of any precise statements about the limits which correspond to the approximations presented.

2.1 Asymptotic Power of a Test

In order to investigate the asymptotic relative efficiency of two tests, it is necessary first to learn something about the asymptotic behavior of each test individually. Accordingly, this subsection is concerned with approximating the power of a single test in large samples. Consider a one- or two-sided test based on a statistic T_n, which rejects for large values of T_n, small values of T_n, or both, as the case may be. (The subscript n is included to indicate the dependence on the sample size.) The power depends on the alternative distribution under discussion, and we shall need to consider more than one alternative at a time. Let us consider a family of distributions depending on a one-dimensional parameter θ, where θ_0 denotes a distribution belonging to the null hypothesis. For instance, we might be interested in the behavior of the ordinary sign test against normal alternatives. Then T_n could be defined as the number of negative observations, θ as the true mean, and θ_0 as 0, with the true standard deviation fixed. (Alternatively, θ could be defined as the mean divided by the standard deviation, since that is all that matters here.)

Suppose that the test statistic T_n is approximately normal with mean $\mu_n(\theta)$ and standard deviation $\sigma_n(\theta)$ when the true distribution is given by θ. In particular, then, under the null hypothesis, T_n is approximately normal with mean $\mu_n(\theta_0)$ and standard deviation $\sigma_n(\theta_0)$. Accordingly, the rejection region of an upper-tailed test based on T_n is approximately

$$T_n \geq \mu_n(\theta_0) + z_\alpha \sigma_n(\theta_0) \tag{2.1}$$

where z_α is the upper α point of the standard normal distribution, the number such that $1 - \Phi(z_\alpha) = \alpha$ for Φ the standard normal c.d.f. (The development for lower-tailed and two-tailed tests is similar and is left as Problem 1.) The power of this test, at any alternative θ, is the probability of rejection, which is approximately the probability of (2.1). Since the test statistic T_n is approximately normal with mean $\mu_n(\theta)$ and standard deviation $\sigma_n(\theta)$, the probability of (2.1) under θ is approximately the probability that a standard normal random variable exceeds the value

$$\frac{\mu_n(\theta_0) + z_\alpha \sigma_n(\theta_0) - \mu_n(\theta)}{\sigma_n(\theta)}. \tag{2.2}$$

In terms of the standard normal c.d.f. Φ, this probability can be written

$$\Phi\left[\frac{\mu_n(\theta) - \mu_n(\theta_0)}{\sigma_n(\theta)} - z_\alpha \frac{\sigma_n(\theta_0)}{\sigma_n(\theta)}\right]. \tag{2.3}$$

Thus the power of the test against θ is given approximately by (2.3).

Now we are ordinarily interested in alternatives against which the test is consistent, so that the power approaches one as $n \to \infty$ for θ fixed. (This presumably restricts θ to one side of θ_0 for a one-tailed test.) Then the approximation (2.3) for fixed θ says only that the power is approximately one. It is therefore appropriate to consider not a particular alternative θ, but a sequence of alternatives θ_n which approach θ_0 as $n \to \infty$. Now if μ_n is differentiable at θ_0 and σ_n is continuous, then for θ_n near θ_0 we have the approximations

$$\mu_n(\theta_n) - \mu_n(\theta_0) \doteq (\theta_n - \theta_0)\mu_n'(\theta_0), \qquad \sigma_n(\theta_n) \doteq \sigma_n(\theta_0) \tag{2.4}$$

where $\mu_n'(\theta_0) = d\mu_n(\theta)/d\theta|_{\theta=\theta_0}$. Substituting (2.4) into (2.3), we find that the power of the test against θ_n is approximately equal to

$$\Phi\left[(\theta_n - \theta_0)\frac{\mu_n'(\theta_0)}{\sigma_n(\theta_0)} - z_\alpha\right]. \tag{2.5}$$

In particular, the rate at which $\theta_n \to \theta_0$ can be adjusted so that the argument of Φ is finite. Furthermore, to a first approximation, the power function of the test is completely determined by the quantity $\mu_n'(\theta_0)/\sigma_n(\theta_0)$. Notice that the variance $\sigma_n^2(\theta)$ is needed only under the null hypothesis $\theta = \theta_0$.

As an example, consider the power function of the ordinary sign test against normal alternatives. Let T_n be the number of negative observations. If the true distribution of the population is normal with mean μ and variance σ^2, then the probability of a negative observation is

$$p = \Phi(-\theta/\sigma). \qquad (2.6)$$

At $\theta = \theta_0 = 0$, the null hypothesis $p = \frac{1}{2}$ is satisfied. T_n is binomial with parameters p and n, and hence is approximately normal with mean and variance given by

$$\mu_n(\theta) = np = n\Phi(-\theta/\sigma) \qquad (2.7)$$
$$\sigma_n^2(\theta) = np(1-p) = n\Phi(-\theta/\sigma)[1 - \Phi(-\theta/\sigma)]. \qquad (2.8)$$

Letting ϕ denote the standard normal density, we calculate

$$\frac{\mu_n'(\theta_0)}{\sigma_n(\theta_0)} = \frac{n(-1/\sigma)\phi(0)}{\sqrt{n/4}} = -2\sqrt{n}\phi(0)/\sigma = -\sqrt{2n/\pi}/\sigma. \qquad (2.9)$$

Substituting (2.9) in (2.5) then gives

$$\Phi[-\sqrt{2n/\pi}(\theta_n - \theta_0)/\sigma - z_\alpha]$$

as an approximation to the power function of the sign test against normal alternatives.

In this example, $\mu_n'(\theta_0)/\sigma_n(\theta_0)$ is of order $\sqrt{n}$. This is typical, as we shall see from other examples. For purposes of relative efficiency, it would be convenient to have a quantity of order n. Accordingly, we introduce the square of the foregoing ratio, namely

$$e_n = [\mu_n'(\theta_0)]^2/\sigma_n^2(\theta_0). \qquad (2.10)$$

The quantity e_n is, in general, called the *efficacy* of the test statistic T_n for the family of distributions in question (at θ_0). As defined here, it depends on the choice of the approximations μ_n and σ_n, which are not unique, and on the choice among equivalent test statistics. As we shall see, however, these choices have only a second-order effect, and the efficacy of a test is uniquely defined to a first order of approximation.

In taking the square, the sign of $\mu_n'(\theta_0)/\sigma_n(\theta_0)$ is lost, but this is unimportant for present purposes. The sign is always the same as the sign of $\mu_n'(\theta_0)$ and merely indicates whether large values of T_n are associated with large or small values of θ. For example, the negative sign in (2.9) corresponds to the fact that for normal distributions the number of negative observations tends to be large when the mean θ is small, which implies that a one-tailed test rejecting when there are too many negative observations is appropriate against the alternative $\theta < 0$, not the alternative $\theta > 0$. Provided the appropriate one-tailed test is used for a one-sided alternative, the sign of $\mu'_n(\theta_0)$ is of no consequence.

The efficacy e_n therefore contains exactly the information we need here, and conveniently is typically of order n. In terms of e_n, by (2.5), the power of the appropriate one-tailed test against the alternative $\theta > \theta_0$, can be expressed approximately as

$$\Phi[(\theta - \theta_0)\sqrt{e_n} - z_\alpha]. \tag{2.11}$$

Similar expressions apply to the alternatives $\theta < \theta_0$ and $\theta \neq \theta_0$.

This indicates that, in large samples, the power functions of all typical tests are approximately the same except for a scale factor $\sqrt{e_n}$ which depends on the test statistic and the family of alternative distributions. Since $e_n \to \infty$, a graph of the approximate power (2.11) as a function of θ without attention to scale would show for large n only that the power is approximately one for the entire alternative hypothesis $\theta > \theta_0$. This would tell us nothing, except that the test is consistent. However, if we rescale θ according to sample size, using $\delta_n = (\theta - \theta_0)\sqrt{e_n}$ for sample size n, and plot the power as a function of δ_n, then we get approximately $\Phi(\delta_n - z_\alpha)$ and we can see what the power function really looks like as $n \to \infty$. This function appears in Fig. 2.1 for $z_\alpha = 1.64$, as the curve labeled $k = 1$. It describes the large-sample power, suitably rescaled, for all typical one-tailed tests with $\alpha = 0.05$.

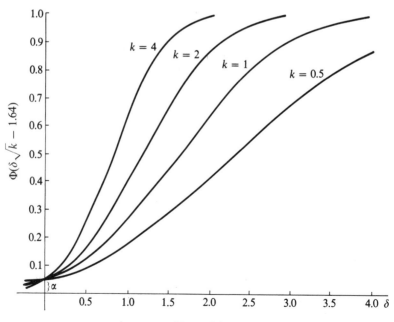

Figure 2.1

We are, of course, not recommending formulas (2.5) and (2.11) as good approximations for actual calculation of the power. Using the normal approximation leads naturally to formula (2.3), but it is seldom necessary to make the additional approximation in (2.4) which leads to (2.5) and (2.11). Equations (2.5) and (2.11) are introduced here primarily to show how the power function behaves as $n \to \infty$. This facilitates an intuitive understanding of the situation and will also be useful later.

We may also mention that, if the exact mean and variance are used in the definition (2.10), the efficacy is less than or equal to the Fisher information, and that this inequality is a form of the famous Cramér–Rao inequality (Problem 2). The asymptotic efficiency of one-parameter maximum likelihood estimators follows heuristically (Problem 3).

2.2 Nuisance Parameters

The problem of nuisance parameters will be introduced by means of an example. Consider a situation where the test statistic T_n is the sample mean, and the population is normal with known variance σ^2. If the null hypothesis is that the mean θ has a specified value θ_0, then the approximations of the previous subsection are exact with

$$\mu_n(\theta) = \theta, \qquad \sigma_n^2(\theta) = \sigma^2/n, \qquad e_n = n_i'\sigma^2. \tag{2.12}$$

The normal approximation is exact because T_n is exactly normal, and (2.4), (2.5) and (2.11) are exact because in addition $\mu_n(\theta)$ is exactly linear and $\sigma_n(\theta)$ is exactly constant.

If the assumption of normality under the null hypothesis is eliminated, the only change is that the level of the test becomes approximate. That is, for the null hypothesis that the population has mean θ_0 and known variance σ^2, the same test has level approximately α and its power against normal alternatives with variance σ^2 is the same as before.

Now suppose we drop the assumption that σ^2 is known under the null hypothesis. In this case, the sample mean itself cannot serve as a test statistic, even for an approximate test, because its distribution under the null hypothesis depends on σ^2, and this dependence does not disappear as $n \to \infty$. Thus σ^2 may be called an unknown "nuisance parameter."

In this situation, the normal-theory test statistic is the t statistic, which is approximately normal with mean $\mu_n(\theta)$ and variance $\sigma_n^2(\theta_0)$, where

$$\mu_n(\theta) = \theta/\sigma, \qquad \sigma_n^2(\theta_0) = 1/n \tag{2.13}$$

(Problem 4). Again we need $\sigma_n^2(\theta)$ only at θ_0. These are not the exact mean and variance of the t statistic, but the distribution of the t statistic can be approximated by a normal distribution with this mean and variance. (This is true whether or not the population is normal.) As a result, the efficacy is the same as before.

The following approach to the problem under discussion is easier to generalize, and also is perhaps more natural when normality is not assumed. Suppose that the critical value of a test based on the sample mean depends on an unknown nuisance parameter σ, and we use an estimate of σ as if it were the true value. As long as this estimator is consistent, the test has approximately the same rejection region as the test with σ known, and hence its size and power are approximately the same. Therefore, the efficacy calculated before still applies for purposes of approximate power.

The t test can be regarded as arising in this way but having its critical value adjusted to make the level exact under normality. Specifically, an upper-tailed test based on the sample mean T_n would reject for

$$T_n \geq \theta_0 + z_\alpha \sigma/\sqrt{n}. \tag{2.14}$$

If the sample standard deviation S is substituted for σ when σ is unknown, the resulting test rejects for

$$T_n \geq \theta_0 + z_\alpha S/\sqrt{n}. \tag{2.15}$$

The previous paragraph states that the test (2.15) is approximately the same as the test (2.14) which requires the true value of σ, and this holds whatever that value may be. The t test is of the form (2.15) except that the constant z_α is adjusted to make the test exact under normality. The amount of the adjustment approaches zero as $n \to \infty$, however.

This illustrates the typical situation where a statistic T_n is selected for a test but the null distribution of T_n depends on one or more nuisance parameters. Then we cannot base the test on T_n in the sense of comparing T_n with a fixed critical value. We can in another sense, however, as follows. Compute the critical value of T_n as a function of the nuisance parameters, substitute consistent estimates of the nuisance parameters, and compare T_n with the resulting critical value. Then the test will be approximately the same as a test which is based on T_n in the ordinary sense but requires knowledge of the values of the nuisance parameters. In particular, the efficacy computed for T_n using (2.10) will relate in the usual way to the power of the test. For instance, Equation (2.11) will apply (Problem 6a). This will still be true if the test is adjusted further to make its level exactly α under some null hypothesis (Problem 6b). In short, the efficacy of T_n will typcially apply to the power of all tests which are based essentially on T_n.

There is an alternative argument that can often be used to justify the approximations of the previous paragraph. For example, since the t test at level $\alpha = 0.50$ can be based on the sample mean, the efficacy computed from the sample mean applies to the t test at level $\alpha = 0.50$. However, since the efficacy does not depend on the value of α, the efficacy of the sample mean must apply to the t test at all levels. This argument is further explained and used in connection with point estimation in Sect. 3.2.

2.3 Asymptotic Relative Efficiency of Two Tests

Now consider two tests at the same level, which are based on statistics $T_{1,n}$ and $T_{2,n}$ with efficacies $e_{1,n}$ and $e_{2,n}$ respectively. The power of each test is given approximately by (2.11). For a given sample size, the two power functions are then approximately the same shape but they differ by a scale factor $\sqrt{e_{2,n}/e_{1,n}}$. That is, the power of the first test at $\theta_0 + \delta$ is approximately the same as the power of the second test at $\theta_0 + \delta\sqrt{e_{2,n}/e_{1,n}}$. Figure 2.1 illustrates this situation with graphs of normal power functions for one-sided tests at level $\alpha = 0.05$, that is, $\Phi(\delta\sqrt{k} - 1.64)$ as a function of δ, for several values of k, where k represents $e_{2,n}/e_{1,n}$.

We have seen that typically $e_{1,n}$ is of order n. More specifically, as $n \to \infty$, the ratio $e_{1,n}/n$ typically approaches[1] some positive constant $e_{1\cdot}$,

$$\lim e_{1,n}/n = e_{1\cdot}. \tag{2.16}$$

We call $e_{1\cdot}$ the *limiting efficacy per observation* or, more briefly, the *asymptotic efficacy* of the first test statistic. If the limiting efficacies per observation exist for both test statistics, then the ratio $e_{1,n}/e_{2,n}$ approaches $E_{1:2}$, where

$$E_{1:2} = \frac{e_{1\cdot}}{e_{2\cdot}} = \frac{\lim e_{1,n}/n}{\lim e_{2,n}/n}$$

$$= \lim \frac{e_{1,n}}{e_{2,n}} = \lim \frac{[\mu'_{1,n}(\theta_0)]^2/\sigma^2_{1,n}(\theta_0)}{[\mu'_{2,n}(\theta_0)]^2/\sigma^2_{2,n}(\theta_0)}. \tag{2.17}$$

Thus, as a large-sample approximation, we may say that, in samples of the same size, tests at the same level based on $T_{1,n}$ and $T_{2,n}$ will have power functions differing by a scale factor $1/\sqrt{E_{1:2}}$. Formula (2.17) is frequently called Pitman's formula, and will be repeated later in Sect. 7.

This scale-factor relationship is an important property of the quantity $E_{1:2}$. In all situations which we discuss, this property occurs together with an even more fundamental property, which is essentially the customary definition of asymptotic relative efficiency and which we now develop.

Let us consider the same two tests and ask when the first test with sample size n_1 will have the same power as the second test with sample size n_2. The power of each test is given approximately by (2.11). From this expression it is clear that the two tests will have approximately the same power against any alternative θ when the scale factors $\sqrt{e_{1,n_1}}$ and $\sqrt{e_{2,n_2}}$ are approximately equal, or

$$e_{1,n_1} \doteq e_{2,n_2}. \tag{2.18}$$

---

[1] All limits in this chapter are to be taken as the relevant sample sizes approach infinity, but this will not be stated explicitly.

Since (2.16) implies that $e_{1,n_1} \doteq n_1 e_{1.}$ and $e_{2,n_2} \doteq n_2 e_{2.}$, substituting these in (2.18) shows that the power functions of the two tests will be equal when

$$n_1 e_{1.} \doteq n_2 e_{2.},$$

and hence when n_1 and n_2 are in the ratio

$$\frac{n_2}{n_1} \doteq \frac{e_{1.}}{e_{2.}} = E_{1:2}. \tag{2.19}$$

Thus $E_{1:2}$ can be interpreted as the limiting ratio of sample sizes for which the tests have the same power. This is the usual definition of asymptotic relative efficiency. The foregoing discussion indicates that both this and the scale-factor interpretation of $E_{1:2}$ mentioned earlier will hold in typical situations.

As an example, consider the asymptotic efficiency of the ordinary sign test relative to the classical normal-theory test for the same situation. The asymptotic relative efficiency depends on the family of distributions under discussion. Let us consider a normal family as one relevant possibility. Let $T_{1.n}$ be the test statistic for the sign test, that is, the number of negative observations. From the results given in (2.9), the efficacy of $T_{1,n}$ for the normal distribution is

$$e_{1,n} = 2n/\pi\sigma^2. \tag{2.20}$$

The normal-theory test is based essentially on the sample mean. The test statistic is the sample mean if σ is known and the t statistic if σ is unknown, but, as explained in the last subsection, we may proceed as if the sample mean were the test statistic in both cases. Accordingly, we let $T_{2,n}$ be the sample mean and obtain its efficacy from (2.12) as

$$e_{2,n} = n/\sigma^2. \tag{2.21}$$

The asymptotic efficiency of the sign test relative to the normal-theory test (for σ either known or unknown), against normal alternatives, is then the ratio

$$E_{1:2} = \lim \frac{e_{1,n}}{e_{2,n}} = \frac{2}{\pi} = 0.64. \tag{2.22}$$

We have given two interpretations of this result, both applying to large samples. First, the power of the sign test against a normal alternative with mean $\theta = \delta$ is approximately the same as the power of the normal-theory test against $\theta = \delta\sqrt{2/\pi} = 0.80\delta$, when the sample size and δ are the same for both tests. Second, the sign test has approximately the same power against normal alternatives as the normal-theory test based on 64% as many observations.

3 Asymptotic Behavior of Point Estimators: Heuristic Discussion

In Sect. 2 we explored heuristically the individual and relative performance of typical tests in large samples. A similar discussion for point estimators is the first subject of this section. We shall see that the asymptotic relative efficiency of two estimators is the same as that of two tests based on those estimators. Most test statistics, however, are not themselves estimators of natural parameters, although they are often derived from such estimators. We shall, therefore, give a general method of obtaining estimators from tests and thereby justify the conclusion that the asymptotic relative efficiencies of tests apply also to suitably related estimators. Finally we shall consider how one might compare two estimators of different quantities. This latter discussion is not really necessary for later purposes, but will provide additional insight into the situation. As in Sect. 2, the approximate statements become exact as the sample sizes approach infinity, but we do not give a precise formulation of limit statements in this section.

3.1 Estimators of the Same Quantity

Suppose we are considering two estimators of the same quantity, say μ. For example, the sample median and the sample mean may be regarded as estimators of the same quantity if the population median and mean are assumed to be equal. Let the two estimators be $T_{1,n}$ and $T_{2,n}$, and suppose that the true distribution belongs to some specified family of distributions indexed by a one-dimensional parameter θ. Since θ determines the true distribution, it also determines the quantity being estimated, which we may write accordingly as $\mu(\theta)$.

Suppose that $T_{1,n}$ and $T_{2,n}$ are approximately normal, as estimators usually are, with mean $\mu(\theta)$ and variances $\sigma_{1,n}^2(\theta)$ and $\sigma_{2,n}^2(\theta)$ respectively. Now if the definitions of Sect. 2 are applied to $T_{1,n}$ and $T_{2,n}$ regarded as test statistics for a particular null value of θ (which they could be at least within our one-parameter family of distributions), then their efficacies and their asymptotic relative efficiency are given by

$$e_{1,n}(\theta) = [\mu'(\theta)]^2/\sigma_{1,n}^2(\theta), \qquad e_{2,n}(\theta) = [\mu'(\theta)]^2/\sigma_{2,n}^2(\theta), \qquad (3.1)$$

$$E_{1:2}(\theta) = \lim \sigma_{2,n}^2(\theta)/\sigma_{1,n}^2(\theta). \qquad (3.2)$$

Note that since $\mu'(\theta)$ is the same for both tests, $E_{1:2}(\theta)$ can be obtained without actually computing $\mu'(\theta)$.

The two interpretations of $E_{1:2}(\theta)$ in estimation are similar to the two in testing. First, the errors of the two estimators $T_{1,n}$ and $T_{2,n}$ have approximately the same distribution except for a scale factor $1/\sqrt{E_{1:2}(\theta)}$, since both estimators are approximately normal with mean $\mu(\theta)$ and the ratio of the variance of the first estimator to that of the second is $1/E_{1:2}(\theta)$.

Second, if the two estimators are based on samples of sizes n_1 and n_2 in the ratio $n_2/n_1 = E_{1:2}(\theta)$, then the distributions of the estimators and hence of the errors will be approximately the same. To see this, recall again that both are approximately normal with mean $\mu(\theta)$. They will, therefore, have approximately the same distribution if their variances are approximately equal, and hence if their efficacies, given by (3.1), are approximately equal. But this is exactly the condition of Equation (2.19) for the tests based on T_{1,n_1} and T_{2,n_2} to have approximately the same power function, except that the dependence on θ_0 was suppressed there while the dependence on θ is not suppressed here. Accordingly, by the same argument as in Sect. 2.3, if the ratio n_2/n_1 is, in the limit, equal to $E_{1:2}(\theta)$, then the two estimators of θ, T_{1,n_1} and T_{2,n_2}, will, in the limit, have the same distribution. Note that in general, this ratio depends on the true value of θ, although this dependence will disappear in our examples.

In typical situations, then, $E_{1:2}(\theta)$ will have the two interpretations above and can be computed for estimators exactly as for test statistics when the estimators are estimating the same quantity.

As an example, suppose θ is the proportion negative in some population, $T_{1,n}$ is the proportion of negative observations in the sample, and $T_{2,n} = \Phi(\overline{X}/S)$, where Φ is the standard normal c.d.f. and $\overline{X}$ and S are the sample mean and standard deviation. The second estimator is appropriate for normal populations with unknown standard deviation. If the population is normal with mean θ and standard deviation σ, then the asymptotic efficiency of the first estimator relative to the second can be computed (Problem 7) as

$$E_{1:2}(\theta) = \frac{(1 + \tfrac{1}{2}\omega^2)\phi^2(\omega)}{p(1 - p)} \tag{3.3}$$

where $\omega = \theta/\sigma$, $p = \Phi(\omega)$, and ϕ is the standard normal density function. $E_{1:2}(\theta)$ is plotted as a function of p in Fig. 3.1. When 25% of the population is negative, for example, $E_{1:2}(\theta) = 0.66$. This implies that the variance of the sample proportion negative is approximately $1/0.66 = 1.51$ times the variance of $\Phi(\overline{X}/S)$, and that their errors have approximately the same distribution except for a scale factor $1/\sqrt{0.66} = 1.23$. It also implies that the sample proportion negative has approximately the same distribution as $\Phi(\overline{X}/S)$ based on 66% as many observations. If $\theta = 0$, then $p = 0.5$ and the asymptotic relative efficiency is the same as for the sign test relative to the t test, namely $2/\pi = 0.64$. As can be seen from Fig. 3.1., the normal-theory estimator, $\Phi(\overline{X}/S)$, is always better under normality and may be much better. Unfortunately, however, if the normality assumption fails, then as discussed in Sect. 3.2 of Chap. 2, $\Phi(\overline{X}/S)$ will not ordinarily be a natural or good estimator of the proportion negative at all.

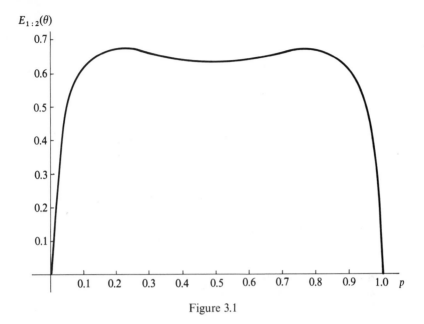

Figure 3.1

3.2 Relation of Estimators and Tests

The previous subsection makes it appear that estimators have the same asymptotic relative efficiency as tests. However, more needs to be said concerning which estimators have the same relative efficiency as which tests.

When we start with tests and look for corresponding estimators, the following difficulty arises. A test can be defined equivalently in terms of many different test statistics. However, if the asymptotic relative efficiency of two tests is to apply to their test statistics when regarded as estimators, then both test statistics must be estimators of the same population quantity, which must not depend on n. Consider, for instance, the sign test and the normal-theory test in a one-sample problem. The sign test was defined earlier in terms of the number of negative observations, but this estimates a quantity depending on n. An equivalent test statistic which eliminates this dependence is the proportion of negative sample observations, which estimates the proportion negative in the population. However, the normal-theory test statistic, either the sample mean or the t statistic, is not an estimator of this parameter, and so we still do not have a pair of estimators which are related in the manner of the previous section and correspond to the two tests, the sign test and the normal-theory test.

If, on the other hand, we start with estimators, the difficulty is that they may not appear to be test statistics for the right kind of test. The sample median, for instance, is not the test statistic for the sign test. Furthermore, since its distribution depends on the shape of the population, it is a possible test statistic only under very restrictive assumptions. How, then, can its efficacy in estimation be related to the efficacy of any interesting test?

One answer is to consider one-tailed tests at the level $\alpha = 0.50$. For example, the upper-tailed sign test at level $\alpha = 0.50$ rejects if the number of negative observations exceeds $n/2$. This is equivalent to rejecting if the sample median is negative. Thus the sample median can serve as the test statistic for a one-tailed sign test at level $\alpha = 0.50$, although not at other levels. As we have seen, however, the efficacy of a test does not depend on the level. Therefore, the efficacy of the sample median must be asymptotically the same as the efficacy of the sign test (Problem 8). While we have been talking about the sign test for the null hypothesis that the population median is 0, the argument extends immediately to any hypothesized value of the median. A sign test of the null hypothesis that the median is μ_0 is ordinarily based on the number of observations less than μ_0, but the sample median can again serve as the test statistic for a one-tailed test at level $\alpha = 0.50$.

Thus the efficacy of the sample median as an estimator of the population median in any family of distributions will be asymptotically the same at each value of the population median as the efficacy of the sign test for that value of the population median. Of course, this can be verified directly (Problem 9). However, the foregoing argument relates the sample median to the sign test explicitly and shows that their efficacies must be asymptotically equal and need not both be computed individually.

A similar argument relates the sample mean to the t test (Problem 10). (Section 2.2 presented another argument, namely that the t test at any level is asymptotically equivalent for present purposes to a test based on the sample mean.)

To generalize the argument relating a family of tests for a parameter μ to an estimator of μ with asymptotically the same efficacy, suppose we perform a one-tailed test at level $\alpha = 0.50$ for each value of μ. Then for any given set of observations there is usually one value of μ such that all values on one side of it would be accepted, and all values on the other side rejected. This point of division, the value of μ which would be "just accepted" at the level $\alpha = 0.50$, may be considered an estimator of μ corresponding to the family of tests in question. It is just the confidence bound for μ at level $1 - \alpha = 0.50$ which corresponds to the family of tests at level $\alpha = 0.50$. Its efficacy might be very difficult to obtain directly. However, since this estimator could be used as a test statistic for any one of the tests at level $\alpha = 0.50$, its efficacy at any value of μ must be asymptotically the same as that of the test for that value of μ.

In summary, the foregoing argument shows that, given a family of tests for μ, the corresponding 50% confidence bound is a naturally related estimator of μ, with asymptotically the same efficacy.

Estimators Derived from One-Sample Tests

We now apply the foregoing method to derive estimators from one-sample tests when the population is assumed symmetric about some point μ. Earlier chapters discussed many tests of the null hypothesis $\mu = 0$, including the sign test and other signed-rank tests. Recall that any such test generates a family of tests, one for each value of μ, as follows. To test a null value of μ, subtract that number from every observation, compute the test statistic, and compare it with an appropriate quantile of its null distribution as critical value. For a one-tailed test at level $\alpha = 0.50$ this quantile is the median. Comparing the test statistic (computed after subtracting μ) with the median of its null distribution is equivalent to comparing μ with the amount that must be subtracted from every observation to make the test statistic equal to the median of its null distribution. Therefore, an estimator of μ corresponding to such a family of tests is the amount that must be subtracted from every observation to make the test statistic equal to the median of its null distribution.

In the case of the one-tailed sign test, for example, a test statistic is the number of negative observations. Its null distribution has median $n/2$, barring complications due to discreteness. If we ignore the problems of zeros, ties, and whether n is even or odd, the amount that must be subtracted from every observation to make the number of negative observations equal to $n/2$ is the sample median. As we have already seen, a one-tailed sign test at level $\alpha = 0.50$ for any μ can be carried out by comparing the sample median with μ.

In the case of the Wilcoxon signed-rank test, it is not quite so obvious what the related estimator is. It is obvious, however, that an estimator can be defined as the amount that must be subtracted from every observation to make the test statistic equal to the median of its null distribution. And of course, a one-tailed Wilcoxon signed-rank test at level $\alpha = 0.50$ for any center of symmetry μ can be carried out by comparing this estimate with μ. It is interesting to note that the estimate is actually the median of the set of all Walsh averages, that is, the median of the set of $n(n + 1)/2$ numbers of the form $(X_i + X_j)/2$ for $i \leq j$ (Problem 11).

Other one-sample tests give rise similarly to estimators of an assumed center of symmetry μ.

Estimators Derived from Two-Sample Tests

In the two-sample shift problem, two-sample tests give rise to estimators of the shift parameter μ in much the same way. Specifically, the estimator is the amount which must be subtracted from every observation of the second sample to make the test statistic equal to the median of its null distribution. For any μ, a one-tailed test at level $\alpha = 0.50$ of the null hypothesis that μ is the amount of the shift by which the populations differ can be performed by comparing this estimate with μ. The estimator corresponding to the two-sample t test in this way is the difference of the sample means. Similarly, the difference of the sample medians corresponds to the two-sample median test, and the median of the set of all differences $Y_j - X_i$ corresponds to the two-sample rank-sum test (Problem 12).

Summary

The foregoing discussion shows how to find estimators which correspond to certain tests even when they are not obvious. This link is illuminating in itself. In addition, though not essential to either the definition or the computation of the asymptotic relative efficiency of estimators, it eliminates the need for such computations if the efficiency of the corresponding tests has already been computed. This is especially advantageous if it is difficult to compute the asymptotic relative efficiency of an estimator, as is true for the estimators corresponding to the Wilcoxon signed-rank test and the two-sample rank-sum test.

In some situations there are estimators which can serve directly as test statistics at all levels. An example of this was given at the end of Sect. 3.1. In such situations, the argument of this subsection is unnecessary, and calculating the asymptotic relative efficiency for the estimators directly is virtually the same as calculating it for the tests.

*3.3 Estimators of Different Quantities

Now let us consider two arbitrary estimators $T_{1,n}$ and $T_{2,n}$ for (presumably) different quantities. There is no natural way to make a direct comparison between $T_{1,n}$ as an estimator of one quantity and $T_{2,n}$ as an estimator of another. If $T_{1,n}$ has smaller variance than $T_{2,n}$, this may only be because the quantity estimated by $T_{1,n}$ is the easier one to estimate. In fact, it could happen at the same time that some function of $T_{2,n}$ estimates the same quantity as $T_{1,n}$ and has smaller variance than $T_{1,n}$. Then $T_{2,n}$ would certainly be more useful than $T_{1,n}$ even though it has larger variance. Two estimators can be compared in a straightforward way only when both are estimating the same quantity.

We could leave the problem here, since there is no really compelling reason to compare two estimators of different quantities. However, an instructive comparison turns out to be possible between any two statistics $T_{1,n}$ and $T_{2,n}$, which need not even be estimates at all. We shall find that there is a very natural way to use functions of them to estimate the same quantity asymptotically, and that the asymptotic relative efficiency of the resulting estimators is the same as the asymptotic relative efficiency of $T_{1,n}$ and $T_{2,n}$ when used for testing in the same situation.

As in earlier sections, consider a family of distributions depending on a one-dimensional parameter θ and suppose that the statistics $T_{1,n}$ and $T_{2,n}$ are approximately normal with means $\mu_{1,n}(\theta)$ and $\mu_{2,n}(\theta)$ and variances $\sigma^2_{1,n}(\theta)$ and $\sigma^2_{2,n}(\theta)$. How can we obtain estimates of θ from $T_{1,n}$ and $T_{2,n}$? Since $T_{1,n}$ is approximately normal with mean $\mu_{1,n}(\theta)$, it is a natural estimator of $\mu_{1,n}(\theta)$. Provided $\mu_{1,n}^{-1}$, the inverse function (not the reciprocal) of $\mu_{1,n}$, is well defined, it is natural to estimate θ correspondingly by

$$\hat\theta_{1,n} = \mu_{1,n}^{-1}(T_{1,n}). \tag{3.4}$$

Equivalently, $\hat\theta_{1,n}$ is the solution of

$$\mu_{1,n}(\hat\theta_{1,n}) = T_{1,n}. \tag{3.5}$$

Define $\hat\theta_{2,n}$ in terms of $T_{2,n}$ similarly.

We now find the asymptotic distribution of $\hat\theta_{1,n}$ using the general procedure known as the δ method. Expanding the right-hand side of (3.4) in a Taylor's series about $\mu_{1,n}(\theta)$, or the left-hand side of (3.5) about θ, gives (Problem 13)

$$\hat\theta_{1,n} - \theta = \frac{T_{1,n} - \mu_{1,n}(\theta)}{\mu'_{1,n}(\theta)} + \text{remainder.} \tag{3.6}$$

Since $T_{1,n}$ is approximately normal with mean $\mu_{1,n}(\theta)$ and variance $\sigma^2_{1,n}(\theta)$, it follows that $\hat\theta_{1,n}$ is approximately normal with mean θ and variance

$$\sigma^2_{1,n}(\theta)/[\mu'_{1,n}(\theta)]^2 = 1/e_{1,n}(\theta). \tag{3.7}$$

Similarly $\hat\theta_{2,n} = \mu_{2,n}^{-1}(T_{2,n})$ is approximately normal with mean θ and variance $1/e_{2,n}(\theta)$.

Therefore, in large samples, there is a natural way to use $T_{1,n}$ and $T_{2,n}$ for purposes of estimating θ, and the resulting estimators are approximately normal with the same mean θ and variances $1/e_{1,n}(\theta)$ and $1/e_{2,n}(\theta)$. The asymptotic efficiency of the estimator of θ based on $T_{1,n}$ relative to that based on $T_{2,n}$ is then

$$E_{1:2}(\theta) = \lim \frac{e_{1,n}(\theta)}{e_{2,n}(\theta)} = \lim \frac{[\mu'_{1,n}(\theta)]^2/\sigma^2_{1,n}(\theta)}{[\mu'_{2,n}(\theta)]^2/\sigma^2_{2,n}(\theta)} \tag{3.8}$$

which is the same as the asymptotic efficiency of $T_{1,n}$ relative to $T_{2,n}$ for testing hypotheses about θ.

Thus $E_{1:2}(\theta)$ gives an assessment of the value of $T_{1,n}$ compared to $T_{2,n}$ asymptotically, both for estimation of θ and for testing hypotheses about θ. Note that this entire analysis depends on the assumption that the true distribution belongs to the family indexed by θ. The meaning of θ does not extend in any automatic way beyond this particular family. If it is extended in some way, then the tests and estimators for θ based on $T_{1,n}$ and $T_{2,n}$ will generally depend on the distribution assumption, because $\mu_{1,n}$ and $\mu_{2,n}$ do (Problem 14).

4 Asymptotic Behavior of Confidence Bounds

We now study the behavior of typical confidence bounds in large samples, particularly their behavior relative to one another. Because of the relation between confidence bounds and tests of hypotheses, we can expect the present investigation to be closely related to that of Sect. 2, and the asymptotic relative efficiency of confidence procedures to be the same as that of the corresponding tests. As before, approximate statements should become exact as the sample sizes approach infinity, but we shall not give a precise formulation of the limit statements here.

Let us consider first an upper confidence bound T_n for a quantity μ at confidence level $1 - \alpha$. Suppose, as usual, that the true distribution belongs to some specified family of distributions indexed by a one-dimensional parameter θ. Since θ determines the true distribution, it also determines the quantity μ, which we may therefore write as $\mu(\theta)$.

Corresponding to the upper confidence bound T_n is a one-tailed test for each μ which rejects the value μ if $T_n < \mu$. Note that this is not a single test, but a family of tests, one for each value of μ. The probability under θ that $T_n < \mu(\theta_0)$ is the power of the test for the null hypothesis $\mu(\theta_0)$ against the alternative given by θ. By Equation (2.11), an approximation to this probability is

$$P_\theta[T_n < \mu(\theta_0)] \doteq \Phi[\pm(\theta - \theta_0)\sqrt{e_n(\theta_0)} - z_\alpha], \tag{4.1}$$

where $e_n(\theta_0)$ is the efficacy of the test at θ_0 and the choice of sign in the argument of Φ depends on whether the test is appropriate against alternatives $\theta < \theta_0$ or against alternatives $\theta > \theta_0$. Substituting $\theta_0 = \theta + \delta/|\mu'(\theta)|$ in (4.1) leads (Problem 15a) to the approximation

$$P_\theta[T_n < \mu(\theta) + \delta] \doteq \Phi\left[\frac{\delta}{|\mu'(\theta)|}\sqrt{e_n(\theta)} - z_\alpha\right]. \tag{4.2}$$

Thus we have obtained an approximation to the c.d.f. of T_n when the true distribution is given by θ.

The efficacy $e_n(\theta)$ need not be computed from the confidence bound T_n, as it can be computed from any test statistic which yields the corresponding test of the null value $\mu(\theta)$. T_n is one such statistic, but is often not the easiest one to use in computing the efficacy, even in the case $\alpha = 0.50$ discussed in Sect. 3.2. Indeed, the natural test statistic for the null value μ usually depends on μ. Consider, for example, the confidence bounds related to the Wilcoxon signed-rank test as in Sect. 4 of Chap. 3. It would be difficult, if not impossible, to compute the mean and variance of this confidence bound directly. However, the corresponding test for any null value of μ can be carried out by subtracting μ from every observation and then computing the signed-rank sum. Note that this test statistic is a function of μ. Its efficacy is easy to compute for any given μ, and is also the efficacy of the confidence bound.

Now suppose we have two upper confidence bounds $T_{1,n}$ and $T_{2,n}$ for the same quantity μ at the same confidence level $1 - \alpha$. The c.d.f. of each is given approximately by (4.2). In addition, as $n \to \infty$, $e_{1,n}(\theta)/e_{2,n}(\theta) \to E_{1:2}(\theta)$, the asymptotic relative efficiency of the corresponding tests. It follows (Problem 15b) that for a given sample size, the c.d.f. of $T_{1,n}$ at $\mu(\theta) + \delta$ is approximately the c.d.f. of $T_{2,n}$ at $\mu(\theta) + \delta\sqrt{E_{1:2}(\theta)}$; that is, $T_{1,n} - \mu(\theta)$ and $T_{2,n} - \mu(\theta)$ have approximately the same c.d.f. except for a scale factor $1/\sqrt{E_{1:2}(\theta)}$. (Compare Fig. 2.1.) In particular, the expectation of $T_{1,n} - \mu(\theta)$ is approximately $1/\sqrt{E_{1:2}(\theta)}$ times the expectation of $T_{2,n} - \mu(\theta)$. The quantity $T_{i,n} - \mu(\theta)$ with its algebraic sign is the amount of overestimation. One might instead be interested in its positive part,

$$[T_{i,n} - \mu(\theta)]^+ = \max\{T_{i,n} - \mu(\theta), 0\},$$

since it is desirable that an upper confidence bound for μ be small as long as it exceeds μ but not when it is smaller than μ. Just as for the signed overestimation, the expectation of $[T_{1,n} - \mu(\theta)]^+$ is approximately $1/\sqrt{E_{1,2}(\theta)}$ times the expectation of $[T_{2,n} - \mu(\theta)]^+$. For other measures of "error" or "loss," $\sqrt{E_{1:2}(\theta)}$ is the scale factor in the random variable but not always in the expected loss (Problem 16).

This provides one interpretation of the asymptotic relative efficiency of tests in connection with the corresponding confidence procedures. We obtain another interpretation by considering the two confidence bounds T_{1,n_1} and T_{2,n_2}, based on samples of sizes n_1 and n_2 respectively. By the same argument as in earlier sections (Problem 17), if the ratio n_2/n_1 is, in the limit, equal to $E_{1:2}(\theta)$, then the confidence bounds T_{1,n_1} and T_{2,n_2} will, in the limit, have the same distribution and hence the same expected loss for any loss function.

These conclusions of course apply to lower as well as upper confidence bounds. As regards two confidence intervals, the previous discussion provides a comparison between the distributions of the two upper bounds, and similarly for the two lower bounds. A full comparison would also require consideration of the joint distribution of the upper and lower bounds, which we have not discussed. What we already know implies, however, that the expected length of the first interval is approximately $1/\sqrt{E_{1:2}(\theta)}$ times that of the second (Problem 18). Furthermore, the length of a typical confidence interval is asymptotically constant, with a standard deviation of smaller order than its mean. Hence, in comparing two confidence intervals, the scale factor $\sqrt{E_{1:2}(\theta)}$ applies asymptotically to both their lengths and the deviation of their endpoints from $\mu(\theta)$, and therefore to all other aspects of their relationships to $\mu(\theta)$. Thus $E_{1:2}(\theta)$ has the same kind of interpretation for confidence intervals as for confidence bounds. A different kind of argument would be required, however, to establish this for confidence intervals whose length is not asymptotically constant.

In typical situations, then, the quantity $E_{1:2}(\theta)$ will have the interpretations given above and will therefore be called the asymptotic efficiency of the confidence procedure $T_{1,n}$ relative to $T_{2,n}$. It is the same as the asymptotic relative efficiency of the corresponding tests of the null value $\mu(\theta)$, which is usually easier to compute directly.

As an example, consider a confidence bound for the population median based on an order statistic as in Sect. 4 of Chap. 2, and the normal-theory confidence bound $\bar{X} + z_\alpha S/\sqrt{n}$ for the population median, where $\bar{X}$ and S are the sample mean and standard deviation and z_α is an appropriate constant. Both are confidence bounds for the same quantity if the population mean and median are assumed equal. (The confidence level of the normal-theory procedure depends on the population shape, but will be correct asymptotically.) Let μ be the common value of the population mean and median. The confidence bound based on the order statistic corresponds to the sign test for each μ, which is based on the number of observations less than μ. The normal-theory confidence bound corresponds to the t test for each μ. Under the normal assumption (the population is normal with mean $\theta = \mu$ and standard deviation σ), the asymptotic efficiency of the first test with respect to the second was found in Sect. 2.3 to be $2/\pi = 0.64$ at $\mu = \theta = 0$, and the same value clearly applies at other values of μ. For the two confidence bounds, this asymptotic efficiency means that under normality the order statistic bound has approximately the same probability of falling below $\mu + \delta$ as the normal-theory bound has of falling below $\mu + \delta\sqrt{0.64} = \mu + 0.80\delta$. In other words, the two confidence bounds differ from μ by amounts having the same distribution except for a scale factor $1/\sqrt{0.64} = 1.25$. In particular, the expected amount by which the order statistic bound exceeds μ is approximately 1.25 times the corresponding expectation for the normal-theory bound, and the same holds for the expected lengths of confidence intervals. Furthermore, the order statistic bound has approximately the same distribution as the normal-theory bound based on 64% as many observations.

5 Example

We have seen that the asymptotic relative efficiency of two tests applies also to the corresponding estimators and confidence bounds, with at least two interpretations in each case. Accordingly, any numerical efficiency value has many meanings. We now illustrate this whole range of ideas for two sets of related procedures in the one-sample problem.

Consider the sign test for the null hypothesis that the population median has a specified value. The natural test statistic is the number of observations falling below the specified value. We thus have a whole family of tests (and test statistics), one for each value which might be specified. The estimator of the population median corresponding to this family of tests is the sample median, as explained in Sect. 3.2. The corresponding confidence bounds are order statistics, as explained in Sect. 4 of Chap. 2. For convenience, all these procedures will be referred to in this section as "median procedures;" they all permit inferences about the population median.

Consider also the family of t tests for null hypotheses specifying the population mean. The corresponding estimator of the population mean is the sample mean, as noted in Sect. 3.2, and the corresponding confidence bounds are the usual ones based on the t distribution. All these procedures will be referred to here as normal-theory procedures, because they are exact under the assumption of normality. They give asymptotically valid inferences about the population mean regardless of the population shape provided that the variance is finite. (If this were not true, comparisons with other procedures under assumptions other than normality would be complicated by the discrepancy between their true level and their nominal level. Such a discrepancy would invalidate all our earlier analysis, and would also bring a new consideration into the problem—the trade-off between level and power. See also the discussion of power comparisons using conservative tests in Sect. 4.3 of Chap. 1.)

Our object here is to compare the median procedures with the normal-theory procedures under the assumption that the population median and mean are equal. Before proceeding, however, let us emphasize that these procedures lead in general to inferences about different quantities. A median procedure provides an inference about the population median, while a normal-theory procedure provides an inference about the population mean. Accordingly, the first question to ask is whether it is really the median or the mean of the population which is of interest. Careful thought may reveal that one or the other (or something else entirely) is really the parameter of interest. If so, and if we we are also unwilling to assume that they differ negligibly, then our choice of procedure will be clear and an efficiency comparison irrelevant. Such considerations are at least as important as efficiency. They receive less attention here simply because they require less explanation.

On the other hand, if we believe that the population median and mean differ negligibly compared to the uncertainty resulting from sampling variability, then the relative efficiency of the median and normal-theory procedures will be of interest. It will also be of interest in situations where we think the population median and mean may well differ appreciably and it is immaterial which one the inference concerns—but then perhaps a more meaningful and useful way of making or scaling the measurements could be found. In either case, the choice among procedures may be facilitated by learning something about their relative efficiency. We shall, however, compute and interpret their asymptotic relative efficiency only in situations where the population median and mean have the same value, say μ. In fact, we shall consider populations which are symmetric about μ. It is difficult to justify the assumption that the population mean and median are equal except when the stronger symmetry assumption can also be justified.

For the median procedures, the estimator is the sample median, say $T_{1,n}$. For large n, this statistic is (Problem 19) approximately normal with mean equal to the population median μ and variance $1/4nd^2$, where d is the population density at μ (provided the population has a positive, continuous density at μ). This result will be used below to obtain the efficacy of the median procedures. Of course, $T_{1,n}$ is not in itself a test statistic or confidence bound for μ (except at the level $\alpha = 0.50$), but we have already seen that efficiencies computed from estimators apply also to the corresponding tests and confidence bounds. Instead of the sample median we could have used the natural test statistic for each null value μ, namely the number of observations less than μ. The reader may verify (Problem 9b) that in what follows this would give the same efficacies to first order as $n \to \infty$. As it happens, the efficacies would be exactly the same if computed in the natural way.

For the normal-theory procedures, the estimator is the sample mean, say $T_{2,n}$. For large n, this statistic is approximately normal with mean μ and variance σ^2/n, where μ and σ^2 are the population mean and variance. This fact will be used to obtain the efficacy of the normal-theory procedures. The t statistic for each value of μ gives the same efficacies to first order as $n \to \infty$.

In order to put the problem in the framework of the discussion of Sects. 2–4, we must restrict consideration to a family of distributions indexed by a single parameter θ. A number of such families are considered below. In each family, the parameter θ is the same as the center of symmetry μ, so that $\mu(\theta) = \theta$.

Suppose first that the population distribution has a Laplace (double exponential) density

$$f(x; \theta) = \frac{1}{2\lambda} e^{-|x-\theta|/\lambda}, \tag{5.1}$$

where λ is an arbitrary but fixed positive number. This distribution is symmetric about θ, with mean θ and variance $2\lambda^2$ (Problem 20a). For the sample median $T_{1,n}$ and the sample mean $T_{2,n}$, we find (Problem 20b)

$$\mu_{1,n}(\theta) = 0, \qquad \sigma^2_{1,n}(\theta) = \lambda^2/n, \qquad e_{1,n}(\theta) = n/\lambda^2, \tag{5.2}$$

$$\mu_{2,n}(\theta) = 0, \qquad \sigma^2_{2,n}(\theta) = 2\lambda^2/n, \qquad e_{2,n}(\theta) = n/2\lambda^2. \tag{5.3}$$

For the Laplace family of distributions, (5.1), the asymptotic efficiency of the median procedures relative to the normal theory procedures is therefore

$$E_{1:2}(\theta) = 2. \tag{5.4}$$

This result does not depend on θ, as could be anticipated from the nature of the procedures and the way θ enters the density (5.1) (Problem 21d). According to Sects. 2–4, the implications of this result under the model (5.1) are as follows.

Consider the median procedure and the normal-theory procedure for testing the hypothesis $\theta = \theta_0$, that is, the sign test and the t test for this null hypothesis. We assume always that the tests have the same level α and are either both one-tailed in the same direction or both two-tailed with the same division of the significance level between the two tails. If the tests are based on samples of the same (large) size, then the power of the sign test at a point δ units away from θ_0 is approximately the same as the power of the t test at a point $\delta\sqrt{2} = 1.41\delta$ units away in the same direction; that is, the sign test gives approximately the same power at any point as the t test gives at a point farther from θ_0 by the factor $\sqrt{2}$. In terms of different sample sizes (still large), we may say that the sign test requires approximately one-half as many observations as the t test to give a specified power at a specified alternative near the null hypothesis. Both these statements apply to Laplace alternative (5.1), of course.

Approximations to the power itself can be given in terms of $e_{1,n}(\theta)$ and $e_{2,n}(\theta)$; by Equation (2.5) the respective powers against θ of the sign test and the t test are approximately $\Phi[\sqrt{n}(\theta - \theta_0)/\lambda - z_\alpha]$ and $\Phi[\sqrt{n/2}(\theta - \theta_0)/\lambda - z_\alpha]$ for one-tailed tests appropriate against the alternative $\theta > \theta_0$. Similar expressions could be written for one-tailed tests appropriate against $\theta < \theta_0$ and for two-tailed tests. One might expect these approximations to be better than usual here, because when the $\mu_{i,n}(\theta)$ are linear in θ and the $\sigma_{i,n}^2(\theta)$ do not depend on θ, as here, (2.4) is exact and (2.5) and (2.7) agree exactly with (2.3) for both the mean and the median. This is misleading, however, as the mean and median can serve as test statistics only at the level $\alpha = 0.50$. At other levels, other test statistics would be required, and, for them, (2.5) and (2.7) would not agree exactly with (2.3).

To estimate μ, the median and normal-theory procedures use the sample median and mean respectively. In large samples from the Laplace distribution (5.1), both are approximately normal with mean $\mu = \theta$, and the variance of the median is approximately one-half of the variance of the mean from a sample of the same size. The estimation error of the median has approximately the same distribution as $1/\sqrt{2}$ times the estimation error of the mean. If the sample size for the median is one-half of the sample size for the mean, their distributions will be approximately the same. The situation is particularly simple in that the factor one-half does not depend on θ. (This simplification occurs frequently, but not always.)

For large samples from the Laplace distribution (5.1), an upper confidence bound for μ computed by the median procedure has approximately the same probability of falling below $\mu + \delta$ as the normal theory bound has of falling below $\mu + \delta\sqrt{2}$. The amount by which the former exceeds μ has approximately the same distribution as $1/\sqrt{2}$ times the amount by which the latter exceeds μ. Similar statements hold for a lower confidence bound. The expectation of the difference between a confidence bound and μ, or of the positive part of this difference, or of the length of a confidence interval, is approximately $1/\sqrt{2}$ times as great for the median procedure as for the normal theory procedure. The median procedure using a sample size one-half as large as the normal theory procedure gives confidence bounds having approximately the same distribution. Again, the implications of these statements are particularly simple because the factor one-half does not depend on θ.

All these statements are implied by the single statement that the asymptotic efficiency of the median relative to the mean is 2 (for all θ). Note that these results apply specifically when the true distribution is Laplace, (5.1), and its center of symmetry θ is the parameter of interest. For a true distribution with a different shape, the asymptotic relative efficiency will generally be different, as we shall illustrate next.

Suppose now that the population is normal with mean θ and variance σ^2, where σ^2 is arbitrary but fixed, like λ above. In this case the asymptotic efficiency of the median procedures relative to the normal-theory procedures is (Problems 7c and 21).

$$E_{1:2}(\theta) = 2/\pi = 0.64. \qquad (5.5)$$

In large samples, then, the sign test of a null hypothesis $\theta = \theta_0$ has approximately the same power at any point as the t test at a point $\sqrt{0.64} = 0.80$ times as far from θ_0. The estimation error of the median has approximately the same distribution as $1/\sqrt{0.64} = 1.25$ times the estimation error of the mean. A confidence interval for θ computed by the median procedure differs from θ by an amount having approximately the same distribution as $1/\sqrt{0.64}$ times the corresponding difference for the normal theory procedure. All these statements apply to samples of the same size. The median procedures for testing, estimation, and setting confidence limits behave approximately like the normal-theory procedures based on a sample 0.64 times as large. Again the factor is independent of θ.

If the population is normal, the median procedures are asymptotically 0.64 times as efficient as the normal theory procedures. On the other hand, if the population is Laplace, we saw earlier that the median procedures are asymptotically twice as efficient as the normal-theory procedures. It is convenient to consider both the normal and Laplace densities as special cases of the general density (which might be called the "double exponential power" or "power Laplace" density) given by

$$f(x; \theta) = \frac{k}{2\lambda\Gamma(1/k)} e^{-|x-\theta|^k/\lambda^k} \qquad (5.6)$$

where Γ denotes the well-known gamma function and k and λ are positive. For any k, the density (5.6) has center of symmetry θ. It is Laplace when

$k = 1$ and normal when $k = 2$. As $k \to \infty$, it approaches the uniform density on the interval $(\theta - \lambda, \theta + \lambda)$ (Problem 22). When the density is of the form (5.6) and θ is the parameter of interest, with k and λ arbitrary but fixed, the asymptotic efficiency of the sample median relative to the sample mean is (Problem 23a) independent of θ and λ, and is given by

$$E_{1:2}(\theta; k) = \frac{k^2 \Gamma(3/k)}{[\Gamma(1/k)]^3}. \tag{5.7}$$

Figure 5.1 is a graph of (5.7) as a function of k. As $k \to \infty$ the limit of (5.7) (Problem 23b) is

$$E_{1:2}(\theta; \infty) = \tfrac{1}{3}, \tag{5.8}$$

which agrees with an independent calculation of the asymptotic relative efficiency for the uniform distribution (Problem 23c). The limit as $k \to 0$ is ∞, but there is no limiting distribution at $k = 0$. Each value of $E_{1:2}(\theta; k)$ has multiple interpretations, as illustrated before for $k = 1$ and $k = 2$. Notice that, while the normal-theory procedure can be only three times as efficient as the median procedure for a family of distributions of the form (5.6), the median procedure may be arbitrarily many times as efficient as the normal-theory procedure. (See also Sect. 8.3.)

We will return to these and other tests and families of probability distributions in Sect. 8, after giving a formal definition of asymptotic relative efficiency in the next section and a formula for its calculation in Sect. 7.

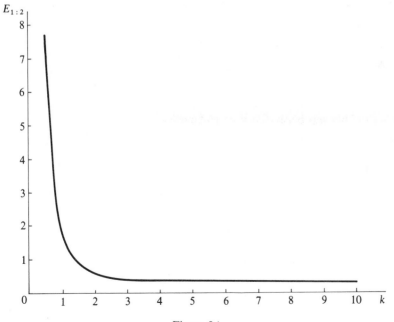

Figure 5.1

Those readers who are satisfied with the informal definitions of asymptotic relative efficiency given in Sects. 2–4 may skip the precise formulation presented in Sect. 6 and go immediately to Sect. 7.

*6 Definitions of Asymptotic Relative Efficiency

6.1 Introduction

In Sect. 1 we defined the relative efficiency of two procedures which are applicable in the same situation as the ratio of the sample sizes needed to achieve the same performance, adding the qualifications necessary to make this a precise definition. We then defined the asymptotic relative efficiency of two procedures as the limit of their relative efficiency as the sample sizes approach infinity, but did not make this definition precise. Instead, in Sects. 2–4 we investigated heuristically and approximately how procedures behave and how they may be compared as sample sizes become large. The investigation revealed that, rather than there being a separate numerical value of the asymptotic relative efficiency for each performance criterion, a great simplification results in large samples, and one number does, in a sense, describe the entire situation asymptotically. This permits much stronger definitions of asymptotic relative efficiency than would otherwise be possible.

In this section, we shall give precise statements of a number of asymptotic properties suggested by the investigation of Sects. 2–4. The statements in this book about asymptotic relative efficiency apply to most of these properties, although we shall not prove this rigorously. In the statistical literature, the standard definition of asymptotic relative efficiency and most statements and proofs about it refer only to tests and only to the property involving the ratio of sample sizes (A(ii) below) or a slightly weaker version of it. Suitably interpreted, the statements ordinarily apply also to the other properties below, including those for estimators and confidence bounds, but this cannot be taken for granted and is seldom proved.

There is also a smaller but significant and growing literature about a kind of asymptotic relative efficiency based on a fundamentally different limiting operation from the set of related properties discussed here. For a fixed alternative, it concerns the rate at which one type of error probability approaches 0 with the other fixed as $n \to \infty$, or the rate at which the maximum error probability can be made to approach 0. It uses the probability theory of "extreme deviations." Some early and other references are Chernoff [1952], [1956], Hodges and Lehmann [1956], Blyth [1958], Bahadur [1960], [1971], and Groeneboom and Oosterhoff [1977].

As already indicated in Sect. 1, asymptotic relative efficiency is a property of two sequences of procedures, each defined for all n. As in Sects. 2–4, we always consider a family of distributions indexed by a one-dimensional parameter θ. The investigations of Sects. 2–4 show that there will typically exist a quantity $E_{1:2}(\theta)$ with the properties described below for tests, estimators, and confidence bounds. This quantity will be called the asymptotic relative efficiency. We assume throughout that $E_{1:2}(\theta)$ is neither 0 nor ∞. If $E_{1:2}(\theta)$ is 0 or ∞, some restatement of the properties is needed (Problem 25).

6.2 Tests

Suppose first that we are comparing two test procedures and that θ_0 gives a distribution belonging to the null hypothesis for each test. If both tests are one-tailed, assume that they are appropriate against the same one-sided alternative and that the exact levels of the tests under θ_0 approach the same positive constant as $n \to \infty$. If both tests are two-tailed, make the same assumption about the level in each tail separately. Then the asymptotic efficiency $E = E_{1:2}(\theta_0)$ of the first test relative to the second can be expected to have the following properties, any one of which could be taken as the definition of asymptotic relative efficiency.

A(i). For two tests based on the sample size n, the difference between the power of the first test at $\theta_0 + \delta$ and the power of the second test at $\theta_0 + \delta\sqrt{E}$ approaches zero uniformly in δ as $n \to \infty$.

If $P_{1,n}$ and $P_{2,n}$ are the power functions of the two tests, the condition is

$$P_{1,n}(\theta_0 + \delta) - P_{2,n}(\theta_0 + \delta\sqrt{E}) \to 0 \quad \text{uniformly in } \delta \text{ as } n \to \infty. \quad (6.1)$$

The statement of (6.1) in ε terminology, directly from the definition of uniform convergence, is that for every $\varepsilon > 0$, there is an N not depending on δ such that, for all $n > N$ and all δ,

$$|P_{1,n}(\theta_0 + \delta) - P_{2,n}(\theta_0 + \delta\sqrt{E})| \le \varepsilon. \quad (6.2)$$

Uniform convergence is equivalent by definition to convergence of the maximum absolute difference. Thus (6.1) is equivalent (Problem 24) to

$$\max_{\delta}|P_{1,n}(\theta_0 + \delta) - P_{2,n}(\theta_0 + \delta\sqrt{E})| \to 0 \quad \text{as } n \to \infty. \quad (6.3)$$

This says that, if the two powers at $\theta_0 + \delta$ and $\theta_0 + \delta\sqrt{E}$ respectively are graphed as functions of δ (Fig. 2.1, for instance), the maximum vertical difference between the graphs approaches zero as $n \to \infty$. Uniform convergence is also equivalent to convergence for every sequence δ_n, so (6.1) is also equivalent (Problem 24) to

$$P_{1,n}(\theta_0 + \delta_n) - P_{2,n}(\theta_0 + \delta_n\sqrt{E}) \to 0 \quad \text{as } n \to \infty \quad (6.4)$$

for all sequences δ_n. These equivalences indicate the significance of the uniformity in (6.1), to be discussed further shortly.

We saw in Sect. 2.1 how the power function of an arbitrary test can typically be approximated by a normal distribution as in Equation (2.11). Using this approximation, along with the fact that the efficacy is typically of order $\sqrt{n}$, we observe that the powers of the two tests at $\theta_0 + \delta_n$ and $\theta_0 + \delta_n\sqrt{E}$ appearing in (6.1)–(6.4) typically behave as follows, for an arbitrary E. If $\delta_n \to 0$ so fast that $\sqrt{n}\delta_n \to 0$, then neither test is effective asymptotically and both powers approach the significance level α. If, at the other extreme, $\sqrt{n}\delta_n \to \pm\infty$, then both powers approach one (or zero for one-tailed tests in the wrong direction). If $\delta_n \to 0$ in such a way that $\sqrt{n}\delta_n \to d$ for d nonzero and finite, then both powers approach limits other than 1, α, or 0 in general. These limits will be equal if $E = E_{1:2}(\theta_0)$, but otherwise will not, with one minor exception. (For two-tailed tests with unequal tails in the limit, for each $E \neq E_{1:2}(\theta_0)$ there is one particular value of d such that the limits are equal and less than α. See Problem 26.)

Notice that, without the condition "uniformly in δ," (6.1) would lose all force, since both terms in (6.1) approach one (or zero for one-tailed tests in the wrong direction) for every fixed $\delta \neq 0$ regardless of the value of E. But with the uniformity, property A(i) implies that the power functions will be the same in the limit except for a scale factor $1/\sqrt{E}$ even when the alternatives are rescaled so that the limits are not degenerate, specifically, when the powers are considered as functions of $\sqrt{n}(\theta - \theta_0)$.

From the foregoing it also follows that the only value of E with property A(i) is $E = E_{1:2}(\theta_0)$. In most of the properties to follow, uniformity is essential to the meaning and can be restated in the style of (6.3) and (6.4); similar comments about degenerate limits and rescaling apply; and similar converses can be stated. This will not be mentioned each time, but left to the reader to fill in (Problems 27–29).

The asymptotic relative efficiency property A(i) concerns the power of the two tests at different alternatives for the same sample size. The next two properties relate to the same alternative but different sample sizes.

A(ii). For two tests with sample sizes n_1 and n_2 respectively, the difference between the powers at the same point θ approaches zero uniformly in θ when n_1 and n_2 approach infinity simultaneously in such a way that $n_2/n_1 \to E$.

In the same notation as before, the condition here is

$$P_{1,n_1}(\theta) - P_{2,n_2}(\theta) \to 0 \quad \text{uniformly in } \theta \text{ if } n_2/n_1 \to E. \tag{6.5}$$

The final property of the asymptotic relative efficiency E of two tests which we state is

A(iii). For two tests, if n_1 is the minimum n for which the first test has power at least $1 - \beta$ against the alternative θ and n_2 is defined similarly for the second test, then $n_2/n_1 \to E$ as $\theta \to \theta_0$.

Here $1 - \beta$ is required to exceed the limiting significance level of the tests and, for one-tailed tests, θ is restricted to the side of θ_0 against which the tests are appropriate. Property A(iii) says that the ratio of the sample sizes required to achieve a specified power against a specified alternative approaches E as the alternative approaches θ_0 and as, consequently, the sample sizes approach infinity.

6.3 Estimators

Consider next two estimators $T_{1,n}$ and $T_{2,n}$ of the same quantity $\mu(\theta)$. Since there is no distinguished value of θ in this context, we return to the notation $E_{1:2}(\theta)$ for the asymptotic relative efficiency, to emphasize its dependence on θ. Continuing our delineation of the properties which we expect $E_{1:2}(\theta)$ to have, we state four such properties, B(i)–B(iv), any one of which could serve to define asymptotic relative efficiency for estimators.

B(i). When the true distribution is given by θ, the difference between the probability of an error of δ or less using $T_{1,n}$ and the probability of an error of $\delta\sqrt{E_{1:2}(\theta)}$ or less using $T_{2,n}$ approaches zero uniformly in δ as the common sample size $n \to \infty$.

In symbols,

$$P_{\theta}[T_{1,n} - \mu(\theta) \le \delta] - P_{\theta}[T_{2,n} - \mu(\theta) \le \delta\sqrt{E_{1:2}(\theta)}] \to 0 \qquad (6.6)$$

uniformly in δ as $n \to \infty$.

This says that the errors of the two estimators have, in an appropriate sense, asymptotically the same distribution except for a scale factor $1/\sqrt{E_{1:2}(\theta)}$ (Problem 30). The errors themselves have distributions which concentrate at zero as $n \to \infty$, so that both terms in (6.6) approach zero for $\delta < 0$ and both approach one for $\delta > 0$. By uniformity, however, Property B(i) implies that even when the errors are scaled up in such a way that their distributions are not degenerate in the limit, the distributions will be the same in the limit except for the scale factor $1/\sqrt{E_{1:2}(\theta)}$. Specifically, when the errors are scaled up by the factor $\sqrt{n}$, their distributions typically converge to normal distributions with mean 0 and variances in the ratio $1/E_{1:2}(\theta)$; equivalently, the distributions of $\sqrt{n}$ times the error of $T_{1,n}$ and $\sqrt{n/E_{1:2}(\theta)}$ times the error of $T_{2,n}$ approach the same normal distribution, with mean zero and positive variance.

Scaling, the importance of uniformity, alternative statements of it, and the uniqueness of E are mostly much the same as in Sect. 6.2 and will not be discussed further here, but they should be borne in mind.

B(ii). When the true distribution is given by θ, the ratio of the variance of $T_{1,n}$ to the variance of $T_{2,n}$ approaches $1/E_{1:2}(\theta)$ as $n \to \infty$. The same is true of the ratio of their mean squared errors. The ratio of the standard deviation of $T_{1,n}$ to that of $T_{2,n}$ approaches $1/\sqrt{E_{1:2}(\theta)}$. The same holds for the ratio of their mean absolute errors.

Property B(ii) does not follow automatically from B(i) because the variance of a limiting distribution, though ordinarily the limit of the variances for finite n, need not be. For instance, $\sqrt{n}[T_{1,n} - \mu(\theta)]$ can have infinite variance for every n even though its limiting distribution has finite variance (Problem 31). Common methods of deriving limiting distributions do not apply to the limit of the variances. Similar remarks hold for mean squared error, standard deviation, and mean absolute error. Accordingly, statements and proofs about asymptotic relative efficiency, in this book and elsewhere, usually apply directly to B(i), but for B(ii), some statements need qualification, especially very general ones, and most proofs need additional justification.

B(iii). For two estimators with sample sizes n_1 and n_2 respectively, if the true distribution corresponds to θ, and n_1 and n_2 approach infinity simultaneously in such a way that $n_2/n_1 \to E_{1:2}(\theta)$, then the difference between the c.d.f.'s of the two estimators converges uniformly to zero.

In symbols,

$$P_\theta(T_{1,n_1} \le t) - P_\theta(T_{2,n_2} \le t) \to 0 \tag{6.7}$$

uniformly in t if $n_2/n_1 \to E_{1:2}(\theta)$.

B(iv). For two estimators with sample sizes n_1 and n_2 respectively, if the true distribution corresponds to θ and n_1 and n_2 approach infinity simultaneously in such a way that $n_2/n_1 \to E_{1:2}(\theta)$, then the ratio of the variances of the estimators approaches one, as does the ratio of their mean squared errors, the ratio of their standard deviations, the ratio of their mean absolute errors, and indeed the ratio of the expectation of any function of their errors.

Comments similar to those following property B(ii) apply to B(iii) and B(iv).

The final phrase of B(iv) implies that no matter what function of the error (and even of the true value as well) is chosen to represent the loss of mis-estimation, the ratio of the expected losses still approaches one if n_1 and n_2 approach infinity in such a way that $n_2/n_1 \to E_{1:2}(\theta)$. The conditions needed to prove that this property holds depend of course on the loss function, and may be strong.

6.4 Confidence Bounds

We consider next two confidence bounds $T_{1,n}$ and $T_{2,n}$ for the same quantity $\mu(\theta)$ and state four properties, which we may expect the asymptotic relative efficiency $E_{1:2}(\theta)$ to have and could use to define asymptotic relative efficiency for confidence bounds.

C(i). When the true distribution is given by θ, the difference between the c.d.f. of $T_{1,n}$ at $\mu(\theta) + \delta$ and the c.d.f. of $T_{2,n}$ at $\mu(\theta) + \delta\sqrt{E_{1:2}(\theta)}$ approaches zero uniformly in δ as the common sample size $n \to \infty$.

In symbols,

$$P_\theta[T_{1,n} \le \mu(\theta) + \delta] - P_\theta[T_{2,n} \le \mu(\theta) + \delta\sqrt{E_{1:2}(\theta)}] \to 0 \qquad (6.8)$$

uniformly in δ as $n \to \infty$.

This implies that, in a relative sense, the two confidence bounds have asymptotically the same distribution except for a scale factor $1/\sqrt{E_{1:2}(\theta)}$. The discussions of scaling, uniformity, and uniqueness of E in Sects. 6.2 and 6.3 apply here with little change and will not be repeated.

C(ii). When the true distribution is given by θ, the ratio of the expectation of $T_{1,n} - \mu(\theta)$ to the expectation of $T_{2,n} - \mu(\theta)$ approaches $1/\sqrt{E_{1:2}(\theta)}$ as $n \to \infty$. The ratio of the expected overestimations, that is, of the expectations of $[T_{1,n} - \mu(\theta)]^+$ and $[T_{2,n} - \mu(\theta)]^+$, and the ratio of expected underestimations $[\mu(\theta) - T_{1,n}]^+$ and $[\mu(\theta) - T_{2,n}]^+$, also approach $1/\sqrt{E_{1:2}(\theta)}$ as $n \to \infty$.

As with property B(ii), C(ii) does not strictly follow from C(i), and statements about asymptotic relative efficiency often need additional justification and may need qualification to apply to property C(ii).

The definition of $[T_{i,n} - \mu(\theta)]^+$ and the reason for considering it when $T_{i,n}$ is an upper confidence bound were given in Section 4. The reason for considering $[\mu(\theta) - T_{i,n}]^+$ when $T_{i,n}$ is a lower confidence bound is analogous.

C(iii). For two confidence bounds with sample sizes n_1 and n_2 respectively, if the true distribution is given by θ and if n_1 and n_2 approach infinity simultaneously in such a way that $n_2/n_1 \to E_{1:2}(\theta)$, then the difference between the c.d.f.'s of the two confidence bounds converges uniformly to zero.

In symbols,

$$P_\theta[T_{1,n_1} \le t] - P_\theta[T_{2,n_2} \le t] \to 0 \qquad (6.9)$$

uniformly in t if $n_2/n_1 \to E_{1:2}(\theta)$.

C(iv). For two confidence bounds for μ with sample sizes n_1 and n_2 respectively, if the true distribution is given by θ, and if n_1 and n_2 approach infinity simultaneously in such a way that $n_2/n_1 \to E_{1:2}(\theta)$, then the ratio of their expected overestimations approaches one, as does the ratio of the expected values of any function of their differences from μ.

Comments similar to those following B(ii), B(iv), and C(ii) apply to C(iii) and C(iv).

6.5 Confidence Intervals

The final properties we introduce apply to two-sided confidence intervals.

D(i). For two confidence interval procedures with the same n, the upper and lower endpoints each have property C(i), and the ratio of the length of the first confidence interval to the length of the second converges in probability to $1/\sqrt{E_{1:2}(\theta)}$ as $n \to \infty$ when the true distribution is given by θ.

If $L_{1,n}$ and $L_{2,n}$ are the respective lengths, the convergence in probability means, by definition, that for every $\varepsilon > 0$,

$$P_\theta\left[\left|\frac{L_{1,n}}{L_{2,n}} - \frac{1}{\sqrt{E_{1:2}(\theta)}}\right| > \varepsilon\right] \to 0 \quad \text{as } n \to \infty. \tag{6.10}$$

Thus, in addition to property C(i) for the endpoints, property D(i) requires that the ratio of the lengths of the intervals has arbitrarily high probability of being arbitrarily near $1/\sqrt{E_{1:2}(\theta)}$ if n is large enough.

D(iii). For two confidence interval procedures with sample sizes n_1 and n_2 respectively, the upper and lower endpoints each have property C(iii), and the ratio of the length of the first confidence interval to the length of the second converges in probability to one if n_1 and n_2 approach infinity simultaneously in such a way that $n_2/n_1 \to E_{1:2}(\theta)$ when the true distribution is given by θ.

It is natural to define properties D(ii) and D(iv) in a similar way, requiring the endpoints to have properties C(ii) and C(iv) but requiring convergence in the ordinary sense of the ratio of the expectations of the lengths (or any functions of the lengths) in place of convergence in probability of the ratio of lengths. Statements about asymptotic relative efficiency often need additional justification and may need qualification to apply to such properties, because they involve expectations.

6.6 Summary

We have now outlined some precise properties of asymptotic relative efficiency. If we state in this book that two tests have asymptotic relative efficiency $E_{1:2}(\theta_0)$ under certain circumstances, then the two tests have the properties A(i)–(iii) under these circumstances. In the same way, the properties B, with the qualifications noted, define asymptotic relative efficiency for estimators, and the properties C and D for confidence procedures. The heuristic discussion of Sects. 2–4 motivated these definitions and suggests that they are satisfied in interesting situations. However, nothing has been proved in this section about their validity in any situation.

The heuristic discussion also suggests that related tests, estimators, and confidence procedures have the same asymptotic relative efficiency. As a result, it is not really misleading to think of the quantity $E_{1:2}(\theta)$ as being the same throughout. However, the definition of the asymptotic relative efficiency of tests, for instance, is quite independent of the definition for estimators and confidence procedures. Hence it is perfectly legitimate to talk about the asymptotic relative efficiency of tests without referring to any estimators or confidence procedures or to their relation to these tests. A rigorous definition of what it means for tests, estimators, and confidence procedures to be "related" has not been given here.

In the statistical literature, asymptotic relative efficiency is usually defined for tests by property A(ii), or a slightly weaker version of it (Problem 32). The other properties for tests are included here because they enrich the meaning of asymptotic relative efficiency and do generally hold, even though they may not be mentioned or their validity proved. Occasionally, reference is made to "Mood's definition" [Mood, 1954] of asymptotic relative efficiency. This applies to one-tailed tests and to two-tailed tests with equal tails. It ordinarily agrees with the usual definition in the case of two-tailed tests with equal tails, but it gives the square root of the usual value in the case of one-tailed tests (Problem 33). Definitions of the asymptotic relative efficiency of point estimators and confidence procedures do not appear frequently in the literature. The properties of these procedures which are most similar to A(ii) for tests are B(iii), C(iii), and D(iii).

7 Pitman's Formula

The asymptotic relative efficiency of two procedures was defined in Sect. 1 as the limit of the relative efficiency, that is, the limit of the ratio of the sample sizes needed to achieve the same performance. The heuristic discussion in Sects. 2–4 indicated that a single number describes the relative efficiency asymptotically for a wide range of measures of performance. This discussion also led to a formula, that given in (2.17) and usually called Pitman's formula, for computing asymptotic relative efficiency as the limiting ratio of two efficacies. We used this formula for calculation in the examples of Sect. 5.

In light of the discussion, we could reasonably define asymptotic relative efficiency as a quantity with the properties indicated in Sects. 2–4 and laid out in Sect. 6, or some specified subset of them. Pitman's formula is not a part of the definition of asymptotic relative efficiency, and numbers computed from it have these properties only under suitable conditions (which, for many of the properties, are very weak). However, the formula provides a convenient and widely applicable method for computation of asymptotic relative efficiencies, and we will use it further below. Nevertheless, we will not give sufficient conditions for the validity of the formula, nor prove formally that numbers computed from it have the properties given in Sect. 6.

For convenience and easier reference, Pitman's formula will now be repeated. Consider a family of distributions indexed by a one-dimensional parameter θ. Let T_n be a one-sample test statistic or point estimator and suppose that when the true distribution is given by θ, T_n is asymptotically normal with mean $\mu_n(\theta)$ and variance $\sigma_n^2(\theta)$, that is, $[T_n - \mu_n(\theta)]/\sigma_n(\theta)$ converges in distribution to the standard normal distribution. Assuming μ_n is differentiable, define the efficacy of T_n by

$$e_n(\theta) = [\mu_n'(\theta)]^2/\sigma_n^2(\theta) \tag{7.1}$$

and the limiting efficacy per observation or asymptotic efficacy by

$$e. = \lim[e_n(\theta)/n]. \tag{7.2}$$

For finite n, $\mu_n(\theta)$ and $\sigma_n^2(\theta)$ need not be the actual mean and variance of T_n, but may be any well-behaved functions of θ for which the asymptotic normality holds. This freedom of choice for μ_n and σ_n means that the efficacy $e_n(\theta)$ is not uniquely defined, but its indeterminacy is of second order only. Hence the asymptotic efficacy $e.$ is uniquely defined, because it depends on only the dominant term of e_n.

Now let $T_{1,n}$ and $T_{2,n}$ be two test statistics or two point estimators with asymptotic efficacies $e_1.$ and $e_2.$ respectively. The asymptotic relative efficiency of $T_{1,n}$ relative to $T_{2,n}$ is the ratio $e_1./e_2.$. However, it is not necessary in practice to divide the respective efficacies $e_{1,n}$ and $e_{2,n}$ by n or to compute the limits $e_1.$ and $e_2.$. The ratio of the limits can usually be found more easily as the limit of the ratio of the efficacies, $e_{1,n}(\theta)/e_{2,n}(\theta)$. Accordingly, we give Pitman's formula for a one-sample problem as follows.

$$E_{1:2}(\theta) = \lim \frac{e_{1,n}(\theta)}{e_{2,n}(\theta)} = \lim \left[\frac{[\mu_{1,n}'(\theta)]^2}{\sigma_{1,n}^2(\theta)} \cdot \frac{\sigma_{2,n}^2(\theta)}{[\mu_{2,n}'(\theta)]^2} \right]. \tag{7.3}$$

For problems with more than one sample, little modification is required beyond replacing n by the vector of sample sizes; for two samples, see Sect. 10.

Applying Pitman's formula to corresponding tests, point estimators, and even confidence bounds gives the same asymptotic relative efficiency. Whichever equivalent statistic is easiest may therefore be used. Most of the asymptotic efficiency results which are reported in the statistical literature are computed by Pitman's formula. Pitman's formula does not apply, however, to definitions of asymptotic relative efficiency based on different limiting operations, such as those mentioned in Sect. 6.1.

8 Asymptotic Relative Efficiencies of One-Sample Procedures for Shift Families

Now that we have a definition of asymptotic relative efficiency for tests, point estimators, and confidence procedures, and a convenient formula for its computation, we can find the actual value of the asymptotic relative efficiency for some procedures in various situations. In this section we consider one-sample procedures in the situation of a large class of densities which satisfy the shift model.

8.1 The Shift Model

A family of densities $f(x; \theta)$ is called a shift family if

$$f(x; \theta) = h(x - \theta) \tag{8.1}$$

for some specified density function h. For example, if h is normal with mean zero and variance σ^2, then $f(x; \theta)$ is normal with mean θ and variance σ^2. If $h(x) = e^{-|x|/\lambda}/(2\lambda)$ with λ specified, then $f(x; \theta)$ is the Laplace density given in Equation (5.1).

In general, f is the same as h except for a shift by the amount θ. The shift is to the right if $\theta > 0$ and to the left if $\theta < 0$. The density h can have any number of parameters, but the shift model is a one-parameter family indexed by θ, with any other parameters fixed. In other words, θ is the parameter of interest, and any other parameters are "nuisance" parameters.

In hypothesis testing, if the density h of a shift model satisfies the null hypothesis, then $\theta = 0$ belongs to the null hypothesis. The alternative $\theta > 0$, for instance, says that the population has the density h but shifted to the right by the amount θ. Hence the shift model is frequently referred to as the shift alternative for hypothesis testing.

The procedures which we shall study below under the shift model are all shift families of procedures. That is, each test has associated with it, for each θ_0, a test performed by subtracting θ_0 from every observation and then applying the original test. The null hypothesis for the second test is the same as that for the first except for a shift by the amount θ_0. Under a shift model, if $\theta = 0$ satisfies the first null hypothesis, then $\theta = \theta_0$ satisfies the second. Similarly, the estimators and confidence bounds will increase by the amount k if every observation is increased by k. Consequently, the asymptotic efficacies and relative efficiencies which we calculate will not depend on θ, and will apply to any value θ_0 (Problem 34; see also Problem 21d).

8.2 Asymptotic Relative Efficiencies for Specific Shift Families

In this subsection, we obtain the asymptotic efficacies and thereby the asymptotic relative efficiencies of various one-sample procedures for some specific families of densities of the shift form. These include the normal, Laplace (double exponential), logistic, Cauchy, and uniform densities and the more general symmetric beta and double exponential power families. The procedures to be considered, tests, estimators and confidence bounds, are the normal-theory procedures, the median procedures, the Wilcoxon signed-rank procedures, the general procedures based on sums of signed constants, including the squared ranks and normal scores procedures, and procedures based on the randomization distribution of the sample mean. We do not consider the Walsh procedures, described in Sect. 7.4 of Chap. 3, because there is no general rule for constructing them for arbitrary sample sizes. They are designed ad hoc for sample sizes of 15 or less and hence their asymptotic efficiency is undefined. Kolmogorov–Smirnov procedures are also omitted for now, since their asymptotic behavior is fundamentally different and neither Pitman's formula nor the usual properties of asymptotic relative efficiency apply to them. They will be treated in Sect. 11.

Procedures Considered and their Asymptotic Efficacies for Shifts

We start by listing the procedures we consider and formulas for their asymptotic efficacies for any shift family h. Thereafter we give numerical values for the shift families just mentioned.

(1) The *normal-theory procedures* include the t test for the null hypothesis that the population mean has a specified value (or the test based on the sample mean if the variance is assumed known), the sample mean as estimator of the population mean, and the confidence bounds for the population mean corresponding to the foregoing tests.

The efficacy is most conveniently computed from the estimator, the sample mean, and is $e_{1,n} = n/\sigma^2$ where σ^2 is the population variance, the variance of the density h (Problem 35a). Thus the asymptotic efficacy is

$$e_{1.} = \lim e_{1,n}/n = 1/\sigma^2. \tag{8.2}$$

(2) The *median procedures* (in the sense of Sect. 5) are the sign test for the null hypothesis that the population median has a specified value, the sample median as estimator of the population median, and the corresponding confidence bounds for the population median which have appropriate order statistics as endpoints.

The efficacy is conveniently computed from the test statistic, the number of observations smaller than the median value specified by the null hypothesis, and is $4nh^2(0)$ if h is a density with median 0 (Problem 35b). The asymptotic efficacy is then

$$e_{2.} = 4h^2(0). \tag{8.3}$$

(3) The *Wilcoxon procedures* are the Wilcoxon signed-rank test for the null hypothesis that the population is symmetric about a specified value, the corresponding estimator of the population center of symmetry, namely the median of the set of Walsh averages (see Sect. 3.2), and the corresponding confidence bounds (Sect. 4, Chap. 3).

The Wilcoxon procedures are not valid in general, even asymptotically, unless h is symmetric about zero. If h is symmetric about zero, the efficacy is most easily computed from the test statistic, and is (Problem 35c)

$$e_{3,n} = \frac{[2n(n-1) \int h^2(x) \, dx + 2nh(0)]^2}{n(n+1)(2n+1)/6}. \tag{8.4}$$

The asymptotic efficacy is therefore

$$e_{3.} = 12\left[\int h^2(x) \, dx\right]^2. \tag{8.5}$$

(4) The *procedures based on sums of signed contants* (Sect. 7, Chap. 3) use, for each n, a set of n constants $c_{n1}, c_{n2}, \ldots, c_{nn}$, where the first subscript n is now included to designate explicitly the dependence of the procedures on n. The test statistic is $\sum_{j=1}^{n} c_{nj} S_j$, where $S_j = \pm 1$ and the sign is the sign of the jth smallest observation in order of absolute value. The relevant procedures are this test and the corresponding estimator and confidence procedures for the population center of symmetry. (The estimator is the amount which must be subtracted from every observation to make the test statistic have the value zero.)

These procedures are generally valid only if h is symmetric about zero. It will be proved at the end of this section that if h is symmetric about zero and sufficiently regular, then the efficacy may be expressed as

$$e_{4,n} = \left\{\sum_{j=1}^{n} c_{nj} E_h[h'(|X|_{(j)})/h(|X|_{(j)})]\right\}^2 \bigg/ \sum_{j=1}^{n} c_{nj}^2 \tag{8.6}$$

where, in the expectation, $|X|_{(1)} < \cdots < |X|_{(n)}$ are the absolute values of a sample of n observations from the density h arranged in increasing order of size. (This notation is imprecise, since the meaning and distribution of $|X|_{(j)}$ depend on n as well as j.) The asymptotic efficacy is $\lim e_{4,n}/n$, and will exist only under some restriction on the c_{nj}.

A common situation is that

$$c_{nj} = b_n c\left(\frac{j - \frac{1}{2}}{n}\right) + \text{remainder}, \tag{8.7}$$

where b_n is a constant for each n, $c(u)$ is a function of u defined for $0 < u < 1$, and the remainders are small enough not to contribute asymptotically. For example, the sign test statistic has this form with $c_{nj} = 1, c(u) = 1$, and $b_n = 1$. The Wilcoxon signed-rank test statistic has this form with $c_{nj} = j, c(u) = u$, and $b_n = n$. The factor b_n has no effect on the test and is included in (8.7) for the convenience of using the test statistics as previously defined. For example, without b_n the Wilcoxon signed-rank test would have $c_{nj} = j/n$, an equivalent but inconvenient form. The reason for evaluating $c(u)$ at $(j - \frac{1}{2})/n$ rather than j/n in (8.7) is to allow the possibility that $c(u) \to \infty$ as $u \to 1$, in which case $c(j/n)$ could not be used for $j = n$.

Under suitable conditions on c and the remainder terms, if h is symmetric about zero and sufficiently regular, then we will show at the end of this section that the asymptotic efficacy when (8.7) holds is

$$e_{4\cdot} = 4\left\{\int_{1/2}^1 c(2u - 1)h'[H^{-1}(u)]/h[H^{-1}(u)]\,du\right\}^2 \Big/ \int_0^1 c^2(u)\,du$$

$$= 4\left\{\int_0^\infty c[2H(x) - 1]h'(x)\,dx\right\}^2 \Big/ \int_0^1 c^2(u)\,du, \qquad (8.8)$$

where H is the c.d.f. of the density h and H^{-1} is its inverse. Alternative expressions appear in Problem 36.

One density h to which (8.6) and (8.8) do not apply is the uniform, for which, if the range is $[-1, 1]$, (Problem 37)

$$e_{4, n} = nc_{nn}^2 \Big/ \sum_{j=1}^n c_{nj}^2. \qquad (8.9)$$

When (8.7) holds with suitable conditions on c, the asymptotic efficacy for the uniform density is (Problem 37)

$$e_{4\cdot} = c^2(1) \Big/ \int_0^1 c^2(u)\,du. \qquad (8.10)$$

(5) The *normal scores procedures* are of the foregoing type and satisfy (8.7) with $c(u) = \Phi^{-1}[(1 + u)/2]$ where Φ is the standard normal c.d.f. For example, we might take c_{nj} as the quantile of order $(j - \frac{1}{2})/n$ or $j/(n + 1)$ of the absolute value of a standard normal random variable, that is, the $(n + j - \frac{1}{2})/2n$ or $(n + j + 1)/2(n + 1)$ quantile of the standard normal distribution, or take $c_{nj} = E[|Z|_{(j)}]$ where $|Z|_{(1)} < \cdots < |Z|_{(n)}$ are the ordered absolute values of a sample of n from the standard normal distribution (Sect. 9, Chap. 3).

When h is symmetric about zero and sufficiently regular, the asymptotic efficacy of any normal scores procedure can be expressed as (Problem 38)

$$e_{5\cdot} = 4\left\{\int_0^\infty \Phi^{-1}[H(x)]h'(x)\,dx\right\}^2 = \left\{\int_{-\infty}^\infty \Phi^{-1\prime}[H(x)]h^2(x)\,dx\right\}^2$$

$$= \left\{\int_0^1 \Phi^{-1\prime}(t)h[H^{-1}(t)]\,dt\right\}^2 \qquad (8.11)$$

where H is the c.d.f. of h, Φ^{-1} is the inverse of the standard normal distribution, and $\Phi^{-1\prime}$ is the derivative of this inverse function.

(6) The *procedures based on the randomization distribution* are the randomization test of the null hypothesis of symmetry around a specified value and the corresponding estimator and confidence procedures, as described in Sects. 2.1 and 2.2 of Chap. 4. With the sample mean as the test statistic, the efficacy and limiting efficacy per observation are the same as for the normal-theory procedures. However, further argument is needed to justify this

method of calculation in the case of the randomization procedures. Since the test compares the sample mean with a quantile of its conditional distribution given the absolute values of the observations, the critical value of the sample mean is not fixed but depends on the sample in a complicated way – more complicated than substituting an estimate for a nuisance parameter and adjusting a constant as in Sect. 2.2.

Asymptotic Relative Efficiencies for Specific Shift Families

The asymptotic efficiency of any one of the foregoing listed procedures relative to any other is now easily obtained by dividing the asymptotic efficacy of the first procedure by that of the second. The resulting asymptotic relative efficiency applies, of course, to the particular shift family specified by the density h, and the two procedures must both be valid for this family, at least asymptotically. If h is symmetric about zero, all the foregoing procedures are asymptotically valid. Otherwise, the circumstances and population parameters for which they are valid vary from one procedure to another. For this reason, all the examples below use densities h which are symmetric about zero.

Unless we are confident of symmetry, the first question in choosing a procedure is not efficiency but whether the parameter of interest is really the mean, the median, or something else, as discussed earlier (Sect. 3.2, Sect. 5).

We also point out that a change of scale, such as a change of standard deviation for the normal distribution, multiplies the efficacies of each of the foregoing procedures by the same factor, and therefore leaves their asymptotic relative efficiencies unchanged. Specifically, if $h(x)$ is symmetric about zero, replacing $h(x)$ by $ah(ax)$ for some $a > 0$ multiplies all the foregoing efficacies by a^2. This can be seen by direct calculation or by consideration of what happens when all observations are divided by the value a (Problem 39). Consequently, the asymptotic efficiency of one procedure relative to another obtained for a particular density h applies also to any rescaling of h, that is, also to $ah(ax)$ for all $a > 0$.

Table 8.1 gives the numerical values of the asymptotic efficacies of the foregoing procedures for the shift families generated by various symmetric densities h. The asymptotic relative efficiencies can be found simply by taking quotients. We now explain this table.

The column labeled "Mean" in Table 8.1 applies to procedures discussed in (1) and (6) earlier, namely, the normal-theory procedures, including those based on the t distribution, and the randomization procedures based on the sample mean. The columns headed "Median," "Wilcoxon," and "Normal scores" apply to the procedures so designated earlier (Numbers (2), (3), and (5)). These three procedures belong to the class of procedures based on sums of signed constants, discussed in Number (4), for the constants $c_{nj} = 1, j$, and normal scores respectively. The column headed "Squared ranks"

Table 8.1 Asymptotic Efficacies and Asymptotic Relative Efficiencies of Some One-Sample Procedures for Some Shift Families

Density $h(x)$		Procedure					
		Mean	Median	Wilcoxon	Squared ranks	Normal scores	Asymptotically efficient
$ce^{-\|ax\|^k}$ (double exponential power) $c = ak/2\Gamma(1/k)$ $a = \Gamma(1/k)k$ $= \Gamma(1+k^{-1})$	$k = 0.5$	0.033	1.00	0.19	0.084	0.11	∞
	$k = 0.75$	0.21	1.00	0.47	0.29	0.39	1.21
	$k = 1$ (Laplace)	0.50	1.00	0.75	0.56	0.64	1.00
	$k = 1.5$	1.10	1.00	1.19	1.08	1.14	1.21
	$k = 2$ (Normal)	1.57	1.00	1.50	1.54	1.57	1.57
	$k = 4$	2.43	1.00	2.12	2.67	2.75	3.33
	$k = 10$	2.88	1.00	2.61	3.84	4.41	9.15
	$k = \infty$ (uniform)	3.00	1.00	3.00	5.00	∞	∞
$c(1 - a^2x^2)^{r-1}$, $\|x\| \le 1/a$; (symmetric beta) $c = a/2^{2r-1}B(r,r)$ $a = 4^{r-1}B(r,r)$ (Normal)	$r = 1$						
	$r = 1.5$	2.47	1.00	2.16	2.90	4.11	∞
	$r = 2$	2.22	1.00	1.92	2.36	2.82	∞
	$r = 3$	1.99	1.00	1.74	2.00	2.18	2.88
	$r = 4$	1.88	1.00	1.67	1.85	1.97	2.20
	$r = \infty$ (Normal)	1.57	1.00	1.50	1.54	1.57	1.57
$ae^{ax}/(e^{ax} + 1)^2$,	$a = 2$ (logistic)	1.22	1.00	1.33	1.25	1.27	1.33
$a/\pi(1 + a^2x^2)$,	$a = \pi/2$ (Cauchy)	0	1.00	0.75	0.44	0.53	1.23

Note: Entries are asymptotic efficacies for values of a specified. They are asymptotic efficiencies relative to the median procedures and their ratios are asymptotic relative efficiencies for all values of a.

applies to procedures based on sums of signed constants for the constants $c_{nj} = j^2$. The last column, headed "Asymptotically efficient," will be explained shortly.

Each row applies to the particular density h which is indicated and gives asymptotic efficacies for the value of a specified. Thus, for instance, the third row gives the asymptotic efficacies of the procedures just listed for the Laplace density $h(x) = e^{-|x|}/2$, that is for the Laplace family $f(x; \theta) = h(x - \theta) = e^{-|x-\theta|}/2$. The fifth row gives asymptotic efficacies for the normal density $h(x) = e^{-[\Gamma(1.5)x]^2}/2 = e^{-\pi x^2/4}/2$, which is not standard normal but has variance $2/\pi$. For each density h in the table, the scale has been chosen in such a way that $h(0) = \frac{1}{2}$ and hence, by (8.3), that the asymptotic efficacy of the median procedures is one.

The asymptotic relative efficiency of any two procedures for any density h in the table is simply the quotient of the appropriate two entries in the row corresponding to h. In particular, the individual entries are asymptotic efficiencies relative to the median procedures. As explained before, the choice of scale affects the asymptotic efficacies but not the asymptotic relative efficiencies, which are therefore valid for all values of a. Thus, for instance, the quotients of the efficacies given in the fifth row are asymptotic relative efficiencies for a family of normal densities with any variance.

Densities of the first set in Table 8.1 are double exponential power densities, as in Equation (5.6) with $\lambda = 1/a$. The cases $k = 1$ and $k = 2$ are the Laplace and normal distributions already mentioned. The uniform distribution is approached as $k \to \infty$. In particular, for the value of a specified, the density approaches the uniform density on the interval $(-1, 1)$ (Problem 40).

The densities of the second set are symmetric beta densities shifted so that they are centered at 0 rather than at $\frac{1}{2}$ and rescaled. For $r = 1$, we have the uniform density, just as for $k \to \infty$ above. The two cases are accordingly combined in the table. The normal distribution is approached as $r \to \infty$ (Problem 41). Hence, it appears in two places in Table 8.1.

The logistic density was defined (without a scale parameter) in Equation (9.11) of Chap. 3. The logistic family $f(x; \theta) = h(x - \theta)$ used in Table 8.1 is the same as the two-parameter form given in Equation (7.2) of Chap. 5 with $\mu = 0$, $\sigma = 1/a$. Alternative expressions in terms of $\cosh z = (e^z + e^{-z})/2$ appear in Problem 42.

The right-hand column of Table 8.1 gives the Fisher information (per observation) for each family $f(x; \theta) = h(x - \theta)$ for the value of a specified. The *Fisher information* is defined as $I(\theta) = E_\theta[\partial \log f(X, \theta)/\partial \theta]^2$ where X has density $f(x; \theta)$, and is in general a function of θ. For the shift family it becomes the constant

$$I = E_h[h'(X)/h(X)]^2 \qquad (8.12)$$

where X has density $h(x)$ (Problem 43). The Cramér-Rao inequality (see also Sect. 2.1 and Problem 2) states that, under certain regularity conditions, the variance of any estimator is at least $[1 + b'(\theta)]^2/nI(\theta)$ where b is the bias of the estimator of θ and b' is its derivative. Since $1 + b'(\theta) = \mu'(\theta)$, this is equivalent to saying that the efficacy per observation cannot exceed $I(\theta)$. Hence the same is true of the asymptotic efficacy, and accordingly the entry in the last column of each row of Table 8.1 is at least as large as every other entry in that row. A procedure with asymptotic efficacy equal to the Fisher information in a particular situation is said to be *asymptotically efficient* in that situation, provided the properties of asymptotic efficacy we have given, such as (2.11), hold. (These properties preclude the phenomenon of "superefficiency" illustrated in Problem 44, and hence so must the regularity conditions which would be required for rigorous proofs of these properties.)

For any "well-behaved" one-parameter family of densities $f(x; \theta)$, asymptotically efficient procedures exist. One is maximum likelihood estimation. The usual tests and confidence procedures based on maximum likelihood are ordinarily invalid outside the family in question, but randomization procedures based on the maximum likelihood estimator for a symmetric shift-parameter family are valid for all symmetric distributions and have the same asymptotic efficacy and hence efficiency (although the properties of efficacy require special justification for randomization procedures, as discussed at (6) earlier in the normal-theory case). Another asymptotically efficient procedure for a symmetric shift-parameter family is the locally most powerful signed-rank test for that family, as discussed in Sect. 8.4; this test and the corresponding confidence procedures are also valid for all symmetric distributions. Note, however, that in either case the procedure depends on h, and the final column of Table 8.1 does not represent any single procedure.

For a given family $f(x; \theta) = h(x - \theta)$, the asymptotic efficiency of any procedure of Table 8.1 relative to a procedure which is asymptotically efficient for that family can be found as the ratio of the appropriate entry in Table 8.1 to the last entry in the same row, but the asymptotic efficiency relative to a procedure which is asymptotically efficient for a different family is not generally available from this table. As before, results on asymptotic efficiency are not affected by a change of scale.

The median, Wilcoxon, and normal-scores procedures correspond to locally most powerful signed-rank tests for the Laplace, logistic, and normal distributions respectively, as we shall see in Sect. 8.4. Each is, therefore, asymptotically efficient for its respective distribution and has efficacy equal to the Fisher information in the appropriate row of Table 8.1.

The asymptotic efficiencies developed in the example of Sect. 5 for the median procedures relative to the mean (normal-theory) procedures can be verified from Table 8.1. For the Laplace family, this asymptotic relative efficiency is $1.00/0.50 = 2.00$, in agreement with (5.4). For the normal family, $1.00/1.57 = 0.64$ agrees with (5.5), and for the uniform family, $1.00/3.00$

agrees with (5.8). The graph in Fig. 5.1 includes these values and several others which can be verified similarly.

This concludes our explanation of Table 8.1. The numbers in the table can all be obtained from Equations (8.2), (8.3), (8.5) and (8.8)–(8.11). These calculations, except those which require numerical integration, are left to the reader as Problems 45–52.

*Derivations

We will first derive formula (8.6) for the efficacy of the procedures based on sums of signed constants c_{nj}, and then formula (8.8) for the asymptotic efficacy. We assume that h is symmetric about zero and sufficiently regular to justify the steps to follow.

The test statistic for the null hypothesis $\theta = 0$ is

$$T_n = \sum_{j=1}^{n} S_j c_{nj} \tag{8.13}$$

where $S_j = 1$ or -1 according as the jth observation in order of absolute value is positive or negative. Under the null hypothesis $\theta = 0$, the S_j are independent and identically distributed (Chap. 3) with

$$P(S_j = 1) = P(S_j = -1) = \tfrac{1}{2}, \tag{8.14}$$

and therefore T_n has null mean and variance (Problem 80, Chap. 3)

$$\mu_n(0) = E_0(T_n) = 0, \qquad \sigma_n^2(0) = \sum_{j=1}^{n} c_{nj}^2. \tag{8.15}$$

The remaining problem is to evaluate

$$\mu_n'(0) = \frac{d}{d\theta} E_\theta(T_n) \bigg|_{\theta=0} = \sum_{j=1}^{n} c_{nj} \frac{d}{d\theta} E_\theta(S_j) \bigg|_{\theta=0}. \tag{8.16}$$

Let $|X|_{(1)}, \ldots, |X|_{(n)}$ be the absolute values of the observations ordered from smallest to largest. Given $|X|_{(1)}, \ldots, |X|_{(n)}$, the conditional probability that $S_j = 1$ is (Problem 53)

$$P_\theta(S_j = 1 | |X|_{(1)}, \ldots, |X|_{(n)}) = \frac{f(|X|_{(j)}; \theta)}{f(|X|_{(j)}; \theta) + f(-|X|_{(j)}; \theta)}. \tag{8.17}$$

Let $f^*(z_1, \ldots, z_n; \theta)$ be the joint density of $|X|_{(1)}, \ldots, |X|_{(n)}$ at $z_1, \ldots, z_n$. Then

$$
\begin{aligned}
E_\theta(S_j) &= P_\theta(S_j = 1) - P_\theta(S_j = -1) = 2P_\theta(S_j = 1) - 1 \\
&= 2E_\theta[P_\theta(S_j = 1 | |X|_{(1)}, \ldots, |X|_{(n)})] - 1 \\
&= 2 \int \frac{f(z_j; \theta)}{f(z_j; \theta) + f(-z_j; \theta)} f^*(z_1, \ldots, z_n; \theta) dz_1 \ldots dz_n - 1. \quad (8.18)
\end{aligned}
$$

Since $f(z; \theta) = h(z - \theta)$ and h is symmetric about 0, the derivative of (8.18) at $\theta = 0$ (provided it is legitimate to differentiate under the integral sign) is

$$\frac{d}{d\theta} E_\theta(S_j)\bigg|_{\theta=0} = -\int \frac{h'(z_j)}{h(z_j)} f^*(z_1, \ldots, z_n; 0)\, dz_1 \ldots dz_n$$

$$+ \frac{d}{d\theta} \int f^*(z_1, \ldots, z_n; \theta)\, dz_1 \ldots dz_n \bigg|_{\theta=0}$$

$$= -E_0[h'(|X|_{(j)})/h(|X|_{(j)})]. \tag{8.19}$$

Substituting this result in (8.16), and then (8.16) and (8.15) in Pitman's Formula (7.1) gives the efficacy in (8.6).

Having the efficacy (8.6), we now seek a formula for the asymptotic efficacy under assumption (8.7). The denominator of (8.6) is

$$\sum_{j=1}^{n} c_{nj}^2 = b_n^2 \sum_{j=1}^{n} c^2 \left(\frac{j - \frac{1}{2}}{n} \right) + \text{remainder}$$

$$= nb_n^2 \int_0^1 c^2(u)\, du + \text{remainder}, \tag{8.20}$$

since the integral can be approximated by the sum above it divided by n. The numerator of (8.6) is a little more complicated to handle, but the same idea can be used in the following manner. When the observations have density h, their absolute values have c.d.f. $2H - 1$ on $x > 0$. Then, for n large, the probability is high that $|X|_{(j)}$ will be close to the pth quantile of this c.d.f., where $p = j/n$ or $(j - \frac{1}{2})/n$ or anything else which is asymptotically equivalent. Since the p-point of $2H - 1$ is $H^{-1}(\frac{1}{2} + \frac{1}{2}p)$, the bracketed quantity in (8.6) is

$$\sum_{j=1}^{n} c_{nj} E_h \left[\frac{h'(|X|_{(j)})}{h(|X|_{(j)})} \right]$$

$$= b_n \sum_{j=1}^{n} c \left(\frac{j - \frac{1}{2}}{n} \right) \frac{h'\{H^{-1}[\frac{1}{2} + \frac{1}{2}(j - \frac{1}{2})/n]\}}{h\{H^{-1}[\frac{1}{2} + \frac{1}{2}(j - \frac{1}{2})/n]\}} + \text{remainder}$$

$$= 2nb_n \int_{1/2}^{1} c(2u - 1) \frac{h'[H^{-1}(u)]}{h[H^{-1}(u)]}\, du + \text{remainder}. \tag{8.21}$$

Dividing the square of this by n times the quantity (8.20) and letting $n \to \infty$ gives the first formula of (8.8). The second formula follows immediately.*

8.3 Bounds on Asymptotic Relative Efficiencies for Shift Families

We now turn to an investigation of bounds on asymptotic relative efficiencies. As before the discussion will be limited to shift families of densities $f(x; \theta) = h(x - \theta)$. We are mainly concerned with symmetric, unimodal densities h. However, all we require, unless further assumptions are specifically stated, is that h be such that those procedures under discussion are valid, at least asymptotically.

We consider first the asymptotic efficiency of the normal-theory (mean) procedures relative to the median procedures. Notice that its value in Table 8.1 ranges from zero for the Cauchy distribution to 3 for the uniform distribution. (Its value is zero for any density h which is positive at its median and has infinite variance.) As it happens, the largest value possible for any shift family given by a density h which has its maximum at its median is 3, although an asymptotic efficiency of infinity is possible for a symmetric, multimodal density h (Problem 54). Thus, for $f(x; \theta) = h(x - \theta)$, the asymptotic efficiency of the normal-theory procedures with respect to the median procedures can be zero, even if h is required to be symmetric and unimodal; it can be infinite if h is unrestricted or merely assumed symmetric; and it can be as large as 3, but no larger, if h has its maximum at its median or, therefore, if h is symmetric and unimodal (because this implies that h has its maximum at its median).

Next we consider the asymptotic efficiency of the Wilcoxon procedures relative to the normal-theory procedures. It can be infinite even if h is symmetric and unimodal (the Cauchy family again provides an example), and it can be as small as but no smaller than $108/125 = 0.864$ if h is symmetric, whether or not it is also required to be unimodal. The value 0.864 is achieved by the symmetric beta density with $r = 2$. This density is a quadratic function with a negative squared term, except that where the quadratic function is negative the density is zero. A proof that 0.864 is minimum appears at the end of this subsection under (1).

The asymptotic efficiency of the Wilcoxon procedures relative to the median procedures can be arbitrarily small, even if h is symmetric and unimodal; it can be infinite if h is unrestricted or merely symmetric; and it can be as large as 3 but not larger if h has its maximum at its median or is symmetric and unimodal (Problem 55). The uniform distribution gives the value 3, while the double exponential power family in Equation (5.6) gives values which approach zero as $k \to 0$ (Problem 55).

The normal scores procedures are at least as efficient asymptotically as the normal-theory procedures for all symmetric h and may be infinitely more so. The asymptotic efficiency of the normal scores procedures relative to the normal-theory procedures is infinite for h either Cauchy or uniform; it is one if h is normal; and it is never less than one if h is symmetric. These results are proved later under (2). Clearly, a requirement of unimodality would not improve the bounds.

The asymptotic efficiency of the normal scores procedures relative to the median procedures can be infinite and it can be arbitrarily small even if h is symmetric and unimodal. The uniform distribution provides an example of the former, and Problem 56 of the latter.

The asymptotic efficiency of the normal scores procedures relative to the Wilcoxon procedures can be infinite, even if h is symmetric and unimodal,

Table 8.2 Bounds on Asymptotic Relative Efficiencies of One-Sample Procedures for Shift of a Symmetric Density

	Mean		Median		Wilcoxon		Normal scores	
Mean	1	1	0	3^a	0	$\frac{125}{108}$	0	1
Median	$\frac{1}{3}{}^a$	∞	1	1	$\frac{1}{3}{}^a$	∞	0	∞
Wilcoxon	$\frac{108}{125}$	∞	0	3^a	1	1	0	$\frac{6}{\pi}$
Normal scores	1	∞	0	∞	$\frac{\pi}{6}$	∞	1	1

^{a} If unimodality is not assumed, $\frac{1}{3}$ must be replaced by 0 and 3 by ∞.

and it must be more than but can come arbitrarily close to $\pi/6 = 0.524$ if h is symmetric, whether or not h is also required to be unimodal. The uniform distribution again provides an example of the former, and the proof of the latter is requested in Problem 57.

Table 8.2 summarizes most of the foregoing bounds. Each cell relates to the procedures designating its row and column and contains the greatest lower bound and the least upper bound on the asymptotic efficiency of the row procedure relative to the column procedure for shifts of a symmetric, unimodal density. If unimodality is not assumed, the bounds are the same except in the cases footnoted. Notice that the table is symmetric, in the sense that, for instance, the greatest lower bound for the median relative to the mean is $\frac{1}{3}$, while the least upper bound for the mean relative to the median is the reciprocal of $\frac{1}{3}$ or 3.

*Proofs

(1) What is the minimum asymptotic efficiency of the Wilcoxon procedures relative to the normal theory procedures for shifts of a density h which is symmetric about zero? By (8.4) and (8.6), the quantity to be minimized is

$$E_{3:1} = 12\sigma^2 \left[\int h^2(x)\, dx \right]^2. \tag{8.22}$$

Since the scale has no effect, we can fix

$$\sigma^2 = \int x^2 h(x)\, dx, \tag{8.23}$$

and then our problem is to minimize $\int h^2(x)\, dx$ subject to the condition (8.23). For any density h satisfying this condition, we have

$$\int h^2(x)\, dx = \int h(x)[h(x) - (a - bx^2)]\, dx + a - b\sigma^2 \tag{8.24}$$

for any values of the "undetermined multipliers" a, b. The integrand on the right-hand side of (8.24) is minimized over all nonnegative $h(x)$ by $h_{a,b}(x) = [a - bx^2]^+/2$, and hence so is the entire expression. We now choose a, b so that $h_{a,b}$ is a density satisfying (8.23), as it is possible to do. For this a, b, the function $h_{a,b}$ still minimizes the right-hand side of (8.24) among nonnegative functions and a fortiori among densities satisfying (8.23). For this $h_{a,b}$ the asymptotic relative efficiency (8.22) is easily calculated as $108/125$. Since $h_{a,b}$ has the symmetric beta form of Table 8.1 with $r = 2$, this value is already available from the table. This result was given by Hodges and Lehmann [1956].

(2) We shall now minimize the asymptotic efficiency of the normal-scores procedures relative to the normal theory procedures for shifts of a density h which is symmetric about zero. The asymptotic relative efficiency by (8.2) and (8.11) is

$$E_{5:1} = \sigma^2 Q^2 \tag{8.25}$$

where

$$Q = \int_{-\infty}^{\infty} \Phi^{-1\prime}[H(x)]h^2(x)\,dx. \tag{8.26}$$

The trivial result

$$z \geq 2 - (1/z) \quad \text{for } z > 0 \tag{8.27}$$

applied to (8.26) with

$$z = \Phi^{-1\prime}[H(x)]h(x) = h(x)/\phi\{\Phi^{-1}[H(x)]\} \tag{8.28}$$

gives

$$Q \geq 2 - \int_{-\infty}^{\infty} \phi\{\Phi^{-1}[H(x)]\}\,dx. \tag{8.29}$$

Integrating by parts then gives

$$Q \geq 2 - x\phi\{\Phi^{-1}[H(x)]\}\Big|_{-\infty}^{\infty} - \int x\Phi^{-1}[H(x)]h(x)\,dx. \tag{8.30}$$

The next-to-last term of Equation (8.30) vanishes if $\sigma^2 < \infty$ (Problem 58a); then applying the Schwarz inequality (or the fact that a correlation is at most one) to the last term gives

$$Q \geq 2 - \left(\sigma^2 \int \left\{\Phi^{-1}[H(x)]\right\}^2 h(x)\,dx\right)^{1/2}. \tag{8.31}$$

The integral in Equation (8.31) is equal to one (Problem 58b), and we may assume without loss of generality that $\sigma^2 = 1$ since we can rescale H if necessary. Then $Q \geq 1$ and

$$E_{5:1} = Q^2 \geq 1; \tag{8.32}$$

equality holds if and only if $H = \Phi$. Therefore $E_{5:1} \geq 1$ with equality if and only if h is normal. $\qquad\square$

Proof (2) was adapted from Chernoff and Savage [1958] and is similar to that of Gastwirth and Wolff [1968] (Problem 59). The reader may wish to consult these papers for further insight.

*8.4 Asymptotically Efficient Signed-Rank Procedures

In this subsection we will give explicitly the locally most powerful signed-rank test for a symmetric shift alternative, show that its asymptotic efficacy equals the Fisher information and hence that the test and related estimators and confidence procedures are asymptotically efficient for the specified alternative, and discuss the possibility of an "adaptive" procedure which would be asymptotically efficient for all symmetric shifts simultaneously.

The signed-rank test which is locally most powerful for a shift family of densities $f(x; \theta) = h(x - \theta)$ with h symmetric about zero is, by (9.8) of Chap. 3, based on the sum of signed constants

$$c_{nj} = -E_h[h'(|X|_{(j)})/h(|X|_{(j)})] \tag{8.33}$$

where the $|X|_{(j)}$ are as usual the ordered absolute values of a sample from the population with distribution h (Problem 60). (The meaning of h is different in Chap. 3.) The efficacy of this procedure for the family $f(x; \theta) = h(x - \theta)$ is, by (8.6), equal to

$$e_{4,n} = \sum_{j=1}^{n} c_{nj}^2. \tag{8.34}$$

It can be shown (see below) that the asymptotic efficacy of this procedure is

$$e_4 = E_h[h'(X)/h(X)]^2 = I, \tag{8.35}$$

where I is the Fisher information given in (8.12) for the family $f(x; \theta) = h(x - \theta)$. As discussed earlier (see Sect. 8.2), no procedure satisfying suitable regularity conditions has asymptotic efficacy greater than the Fisher information. Therefore the locally most powerful signed-rank procedure for the family $f(x; \theta) = h(x - \theta)$ is asymptotically efficient for this family, and it has asymptotic efficiency at least one relative to every other procedure satisfying the regularity conditions. This applies, of course, to the corresponding estimators and confidence procedures as well as tests. Exact regularity conditions which are also simple are hard to find, but the foregoing statements certainly apply to most standard nonparametric and parametric procedures.

We now have an asymptotically efficient procedure for any given h. Is there a procedure which is asymptotically efficient for all h simultaneously?

Apparently so, if we first estimate h and then use a procedure which is asymptotically efficient for the estimated h. Two difficulties arise here, namely, how to estimate the density h, and how the preliminary estimation affects the overall properties of the procedure, such as the level and power of a test. Suppose that we always use a test of $\theta = 0$ which has conditional level α given the absolute values $|X_1|, \ldots, |X_n|$. Then the test chosen may depend on the absolute values in any way we like without affecting the level α. We can take advantage of this by using the absolute values to estimate h. If the observations have a density h which is symmetric about zero for all x, then their absolute values have density $2h$ for $x > 0$. Suppose we estimate h from the absolute values as if they had density $2h$, and then find the locally most powerful signed-rank test for the estimated h at level α. The overall level will be α. Under the alternative $f(x; \theta) = h(x - \theta)$, the absolute values will have density $h(x - \theta) + h(x + \theta)$, so we will be treating an estimate of this as if it were an estimate of $2h(x)$ in choosing our test. Presumably, this will make little difference if $|\theta|$ is small, while otherwise the power will be close to one in large samples anyway. Our procedure will therefore have asymptotically as good power everywhere as the locally most powerful signed-rank test for the h which actually obtains, whatever it may be. Thus we have obtained a test which is asymptotically efficient for all h simultaneously. The corresponding estimators and confidence procedures would be even more complicated to compute but would also be asymptotically efficient.

It sounds marvelous to have a procedure which is asymptotically efficient for all h simultaneously, but there are severe limitations. The density h cannot be estimated at all without some assumption about its smoothness. Even under such an assumption, it cannot be estimated reliably except in very large samples. The density is especially hard to estimate in the tails, which have a strong influence on the procedure to be used. It appears, therefore, that the sample size must be extremely large before the simultaneous asymptotic efficiency of the foregoing procedure for all h will really make itself felt. Aside from this, it should be noted that the efficiency property applies only to shift parameter families. The foregoing procedure will not ordinarily be asymptotically efficient for a family not of the form $f(x; \theta) = h(x - \theta)$. For example, the locally most powerful signed-rank test for such a family will ordinarily be asymptotically more efficient. Presumably, no procedure is asymptotically efficient for all families $f(x; \theta)$ simultaneously, as the development of an asymptotically efficient procedure depends on making some assumption about the way in which the alternatives "close to" any particular null distribution approach that distribution.

These and other procedures using some sample information to choose among more "elementary," conditionally valid procedures have come to be called "adaptive." See also Stein [1956], Hogg [1974], and Sect. 3.2 of Chap. 4.

PROOF. The asymptotic efficacy of the locally most powerful signed-rank test given by (8.33) can be obtained by the argument used at (8.21). Since

c_{nj} is now equal to the expectation multiplying it in (8.21), this argument gives, for the efficacy (8.34),

$$\sum_{j=1}^{n} c_{nj}^2 = 2n \int_{1/2}^{1} \left[\frac{h'[H^{-1}(u)]}{h[H^{-1}(u)]} \right]^2 du + \text{remainder}$$

$$= 2n \int_{0}^{\infty} \left[\frac{h'(x)}{h(x)} \right]^2 h(x)\, dx + \text{remainder}$$

$$= nE_h \left[\frac{h'(X)}{h(X)} \right]^2 + \text{remainder}. \tag{8.36}$$

Formula (8.35) for the asymptotic efficacy follows. An essentially equivalent proof can be based on (8.7) and (8.8) (Problem 61).

Still another proof can be based on the following expression for the efficacy (Problem 62), which also provides some insight in itself:

$$\sum_{j=1}^{n} c_{nj}^2 = \sum_{j=1}^{n} E_h \left[\frac{h'(|X|_{(j)})}{h(|X|_{(j)})} \right]^2 - \text{var}_h \left[\frac{h'(|X|_{(j)})}{h(|X|_{(j)})} \right]$$

$$= nE_h \left[\frac{h'(X)}{h(X)} \right]^2 - \sum_{j=1}^{n} \text{var}_h \left[\frac{h'(|X|_{(j)})}{h(|X|_{(j)})} \right]. \tag{8.37}$$

The last term here represents the amount by which the efficacy falls short of the information in n observations.*   □

9 Asymptotic Relative Efficiency of Procedures for Matched Pairs

This section concerns observations which occur as matched pairs. As usual, we let X_j be the treatment-control difference in the jth pair and apply one-sample procedures and results to the n differences, $X_1, \ldots, X_n$. Thus the relevant distribution is now the distribution of the treatment-control differences. This affects the assumptions it is natural to make and hence the resulting consequences, as we will now discuss. The earlier discussions of matched pairs in Sect. 7 of Chap. 2 and Sect. 2 of Chap. 3 should be kept in mind and will not be repeated, although some overlap is inevitable.

9.1 Assumptions

To be specific, let the jth pair consist of a "control" measurement V_j and a "treatment" measurement W_j. Then if $X_j = W_j - V_j$, asymptotic relative efficiencies (and asymptotic power, etc.) can be obtained exactly as before from the distribution of the X_j. In particular, the appropriate probability distribution $f(x; \theta)$ to use in Sects. 5 and 8 is that of $X_j = W_j - V_j$, not that of W_j or V_j individually.

The question then is what assumptions the distribution of X_j might satisfy. Any family of densities $f(x; \theta)$ can arise from some family of joint distributions of (V_j, W_j), and any shift family $f(x; \theta) = h(x - \theta)$ arises from what might naturally be called a shift family of joint distributions of (V_j, W_j) (Problems 63a,b). The consequences of other assumptions about the distribution of (V_j, W_j) for the differences X_j are not entirely obvious however. Suppose, for example, that for some constant θ, all the V_j and $W_j - \theta$ are independent and identically distributed. Then varying θ gives rise to a shift parameter family of distributions for X_j, but in addition,

$$X_j - \theta = [(W_j - \theta) - V_j]$$

is the difference of two independent, identically distributed random variables. By no means every shift family can arise in this way. In particular, X_j must be symmetrically distributed around θ. This is not the only restriction, however, and several symmetric densities in Table 8.1, for example, including the uniform and quadratic (beta with $r = 2$), are excluded. The same restrictions arise under broader assumptions, as will be discussed further below.

The assumption that all the V_j and $W_j - \theta$ are independently, identically distributed, although a special case of assumptions to follow, is unrealistic in the context of matched pairs, because it implies that the pairing accounts for none of the variability, contrary to what we expect when we pair.

We now consider several more natural assumptions about V_j and W_j and see what they imply about X_j. In each case, we assume that the V_i and W_j are independent between pairs, but not necessarily within pairs; that is, the pairs (V_1, W_1), $(V_2, W_2), \ldots, (V_n, W_n)$ are mutually independent, but V_1 may or may not be independent of W_1, V_2 of W_2, etc. It follows that the n differences $X_1 = W_1 - V_1, \ldots, X_n = W_n - V_n$ are mutually independent, as we have assumed the X_i to be throughout this chapter.

Suppose first that (V_j, W_j) is bivariate normal with the same distribution for each j, so that (V_1, W_1), $(V_2, W_2), \ldots, (V_n, W_n)$ are independent and identically distributed according to the bivariate normal distribution. Then the differences $X_j = W_j - V_j$ are independently, identically, normally distributed with mean θ equal to the treatment population mean minus the control population mean. The situation then reduces to that of the earlier sections with normal alternatives, and the results given there apply here. The relevant variance is of course the variance of the X_j, which is

$$\sigma^2 = \sigma_x^2 = \sigma_v^2 + \sigma_w^2 - 2\sigma_{vw} \tag{9.1}$$

in standard notation. Alternatives with $\sigma_v^2 = \sigma_w^2$ are usual but not necessary.

Dropping the bivariate normal assumption, now suppose that there is an "effect" associated with each pair, which may be regarded as either fixed or random, and in addition a fixed treatment effect, but no treatment-pair interaction. More specifically, suppose that

$$V_j = \mu_j + V'_j, \tag{9.2}$$

$$W_j = \mu_j + \theta + W'_j \tag{9.3}$$

where μ_j is the effect associated with the jth pair, θ is the treatment effect, and V'_j and W'_j are "disturbances" or "errors." If the V'_j and W'_j are mutually independent and normally distributed with mean 0 and common variance σ^2, this is a normal, fixed-effects or mixed model for two-way analysis of variance with no interaction; the design consists of two treatments (one is the control) in blocks of size two. If, more generally, the n pairs of errors $(V'_1, W'_1), (V'_2, W'_2), \ldots, (V'_n, W'_n)$ are independently, identically and symmetrically[2] distributed, then the random variables

$$X_j = W_j - V_j = \theta + W'_j - V'_j \tag{9.4}$$

are independently, identically, symmetrically distributed about θ (Problem 63). The one-sample tests based on symmetry apply to the null hypothesis $\theta = 0$, and the alternative $\theta \neq 0$ is a shift alternative provided, as we assume, the distribution of (V'_j, W'_j) is the same for all θ.

Specifically, if V'_j and W'_j have a joint density, then X_j will have a density $f(x; \theta) = h(x - \theta)$ where h is the density of $W'_j - V'_j$. Any density h which is symmetric around zero, like those in Table 8.1, arises from some symmetric joint density for V'_j and W'_j (Problem 63d).

If we assume, however, that the two "errors" V'_j and W'_j are themselves independently, identically distributed, with density q, say, then h is given by

$$h(x) = \int q(x + t)q(t) \, dt = \int q(t)q(t - x) \, dt. \tag{9.5}$$

This density is automatically symmetric about zero, but it is sometimes not trivial to determine whether a given symmetric density has this form or not. Many symmetric densities do not, including some of those in Table 8.1, as mentioned earlier (Problem 64). Furthermore, if the density q of the errors is not arbitrary but is one of those in Table 8.1, then the density h of the differences of the errors will not be one of those in Table 8.1 except in two special cases. If q is normal, then h is normal; if q is Cauchy, then h is Cauchy; but if q is any other distribution of Table 8.1, then h is not a distribution of Table 8.1. Thus, Table 8.1 fails to apply generally to the present situation in two respects: first, some of the distributions in Table 8.1 cannot arise at all; second, if the distribution of the errors is one of those in the table, the relative efficiencies are generally different from those given in the table.

---

[2] The pair of random variables (V'_j, W'_j) is said to be distributed symmetrically ("permutationally symmetrically" is more specific but rarely used) if (V'_j, W'_j) has the same joint distribution as (W'_j, V'_j). This contrasts with the definition of symmetry around θ for a single random variable.

The relative efficiencies in this situation can, of course, be computed. The efficacy of the mean is much as before (Problem 66a). As it happens, the efficacy of the median test here is closely related to that of the Wilcoxon procedure in the situations of previous sections (Problem 66b). Unfortunately, however, the efficacies of the Wilcoxon, squared ranks, and normal scores procedures do not seem to be obtainable in closed form for most of the distributions of Table 8.1 in the present situation.

To summarize, for procedures based on the differences $X_j = W_j - V_j$, the consequences for X_j of any assumption about the paired observations (V_j, W_j) can be traced, and asymptotic relative efficiencies can be computed from the distribution of the X_j under these assumptions. However, the distribution of the X_j does not usually satisfy the same assumptions as that of V_j and W_j, or have a related univariate form. It is true, however, that X_j is normal under common normal models for (V_j, W_j).

9.2 Bounds for Asymptotic Efficiencies for Matched Pairs

The situation here as regards the bounds on asymptotic relative efficiencies is similar to that of Sect. 8.3. All of the statements there apply equally to procedures based on the X_j in any situation where the X_j are independent and identically distributed with density $f(x; \theta) = h(x - \theta)$. However, some of the restrictions on h discussed there are less natural here, and some restrictions which are natural here were not discussed there.

Suppose, in particular, that the model given by (9.2) and (9.3) holds and that the $2n$ errors V'_j and W'_j are independent and identically distributed with density q. Then the X_j have density $f(x; \theta) = h(x - \theta)$ where h is given by (9.5) and is automatically symmetric. As in Sect. 8.3, the asymptotic efficiency of the normal-theory procedures relative to the median procedures can be zero, but now its maximum value is $125/72 = 1.74$ (Problem 67a), rather than 3. Several other bounds of Sect. 8.3 cannot be improved here: the asymptotic efficiency of the normal scores procedures relative to the normal-theory procedures can be infinite and can be as small as one but no smaller; the asymptotic efficiency of the Wilcoxon procedures relative to the normal-theory procedures can be infinite; the asymptotic efficiency of either the Wilcoxon or normal scores procedures relative to the median procedures can be arbitrarily small; and the asymptotic efficiency of the normal scores procedures relative to the Wilcoxon procedures must be more than but can come arbitrarily close to $\pi/6$. None of these statements is altered by requiring q to be symmetric, unimodal, or both. The proofs are requested in Problem 67. For the remaining cases, sharp bounds appear to be unknown in the present situation. Of course all the bounds given in Sect. 8.3 are still valid here, but the added restrictions may make better bounds possible.

The foregoing bounds are summarized in Table 9.1. Each cell contains lower and upper bounds on the asymptotic efficiency of the procedure for

Table 9.1 Bounds on Asymptotic Relative Efficiencies for Shifts under (9.5)

	Mean		Median		Wilcoxon		Normal scores	
Mean	1	1	0	$\frac{125}{72}$	0	$\frac{125^a}{108}$	1	0
Median	$\frac{72}{125}$	∞	1	1	$\frac{1}{3}^a$	∞	0^a	∞
Wilcoxon	$\frac{108^a}{125}$	∞	0	3^a	1	1	0^a	$\frac{6}{\pi}$
Normal scores	1	∞	0	∞^a	$\frac{\pi}{6}$	∞^a	1	1

[a] Bound taken from Table 8.2. A better bound may be possible.

that row relative to the procedure for that column for shifts of a density of the form (9.5). Except as indicated by a footnote, the bounds are greatest lower and least upper bounds, whether or not q is symmetric and/or unimodal.

10 Asymptotic Relative Efficiency of Two-Sample Procedures for Shift Families

The specific results of Sects. 5 and 8 all related to one-sample procedures. In Sect. 9 we discussed the applicability of these results to matched pairs. We now turn to two-sample procedures, such as those of Chaps. 5 and 6. As we shall see, the heuristic discussion and properties of asymptotic relative efficiency carry over easily to this situation, and Pitman's formula can again be used to compute asymptotic efficiencies, and the special formulas and numerical results of Sect. 8 have natural counterparts here.

The development given in Sects. 2–4 applies here to two independent samples of sizes m and n with essentially only the changes necessitated by replacing a single sample size n by a pair of sample sizes m, n. Thus a procedure based on a statistic $T_{m,n}$ with mean $\mu_{m,n}(\theta)$ and variance $\sigma_{m,n}^2(\theta)$ will have an efficacy

$$e_{m,n} = e_{m,n}(\theta) = [\mu'_{m,n}(\theta)]^2/\sigma_{m,n}^2(\theta) \tag{10.1}$$

which relates in exactly the same way as the one-sample efficacy to the power of a test or the distribution of an estimator or confidence limit. A specified performance (power, for instance) can now, however, be achieved by various combinations of m and n. Both the total sample size $N = m + n$ and the allocation, say $\lambda = m/N$, are important. It turns out, however, as one might expect, that the influence of the allocation is the same for all procedures of the types we have considered. The role of the limiting efficacy per observation can be played here by

$$e. = \lim \frac{N}{mn} e_{m,n}, \tag{10.2}$$

because this limit is the same however m and n approach ∞. Equivalently, $e_{m,n} = (mn/N)e. = N\lambda(1 - \lambda)e.$ except for an error of smaller order. In particular, if m/N approaches a limit λ', the limit in (10.2) does not depend on λ'.

The asymptotic relative efficiency of two procedures is again given by Pitman's formula

$$E_{1:2} = \frac{e_1.}{e_2.} = \lim \frac{e_{1,m,n}}{e_{2,m,n}} = \lim \frac{[\mu'_{1,m,n}(\theta)]^2}{\sigma^2_{1,m,n}(\theta)} \frac{\sigma^2_{2,m,n}(\theta)}{[\mu'_{2,m,n}(\theta)]^2}. \tag{10.3}$$

In comparing the performance of two procedures, the comparison now is between a procedure Ω_1 based on two mutually independent samples of sizes m_1 and n_1, and another procedure Ω_2 based on samples of sizes m_2 and n_2. To exemplify the use of Pitman's formula and the meaning of asymptotic relative efficiency here, suppose that we want to compare two sequences of test procedures T_1 and T_2 by considering the sample sizes m_2 and n_2 required by test T_2 to achieve the same power against the same alternative as the test T_1 based on m_1 and n_1 observations. Let m_2 and n_2 be the same multiples of m_1 and n_1, that is, $m_2/m_1 = n_2/n_1 = E$, say, except for rounding. (Equivalently, $E = N_2/N_1$ and the allocations m_2/N_2 and m_1/N_1 are equal.) Then the multiple E required approaches the asymptotic relative efficiency $E_{1:2}$ as the sample sizes approach infinity and the alternative approaches the null hypothesis. This limit $E_{1:2}$ is independent of the level and power specified and the allocation between samples, as long as the allocation is the same for both procedures.

The definitions of asymptotic relative efficiency in Sect. 6 can all be adapted to the two-sample case in a similar manner, and comments like those of Sect. 6 still apply. Pitman's formula in its two-sample form (10.3) can be used to compute asymptotic relative efficiencies of two-sample procedures for families of alternative distributions indexed by a single parameter θ, and the same numerical values hold for corresponding tests, point estimators, and confidence procedures.

For two-sample shift alternatives, specifically, let us write the respective population densities in the form

$$f(x; \theta) = h(x) \quad \text{and} \quad g(x; \theta) = h(x - \theta), \tag{10.4}$$

so that the null hypothesis $\theta = 0$ corresponds to the hypothesis of identical distributions. Then we can calculate the value of the efficacy of two-sample procedures for shift families exactly as we did in Sect. 8 for the one-sample procedures.

Since the limit (10.2) is the same however m and n approach infinity, that is, whatever the allocation m/N, we need not carry out additional computations for the specific procedures discussed in Chaps. 5 and 6 and the distributions h considered in Sect. 8, because each of the two-sample tests has a corresponding one-sample limit for which we have already computed the efficacy. A two-sample test for identical but unspecified populations, based on samples of sizes m and n, reduces as $m \to \infty$ with n fixed to a one-sample test of a hypothesis about the distribution of the Y sample, which is of finite size n. In particular, the two-sample t test for equality of means approaches the standard t test for the hypothesis that the Y mean equals a given value, namely the X population mean as determined from the infinite X sample; the two-sample median test for identical populations reduces to the sign test of the hypothesis that the probability is $\frac{1}{2}$ that a Y observation exceeds a given value, namely the X population median (see Problem 27, Chap. 5); and Problem 69 shows that the asymptotic properties of the two-sample Wilcoxon rank-sum statistic for identical populations are the same as the asymptotic properties of the one-sample Wilcoxon rank-sum statistic for the null hypothesis of symmetry about a given point. Similar reductions hold for the two-sample normal scores test and the randomization test based on the difference of the sample means. Accordingly, the results for efficacy given in Table 8.1 and for bounds on asymptotic efficiency given in Table 8.2 apply to the corresponding two-sample tests here.

Unfortunately, there seems to be no easy way to see that the limit (10.2) does not depend on how m and n approach infinity except by calculation in special cases or appeal to powerful theorems. Problem 70a requests such calculation for the two-sample normal-theory, median, and Wilcoxon procedures. For a procedure based on a sum of scores c_{mnk} (Sect. 5, Chap. 5), the efficacy for the shift alternative (10.4) is (Problem 71a)

$$e_{m,n} = mn \frac{N-1}{N^3} \left\{ \sum_{k=1}^{N} c_{mnk} E_h[h'(X_{(k)})/h(X_{(k)})] \right\}^2 \Bigg/ \left[N \sum_{k=1}^{N} c_{mnk}^2 - \left(\sum_{k=1}^{N} c_{mnk} \right)^2 \right].$$

(10.5)

If the scores are of the form

$$c_{mnk} = b_{mn} c\left(\frac{k - \frac{1}{2}}{N} \right) + \text{remainder},$$

(10.6)

then under suitable regularity conditions the limit (10.2) exists and is (Problem 71b)

$$e. = \left\{ \int_0^1 c(u) h'[H^{-1}(u)]/h[H^{-1}(u)] \, du \right\}^2 \Bigg/ \left\{ \int_0^1 c^2(u) \, du - \left[\int_0^1 c(u) \, du \right]^2 \right\}$$

$$= \left\{ \int_{-\infty}^{\infty} c[H(x)] h'(x) \, dx \right\}^2 \Bigg/ \left\{ \int_0^1 c^2(u) \, du - \left[\int_0^1 c(u) \, du \right]^2 \right\}.$$

(10.7)

If h is symmetric and $c(1 - u) = -c(u)$ for all u, this reduces to (8.8) with $c(u)$ and $c[H(x)]$ in place of $c(2u - 1)$ and $c[2H(x) - 1]$. Thus the one-sample and two-sample asymptotic efficacies are the same for shifts of symmetric densities h if the one-sample function $c_{(1)}(u)$ and the two-sample function $c_{(2)}(u)$ satisfy

$$c_{(2)}(u) = -c_{(2)}(1 - u) = c_{(1)}(2u - 1) \quad \text{for } \tfrac{1}{2} < u < 1. \tag{10.8}$$

These conditions on $c_{(2)}$ and $c_{(1)}$ are actually quite natural. For instance, the one- and two-sample normal scores procedures have

$$c_{(1)}(u) = \Phi^{-1}[(1 + u)/2] \quad \text{and} \quad c_{(2)}(u) = \Phi^{-1}(u),$$

which satisfy (10.8). The one- and two-sample median and Wilcoxon procedures also satisfy (10.8), and a two-sample sum-of-scores procedure corresponding similarly to the one-sample squared rank procedure of Sect. 8 has scores $c_{mnk} = \pm(k - N/2)^2$, one sign applying for $k > N/2$ and the other for $k < N/2$ (Problem 72).

Thus it can be shown that the efficiencies of appropriately corresponding one- and two-sample procedures are the same for symmetric densities h, and that Tables 8.1 and 8.2 apply to two-sample procedures as well as to one-sample procedures. For asymmetric densities, where the one-sample procedures are generally not valid, the two-sample procedures are valid and could also be compared using the formulas above. For the normal theory procedures we still have $e. = 1/\sigma^2$ and for the median procedures $e. = 4h^2(\xi_{0.5})$ where $\xi_{0.5}$ is the median of h, but for the Wilcoxon procedure we must use

$$e. = 12\left[\int h(x)h(-x)\,dx\right]^2. \tag{10.9}$$

Numerical results will not be given here.

The two-sample tests we have discussed become invalid, in general, if the population variances are unequal, even if the populations are symmetric about the same point. Their sensitivity to this departure from assumptions can be studied asymptotically by methods similar to those discussed here. See Problems 73–78 and Pratt (1964).

11 Asymptotic Efficiency of Kolmogorov–Smirnov Procedures

The asymptotic behavior of the Kolmogorov–Smirnov statistics is fundamentally different from that of the other test statistics we have considered, and hence these statistics require separate discussion. We have already seen, in Sect. 3.2 of Chap. 7, that their asymptotic distributions have a different shape under the null hypothesis. This is also true under alternative hypotheses, as we shall see below. Accordingly, the asymptotic power function of a Kolmogorov–Smirnov test has a different shape from that of the other tests, and its dependence on the level is also different. The sample size required to achieve a given level and a given power at a given alternative therefore depends on the level, power, and alternative in a fundamentally different way. Consequently, the asymptotic efficiency of the Kolmogorov–Smirnov procedures relative to other procedures depends on the Type I and Type II errors, α and β, and therefore will have a much more restrictive meaning in this section than in the rest of this chapter.

Since the asymptotic distributions of $\sqrt{mn/N}\,D_{mn}$ and $\sqrt{mn/N}\,D_{mn}^{+}$ are independent of the way m and n approach infinity under both the null hypothesis and nearby alternatives, it is true here as in the previous section that the asymptotic efficiency of the Kolmogorov–Smirnov procedures relative to other procedures with the same allocation m/N does not depend on what that allocation is. Therefore, two-sample efficiencies are again the same as one-sample efficiencies. We shall treat the one-sample case here since the notation is simpler, but the results are applicable to the two-sample case once $\sqrt{n}$ is replaced by $\sqrt{mn/N}$. One sided and two-sided tests are not simply related however. We shall discuss one-sided tests first.

11.1 One-Sided Tests

An effective method of computing the asymptotic power of the Kolmogorov–Smirnov procedures is not known in general. It is easily computed for a uniform shift alternative, however. If a null distribution which is uniform with range 1 is shifted by an amount θ, or more generally is shifted by θ times its range, then the power of the one-sided Kolmogorov–Smirnov test is approximately (Problem 79a; Quade, 1965, Theorem 4.1)

$$
\begin{cases}
e^{-2(c_\alpha - \sqrt{n}\theta)^2} & \text{if } \sqrt{n}\theta < c_\alpha, \\
1 & \text{if } \sqrt{n}\theta \geq c_\alpha,
\end{cases}
\tag{11.1}
$$

where c_α is the upper α point of the asymptotic null distribution of $\sqrt{n}D_n^{+}$ given in (6.3) of Chap. 7 as

$$
c_\alpha = \sqrt{-(\log_e \alpha)/2}.
\tag{11.2}
$$

The median test turns out to be a particularly appropriate comparison procedure here, as we shall see. Its power against the same alternative is approximately

$$
\Phi(2\sqrt{n}\theta - z_\alpha),
\tag{11.3}
$$

where z_α is the upper α point of the standard normal distribution, as follows (Problem 79b) from (2.11) and (8.3). The ratio of sample sizes needed to achieve level α and power $1 - \beta$ at the same alternative θ therefore approaches (Problem 79c)

$$E_{KS:2} = (z_\alpha + z_\beta)^2 / (2c_\alpha - 2c_{1-\beta})^2 \qquad (11.4)$$

as $\theta \to 0$ and the sample sizes approach infinity. The dependence of the asymptotic efficiency on α and β is immediately evident. Some numerical values are given in Table 11.1, which is explained further below. The efficiency approaches 1 as $\alpha \to 0$ for fixed β, it approaches ∞ as $\beta \to 0$ for fixed α, and it approaches (Problem 79d)

$$[2\alpha c_\alpha/\phi(z_\alpha)]^2 = 4\pi\alpha^2 \left(\log \frac{1}{\alpha}\right) e^{z_\alpha^2} \qquad (11.5)$$

as $\beta \to 1 - \alpha$, that is, when the alternative approaches the null hypothesis faster than $1/\sqrt{n} \to 0$.

Equation (11.4) gives the asymptotic efficiency of the one-sided, one- or two-sample Kolmogorov–Smirnov test relative to the one- or two-sample median test, for uniform shift alternatives. Its asymptotic efficiency relative to any other test can be obtained by dividing (11.4) by the asymptotic efficiency of the other test relative to the median test for the same alternatives, for instance, if the other test appears in Table 8.1, by an entry in the line labeled uniform.

For other alternatives, the dependence on α and β will generally differ from that in (11.4). While no simple, general method of computation is known, it is possible to obtain bounds which provide some insight.

First, for any symmetric, unimodal shift alternative, (11.1) is an asymptotic upper bound on the power of the Kolmogorov–Smirnov test (Problem 79e), where θ is now the difference between the true and hypothesized c.d.f. at the median. (In the case of a uniform shift, this agrees with the previous definition of θ, and the uniform is the "most favorable" symmetric unimodal shift.) Therefore (11.4) is an upper bound on the asymptotic efficiency of the Kolmogorov–Smirnov test relative to the median test. If (11.4) is divided by the asymptotic efficiency of any other test relative to the median test for any particular symmetric, unimodal shift family, we obtain an upper bound on the asymptotic efficiency of the Kolmogorov–Smirnov test relative to the other test for this family.

Second, it is easy to obtain a lower bound (unfortunately very weak) on the power by observing that the Kolmogorov–Smirnov test will certainly reject H_0 if $F_n(\mu) - F(\mu)$ exceeds the Kolmogorov–Smirnov critical value at the median μ, or equivalently, if the median test rejects even when the Kolmogorov–Smirnov critical value is used in place of the (smaller) median test critical value for $F_n(\mu) - F(\mu)$. Approximating the relevant binomial probability by a normal probability in the usual way, a lower bound on the approximate power is obtained (Problem 80a) as

$$\Phi[2(\sqrt{n}\theta - c_\alpha)]. \qquad (11.6)$$

The power of the median test with its own critical value is again approximated by (11.3), and it follows (Problem 80b) that

$$E_{KS:2} \geq (z_\alpha + z_\beta)^2/(2c_\alpha + z_\beta)^2. \tag{11.7}$$

Lower bounds relative to tests other than the median test are obtained by division, as the upper bounds in the previous paragraph were.

Unfortunately, for alternatives very near the null hypothesis, (11.6) does not even say that the power exceeds α, and (11.7) is correspondingly poor. Furthermore, the right-hand side of (11.7) is always less than 1, while we could hope to be able to prove that the Kolmogorov–Smirnov test is asymptotically *more* efficient than the median test for at least some alternatives other than the uniform shift. It is worth noting, however, that the right-hand side of (11.7) approaches 1 as $\alpha \to 0$ [Capon, 1965] or $\beta \to 0$, so that the Kolmogorov–Smirnov test is asymptotically almost as efficient as the median test if α or β is small enough. Unfortunately the approach is very slow, as is evident from Table 11.2.

A lower bound which improves on the previous one for any alternative $G \geq F$ can be obtained as follows.

$$P_G\{\sup[F_n(t) - F(t)] \geq c\} = 1 - P_G\{\sup[F_n(t) - G(t) + G(t) - F(t)] \leq c\}$$
$$\geq 1 - P_G\{\sup[F_n(t) - G(t)] \leq c \text{ and } F_n(\mu) - G(\mu) \leq c - \theta\},$$

$$\tag{11.8}$$

where $\theta = G(\mu) - F(\mu)$ as before. This last probability, conditional on $F_n(\mu)$, can be evaluated asymptotically by arguments like those leading to the asymptotic null distribution, and the expectation of the result over the distribution of $F_n(\mu)$ can be found (see below). The resulting lower bound on the asymptotic power is (Problem 82a; Quade, [1965, Theorem 4.2(d) with $\tau = \frac{1}{2}$])

$$\Phi[2(\sqrt{n}\theta - c_\alpha)] + 2\alpha\Phi(-2\sqrt{n}\theta) - \Phi[-2(\sqrt{n}\theta + c_\alpha)], \tag{11.9}$$

which in turn gives the lower bound on asymptotic efficiency as (Problem 82b)

$$E_{KS:2} \geq (z_\alpha + z_\beta)^2/d^2 \tag{11.10}$$

where d is the solution of

$$\beta = \Phi(-d + 2c_\alpha) - 2\alpha\Phi(-d) + \Phi(-d - 2c_\alpha). \tag{11.11}$$

While (11.11) cannot be solved explicitly, a table of β as a function of d can be generated directly, and a table of d as a function of β can be obtained by inverse interpolation. Some values resulting for the lower bound (11.10) are given in Table 11.1. Lower bounds relative to tests other than the median test are obtained as before. When $\alpha \to 0$ or $\beta \to 0$, this lower bound also approaches 1 since the first term of (11.9) dominates and (11.10) behaves like (11.7). As $\beta \to 1 - \alpha$, (11.9) approaches α as it should (unlike (11.6)) but the approach is so rapid that the right-hand side of (11.10) approaches 0 (Problem

Table 11.1 Bounds on Asymptotic Efficiency of the One-Sided Kolmogorov–Smirnov Test Relative to the Median Test

Power 1 − β:	*	0.1	0.3	0.5	0.7	0.9	0.99	0.999	0.9999	1.0
Type II Error β:	*	0.9	0.7	0.5	0.3	0.1	0.01	0.001	0.0001	0
α										
0.9						5.54	10.85	19.06	30.09	∞
						0.44	0.53	0.57	0.61	1
						0	0.17	0.28	0.35	1
0.5				2.18	2.48	3.18	5.05	7.44	10.24	∞
				0.59	0.63	0.67	0.72	0.74	0.76	1
				0	0.12	0.28	0.44	0.53	0.58	1
0.1		1.50	1.62	1.75	1.93	2.31	3.24	4.33	5.51	∞
		0.69	0.74	0.76	0.78	0.80	0.82	0.84	0.85	1
		0	0.23	0.36	0.46	0.56	0.65	0.70	0.73	1
0.05	1.41	1.45	1.56	1.68	1.83	2.17	2.97	3.88	4.86	∞
	0.71	0.74	0.77	0.79	0.80	0.82	0.84	0.86	0.87	1
	0	0.11	0.34	0.45	0.53	0.62	0.69	0.73	0.76	1
0.025	1.35	1.42	1.52	1.62	1.76	2.06	2.77	3.57	4.42	∞
	0.73	0.77	0.79	0.81	0.82	0.84	0.86	0.87	0.88	1
	0	0.23	0.43	0.52	0.59	0.66	0.72	0.76	0.78	1
0.01	1.30	1.38	1.48	1.57	1.69	1.96	2.59	3.28	4.01	∞
	0.75	0.79	0.82	0.83	0.84	0.86	0.87	0.88	0.89	1
	0	0.36	0.52	0.59	0.64	0.70	0.75	0.78	0.80	1
0.005	1.28	1.36	1.45	1.54	1.65	1.90	2.48	3.12	3.78	∞
	0.76	0.81	0.83	0.84	0.85	0.87	0.88	0.89	0.90	1
	0	0.43	0.56	0.63	0.67	0.72	0.77	0.80	0.81	1
0.001	1.22	1.33	1.40	1.48	1.58	1.80	2.30	2.83	3.38	∞
	0.78	0.84	0.86	0.87	0.88	0.89	0.90	0.90	0.91	1
	0	0.55	0.65	0.69	0.73	0.77	0.80	0.82	0.84	1
0.0001	1.20	1.29	1.36	1.43	1.52	1.70	2.12	2.57	3.03	∞
	0.81	0.87	0.88	0.89	0.89	0.90	0.91	0.92	0.92	1
	0	0.66	0.72	0.75	0.78	0.81	0.83	0.85	0.86	1
0		1	1	1	1	1	1	1	1	

Note: For each α and β pair, the entry in the first row is the value for uniform shifts, (11.4), which is also an upper bound for alternatives G most distant from F at the median; the entry in the second row is the value for Laplace shifts, (11.19), which is also a lower bound for shifts of a density h satisfying $h(x) \geq 2h(\mu)\min\{H(x), 1 - H(x)\}$ where μ is the median; and the entry in the third row is the lower bound (11.10) for stochastically one-sided alternatives. All bounds are valid for symmetric unimodal shifts. Column* applies as $\beta \to 1 - \alpha$.

82c). The improvement of (11.10) over (11.7) is revealed numerically by comparing the results given in Tables 11.1 and 11.2; it is very small.

Symmetric, unimodal shift alternatives which come arbitrarily close to achieving equality in the bounds given in (11.8)–(11.11) are described in Problem 83. We have thus found the maximum and minimum asymptotic efficiency of the one-sided Kolmogorov–Smirnov test relative to the median test for such alternatives (Equations (11.4) and (11.10) and Table 11.1). Bounds on asymptotic efficiency relative to other tests could be obtained by dividing by the minimum and maximum asymptotic efficiency of the other tests relative to the median test, but these bounds are probably very poor because different kinds of alternatives are "least favorable" for the two factors.

* Better bounds on the asymptotic power and relative efficiency of the Kolmogorov–Smirnov test for specific alternatives other than those achieving equality in (11.1) and (11.4) or (11.9) and (11.10) can be obtained as follows. Let $\mu_1, \mu_2, \ldots, \mu_r$ be any r quantiles of G, with $\mu_1 < \mu_2 < \cdots < \mu_r$, $\mu_0 = -\infty$ and $\mu_{r+1} = \infty$, and suppose that

$$G(x) - F(x) \geq \delta_i \quad \text{for} \quad \mu_{i-1} < x < \mu_i, i = 1, 2, \ldots, r+1 \quad (11.12a)$$

$$G(\mu_i) - F(\mu_i) \geq \varepsilon_i \quad \text{for } i = 1, 2, \ldots, r \quad (11.12b)$$

with $\varepsilon_i \geq \max(\delta_i, \delta_{i+1})$. (This bounds $G - F$ from below by a step function. Spikes are allowed at the steps, but presumably one would take $\varepsilon_i = \delta_i$ or δ_{i+1} except in peculiar cases. Then

$$\begin{aligned}
\beta &= P_G\{\sup[F_n(x) - F(x)] \leq c\} \\
&= P_G\{\sup[F_n(x) - G(x) + G(x) - F(x)] \leq c\} \\
&\leq P_G[F_n(x) - G(x) \leq c - \delta_i \quad \text{for } \mu_{i-1} < x < \mu_i, i = 1, \ldots, r+1, \\
&\qquad \text{and } F_n(\mu_i) - G(\mu_i) \leq c - \varepsilon_i, i = 1, \ldots, r]. \quad (11.13)
\end{aligned}$$

Conditional on $F_n(\mu_i) - G(\mu_i) = u_i, i = 1, \ldots, r$, the $r+1$ events in the first line of (11.13) are independent and their probabilities can be evaluated asymptotically by the reflection principle and a limiting argument as in Sect. 3, Chap. 7. For $u_i \leq c - \varepsilon_i$, we have $u_{i-1} \leq c - \delta_i$ and $u_i \leq c - \delta_i$ and the ith conditional probability is

$$P_G[F_n(x) - G(x) \leq c - \delta_i \quad \text{for } \mu_{i-1} < x < \mu_i | u_1, u_2, \ldots, u_r]$$

$$= 1 - \exp[-(2n/v_i)(c - \delta_i - u_{i-1})(c - \delta_i - u_i)], \quad i = 1, 2, \ldots, r+1, \quad (11.14)$$

where $v_i = G(\mu_i) - G(\mu_{i-1})$ and $u_0 = u_{r+1} = 0$. The asymptotic value of the right-hand side of (11.13) is

$$\int_0^{c-\varepsilon_1} \cdots \int_0^{c-\varepsilon_r} \prod_{i=1}^{r+1} \{1 - \exp[-(2n/v_i)(c - \delta_i - u_{i-1})(c - \delta_i - u_i)]\}$$

$$g(u_1, \ldots, u_r) \, du_1, \ldots, du_r \quad (11.15)$$

Table 11.2 Weak Lower Bound (11.7) on Asymptotic Efficiency of One-Sided Kolmogorov–Smirnov Test Relative to Median Test

Power $1 - \beta$:	*	0.1	0.3	0.5	0.7	0.9	0.99	0.999	0.9999	1.0
Type II Error β:	*	0.9	0.7	0.5	0.3	0.1	0.01	0.001	0.0001	0
α										
0.9						0	0.14	0.26	0.34	1
0.5				0	0.10	0.27	0.44	0.52	0.58	1
0.1		0	0.22	0.36	0.46	0.56	0.65	0.70	0.73	1
0.05	0	0.10	0.34	0.45	0.53	0.62	0.69	0.73	0.76	1
0.025	0	0.22	0.43	0.52	0.59	0.66	0.72	0.76	0.78	1
0.01	0	0.36	0.52	0.59	0.64	0.70	0.75	0.78	0.80	1
0.005	0	0.43	0.56	0.63	0.67	0.72	0.77	0.80	0.81	1
0.001	0	0.55	0.65	0.69	0.73	0.77	0.80	0.82	0.84	1
0.0001	0	0.66	0.72	0.75	0.78	0.81	0.84	0.85	0.86	1
0		1	1	1	1	1	1	1	1	1

Column * applies as $\beta \to 1 - \alpha$. (Bound approaches 0.)

where g is the asymptotic joint density of $u_1, \ldots, u_r$. The product in (11.15) has 2^{r+1} terms in general, but each term is the exponential of a quadratic (or lower) polynomial in $u_1, \ldots, u_r$, and g is of the same form since $u_1, \ldots, u_r$ are asymptotically multivariate normal. Therefore, each term to be integrated is the exponential of a quadratic form, and completing the square leads to an expression for its integral as a multiple of an r-variate normal c.d.f. Hence (11.15) can be expressed as a linear combination of r-variate normal c.d.f.'s, with 2^{r+1} different terms in general. Thus we have obtained an upper bound on β and hence a lower bound on the asymptotic power and relative efficiency of the one-sided Kolmogorov–Smirnov tests. Upper bounds could be obtained similarly by bounding $G - F$ from above instead of from below in (11.12).

The bound in (11.9) was obtained in this way with $r = 1$, $\delta_1 = \delta_2 = 0$, $\varepsilon = 0$. For $r = 2$, the computation is not difficult since the bivariate normal c.d.f. can be computed quite easily and has even been tabled. Since the algebraic expressions are lengthy and unrevealing and the calculations have not been carried out, further details will not be given here. For $r \geq 3$, the computation is more difficult because of the difficulty of computing the multivariate normal c.d.f., but certainly feasible for small r.

The following further improvement leads to bounds of the same form, without any additional complication of the algebra or calculation. However, somewhat more analytical grasp on $G - F$ is required, because a piecewise linear bound is used instead of a step function. Specifically, we replace the right-hand side of (11.12a) by a linear function of G with values γ_i and δ_i at the endpoints μ_{i-1} and μ_i of the x-interval in question; that is, suppose that

$$G(x) - F(x) \geq \gamma_i[\mu_i - G(x)]/v_i + \delta_i[G(x) - \mu_{i-1}]/v_i$$
$$\text{for } \mu_{i-1} < x < \mu_i, \; i = 1, 2, \ldots, r + 1 \tag{11.16}$$

and that (11.12b) still holds. Then (11.13) changes correspondingly, but the only effect in (11.14) and (11.15) is to replace the first δ_i by γ_i. (This follows because, given u_{i-1} and u_i and a linear boundary under the limiting distribution of F_n, only the distances from u_{i-1} and u_i to the boundary matter.) The nature of the integrand in (11.15) as a function of $u_1, \ldots, u_r$ is thus unchanged.

For $r = 1$, $\delta_1 = \varepsilon_1 = \gamma_2 = \theta$, $\gamma_1 = \delta_2 = 0$, and $\mu_1 = \mu$ (the median), the condition in (11.16) becomes

$$G(x) - F(x) \geq \begin{cases} 2\theta G(x) & \text{for } x < \mu \\ 2\theta[1 - G(x)] & \text{for } x > \mu. \end{cases} \tag{11.17}$$

Under this condition the asymptotic power is at least

$$\Phi[2(\sqrt{n}\theta - c_\alpha)] + 2\alpha\Phi(-2\sqrt{n}\theta)e^{4\sqrt{n}\theta c_\alpha} - \Phi[-2(\sqrt{n}\theta + c_\alpha)]e^{8\sqrt{n}\theta c_\alpha} \tag{11.18}$$

and the improved lower bound on asymptotic efficiency is

$$E_{KS:2} \geq (z_\alpha + z_\beta)^2/d^2 \tag{11.19}$$

where d is the solution of

$$\beta = \Phi(-d + 2c_\alpha) - 2\alpha\Phi(-d)e^{2dc_\alpha} + \Phi(-d - 2c_\alpha)e^{4dc_\alpha}. \tag{11.20}$$

Some values for the lower bound in this case are given in Table 11.1. When $\alpha \to 0$ or $\beta \to 0$, the lower bound (11.19) again approaches 1, but as $\beta \to 1 - \alpha$, it approaches (Problem 84)

$$\{2c_\alpha[\alpha - 2\Phi(-2c_\alpha)]/\phi(z_\alpha)\}^2 = 4\pi[\alpha - 2\Phi(-2c_\alpha)]^2\left(\log\frac{1}{\alpha}\right)e^{z_\alpha^2} \tag{11.21}$$

rather than 0. Thus, this lower bound is effective in the neighborhood of the null hypothesis, while (11.7) and (11.10) were not.

For a symmetric shift family with density $h(x - \mu)$, as $\mu \to 0$, the condition (11.17) under which the foregoing bound applies becomes

$$h(x) \geq 2h(0)H(x) \quad \text{for } x < 0. \tag{11.22}$$

A sufficient condition for this to hold is that

$$h(x)/H(x) \text{ is monotonically decreasing for } x < 0, \tag{11.23}$$

or, in terms of the upper tail, that the hazard function $h(x)/[1 - H(x)]$ be monotonically increasing for $x > 0$ (increasing failure rate). A sufficient condition for this, in turn, is that

$$h'(x)/h(x) \text{ is monotonically decreasing.} \tag{11.24}$$

Problem 85 gives further conditions equivalent to these, which help clarify their relationship.

Condition (11.24) is satisfied, and hence so are (11.23) and (11.22), and the bound in (11.19) applies, for the normal, logistic, Laplace, and uniform distributions, and more generally for the double exponential power distribution (5.6) with $k \geq 1$ and the symmetric beta distribution with $r \geq 1$. Conditions (11.22) and (11.23) also hold, although (11.24) does not, for the symmetric beta with $r < 1$ and the Cauchy distributions. For the double exponential power distribution with $k < 1$ (high-tailed distributions), however, not even condition (11.22) holds, and the bound in (11.19) has not been proved. Derivation of these results is requested in Problem 86.

For the Laplace distribution, equality holds in (11.19), which therefore gives the actual minimum, not just a lower bound, under any one of the conditions (11.17), (11.22), (11.23), or (11.24) (Problem 87).

For shifts of a density h, as $\beta \to 1 - \alpha$, that is, in the asymptotic neighborhood of the null hypothesis, the quantity

$$
\left\{ [2\alpha c_\alpha / \phi(z_\alpha)] \int_0^1 \{2\phi[T(u)c_\alpha] - 1\} h'[H^{-1}(u)]/h[H^{-1}(u)]\, du \right\}^2
$$

$$
= \left\{ [4\alpha c_\alpha / \phi(z_\alpha)] \int_0^1 \phi[T(u)c_\alpha] h[H^{-1}(u)] T'(u)\, du \right\}^2, \tag{11.25}
$$

where

$$
T(u) = (2u - 1)/[u(1 - u)]^{1/2}, \tag{11.26}
$$

plays the role of the asymptotic efficacy of the one-sided Kolmogorov–Smirnov test. In other words, (11.25) can be divided by the limiting efficacy per observation of another test to obtain the asymptotic efficiency of the Kolmogorov–Smirnov test relative to the other test when β is very close to $1 - \alpha$. Equation (11.25) may be compared to (8.8) for tests based on sums of signed constants or scores. Unfortunately, it is typically difficult to evaluate, as well as being applicable only as $\beta \to 1 - \alpha$. For proof, see Hájek and Šidák [1967].*

11.2 Two-Sided Tests

We turn now to two-sided tests. One might expect the results for a one-sided test at level $\alpha/2$ to apply to the two-sided test at level α, and some of them do approximately in some situations, as we shall see. They do not apply directly or exactly, however, and may not apply even approximately, because they do not take account of either rejection in the "wrong" direction or rejection in both directions at once.

The possibility of rejection in both directions at once implies that the rejection region for a two-sided Kolmogorov–Smirnov test is the union of two one-sided regions which are neither mutually exclusive nor each at level exactly $\alpha/2$. As far as the null hypothesis is concerned, however, as observed in Sect. 4.2 of Chap. 7, the numerical difference between the two-sided P-value and twice the one-sided P-value is negligible whenever the latter is no larger than 0.1. Thus, at usual levels α, the error in the level of a two-sided test will be negligible if we use the one-sided regions for level $\alpha/2$. Specifically, the error is the probability under the null hypothesis of the intersection of the two one-sided regions, that is, the probability of rejection in both directions at once, and this is negligible for $\alpha \le 0.2$.

Under alternatives, however, this probability may not be negligible. For instance, it is far from negligible under changes of scale. Even if it were negligible, there would remain the problem that the contribution to power from rejection in the wrong direction may be different for the Kolmogorov–Smirnov test than for other tests, especially as its power function has a different shape anyway. For some alternatives (such as scale) against which the Kolmogorov–Smirnov test might be used, it is not even clear which direction is "wrong."

There is one set of assumptions which leads to a simple result. Suppose that it is clear which region is "wrong" for rejection, as it is in the situations we have been concerned with, and in particular for shift alternatives. Suppose also that we adopt the three-conclusion interpretation of two-sided tests (see Sect. 4.6, Chap. 1) and interpret rejection in both directions at once as a correct decision. Then the probability of a correction decision is just the probability of rejection in the right direction, by the corresponding one-sided test, for the Kolmogorov–Smirnov test as well as for the other tests we have been considering. This makes the earlier results for one-sided tests at level $\alpha/2$ numerically applicable with negligible error to two-sided tests at level α, with the same β, provided that $\alpha \le 0.2$. This applies to both the bounds and the maximum and minimum efficiencies. Furthermore, one could obtain theoretically and numerically exact results, for all α, by substituting the exact two-sided critical value rather than $c_{\alpha/2}$ for c_α, while still substituting $\alpha/2$ for α and $z_{\alpha/2}$ for z_α, but leaving those quantities involving β unchanged.

Things are unfortunately less simple under the two-conclusion interpretation of tests, which we consider next. Rejection in the "wrong" direction is now a correct decision under the alternative hypothesis, and must be taken into account. Near the null hypothesis, it represents an appreciable fraction of the power, for both the Kolmogorov–Smirnov tests and asymptotically normal tests. Far from the null hypothesis, it represents a negligible fraction of the power for asymptotically normal tests, but the situation is more complicated for the Kolmogorov–Smirnov tests. Of course, since ignoring rejection in the "wrong" direction reduces power, the lower bounds on the power of the one-sided test at level $\alpha/2$ still apply. They may be weak, however. Furthermore, the upper bounds on power are invalid; both upper and lower bounds on relative efficiency are also invalid since they require upper bounds on the power of one of the tests being compared. The theory of maximum and minimum relative efficiencies is affected not only by this but also by the fact that the "least favorable" alternatives in the one-sided case are not or may not be least favorable in the two-sided case.

For a normal-theory test, numerical investigation indicates that ignoring rejection in the wrong direction understates power little enough so that the sample size required to achieve a given β is overstated by no more than 2% provided that $\alpha \leq 0.1$ and $\beta \leq 0.74$, or $\alpha \leq 0.05$ and $\beta \leq 0.85$, etc. (Problem 88). In this region, therefore, the lower bounds on the efficiency of the one-sided Kolmorogov–Smirnov test at level $\alpha/2$ relative to an asymptotically normal test apply to the two-sided test at level α after multiplication by 0.98. However, such lower bounds, whether based on the exact normal power or an upper bound, are not tight and may be appreciably lower than necessary because they ignore the probability of rejection in the "wrong" direction by the Kolmogorov–Smirnov test; the nature of the tests suggests that this probability may be larger than for asymptotically normal tests.

An upper bound on the power or efficiency of the Kolmogorov–Smirnov test requires an upper bound on its probability of rejection in the "wrong" direction but not also the "right" direction. As long as the alternative is one-sided in the sense $G(x) \geq F(x)$ for all x, then the probability of rejection in the "wrong" but not the "right" direction does not exceed $\alpha/2$ (Problem 89a). Although this bound seems rather crude, without further assumptions it cannot be improved asymptotically, even for shift alternatives, except to account for rejection in both directions at once. Accounting for this, we have [Quade, 1965, Theorem 4.3(b)], for all G such that $F(x) \leq G(x) \leq F(x) + \theta$ for all x,

$$1 - \beta = P_G[\sup(F_n - F) \geq c \quad \text{or} \quad \inf(F_n - F) \leq -c]$$

$$\leq P_G[\sup(F_n - G) \geq c - \theta \quad \text{or} \quad \inf(F_n - G) \leq -c]. \quad (11.27)$$

There are alternatives, including shift alternatives, which give results arbitrarily close to equality asymptotically (Problem 89b). The upper bound can be evaluated asymptotically (Problem 90), and a solution for θ given β can then be found by inverse interpolation or search. The same is true for the normal power. Thus it would be feasible to calculate a tight upper bound on $E_{KS:2}$ for alternatives G such that

$$F(x) \leq G(x) \leq F(x) + \theta \quad \text{for all } x \text{ and } G(\mu) = F(\mu) + \theta \text{ at the median } \mu.$$

$$(11.28)$$

We have not done so, however, because the calculation is rather complicated and the distributions coming close to equality are multimodal in a pathological way. When additional conditions on G are imposed, it is not known what form the most favorable distributions or maximum power or efficiency have, and they are probably of a form leading to even more difficult calculation.

A simple upper bound is obtained, however, by bounding the probability of rejection in the "wrong" direction by $\alpha/2$ for the Kolmogorov–Smirnov test and ignoring it for the median test. Thus, under (11.28) the power of the Kolmogorov–Smirnov test satisfies (11.27) and hence

$$1 - \beta \leq P_G[\sup(F_n - G) \geq c - \theta] + \alpha/2 \doteq e^{-n(c-\theta)^2} + \alpha/2, \quad (11.29)$$

where c is the critical value of the two-tailed test. Ignoring the probability of rejection in the "wrong" direction gives $\Phi(2\sqrt{n}\,\theta - z_{\alpha/2})$ as a lower bound on the asymptotic power of the median test (compare (11.3)). Therefore, under (11.28),

$$E_{KS:2} \leq (z_{\alpha/2} + z_\beta)^2/(2c - 2c_{1-\beta-\alpha/2})^2, \quad (11.30)$$

where the subscripted c is defined by (11.2) and the unsubscripted c is the upper α point of the asymptotic distribution of the two-sided Kolmogorov–Smirnov statistic and hence is negligibly different from $c_{\alpha/2}$ for $\alpha \leq 0.1$, as mentioned earlier.

*The general methods described after (11.2) for bounding the asymptotic power and relative efficiency of the one-sided Kolmogorov–Smirnov tests under specific alternatives can be extended to the two-sided tests as follows. The difference $G - F$ must be bounded above and below by a step function or a piecewise linear function of G. Then on the left-hand side of (11.14), $F_n - G$ is bounded above and below by constants or linear functions of G, and the right-hand side has an infinite series of exponential terms. Thus the integrand in (11.15) becomes a product of infinite series of the same type, and the bound on β is, therefore, an infinite series of r-variate normal c.d.f.'s. Unless the region being used to bound the acceptance region is very small, that is, unless either α is very large or the bound is very poor, the series converges rapidly and calculation would be feasible for $r = 2$, as in the one-sided case.

The two-sided problem presents one additional difficulty not present in the one-sided case, namely that neither the general method just described nor any of the similar methods yields effective lower bounds on the power or efficiency of the Kolmogorov–Smirnov two-sided test near the null hypothesis, that is, as $\beta \rightarrow 1 - \alpha$. The reason is that, as $\theta \rightarrow 0$, the true power is, to first order, α plus a term of order θ^2 while the bound is less than the exact value by a term of order θ. Accordingly, for sufficiently small θ, the bound is actually less than α. While this may be unimportant from the point of view of power, the resulting lower bound on the efficiency is unfortunately 0 for small θ. To avoid this, more refined methods seem necessary.*

PROBLEMS

1. Develop an expression similar to (2.11) for the approximate power of lower-tailed and two-tailed tests and show that the interpretations of $E_{1:2}$ in terms of the scale factor and relative sample sizes continue to apply.

2. Let $\mu_n(\theta)$ and $\sigma_n^2(\theta)$ be the exact mean and variance of a statistic T_n for some one-parameter family of densities $f(x_n; \theta)$. Derive the Cramér-Rao inequality $[\mu_n'(\theta)]^2 \leq \sigma_n^2(\theta)I_n(\theta)$, where $I_n(\theta) = -E[(\partial^2/\partial\theta^2)\log f(x_n; \theta)]$ is the Fisher information, by showing that $\mu_n'(\theta) = \text{cov}(T_n, U_n)$ and $I_n(\theta) = \text{var}(U_n)$ where $U_n = (\partial/\partial\theta)\log f(x_n; \theta)$. Conclude that the efficacy satisfies $e_n(\theta) \leq I_n(\theta)$. Note that, if T_n is an estimator of θ, then $\mu_n(\theta) = 1 + b_n'(\theta)$ where $b_n(\theta)$ is the bias of T_n. See also Problem 43.

*3. Relate the asymptotic efficiency of maximum likelihood estimators to the development in this chapter.

4. Show that, for a sample of independently, identically distributed observations with mean θ and variance σ^2, the t statistic for $H_0: \theta = \theta_0$ is approximately normal with mean θ/σ and variance $1/n$ if n is large and θ is close to θ_0.

*5. Trace the development of Sect. 2 and give the leading error terms in (2.11) and the approximations leading to it when T_n is the sample mean, the population is normal with variance $\sigma^2(\theta)$ depending on the mean, and the null hypothesis is given by $\theta_0, \sigma^2(\theta_0)$.

*6. (a) Derive the order of magnitude of the error introduced in (2.11) and the approximations leading to it when a consistent estimate is substituted for a nuisance parameter as described in Sect. 2.2.
   (b) Is the order of magnitude of this error changed if the test is adjusted to make its level exact under some null hypothesis?

7. (a) Derive the asymptotic efficiency (3.3) of the sample proportion negative relative to the normal-theory estimator of p, the proportion negative in the population, for a normal population.
   (b) Show that this efficiency depends only on p.
   (c) Show that this efficiency is the same at $1 - p$ as at p.
   (d) Show that this efficiency is $2/\pi$ when $p = 0.5$.
   (e) Evaluate this efficiency at $p = 0.01, 0.05, 0.10, 0.25$, and 0.50.
   (f) Which value of p gives the largest efficiency? (See also Problem 38(c) of Chap. 2.)

8. Show that the efficacies of equivalent test statistics, as defined by (2.10), must be asymptotically the same, but need not be identical in finite samples, even if $\mu_n(\theta)$ is defined as the exact finite-sample mean.

9. (a) For a sample of n independently, identically distributed observations with density $f(x; \theta)$, find the limiting efficacy per observation of the sample median by direct calculation, using the fact that the median is asymptotically normal with mean equal to the population median $\mu(\theta)$ and variance $1/4nf^2[\mu(\theta); \theta]$.
   *(b) Find the efficacy of the number of negative observations and verify that it agrees with the limiting efficacy in (a).
   (Hint: Show that $F[\mu(\theta); \theta] = 0.5$ and find $\mu'(\theta)$ by differentiating.)

10. Use tests at level $\alpha = 0.50$ to argue that the sample mean and the t statistic have asymptotically the same efficacy in samples from a distribution with finite variance.

11. Show that application of the method of Sect. 3.2 to the Wilcoxon signed-rank test for an assumed center of symmetry μ gives the median of the Walsh averages as an estimator of μ with the same asymptotic efficacy.

12. In the two-sample shift problem, apply the method of Sect. 3.2 to obtain estimators that have the same asymptotic efficacy as
    (a) The two-sample t test.
    (b) The two-sample median test.
    (c) The two-sample rank-sum test.

13. Derive Equation (3.6) by expanding the right-hand side of (3.4) in a Taylor's series about $\mu_{1,n}(\theta)$, or the left-hand side of (3.5) about θ.

*14. (a) Consider a family of normal distributions with mean μ and standard deviation σ that are indexed by the 90th percentile value θ and σ. Given that $\sigma = 1$, how would tests and estimators for θ be based on the sample mean, the sample median, or the 90th percentile of the sample?
    (b) What would be the asymptotic efficacies and relative efficiencies of the tests and estimators in (a)?
    (c) Consider the following three possible definitions of θ for arbitrary distributions:
        (i) $\theta = \mu + 1.645\sigma$
        (ii) $\theta = \xi_{0.5} + 1.645\sigma$
        (iii) $\theta = \xi_{0.9}$
        where ξ_p is the quantile of order p in the distribution. Show that each of these definitions agrees with the definition given for normal distributions with $\sigma = 1$.
    (d) How might tests and estimators based on the sample mean, the sample median, or the 90th percentile of the sample be developed for each of the extended definitions of θ in (c)?

15. (a) Derive the approximate c.d.f. (4.2) of a confidence bound T_n from the approximate power (4.1) of the corresponding test.
    (b) Show using (a) that the c.d.f.'s of two confidence bounds $T_{1,n}$ and $T_{2,n}$ are approximately the same except for a scale factor $1/\sqrt{E_{1:2}(\theta)}$.

16. Let $L(t - \mu, \theta)$ be the "loss" incurred when the confidence bound t is given for the parameter $\mu(\theta)$. Show that, in the situations of Sect. 4,
    (a) The distribution of $L(T_{1,n} - \mu(\theta), \theta)$ is approximately the same as that of $L\{[T_{2,n} - \mu(\theta)]/\sqrt{E_{1:2}(\theta)}, \theta\}$.
    (b) If $L(z, \theta)$ is homogeneous of degree k in z, that is, $L(az, \theta) = a^k L(z, \theta)$, then $E_\theta\{L[T_{1,n} - \mu(\theta), \theta]\}$ is approximately $[E_{1:2}(\theta)]^{-k/2} E_\theta\{L[T_{2,n} - \mu(\theta), \theta]\}$.
    (c) Apply (b) to $L(z, \theta) = z$ and $L(z, \theta) = \max(z, 0)$.
    *(d) Show that if $\partial L(z, \theta)/\partial z$ exists and does not vanish at $z = 0$, then the same conclusion as for $k = 1$ in (b) holds.
    (e) If $T_{1,n}$ and $T_{2,n}$ are estimators rather than confidence bounds, what changes are necessary in (a)–(d)?
    (f) Apply (b) for estimators to the loss functions
    $$|z|, z^2, \text{ and } c\max(z, 0) + d\max(-z, 0).$$

17. Show that the interpretation of asymptotic relative efficiency in terms of sample sizes applies also to confidence bounds.

18. If $T'_{i,n}$ and $T''_{i,n}$ are respectively lower and upper confidence bounds for $\mu(\theta)$, $i = 1, 2$, and if the asymptotic relative efficiencies of $T'_{1,n}$ with respect to $T'_{2,n}$, and of $T''_{1,n}$ with respect to $T''_{2,n}$, are both $E_{1:2}(\theta)$, show that
$$E_\theta(T''_{1,n} - T'_{1,n}) = E_\theta(T''_{2,n} - T'_{2,n})/\sqrt{E_{1:2}(\theta)}$$
asymptotically.

19. For a sample from a population with media μ and a density positive and continuous at μ, show that the sample median is asymptotically normal with mean μ and variance $1/(4nd^2)$, where d is the density at μ.

20. For the Laplace density (5.1),
    (a) Show that the mean is θ and the variance is $2\lambda^2$.
    (b) Show that the efficacies of the sample median and the sample mean are as given in (5.2) and (5.3) respectively.

21. For a sample of n from a density of the form $f[(x - \theta)/\lambda]$,
    (a) Show that the efficacy of the normal-theory procedures satisfies $e_n(\theta; \lambda) = ne_1(0; 1)/\lambda^2$ and in particular is independent of θ and inversely proportional to λ^2.
    (b) Show that the results in (a) hold also for the median procedures.
    (c) Show that the asymptotic relative efficiency of the median procedures relative to the normal-theory procedures is independent of both θ and λ and depends only on f.
    (d) Generalize the results in (a)–(c).

*22. Show that the double exponential power density (5.6) approaches the uniform density on $(\theta - \lambda, \theta + \lambda)$ as $k \to \infty$.

*23. (a) Find the efficacies of the sample median and mean and their asymptotic relative efficiency for the double exponential power density (5.6).
    (b) Show that the asymptotic relative efficiency in (a) approaches $\frac{1}{3}$ as $k \to \infty$.
    (c) Show that the result in (b) agrees with that of a direct calculation for the uniform distribution.

24. Show that (6.1)–(6.4) are equivalent, that is, that (6.3) and (6.4) are equivalent to uniform convergence.

25. How might the various properties of asymptotic relative efficiency be restated for the case where it is 0 or ∞?

26. In typical testing situations as described in Section 2, show that
    (a) The powers $P_{1,n}(\theta_0 + \delta_n)$ and $P_{2,n}(\theta_0 + \delta_n\sqrt{E})$ appearing in (6.4) approach limits other than 1, α, or 0 if $\sqrt{n}\delta_n \to d \neq 0$, ∞, and $E > 0$.
    (b) If $E \neq E_{1:2}(\theta_0)$, the limits in (a) are different except in the situation mentioned in Sect. 6.2.

27. Restate the uniform convergence conditions (6.5)–(6.9) in the style of (6.3) and (6.4).

28. Rescale the variables in the uniform convergence conditions (6.5)–(6.9) in such a way that the terms have nondegenerate limits and uniformity is no longer essential to the meaning. What are the limits? State the corresponding properties of asymptotic relative efficiency in terms of these limits.

29. State and justify converses to the properties of asymptotic relative efficiency, according to which each property determines the asymptotic relative efficiency uniquely.

30. Let $T_{1,n}$ and $T_{2,n}$ be estimators of the same quantity $\mu = \mu(\theta)$. If $\sqrt{n}(T_{1,n} - \mu)$ has a nondegenerate limiting distribution and property $B(i)$ holds, then $\sqrt{n}(T_{2,n} - \mu)$ has the same nondegenerate limiting distribution except for the scale factor $1/\sqrt{E_{1:2}(\theta)}$. Show this.

*31. (a) Give an example of an estimator $T_{1,n}$ and a population distribution such that $T_{1,n}$ has infinite variance for every n but its limiting distribution has finite variance.

(b) Can the variance of the limiting distribution exceed the limit (or lim inf or lim sup) of the variance of $\sqrt{n}\,T_{1,n}$?

32. (a) Show that Property $A(ii)$ implies that, for all δ, the difference between the two powers $P_{1,n_1}(\theta_0 + \sqrt{n_1}\,\delta) - P_{2,n_2}(\theta_0 + \sqrt{n_1}\,\delta)$ approaches zero if $n_2/n_1 \to E$. This property is sometimes taken as the definition of asymptotic relative efficiency.

*(b) Show the converse of (a) under the additional condition that the power functions $P_{1,n}(\theta)$ and $P_{2,n}(\theta)$ are both monotone functions of θ for all n.

(c) Give statements analogous to (a) and (b) for estimators.

(d) Give statements analogous to (a) and (b) for confidence bounds.

33. (a) For one-tailed tests, choose d_n such that the powers $P_{1,n}(\theta_0 + t)$ and $P_{2,n}(\theta_0 + d_n t)$ have the same derivative with respect to t at $t=0$. Argue that $d_n \to \sqrt{E_{1:2}(\theta_0)}$ as $n \to \infty$.

(b) For unbiased two-tailed tests, choose d_n such that the powers in (a) have the same second derivative with respect to t at $t = 0$. Argue that $d_n \to E_{1:2}(\theta_0)$ as $n \to \infty$.

(c) Interpret these results.

34. Show that the efficacy of a shift family of procedures under a shift model is constant.

35. For a sample of n observations on a shift family $f(x; \theta) = h(x - \theta)$, derive the efficacy of

(a) The sample mean.

(b) The sample median when h has median 0.

(c) The Wilcoxon signed-rank test statistic when h is symmetric about zero.

(d) Interpret $\int h^2(x)dx$ in terms of the density of a sum or difference of random variables with density h.

36. Define Q as the quantity in braces in either line of Equation (8.8) so that the asymptotic efficacy of procedures based on sums of signed constants is $e_4 = 4Q^2/\int_0^1 c^2(u)du$.

(a) Show that Q has the alternative expressions

$$Q = c(0)h(0) + 2\int_{1/2}^1 c'(2u - 1)h[H^{-1}(u)]du$$

$$= c(0)h(0) + 2\int_0^\infty c'[2H(x) - 1]h^2(x)dx$$

if $c[2H(x) - 1]h(x) \to 0$ as $x \to \infty$.

(b) Use the result in (a) to derive the asymptotic efficacy of the median and Wilcoxon signed-rank procedures as given in Equations (8.3) and (8.5).

(c) Use the result in (a) to derive expressions for the asymptotic efficacy of the normal scores and squared rank procedures.

37. For the uniform shift family, $f(x - \theta) = 1/2R$ for $|x - \theta| < R$, derive the efficacy and, under (8.7), the asymptotic efficacy of the procedures based on sums of signed constants. Verify the results given in (8.9) and (8.10).

38. (a) Derive the asymptotic efficacy of the normal scores procedures given in (8.11) for a symmetric shift.
    (b) Show that the result in (a) can also be written as $e_5 = [\int_0^1 \{\Phi^{-1\prime}(t)/H^{-1\prime}(t)\}dt]^2$.

39. Determine the effect of a change of scale in the model on the efficacies of the procedures in Sect. 8.2.

*40. (a) Show that the double exponential power density with the value of a specified in Table 8.1 approaches $\frac{1}{2}$ for $|x| < 1$ and 0 for $|x| > 1$ as $k \to \infty$.
    (b) What happens to the limit in (a) at $x = 1$? Is it relevant?
    (c) What happens to the limit in (a) if a is kept fixed as $k \to \infty$?

*41. (a) Show that the symmetric beta density with the value of a specified in Table 8.1 approaches a normal density for all x as $r \to \infty$.
    (b) What happens to the limit in (a) if a is kept fixed as $r \to \infty$?

42. Show that the logistic density appearing in Table 8.1 can be written in the alternative forms
$$h(x) = a/2(1 + \cosh ax) = a/4 \cosh^2(ax/2)$$
where $\cosh z = (e^z + e^{-z})/2$.

43. For the shift family $f(x; \theta) = h(x - \theta)$, show that the Fisher information is
$$I(\theta) = E[h'(X)/h(X)]^2 = -E[\phi''(X)]$$
where X has density h and $\phi(x) = \log h(x)$. (See also Problem 2.)

44. Let U_n be the sample mean and let $T_n = U_n$ if $|U_n| > n^{-1/4}$ and $T_n = 0$ otherwise. For a population with mean μ and finite variance σ^2,
    (a) Show that T_n is asymptotically $N(\mu, \sigma^2/n)$ if $\mu \neq 0$.
    (b) What is the asymptotic distribution of T_n if $\mu = 0$?
    (c) What is the asymptotic efficacy of T_n? Is it continuous at $\mu = 0$?

*45. For the double exponential power density $h(x)$ of Table 8.1, show that
    (a) $\sigma^2 = \Gamma(3/k)/a^2\Gamma(1/k)$ and the asymptotic efficacy of the mean procedures is $1/\sigma^2$.
    (b) $h(0) = ak/2\Gamma(1/k)$ and the asymptotic efficacy of the median procedures is $[ak/\Gamma(1/k)]^2$.
    (c) $\int h^2(x)dx = ak/2^{1+(1/k)}\Gamma(1/k)$ and the asymptotic efficacy of the Wilcoxon procedures is $3[ak/2^{1/k}\Gamma(1/k)]^2$.
    (d) $I = a^2k^2\Gamma(2 - 1/k)/\Gamma(1/k)$.
    (e) The entries in Table 8.1 corresponding to (a)–(d) are correct.
    (f) The asymptotic efficiencies of the mean and Wilcoxon procedures relative to the median procedures both approach 3 as $k \to \infty$.

46. For the normal scores procedures,
    (a) Show that the asymptotic efficacy is $2/\pi$ for the Laplace distribution and 1 for the standard normal distribution.
    (b) Verify the entries in Table 8.1 corresponding to (a). (Note that the variance of the normal density there is $\pi/2$.)

*47. For the one-sample squared rank procedures,
    (a) Show that the asymptotic efficacy is $\frac{5}{9}$ for the Laplace distribution and $80(\arctan\sqrt{\frac{1}{2}})^2/\pi^3$ (with arctan in radians) for the standard normal distribution.

(b) Verify the entries in Table 8.1 corresponding to (a). (Note that the variance of the normal density there is $\pi/2$.)

48. For the uniform distribution on $[-1, 1]$, verify the asymptotic efficacies given in Table 8.1.

*49. For the symmetric beta density $h(x)$ of Table 8.1, show that
  (a) $\sigma^2 = 1/(2r + 1)a^2$ and the asymptotic efficacy of the mean procedures is $(2r + 1)a^2$.
  (b) $h(0) = a/2^{2r-1}B(r, r)$ and the asymptotic efficacy of the median procedures is $[a/4^{r-1}B(r, r)]^2$.
  (c) $\int h^2(x)dx = aB(2r - 1, 2r - 1)/2B^2(r, r)$ and the asymptotic efficacy of the Wilcoxon procedures is $3a^2B^2(2r - 1, 2r - 1)/B^4(r, r)$.
  (d) $I = a^2(r - 1)(2r - 1)/(r - 2)$ for $r > 2$. What happens if $r \le 2$?
  (e) The entries in Table 8.1 corresponding to (a)–(d) are correct.

50. For the one-sample squared rank procedures for shifts, show that
  (a) The asymptotic efficacy is

$$20\left[\int_0^\infty [2H(x) - 1]^2h'(x)dx\right]^2 = 320\left\{\int_0^\infty [2H(x) - 1]h^2(x)dx\right\}^2.$$

  *(b) The result in (a) for the symmetric beta density of Table 8.1 with $a = 1$ is $5(33/32)^2$ if $r = 2$; $320(2/\pi)^6(\pi/3 + 1/5 - 7/9)^2$ if $r = 1.5$; $5(283 \cdot 15/7 \cdot 2^9)^2$ if $r = 3$; $5(187 \cdot 175/3 \cdot 2^{13})^2$ if $r = 4$.
  *(c) Verify the entries in Table 8.1 corresponding to (b).

*51. For the cumulative logistic distribution $H(x) = 1/(1 + e^{-x})$, show that (Hint: The relation $h(x) = H(x)[1 - H(x)]$ and the substitution $y = H(x)$ may sometimes be helpful.)
  (a) $\sigma^2 = \pi^2/3$ and the asymptotic efficacy of the mean procedures is $3/\pi^2$.
  (b) $h(0) = \frac{1}{4}$ and the asymptotic efficacy of the median procedures is $\frac{1}{4}$.
  (c) $\int h^2(x)dx = \frac{1}{6}$ and the asymptotic efficacy of the Wilcoxon procedures is $\frac{1}{3}$.
  (d) The asymptotic efficacy of procedures based on sums of signed constants satisfying (8.7) is $e_4 = [\int_0^1 uc(u)du]^2/\int_0^1 c^2(u)du$.
  (e) For $c(u) = u^2$, $e_4 = \frac{5}{16}$.
  (f) The efficacy of the normal scores procedures is $1/\pi$.
  (g) $I = \frac{1}{3}$.
  (h) The entries in Table 8.1 corresponding to (a)–(g) are correct.

52. For the Cauchy density $h(x) = 1/\pi(1 + x^2)$, show that
  (a) $\sigma^2 = \infty$ and the asymptotic efficacy of the mean procedures is 0.
  (b) $h(0) = 1/\pi$ and the asymptotic efficacy of the median procedures is $4/\pi^2$.
  (c) $\int h^2(x)dx = 1/2\pi$ and the asymptotic efficacy of the Wilcoxon procedures is $3/\pi^2$.
  *(d) The asymptotic efficacy of procedures based on sums of signed constants satisfying (8.7) is

$$e_4 = \left[\int_0^1 c(u)\sin(\pi u)du\right]^2 / \int_0^1 c^2(u)du.$$

  *(e) For $c(u) = u^2$, $e_4 = (5/\pi^2)(1 - 4/\pi^2)^2$.
  *(f) $I = \frac{1}{2}$.
  (g) The entries in Table 8.1 corresponding to (a)–(f) are correct.

53. Derive Equation (8.17) for the conditional probability, given the absolute values of a sample, that the jth smallest belongs to a positive observation.

54. Show that the asymptotic efficiency of the normal-theory procedures relative to the median procedures for shifts $h(x - \theta)$ can be infinite even if h is symmetric, but cannot exceed 3 if h has its maximum at its median.

55. Justify the bounds given in Sect. 8.3 for the asymptotic efficiency of the Wilcoxon procedures relative to the median procedures.

56. Show that the asymptotic efficiency of the normal scores procedures relative to the median procedures can be arbitrarily small for shifts of a symmetric density with a spike of height $1/\varepsilon$ and width ε^2 at the median.

*57. (a) Show that the asymptotic efficiency of the Wilcoxon procedures relative to the normal scores procedures for shifts of a symmetric density h cannot exceed $6/\pi$. (Hint: Use the second expression in (8.11).)
   (b) Show that the value in (a) is approached for a suitable symmetric unimodal density with a spike of height $1/\varepsilon^2$ and width ε^3 at the median.

*58. (a) If H is a c.d.f. with finite variance, show that $x\phi\{\Phi^{-1}[H(x)]\} \to 0$ as $x \to \pm\infty$, where Φ and ϕ are the standard normal cumulative and density functions. (Hint: Tchebycheff's inequality can be used [Gastwirth and Wolff, 1968].)
   (b) If H is continuous, show that $\{\int \Phi^{-1}[H(x)]h(x)dx\}^2 = 1$.

59. (a) Show that $E(1/Z) \geq 1/E(Z)$ if the random variable Z is positive with probability one. (One proof uses Jensen's inequality; another uses (8.27) or a similar inequality. For others, see Gurland [1967].)
   *(b) Use the result in (a) instead of (8.27) to show that the asymptotic efficiency of the normal scores procedures relative to the normal-theory procedures for symmetric shifts is at least one [Gastwirth and Wolff, 1968].)

*60. Derive the locally most powerful signed-rank test given by (8.33) for a symmetric shift family.

*61. (a) Show that the test based on the sum of signed constants (8.33) satisfies (8.7) with

$$c(u) = h'\left[H^{-1}\left(\frac{1+u}{2}\right)\right] \bigg/ h\left[H^{-1}\left(\frac{1+u}{2}\right)\right].$$

   (b) Verify that substitution of the result in (a) in (8.8) gives $e_4 = I$.

*62. Derive the expression (8.37) relating the efficacy of the signed-rank test given by (8.33) to the Fisher information.

63. Let the joint density of (V, W) belong to a shift family $g(v, w; \theta) = g(v, w - \theta; 0)$ and let $X = W - V$.
   (a) Express the density of X in terms of g and show that it can be written as a shift family $f(x; \theta) = h(x - \theta)$.
   (b) Show that every univariate shift family of densities can arise as in (a).
   (c) If (V, W) is permutationally symmetrically distributed when $\theta = 0$, show that X is symmetrically distributed about θ.
   (d) Show that every symmetric density h can arise in the manner of (c).
   (e) What role does the restriction to densities play in (a)–(d)?

64. Let ϕ be the characteristic function of X. Show that
    (a) If V and W are independently, identically distributed and $X = W - V$, then ϕ is real and nonnegative everywhere.
    (b) If X has the rectangular distribution on $[-1, 1]$, then $\phi(t) = (\sin t)/t$, which can take on negative values.
    *(c) If X has the quadratic density $h(x) = 3(1 - x^2)^+/4$ (symmetric beta with $r = 2$, $a = 1$), then $\phi(t) = 3(\sin t - t \cos t)/t^3$, which can take on negative values. (See Corollary 4.6 of Hollander [1967].)
    *(d) If X has the symmetric beta density with $r = 3$, $a = 1$, then

$$\phi(t) = 15[(3 - t^2)\sin t - 3t \cos t]/t^5,$$

    which can take on negative values.
    (e) $\phi(t; r) = \int_0^1 (1 - y)^{r-2} \sin(t\sqrt{y})dy/tB(1.5, r - 1)$ for $r > 1$, where $\phi(t; r)$ is the characteristic function of the symmetric beta density with $a = 1$ for arbitrary r.
    *(f) If $\phi(t; r) < 0$ for some $t \in (0, 2\pi]$ and some $r > 1$, then $\phi(t; s) < 0$ for the same t and all $s \in (1, r]$. (Hint: Use the result in (e) and the fact that $(1 - y)^{r-2}/(1 - y)^{s-2}$ is decreasing in y while $\sin(t\sqrt{y})$ changes sign at most once for $y \in [0, 1]$.)

65. Let V and W be independently, identically distributed with central moments μ_j, and let $X = W - V$ have central moments $\theta_j, j = 1, 2, \ldots$. Show that
    (a) $\theta_j = 0$ for j odd, $\theta_2 = 2\mu_2$, and $\theta_4 = 2\mu_4 + 6\mu_2^2$.
    (b) $\theta_4 \geq 2\theta_2^2$.
    (c) X cannot have a rectangular distribution because this would contradict (b).
    (d) $\theta_6 = 2\mu_6 + 30\mu_2\mu_4 - 20\mu_3^2$ and $\theta_8 = 2\mu_8 + 56\mu_2\mu_6 - 112\mu_3\mu_5 + 70\mu_4^2$.
    *(e) X cannot have a symmetric beta distribution with $r = 2$ because this would contradict known inequalities among the moments.

66. Under the matched pairs model with independent errors leading to (9.5), show that
    (a) The efficacy of the mean of the differences X_j is $n/2\sigma^2$, where σ^2 is the variance of the errors.
    (b) The efficacy of the median of the differences X_j is $4n[\int q^2(x)dx]^2$, where q is the density of the errors.

67. Under the matched pairs model with independent errors leading to (9.5), show that
    (a) The bounds of Table 8.2 on asymptotic relative efficiencies of one-sample procedures which can be achieved by normal and Cauchy distributions cannot be improved.
    (b) The maximum asymptotic efficiency of the normal-theory procedures relative to the median procedures is $125(72$, achieved when q is the symmetric beta density with $r = 2$. (Hint: Use Problem 66.)
    *(c) The Wilcoxon and normal scores procedures can have arbitrarily small asymptotic efficiency relative to the median procedures. (Mixtures of normal distributions can be used for q.)
    *(d) The asymptotic efficiency of the normal scores procedures relative to the Wilcoxon procedures must be more than but can be arbitrarily close to $\pi/6$. (Mixtures of normal distributions can again be used for q.)
    (e) In each case (a)–(d), q is or can be chosen symmetric and unimodal.

68. (a) Give a two-sample form of the development leading to (2.11).
    (b) Which steps in the development in (a) are exact for two-sample normal-theory tests of equality of population means when the variances are known but possibly unequal?

69. Let $X_1, \ldots, X_m$ have c.d.f. F and $Y_1, \ldots, Y_n$ have c.d.f. G; let $Z_1, Z_2, \ldots$ be independently, identically distributed according to H which is symmetric about 0; and define

$$nG_n^*(y) = \begin{cases} \text{the number of } Y_j < y & \text{if } y < 0, \\ \text{the number of } Y_j \le y & \text{if } y \ge 0. \end{cases}$$

Show that
(a) The two-sample Wilcoxon test statistic is equivalent to $\sum_j F_m(Y_j)$, which approaches $\sum_j F(Y_j)$ as $m \to \infty$.
(b) The one-sample Wilcoxon test statistic for the Y sample is equivalent to $\sum_j [G_n^*(Y_j) - G_n^*(-Y_j)]$.
(c) For shifts of H, each of the statistics in (a) and (b) has the same asymptotic behavior as a linear function of $\sum_j H(Z_j + \theta)$.
(d) A two-sample sum of scores test statistic satisfying (10.6) is asymptotically equivalent to $\sum_j c[F(Y_j)]$ as first $m \to \infty$ and then $n \to \infty$.
(e) A one-sample sum of scores test statistic satisfying (8.7) is asymptotically equivalent to $\sum_j c[G_n^*(Y_j) - G_n^*(-Y_j)]$ as $n \to \infty$.
(f) In the situation of (c), the sum in (d) is distributed as $\sum_j c[H(Z_j + \theta)]$ while the sum in (e) is distributed as $\sum_j c[2H(Z_j + \theta) - 1]$.
(g) With $c = c_{(1)}$ or $c_{(2)}$ as appropriate, the sums in (d)–(f) are the same under condition (10.8).

*70. For each of the two-sample normal-theory, median, and Wilcoxon procedures for shift alternatives (10.4),
(a) Derive directly the asymptotic efficacy.
(b) Verify that the efficacy does not depend on the ratio of the sample sizes and agrees with the corresponding one-sample efficacy.

71. For a two-sample procedure based on a sum of scores,
(a) Derive the efficacy (10.5).
(b) Derive the asymptotic efficacy (10.7).

72. (a) Show that the one- and two-sample median and Wilcoxon procedures have corresponding scores functions in the sense of (10.8).
(b) Derive the two-sample procedure that corresponds similarly to the one-sample squared ranks procedure.

73. Problems 73–78 all concern the asymptotic robustness of two-sample location tests against inequality of scale. See also Pratt [1964]. Consider a test at level α for the null hypothesis that the densities f and g, with c.d.f.'s F and G respectively, are equal. Suppose, for definiteness, that α is one-tailed, and let $K = \Phi^{-1}(\alpha)$ be the corresponding standard normal deviate. Assume that $m/N \to \lambda$ (which now matters). For the two-sample t test, show that
(a) The probability of rejection approaches 0 or 1 if f and g have different means (and finite variances).
(b) The probability of rejection approaches $\Phi(Kd)$ if f and g have equal means, where $d = [\lambda + (1 - \lambda)\theta^2]^{1/2}/(1 - \lambda + \lambda\theta^2)^{1/2}$ and $\theta = \text{var}(Y)/\text{var}(X)$.
(c) For fixed λ, the range of possible values of d is an interval with endpoints $(\lambda^{-1} - 1)^{1/2}$ and $(\lambda^{-1} - 1)^{-1/2}$.

*74. For the two-sample median test, show that results like those in Problem 73 hold with medians replacing means, $d = (1 - \lambda + \lambda\theta)/(1 - \lambda + \lambda\theta^2)^{1/2}$, and $\theta = f/g$ at the common median, and the endpoints of the range of d are $[\min(\lambda, 1 - \lambda)]^{1/2}$ and 1. (Hint: Use the corresponding confidence bound given in (3.6) of Chap. 5.)

*75. For the two-sample Wilcoxon test, show that results like those of Problem 73 hold but $p_1 = 1/2$ replaces equal means, $d = [12\lambda p_2 + 12(1 - \lambda)p_3 - 3]^{-1/2}$, and the range of d has endpoints $(3 - 3\lambda)^{-1/2}$ and $(1 - \lambda)^{1/2}/(3\lambda - 4\lambda^2)^{1/2}$ if $\lambda \le \frac{1}{2}$ and the same with λ replaced by $(1 - \lambda)$ otherwise, where p_1, p_2, and p_3 are defined by (4.12)–(4.14) of Chap. 5. For the range of d, use $\text{var}(U) \ge mn[3m - (m-1)^2/(n-1)]/12$ if $p_1 = \frac{1}{2}$ where U is the Mann–Whitney statistic defined in (4.5) of Chap. 5 [Birnbaum and Klose, 1957]. (Hint: Equation (4.26) of Chap. 5 gives an upper bound for $\text{var}(U)$ which is achieved for constant Y.)

*76. In Problem 75, if f and g have the same center of symmetry and variances in the ratio $\theta^2 = \text{var}(Y)/\text{var}(X)$, show that
   (a) $p_2 = \frac{1}{4} + (1/2\pi) \arcsin [\theta^2/(\theta^2 + 1)]$ for f and g normal, where the arcsin is in radians.
   (b) $p_2 = \frac{1}{4} + \theta^2/2(\theta + 1)(2\theta + 1)$ for f and g the Laplace density (5.6).
   (c) $p_2 = \frac{1}{4} + \theta^2/12$ if $\theta \le 1$ and $p_2 = \frac{1}{2} - 1/6\theta$ if $\theta \ge 1$ for f and g uniform.
   (d) p_3 is given by the same formulas but with θ replaced by $1/\theta$.

*77. For the two-sample normal scores test, show that results like those of Problem 73(a) and (b) hold but $\gamma = 0$ replaces equal means and

$$d = [2\lambda I(F, G, \lambda) + 2(1 - \lambda)I(G, F, 1 - \lambda)]^{-1/2}$$

where $\gamma = \int \Phi^{-1}[H(x)]f(x)dx$, $H(x) = \lambda F(x) + (1 - \lambda)G(x)$, and

$$I(F, G, \lambda) = \int\!\!\int_{x<y} G(x)[1 - G(y)]\Phi^{-1'}[H(x)]\Phi^{-1'}[H(y)]f(x)f(y)dxdy.$$

*78. What are the implications of Problems 73–77 for confidence procedures?

79. (a) Derive the approximation (11.1) for the power of the one-sided, one-sample Kolmogorov–Smirnov test against shifts of the uniform distribution.
   (b) Derive the approximate power (11.3) of the median test against the alternative in (a).
   (c) Derive the asymptotic relative efficiency (11.4) of the one-sided Kolmogorov–Smirnov test relative to the median test for uniform shift alternatives.
   *(d) Find the limits of the asymptotic relative efficiency as $\alpha \to 0$ for fixed β, and as $\beta \to 0$ for fixed α, and as $\beta \to 1 - \alpha$.
   *(e) Show that (a) and (c) provide upper bounds for any symmetric, unimodal shift alternative.
   *(f) For what other alternatives do (a) and (c) provide upper bounds?

80. For the one-sided, one-sample Kolmogorov–Smirnov test, derive
   (a) The lower bound (11.6) for its power.
   *(b) The lower bound (11.7) for its asymptotic efficiency relative to the median test. (Hint: Use (11.2) and (11.3).)

81. (a) Let f be a density on $(-\infty, B)$ where B is finite and let ξ_θ be its quantile of order θ. Let g be a density on (ξ_θ, ∞) such that $g(y) = f(y)$ for $\xi_\theta \le y \le B$. Show that the power of a Kolmogorov–Smirnov test of the null hypothesis f against the alternative g depends only on θ, and not on f, and in particular, it is the same for any f as for a uniform distribution on $(0, 1)$.
   *(b) What happens in (a) if no such finite limit B exists?

*82. For the one-sided, one-sample Kolmogorov–Smirnov test,
  (a) Derive the lower bound (11.9) for its power.
  (b) Derive the lower bound (11.10) for its asymptotic efficiency relative to the median test.
  (c) Show that the bound in (b) approaches 0 as $\beta \to 1 - \alpha$.

*83. (a) Let $f(x) = \frac{1}{2} - \varepsilon + 1/\varepsilon$ for $|x| < \varepsilon^2$, $f(x) = \frac{1}{2} - \varepsilon$ for $\varepsilon^2 \leq |x| \leq 1$, and $f(x) = 0$ otherwise. Show that, for sufficiently small ε, shifts of $f(x)$ come arbitrarily close to achieving the lower bounds (11.8)–(11.10).
  (b) Find the limits of the lower bound (11.10) as $\alpha \to 0$ for fixed β, as $\beta \to 0$ for fixed α, and as $\beta \to 1 - \alpha$.

*84. Find the limits of the improved lower bound (11.19) for the asymptotic efficiency of the Kolmogorov–Smirnov test relative to the median test as $\alpha \to 0$ for fixed β, as $\beta \to 0$ for fixed α, and as $\beta \to 1 - \alpha$.

85. Let h be a density with c.d.f. H and median 0. Show that
  (a) The following conditions are equivalent:
    (i) $h(x)/H(x) \geq 2h(0)$ for $x \leq 0$;
    (ii) $h[H^{-1}(u)]/u \geq 2h(0)$ for $u \leq \frac{1}{2}$;
    (iii) $h(x)/H(x)$ is minimized for $x \leq 0$ by $x = 0$.
  (b) The following conditions are equivalent:
    (iv) $h(x)/H(x)$ is decreasing in x for $x \leq 0$;
    (v) $h[H^{-1}(u)]/u$ is decreasing in u for $u \leq \frac{1}{2}$;
    (vi) $\log H(x)$ is strictly concave in x for $x \leq 0$.
  (c) The following conditions are equivalent:
    (vii) $h'(x)/h(x)$ is decreasing in x for $x \leq 0$;
    (viii) $\log h(x)$ is strictly concave in x for $x \leq 0$;
    (ix) $h[H^{-1}(u)]$ is strictly concave in u for $u \leq \frac{1}{2}$.
  (d) (iv)–(vi) imply (i)–(iii).
  *(e) (vii)–(ix) imply (iv)–(vi). (One proof uses Cauchy's theorem of elementary calculus. Another uses the theorem that $\log \int f(x, y)dy$ is a concave function of x if $\log f(x, y)$ is a concave function of (x, y) [Brascamp and Lieb, 1976].)
  (f) The relations in (b)–(e) hold if "decreasing" is replaced by "nonincreasing" and "strictly concave" by "concave" throughout.

86. Show that [Pratt, 1981]
  (a) The double exponential power distribution satisfies (i)–(ix) of Problem 85 if $k \geq 1$ and none of these conditions otherwise.
  (b) The symmetric beta distribution satisfies (i)–(vi) for all r and (vii)–(ix) if and only if $r \geq 1$.
  (c) The logistic distribution satisfies (i)–(ix).
  (d) The Cauchy distribution satisfies (i)–(vi) but not (vii)–(ix).
  *(e) The t distribution satisfies (i)–(vi) but not (vii)-(ix).
  (f) The density $h(x) = \frac{1}{2} \cos x$ for $|x| \leq \pi/2$ and 0 elsewhere satisfies (i)–(ix).

*87. Show that the Laplace distribution achieves equality in the lower bounds (11.18) and (11.19).

88. Let n_1 and n_2 be the sample sizes required to achieve a given β by a one-sided normal-theory test at level $\alpha/2$ and by a two-sided normal-theory test at level α for the one-sample shift problem.

(a) Give expressions that determine n_1 and n_2 for alternatives near the null hypothesis.

(b) Show that $n_1/n_2 \leq 1.02$ asymptotically for $\alpha \leq 0.1$ and $\beta \leq 0.74$; also for $\alpha \leq 0.05$ and $\beta \leq 0.85$; also for other combinations of α and β.

*89. For the Kolmogorov–Smirnov test with one-sided critical value c at level $\alpha/2$, show that

(a) $P_G[\inf(F_n - F) \leq -c$ and $\sup(F_n - F) \leq c] \leq \alpha/2$ if $G(x) \geq F(x)$ for all x.

(b) Asymptotically the probability in (a) comes arbitrarily close to

$$\alpha/2 - P_G[\inf(F_n - G) \leq -c \quad \text{and} \sup(F_n - G) \geq c]$$

and the inequality (11.27) comes arbitrarily close to equality for shifts of a distribution which is uniform on $\bigcup_{i=1}^{m} (i, i + \frac{1}{2})$ for sufficiently large m.

*90. Show that, in the notation of Problem 89, for G continuous, as $n \to \infty$

$$P_G[\sup\sqrt{n}(F_n - G) \geq c - \theta \quad \text{or} \inf\sqrt{n}(F_n - G) \leq -c]$$

$$\to \sum_{i=1}^{\infty} e^{-2[(2i-1)c-(i-1)\theta]^2} + e^{-2[(2i-1)c-i\theta]^2} - 2e^{-2i^2(2c-\theta)^2}.$$

Tables

Table A Cumulative Standard Normal Distribution

Each table entry is the cumulative probability of a standardized normal variable $z = (x - \mu)/\sigma$, right tail from the value of z to plus infinity, and also left tail from minus infinity to the value of $-z$, for all $P \leq 0.50$. Read down the first column to the correct first decimal value of z, and over to the correct column for the second decimal value. The number at the intersection is the value of P.

z		0.00	0.01	0.02	0.03	0.04	0.05	0.06	0.07	0.08	0.09
0.0	0.	50000	49601	49202	48803	48405	48006	47608	47210	46812	46414
0.1		46017	45620	45224	44828	44433	44038	43644	43251	42858	42465
0.2		42074	41683	41294	40905	40517	40129	39743	39358	38974	38591
0.3		38209	37828	37448	37070	36693	36317	35942	35569	35197	34827
0.4		34458	34090	33724	33360	32997	32636	32276	31918	31561	31207
0.5		30854	30503	30153	29806	29460	29116	28774	28434	28096	27760
0.6		27425	27093	26763	26435	26109	25785	25463	25143	24825	24510
0.7		24196	23885	23576	23270	22965	22663	22363	22065	21770	21476
0.8		21186	20897	20611	20327	20045	19766	19489	19215	18943	18673
0.9		18406	18141	17879	17619	17361	17106	16853	16602	16354	16109
1.0		15866	15625	15386	15151	14917	14686	14457	14231	14007	13786
1.1		13567	13350	13136	12924	12714	12507	12302	12100	11900	11702
1.2		11507	11314	11123	10935	10749	10565	10383	10204	10027	09853
1.3	0.0	96800	95098	93418	91759	90123	88508	86915	85343	83793	82264
1.4		80757	79270	77804	76359	74934	73529	72145	70781	69437	68112
1.5		66807	65522	64255	63008	61780	60571	59380	58208	57053	55917
1.6		54799	53699	52616	51551	50503	49471	48457	47460	46479	45514
1.7		44565	43633	42716	41815	40930	40059	39204	38364	37538	36727
1.8		35930	35148	34380	33625	32884	32157	31443	30742	30054	29379
1.9		28717	28067	27429	26803	26190	25588	24998	24419	23852	23295
2.0		22750	22216	21692	21178	20675	20182	19699	19226	18763	18309
2.1		17864	17429	17003	16586	16177	15778	15386	15003	14629	14262
2.2		13903	13553	13209	12874	12545	12224	11911	11604	11304	11011
2.3		10724	10444	10170	09903	09642	09387	09137	08894	08656	08424
2.4	0.0^2	81975	79763	77603	75494	73436	71428	69469	67557	65691	63872
2.5		62097	60366	58677	57031	55426	53861	52336	50849	49400	47988
2.6		46612	45271	43965	42692	41453	40246	39070	37926	36811	35726
2.7		34670	33642	32641	31667	30720	29798	28901	28028	27179	26354
2.8		25551	24771	24012	23274	22557	21860	21182	20524	19884	19262
2.9		18658	18071	17502	16948	16411	15889	15382	14890	14412	13949
3.0		13499	13062	12639	12228	11829	11442	11067	10703	10350	10008
3.1	0.0^3	96760	93544	90426	87403	84474	81635	78885	76219	73638	71136
3.2		68714	66367	64095	61895	59765	57703	55706	53774	51904	50094
3.3		48342	46648	45009	43423	41889	40406	38971	37584	36243	34946
3.4		33693	32481	31311	30179	29086	28029	27009	26023	25071	24151
3.5		23263	22405	21577	20778	20006	19262	18543	17849	17180	16534
3.6		15911	15310	14730	14171	13632	13112	12611	12128	11662	11213
3.7		10780	10363	09961	09574	09201	08842	08496	08162	07841	07532
3.8	0.0^4	72348	69483	66726	64072	61517	59059	56694	54418	52228	50122
3.9		48096	46148	44274	42473	40741	39076	37475	35936	34458	33037
4.0		31671	30359	29099	27888	26726	25609	24536	23507	22518	21569
4.1		20658	19783	18944	18138	17365	16624	15912	15230	14575	13948
4.2		13346	12769	12215	11685	11176	10689	10221	09774	09345	08934
4.3	0.0^5	85399	81627	78015	74555	71241	68069	65031	62123	59340	56675
4.4		54125	51685	49350	47117	44979	42935	40980	39110	37322	35612

Table A (continued)

z		0.00	0.01	0.02	0.03	0.04	0.05	0.06	0.07	0.08	0.09
4.5		33977	32414	30920	29492	28127	26823	25577	24386	23249	22162
4.6		21125	20133	19187	18283	17420	16597	15810	15060	14344	13660
4.7		13008	12386	11792	11226	10686	10171	09680	09211	08765	08339
4.8	0.0^6	79333	75465	71779	68267	64920	61731	58693	55799	53043	50418
4.9		47918	45538	43272	41115	39061	37107	35247	33476	31792	30190

For larger values of z, P can be approximated by $(2\pi)^{-1/2} M e^{-z^2/2} = 0.398942\, M e^{-z^2/2}$ where

$$M = (1 - \{z^2 + 3 - [0.22(z^2 + 3.2)]^{-1}\}^{-1})/z$$

The error is less than $0.0005P$ for $z \geq 1.2$. See Peizer and Pratt [1968].

Source: Taken from Table IIi of Fisher and Yates: *Statistical Tables for Biological, Agricultural and Medical Research*, published by Longman Group Ltd., London (6th edition, 1974, page 45) (previously published by Oliver & Boyd, Ltd., Edinburgh), by permission of the authors and publishers.

Table B Cumulative Binomial Distribution

Each table entry is the left-tail binomial cumulative probability of s or less successes in n trials with probability p of success, for $1 \leq s < n/2, 4 \leq n \leq 20$ and $p = 0.5, 21 \leq n \leq 30$. Each entry is also the right-tail cumulative probability of $n - s$ or more successes for probability $1 - p$, as given in the right column and last row for $n \leq 20$. For $s = 0$, use $P(S = 0) = (1 - p)^n$. For other values of s, consider the complementary tail. For other values of p and/or n, Table A can be used with the approximate standard normal deviate

$$z = \frac{d}{|s' - np|} \left\{ \frac{12n}{6n + 1} \left(s' \ln \frac{s'}{np} + t' \ln \frac{t'}{nq} \right) \right\}^{1/2}$$

where $s' = s + 1/2$, $t' = n - s - 1/2$, $d = s + 2/3 - (n + 1/6)p + 0.02 \, [q/(s + 1) - p/(n - s) + (q - 0.5)/(n + 1)]$. The error is less than 1% of the smaller tail probability if $s, n - s - 1 \geq 2$ and $0.19 \leq s'q/t'p \leq 5.3$. It is less than 0.2% if $s, n - s - 1 \geq 4$ and $0.40 \leq s'q/t'p \leq 2.5$. See Peizer and Pratt [1968]. It is less than 0.00082 if $s, n - s - 1 \geq 1$, than 0.00012 if $s, n - s - 1 \geq 4$. See Ling [1978].

p		0.05	0.10	0.20	0.30	0.40	0.50	0.60	0.70	0.80	0.90	0.95	
n	s												$n - s$
4	1	0.9860	0.9477	0.8192	0.6517	0.4752	0.3125	0.1792	0.0837	0.0272	0.0037	0.0005	3
5	1	0.9774	0.9185	0.7373	0.5282	0.3370	0.1875	0.0870	0.0308	0.0067	0.0005	0.0000	4
	2	0.9988	0.9914	0.9421	0.8369	0.6826	0.5000	0.3174	0.1631	0.0579	0.0086	0.0012	3
6	1	0.9672	0.8857	0.6554	0.4202	0.2333	0.1094	0.0410	0.0109	0.0016	0.0001	0.0000	5
	2	0.9978	0.9842	0.9011	0.7443	0.5443	0.3438	0.1792	0.0705	0.0170	0.0013	0.0001	4
7	1	0.9556	0.8503	0.5767	0.3294	0.1586	0.0625	0.0188	0.0038	0.0004	0.0000	0.0000	6
	2	0.9962	0.9743	0.8520	0.6471	0.4199	0.2266	0.0963	0.0288	0.0047	0.0002	0.0000	5
	3	0.9998	0.9973	0.9667	0.8740	0.7102	0.5000	0.2898	0.1260	0.0333	0.0027	0.0002	4
8	1	0.9428	0.8131	0.5033	0.2553	0.1064	0.0352	0.0085	0.0013	0.0001	0.0000	0.0000	7
	2	0.9942	0.9619	0.7969	0.5518	0.3154	0.1445	0.0498	0.0113	0.0012	0.0000	0.0000	6
	3	0.9996	0.9950	0.9437	0.8059	0.5941	0.3633	0.1737	0.0580	0.0104	0.0004	0.0000	5
9	1	0.9288	0.7748	0.4362	0.1960	0.0705	0.0195	0.0038	0.0004	0.0000	0.0000	0.0000	8
	2	0.9916	0.9470	0.7382	0.4628	0.2318	0.0898	0.0250	0.0043	0.0003	0.0000	0.0000	7
	3	0.9994	0.9917	0.9144	0.7297	0.4826	0.2539	0.0994	0.0253	0.0031	0.0001	0.0000	6
	4	1.0000	0.9991	0.9804	0.9012	0.7334	0.5000	0.2666	0.0988	0.0196	0.0009	0.0000	5
10	1	0.9139	0.7361	0.3758	0.1493	0.0464	0.0107	0.0017	0.0001	0.0000	0.0000	0.0000	9
	2	0.9885	0.9298	0.6778	0.3828	0.1673	0.0547	0.0123	0.0016	0.0001	0.0000	0.0000	8
	3	0.9990	0.9872	0.8791	0.6496	0.3823	0.1719	0.0548	0.0106	0.0009	0.0000	0.0000	7
	4	0.9999	0.9984	0.9672	0.8497	0.6331	0.3770	0.1662	0.0473	0.0064	0.0001	0.0000	6
11	1	0.8981	0.6974	0.3221	0.1130	0.0302	0.0059	0.0007	0.0000	0.0000	0.0000	0.0000	10
	2	0.9848	0.9104	0.6174	0.3127	0.1189	0.0327	0.0059	0.0006	0.0000	0.0000	0.0000	9
	3	0.9984	0.9815	0.8389	0.5696	0.2963	0.1133	0.0293	0.0043	0.0002	0.0000	0.0000	8
	4	0.9999	0.9972	0.9496	0.7897	0.5328	0.2744	0.0994	0.0216	0.0020	0.0000	0.0000	7
	5	1.0000	0.9997	0.9883	0.9218	0.7535	0.5000	0.2465	0.0782	0.0117	0.0003	0.0000	6
12	1	0.8816	0.6590	0.2749	0.0850	0.0196	0.0032	0.0003	0.0000	0.0000	0.0000	0.0000	11
	2	0.9804	0.8891	0.5583	0.2528	0.0834	0.0193	0.0028	0.0002	0.0000	0.0000	0.0000	10
	3	0.9978	0.9744	0.7946	0.4925	0.2253	0.0730	0.0153	0.0017	0.0001	0.0000	0.0000	9
	4	0.9998	0.9957	0.9274	0.7237	0.4382	0.1938	0.0573	0.0095	0.0006	0.0000	0.0000	8
	5	1.0000	0.9995	0.9806	0.8822	0.6652	0.3872	0.1582	0.0386	0.0039	0.0001	0.0000	7
13	1	0.8646	0.6213	0.2336	0.0637	0.0126	0.0017	0.0001	0.0000	0.0000	0.0000	0.0000	12
	2	0.9755	0.8661	0.5017	0.2025	0.0579	0.0112	0.0013	0.0001	0.0000	0.0000	0.0000	11
	3	0.9969	0.9658	0.7473	0.4206	0.1686	0.0461	0.0078	0.0007	0.0000	0.0000	0.0000	10
	4	0.9997	0.9935	0.9009	0.6543	0.3530	0.1334	0.0321	0.0040	0.0002	0.0000	0.0000	9
	5	1.0000	0.9991	0.9700	0.8346	0.5744	0.2905	0.0977	0.0182	0.0012	0.0000	0.0000	8
	6	1.0000	0.9999	0.9930	0.9376	0.7712	0.5000	0.2288	0.0624	0.0070	0.0001	0.0000	7
$1 - p$		0.95	0.90	0.80	0.70	0.60	0.50	0.40	0.30	0.20	0.10	0.05	

Table B (*continued*)

	p	0.05	0.10	0.20	0.30	0.40	0.50	0.60	0.70	0.80	0.90	0.95	
n	s												n − s
14	1	0.8470	0.5848	0.1979	0.0475	0.0081	0.0009	0.0001	0.0000	0.0000	0.0000	0.0000	13
	2	0.9699	0.8416	0.4481	0.1608	0.0398	0.0065	0.0006	0.0000	0.0000	0.0000	0.0000	12
	3	0.9958	0.9559	0.6982	0.3552	0.1243	0.0287	0.0039	0.0002	0.0000	0.0000	0.0000	11
	4	0.9996	0.9908	0.8702	0.5842	0.2793	0.0898	0.0175	0.0017	0.0000	0.0000	0.0000	10
	5	1.0000	0.9985	0.9561	0.7805	0.4859	0.2120	0.0583	0.0083	0.0004	0.0000	0.0000	9
	6	1.0000	0.9998	0.9884	0.9067	0.6925	0.3953	0.1501	0.0315	0.0024	0.0000	0.0000	8
15	1	0.8290	0.5490	0.1671	0.0353	0.0052	0.0005	0.0000	0.0000	0.0000	0.0000	0.0000	14
	2	0.9638	0.8159	0.3980	0.1268	0.0271	0.0037	0.0003	0.0000	0.0000	0.0000	0.0000	13
	3	0.9945	0.9444	0.6482	0.2969	0.0905	0.0176	0.0019	0.0001	0.0000	0.0000	0.0000	12
	4	0.9994	0.9873	0.8358	0.5155	0.2173	0.0592	0.0093	0.0007	0.0000	0.0000	0.0000	11
	5	0.9999	0.9978	0.9389	0.7216	0.4032	0.1509	0.0338	0.0037	0.0001	0.0000	0.0000	10
	6	1.0000	0.9997	0.9819	0.8689	0.6098	0.3036	0.0950	0.0152	0.0008	0.0000	0.0000	9
	7	1.0000	1.0000	0.9958	0.9500	0.7869	0.5000	0.2131	0.0500	0.0042	0.0000	0.0000	8
16	1	0.8108	0.5147	0.1407	0.0261	0.0033	0.0003	0.0000	0.0000	0.0000	0.0000	0.0000	15
	2	0.9571	0.7892	0.3518	0.0994	0.0183	0.0021	0.0001	0.0000	0.0000	0.0000	0.0000	14
	3	0.9930	0.9316	0.5981	0.2459	0.0651	0.0106	0.0009	0.0000	0.0000	0.0000	0.0000	13
	4	0.9991	0.9830	0.7982	0.4499	0.1666	0.0384	0.0049	0.0003	0.0000	0.0000	0.0000	12
	5	0.9999	0.9967	0.9183	0.6598	0.3288	0.1051	0.0191	0.0016	0.0000	0.0000	0.0000	11
	6	1.0000	0.9995	0.9733	0.8247	0.5272	0.2272	0.0583	0.0071	0.0002	0.0000	0.0000	10
	7	1.0000	0.9999	0.9930	0.9256	0.7161	0.4018	0.1423	0.0257	0.0015	0.0000	0.0000	9
17	1	0.7922	0.4818	0.1182	0.0193	0.0021	0.0001	0.0000	0.0000	0.0000	0.0000	0.0000	16
	2	0.9497	0.7618	0.3096	0.0774	0.0123	0.0012	0.0001	0.0000	0.0000	0.0000	0.0000	15
	3	0.9912	0.9174	0.5489	0.2019	0.0464	0.0064	0.0005	0.0000	0.0000	0.0000	0.0000	14
	4	0.9988	0.9779	0.7582	0.3887	0.1260	0.0245	0.0025	0.0001	0.0000	0.0000	0.0000	13
	5	0.9999	0.9953	0.8943	0.5968	0.2639	0.0717	0.0106	0.0007	0.0000	0.0000	0.0000	12
	6	1.0000	0.9992	0.9623	0.7752	0.4478	0.1662	0.0348	0.0032	0.0001	0.0000	0.0000	11
	7	1.0000	0.9999	0.9891	0.8954	0.6405	0.3145	0.0919	0.0127	0.0005	0.0000	0.0000	10
	8	1.0000	1.0000	0.9974	0.9597	0.8011	0.5000	0.1989	0.0403	0.0026	0.0000	0.0000	9
18	1	0.7735	0.4503	0.0991	0.0142	0.0013	0.0001	0.0000	0.0000	0.0000	0.0000	0.0000	17
	2	0.9419	0.7338	0.2713	0.0600	0.0082	0.0007	0.0000	0.0000	0.0000	0.0000	0.0000	16
	3	0.9891	0.9018	0.5010	0.1646	0.0328	0.0038	0.0002	0.0000	0.0000	0.0000	0.0000	15
	4	0.9985	0.9718	0.7164	0.3327	0.0942	0.0154	0.0013	0.0000	0.0000	0.0000	0.0000	14
	5	0.9998	0.9936	0.8671	0.5344	0.2088	0.0481	0.0058	0.0003	0.0000	0.0000	0.0000	13
	6	1.0000	0.9988	0.9487	0.7217	0.3743	0.1189	0.0203	0.0014	0.0000	0.0000	0.0000	12
	7	1.0000	0.9998	0.9837	0.8593	0.5634	0.2403	0.0576	0.0061	0.0002	0.0000	0.0000	11
	8	1.0000	1.0000	0.9957	0.9404	0.7368	0.4073	0.1347	0.0210	0.0009	0.0000	0.0000	10
19	1	0.7547	0.4203	0.0829	0.0104	0.0008	0.0000	0.0000	0.0000	0.0000	0.0000	0.0000	18
	2	0.9335	0.7054	0.2369	0.0462	0.0055	0.0004	0.0000	0.0000	0.0000	0.0000	0.0000	17
	3	0.9869	0.8850	0.4551	0.1332	0.0230	0.0022	0.0001	0.0000	0.0000	0.0000	0.0000	16
	4	0.9980	0.9648	0.6733	0.2822	0.0696	0.0096	0.0006	0.0000	0.0000	0.0000	0.0000	15
	5	0.9998	0.9914	0.8369	0.4739	0.1629	0.0318	0.0031	0.0001	0.0000	0.0000	0.0000	14
	6	1.0000	0.9983	0.9324	0.6655	0.3081	0.0835	0.0116	0.0006	0.0000	0.0000	0.0000	13
	7	1.0000	0.9997	0.9767	0.8180	0.4878	0.1796	0.0352	0.0028	0.0000	0.0000	0.0000	12
	8	1.0000	1.0000	0.9933	0.9161	0.6675	0.3238	0.0885	0.0105	0.0003	0.0000	0.0000	11
	9	1.0000	1.0000	0.9984	0.9674	0.8139	0.5000	0.1861	0.0326	0.0016	0.0000	0.0000	10
	1 − p	0.95	0.90	0.80	0.70	0.60	0.50	0.40	0.30	0.20	0.10	0.05	

Table B (*continued*)

p	0.05	0.10	0.20	0.30	0.40	0.50	0.60	0.70	0.80	0.90	0.95	
n s												n − s
20 1	0.7358	0.3917	0.0692	0.0076	0.0005	0.0000	0.0000	0.0000	0.0000	0.0000	0.0000	19
2	0.9245	0.6769	0.2061	0.0355	0.0036	0.0002	0.0000	0.0000	0.0000	0.0000	0.0000	18
3	0.9841	0.8670	0.4114	0.1071	0.0160	0.0013	0.0000	0.0000	0.0000	0.0000	0.0000	17
4	0.9974	0.9568	0.6296	0.2375	0.0510	0.0059	0.0003	0.0000	0.0000	0.0000	0.0000	16
5	0.9997	0.9887	0.8042	0.4164	0.1256	0.0207	0.0016	0.0000	0.0000	0.0000	0.0000	15
6	1.0000	0.9976	0.9133	0.6080	0.2500	0.0577	0.0065	0.0003	0.0000	0.0000	0.0000	14
7	1.0000	0.9996	0.9679	0.7723	0.4159	0.1316	0.0210	0.0013	0.0000	0.0000	0.0000	13
8	1.0000	0.9999	0.9900	0.8867	0.5956	0.2517	0.0565	0.0051	0.0001	0.0000	0.0000	12
9	1.0000	1.0000	0.9974	0.9520	0.7553	0.4119	0.1275	0.0171	0.0006	0.0000	0.0000	11
1 − p	0.95	0.90	0.80	0.70	0.60	0.50	0.40	0.30	0.20	0.10	0.05	

Supplementary Table for $p = 0.5$

s n	21	22	23	24	25	26	27	28	29	30
1	0.0000	0.0000	0.0000	0.0000	0.0000	0.0000	0.0000	0.0000	0.0000	0.0000
2	0.0001	0.0001	0.0000	0.0000	0.0000	0.0000	0.0000	0.0000	0.0000	0.0000
3	0.0007	0.0004	0.0002	0.0001	0.0001	0.0000	0.0000	0.0000	0.0000	0.0000
4	0.0036	0.0022	0.0013	0.0008	0.0005	0.0003	0.0002	0.0001	0.0001	0.0000
5	0.0133	0.0085	0.0053	0.0033	0.0020	0.0012	0.0008	0.0005	0.0003	0.0002
6	0.0392	0.0262	0.0173	0.0113	0.0073	0.0047	0.0030	0.0019	0.0012	0.0007
7	0.0946	0.0669	0.0466	0.0320	0.0216	0.0145	0.0096	0.0063	0.0041	0.0026
8	0.1917	0.1431	0.1050	0.0758	0.0539	0.0378	0.0261	0.0178	0.0121	0.0081
9	0.3318	0.2617	0.2024	0.1537	0.1148	0.0843	0.0610	0.0436	0.0307	0.0214
10	0.5000	0.4159	0.3388	0.2706	0.2122	0.1635	0.1239	0.0925	0.0680	0.0494
11	0.6692	0.5841	0.5000	0.4194	0.3450	0.2786	0.2210	0.1725	0.1325	0.1002
12	0.8083	0.7383	0.6612	0.5806	0.5000	0.4225	0.3506	0.2858	0.2291	0.1808
13	0.9054	0.8569	0.7976	0.7294	0.6550	0.5775	0.5000	0.4253	0.3555	0.2923
14	0.9608	0.9331	0.8950	0.8463	0.7878	0.7214	0.6494	0.5747	0.5000	0.4278

Table C Binomial Confidence Limits

Each table entry is n times the lower or upper confidence bound of the binomial parameter p for an observed number s of successes in n trials when $n \geq 5$ and $s/n < 0.50$ at typical levels $1 - \alpha$. For $s/n > 0.50$, enter the table with $(n - s)/n$, interchange Upper and Lower and subtract the table entry from n. In either case, divide the result by n to obtain the confidence limit. For $s = 0$, the lower limit is 0 and the upper limit is $1 - \alpha^{1/n}$ for all n. For $s > 14$, use the approximation indicated at the end of the table.

s	s/n	Lower Tail α					Upper Tail α				
		0.005	0.010	0.025	0.050	0.100	0.100	0.050	0.025	0.010	0.005
1	0.00	0.005	0.010	0.025	0.051	0.105	3.89	4.74	5.57	6.64	7.43
1	0.05	0.005	0.010	0.025	0.051	0.105	3.62	4.32	4.97	5.78	6.34
1	0.10	0.005	0.010	0.025	0.051	0.105	3.37	3.94	4.45	5.04	5.44
1	0.15	0.005	0.010	0.025	0.051	0.105	3.14	3.60	3.99	4.42	4.70
1	0.20	0.005	0.010	0.025	0.051	0.104	2.92	3.29	3.58	3.89	4.07
2	0.00	0.103	0.149	0.242	0.355	0.532	5.32	6.30	7.22	8.41	9.27
2	0.05	0.105	0.150	0.245	0.358	0.535	5.11	5.97	6.77	7.76	8.47
2	0.10	0.106	0.152	0.247	0.361	0.538	4.90	5.65	6.34	7.17	7.74
2	0.15	0.107	0.154	0.249	0.364	0.542	4.69	5.35	5.94	6.62	7.08
2	0.20	0.109	0.155	0.252	0.368	0.545	4.50	5.07	5.56	6.12	6.48
2	0.25	0.110	0.157	0.255	0.371	0.549	4.31	4.80	5.21	5.65	5.94
2	2\7	0.111	0.159	0.257	0.374	0.552	4.17	4.61	4.97	5.35	5.58
2	2\6	0.112	0.161	0.260	0.377	0.556	4.00	4.37	4.66	4.96	5.14
2	0.40	0.114	0.163	0.264	0.382	0.561	3.77	4.05	4.27	4.47	4.59
3	0.00	0.338	0.436	0.619	0.818	1.102	6.68	7.75	8.77	10.05	10.98
3	0.10	0.348	0.448	0.634	0.834	1.119	6.28	7.16	7.96	8.93	9.61
3	0.20	0.358	0.461	0.650	0.853	1.138	5.89	6.60	7.21	7.93	8.41
3	0.30	0.370	0.475	0.667	0.873	1.158	5.52	6.07	6.52	7.03	7.35
3	0.40	0.383	0.491	0.687	0.895	1.181	5.15	5.56	5.89	6.22	6.42
3	0.50	0.398	0.508	0.709	0.919	1.205	4.79	5.08	5.29	5.49	5.60
4	0.00	0.672	0.823	1.090	1.366	1.745	7.99	9.15	10.24	11.60	12.59
4	0.10	0.693	0.847	1.117	1.395	1.773	7.60	8.58	9.47	10.54	11.30
4	0.20	0.715	0.872	1.147	1.427	1.804	7.21	8.02	8.73	9.57	10.13
4	0.30	0.740	0.901	1.179	1.462	1.838	6.83	7.48	8.04	8.66	9.06
4	0.40	0.768	0.932	1.216	1.500	1.876	6.46	6.96	7.38	7.82	8.09
4	0.50	0.799	0.968	1.256	1.543	1.917	6.08	6.46	6.74	7.03	7.20
5	0.00	1.078	1.279	1.623	1.970	2.433	9.27	10.51	11.67	13.11	14.15
5	0.10	1.111	1.315	1.664	2.012	2.472	8.88	9.94	10.91	12.08	12.90
5	0.20	1.147	1.355	1.708	2.057	2.515	8.49	9.39	10.18	11.11	11.74
5	0.30	1.187	1.400	1.756	2.107	2.563	8.10	8.84	9.47	10.19	10.67
5	0.40	1.232	1.449	1.810	2.162	2.615	7.72	8.31	8.79	9.32	9.66
5	0.50	1.283	1.504	1.871	2.224	2.673	7.33	7.78	8.13	8.50	8.72
6	0.00	1.537	1.785	2.202	2.613	3.152	10.53	11.84	13.06	14.57	15.66
6	0.10	1.583	1.835	2.255	2.667	3.202	10.14	11.27	12.30	13.55	14.44
6	0.20	1.634	1.890	2.314	2.726	3.257	9.74	10.71	11.57	12.59	13.28
6	0.30	1.691	1.951	2.379	2.791	3.317	9.35	10.16	10.86	11.66	12.19
6	0.40	1.754	2.019	2.450	2.863	3.384	8.95	9.61	10.16	10.77	11.16
6	0.50	1.826	2.095	2.531	2.944	3.458	8.54	9.06	9.47	9.90	10.17
7	0.00	2.037	2.330	2.814	3.285	3.895	11.77	13.15	14.42	16.00	17.13
7	0.10	2.098	2.394	2.881	3.352	3.956	11.37	12.57	13.67	14.99	15.92
7	0.20	2.164	2.464	2.954	3.424	4.022	10.97	12.01	12.93	14.02	14.77
7	0.30	2.238	2.542	3.035	3.503	4.094	10.56	11.44	12.20	13.08	13.67
7	0.40	2.320	2.628	3.124	3.591	4.174	10.16	10.88	11.49	12.17	12.61
7	0.50	2.414	2.726	3.225	3.690	4.264	9.74	10.31	10.77	11.27	11.59

Table C (*continued*)

s	s/n	|	Lower Tail α				|	Upper Tail α			
		0.005	0.010	0.025	0.050	0.100	0.100	0.050	0.025	0.010	0.005
8	0.00	2.571	2.906	3.454	3.981	4.656	12.99	14.43	15.76	17.40	18.58
8	0.10	2.646	2.984	3.534	4.059	4.727	12.59	13.86	15.01	16.40	17.37
8	0.20	2.727	3.069	3.621	4.144	4.804	12.18	13.28	14.26	15.42	16.22
8	0.30	2.818	3.163	3.717	4.238	4.888	11.77	12.71	13.52	14.46	15.10
8	0.40	2.920	3.268	3.824	4.341	4.981	11.35	12.13	12.79	13.53	14.02
8	0.50	3.035	3.387	3.944	4.458	5.085	10.91	11.54	12.06	12.61	12.97
9	0.00	3.132	3.507	4.115	4.695	5.432	14.21	15.71	17.08	18.78	20.00
9	0.10	3.221	3.599	4.208	4.785	5.513	13.79	15.12	16.32	17.77	18.80
9	0.20	3.318	3.699	4.309	4.883	5.600	13.38	14.54	15.57	16.79	17.63
9	0.30	3.426	3.810	4.420	4.990	5.696	12.96	13.95	14.82	15.82	16.50
9	0.40	3.547	3.933	4.544	5.108	5.801	12.53	13.36	14.07	14.87	15.40
9	0.50	3.684	4.073	4.683	5.242	5.919	12.08	12.76	13.32	13.93	14.32
10	0.00	3.717	4.130	4.795	5.425	6.221	15.41	16.96	18.39	20.14	21.40
10	0.10	3.820	4.235	4.900	5.526	6.311	14.99	16.37	17.62	19.13	20.20
10	0.20	3.932	4.350	5.015	5.636	6.409	14.56	15.78	16.86	18.14	19.02
10	0.30	4.057	4.477	5.141	5.756	6.515	14.13	15.18	16.10	17.16	17.88
10	0.40	4.197	4.619	5.281	5.890	6.632	13.69	14.58	15.33	16.19	16.76
10	0.50	4.355	4.779	5.439	6.039	6.763	13.24	13.96	14.56	15.22	15.65
11	0.00	4.321	4.771	5.491	6.169	7.021	16.60	18.21	19.68	21.49	22.78
11	0.10	4.438	4.890	5.608	6.281	7.120	16.17	17.61	18.91	20.47	21.57
11	0.20	4.566	5.019	5.736	6.402	7.227	15.74	17.01	18.14	19.47	20.39
11	0.30	4.707	5.162	5.877	6.535	7.343	15.30	16.40	17.36	18.48	19.23
11	0.40	4.866	5.322	6.033	6.683	7.472	14.85	15.79	16.58	17.49	18.09
11	0.50	5.045	5.502	6.209	6.848	7.616	14.38	15.15	15.79	16.50	16.96
12	0.00	4.943	5.428	6.201	6.924	7.829	17.78	19.44	20.96	22.82	24.14
12	0.10	5.073	5.560	6.330	7.047	7.937	17.35	18.84	20.18	21.80	22.93
12	0.20	5.216	5.703	6.470	7.179	8.053	16.91	18.23	19.40	20.79	21.75
12	0.30	5.374	5.862	6.625	7.325	8.180	16.46	17.61	18.61	19.78	20.57
12	0.40	5.551	6.039	6.797	7.486	8.320	16.00	16.98	17.82	18.77	19.41
12	0.50	5.751	6.239	6.990	7.666	8.476	15.52	16.33	17.01	17.76	18.25
13	0.00	5.580	6.099	6.922	7.690	8.646	18.96	20.67	22.23	24.14	25.50
13	0.10	5.724	6.243	7.063	7.822	8.762	18.52	20.06	21.44	23.11	24.28
13	0.20	5.882	6.401	7.216	7.966	8.887	18.07	19.44	20.65	22.09	23.08
13	0.30	6.056	6.575	7.384	8.124	9.024	17.62	18.81	19.85	21.07	21.89
13	0.40	6.250	6.769	7.571	8.298	9.174	17.15	18.17	19.04	20.05	20.71
13	0.50	6.471	6.988	7.781	8.493	9.342	16.66	17.51	18.22	19.01	19.53
14	0.00	6.231	6.782	7.654	8.464	9.470	20.13	21.89	23.49	25.45	26.84
14	0.10	6.388	6.939	7.806	8.606	9.594	19.68	21.27	22.69	24.41	25.61
14	0.20	6.560	7.111	7.972	8.761	9.728	19.23	20.64	21.89	23.38	24.40
14	0.30	6.750	7.300	8.153	8.930	9.874	18.77	20.00	21.08	22.34	23.20
14	0.40	6.962	7.510	8.355	9.117	10.035	18.29	19.35	20.26	21.31	22.00
14	0.50	7.203	7.748	8.581	9.327	10.214	17.79	18.67	19.42	20.25	20.80

Other values can be approximated as follows. For lower limits, let $a = n - s$, $b = s + 1$, z = positive normal quantile (Table A). For upper limits, let $a = n - s + 1$, $b = s$, z = negative normal quantile. Calculate $A = 9a - 1$, $B = 9b - 1$, $C = 3z$, $D = B^2 - bC^2$, $E = AB + C(aD + bA^2)^{1/2}$. The confidence limit is $1/[1 + (b/a)^2(E/D)^3]$. For $\alpha \geq 0.005$, the error is less than 1% if $s, n - s \geq 9$ and less than 0.5% if $s, n - s \geq 12$. See Pratt [1968].

Table D Cumulative Probabilities for Wilcoxon Signed Rank Statistic

Each table entry is the cumulative probability of t or less (left-tail) under the null distribution of T^+(or equivalently T^-).

$t\backslash n$	5	6	7	8	9	10	$t\backslash n$	8	9	10
0	0.0313	0.0156	0.0078	0.0039	0.0020	0.0010	14	0.3203	0.1797	0.0967
1	0.0625	0.0313	0.0156	0.0078	0.0039	0.0020	15	0.3711	0.2129	0.1162
2	0.0938	0.0469	0.0234	0.0117	0.0059	0.0029	16	0.4219	0.2480	0.1377
3	0.1563	0.0781	0.0391	0.0195	0.0098	0.0049	17	0.4727	0.2852	0.1611
4	0.2188	0.1094	0.0547	0.0273	0.0137	0.0068	18	0.5273	0.3262	0.1875
5	0.3125	0.1563	0.0781	0.0391	0.0195	0.0098	19		0.3672	0.2158
6	0.4063	0.2188	0.1094	0.0547	0.0273	0.0137	20		0.4102	0.2461
7	0.5000	0.2813	0.1484	0.0742	0.0371	0.0186	21		0.4551	0.2783
8		0.3438	0.1875	0.0977	0.0488	0.0244	22		0.5000	0.3125
9		0.4219	0.2344	0.1250	0.0645	0.0322	23			0.3477
10		0.5000	0.2891	0.1563	0.0820	0.0420	24			0.3848
11			0.3438	0.1914	0.1016	0.0527	25			0.4229
12			0.4063	0.2305	0.1250	0.0654	26			0.4609
13			0.4688	0.2734	0.1504	0.0801	27			0.5000

$t\backslash n$	11	12	13	14	15	16	17	18	19	20
0	0.0005	0.0002	0.0001	0.0001	0.0000	0.0000	0.0000	0.0000	0.0000	0.0000
1	0.0010	0.0005	0.0002	0.0001	0.0001	0.0000	0.0000	0.0000	0.0000	0.0000
2	0.0015	0.0007	0.0004	0.0002	0.0001	0.0000	0.0000	0.0000	0.0000	0.0000
3	0.0024	0.0012	0.0006	0.0003	0.0002	0.0001	0.0000	0.0000	0.0000	0.0000
4	0.0034	0.0017	0.0009	0.0004	0.0002	0.0001	0.0001	0.0000	0.0000	0.0000
5	0.0049	0.0024	0.0012	0.0006	0.0003	0.0002	0.0001	0.0001	0.0000	0.0000
6	0.0068	0.0034	0.0017	0.0009	0.0004	0.0002	0.0001	0.0001	0.0000	0.0000
7	0.0093	0.0046	0.0023	0.0012	0.0006	0.0003	0.0001	0.0001	0.0000	0.0000
8	0.0122	0.0061	0.0031	0.0015	0.0008	0.0004	0.0002	0.0001	0.0000	0.0000
9	0.0161	0.0081	0.0040	0.0020	0.0010	0.0005	0.0003	0.0001	0.0001	0.0000
10	0.0210	0.0105	0.0052	0.0026	0.0013	0.0007	0.0003	0.0002	0.0001	0.0000
11	0.0269	0.0134	0.0067	0.0034	0.0017	0.0008	0.0004	0.0002	0.0001	0.0001
12	0.0337	0.0171	0.0085	0.0043	0.0021	0.0011	0.0005	0.0003	0.0001	0.0001
13	0.0415	0.0212	0.0107	0.0054	0.0027	0.0013	0.0007	0.0003	0.0002	0.0001
14	0.0508	0.0261	0.0133	0.0067	0.0034	0.0017	0.0008	0.0004	0.0002	0.0001
15	0.0615	0.0320	0.0164	0.0083	0.0042	0.0021	0.0010	0.0005	0.0003	0.0001
16	0.0737	0.0386	0.0199	0.0101	0.0051	0.0026	0.0013	0.0006	0.0003	0.0002
17	0.0874	0.0461	0.0239	0.0123	0.0062	0.0031	0.0016	0.0008	0.0004	0.0002
18	0.1030	0.0549	0.0287	0.0148	0.0075	0.0038	0.0019	0.0010	0.0005	0.0002
19	0.1201	0.0647	0.0341	0.0176	0.0090	0.0046	0.0023	0.0012	0.0006	0.0003
20	0.1392	0.0757	0.0402	0.0209	0.0108	0.0055	0.0028	0.0014	0.0007	0.0004
21	0.1602	0.0881	0.0471	0.0247	0.0128	0.0065	0.0033	0.0017	0.0008	0.0004
22	0.1826	0.1018	0.0549	0.0290	0.0151	0.0078	0.0040	0.0020	0.0010	0.0005
23	0.2065	0.1167	0.0636	0.0338	0.0177	0.0091	0.0047	0.0024	0.0012	0.0006
24	0.2324	0.1331	0.0732	0.0392	0.0206	0.0107	0.0055	0.0028	0.0014	0.0007
25	0.2598	0.1506	0.0839	0.0453	0.0240	0.0125	0.0064	0.0033	0.0017	0.0008
26	0.2886	0.1697	0.0955	0.0520	0.0277	0.0145	0.0075	0.0038	0.0020	0.0010
27	0.3188	0.1902	0.1082	0.0594	0.0319	0.0168	0.0087	0.0045	0.0023	0.0012
28	0.3501	0.2119	0.1219	0.0676	0.0365	0.0193	0.0101	0.0052	0.0027	0.0014
29	0.3823	0.2349	0.1367	0.0765	0.0416	0.0222	0.0116	0.0060	0.0031	0.0016
30	0.4155	0.2593	0.1527	0.0863	0.0473	0.0253	0.0133	0.0069	0.0036	0.0018
31	0.4492	0.2847	0.1698	0.0969	0.0535	0.0288	0.0153	0.0080	0.0041	0.0021
32	0.4829	0.3110	0.1879	0.1083	0.0603	0.0327	0.0174	0.0091	0.0047	0.0024
33	0.5171	0.3386	0.2072	0.1206	0.0677	0.0370	0.0198	0.0104	0.0054	0.0028
34		0.3667	0.2274	0.1338	0.0757	0.0416	0.0224	0.0118	0.0062	0.0032
35		0.3955	0.2487	0.1479	0.0844	0.0467	0.0253	0.0134	0.0070	0.0036
36		0.4250	0.2709	0.1629	0.0938	0.0523	0.0284	0.0152	0.0080	0.0042
37		0.4548	0.2939	0.1788	0.1039	0.0583	0.0319	0.0171	0.0090	0.0047

Table D (continued)

$t\backslash n$	11	12	13	14	15	16	17	18	19	20
38		0.4849	0.3177	0.1955	0.1147	0.0649	0.0357	0.0192	0.0102	0.0053
39		0.5151	0.3424	0.2131	0.1262	0.0719	0.0398	0.0216	0.0115	0.0060
40			0.3677	0.2316	0.1384	0.0795	0.0443	0.0241	0.0129	0.0068
41			0.3934	0.2508	0.1514	0.0877	0.0492	0.0269	0.0145	0.0077
42			0.4197	0.2708	0.1651	0.0964	0.0544	0.0300	0.0162	0.0086
43			0.4463	0.2915	0.1796	0.1057	0.0601	0.0333	0.0180	0.0096
44			0.4730	0.3129	0.1947	0.1156	0.0662	0.0368	0.0201	0.0107
45			0.5000	0.3349	0.2106	0.1261	0.0727	0.0407	0.0223	0.0120
46				0.3574	0.2271	0.1372	0.0797	0.0449	0.0247	0.0133
47				0.3804	0.2444	0.1489	0.0871	0.0494	0.0273	0.0148
48				0.4039	0.2622	0.1613	0.0950	0.0542	0.0301	0.0164
49				0.4276	0.2807	0.1742	0.1034	0.0594	0.0331	0.0181
50				0.4516	0.2997	0.1877	0.1123	0.0649	0.0364	0.0200
51				0.4758	0.3193	0.2019	0.1218	0.0708	0.0399	0.0220
52				0.5000	0.3394	0.2166	0.1317	0.0770	0.0437	0.0242

$t\backslash n$	15	16	17	18	19	20
53	0.3599	0.2319	0.1421	0.0837	0.0478	0.0266
54	0.3808	0.2477	0.1530	0.0907	0.0521	0.0291
55	0.4020	0.2641	0.1645	0.0982	0.0567	0.0319
56	0.4235	0.2809	0.1764	0.1061	0.0616	0.0348
57	0.4452	0.2983	0.1889	0.1144	0.0668	0.0379
58	0.4670	0.3161	0.2019	0.1231	0.0723	0.0413
59	0.4890	0.3343	0.2153	0.1323	0.0782	0.0448
60	0.5110	0.3529	0.2293	0.1419	0.0844	0.0487
61		0.3718	0.2437	0.1519	0.0909	0.0527
62		0.3910	0.2585	0.1624	0.0978	0.0570
63		0.4104	0.2738	0.1733	0.1051	0.0615
64		0.4301	0.2895	0.1846	0.1127	0.0664
65		0.4500	0.3056	0.1964	0.1206	0.0715
66		0.4699	0.3221	0.2086	0.1290	0.0768
67		0.4900	0.3389	0.2211	0.1377	0.0825
68		0.5100	0.3559	0.2341	0.1467	0.0884
69			0.3733	0.2475	0.1562	0.0947
70			0.3910	0.2613	0.1660	0.1012
71			0.4088	0.2754	0.1762	0.1081
72			0.4268	0.2899	0.1868	0.1153
73			0.4450	0.3047	0.1977	0.1227
74			0.4633	0.3198	0.2090	0.1305
75			0.4816	0.3353	0.2207	0.1387
76			0.5000	0.3509	0.2327	0.1471
77				0.3669	0.2450	0.1559
78				0.3830	0.2576	0.1650

$t\backslash n$	18	19	20
79	0.3994	0.2706	0.1744
80	0.4159	0.2839	0.1841
81	0.4325	0.2974	0.1942
82	0.4493	0.3113	0.2045
83	0.4661	0.3254	0.2152
84	0.4831	0.3397	0.2262
85	0.5000	0.3543	0.2375
86		0.3690	0.2490
87		0.3840	0.2608
88		0.3991	0.2729
89		0.4144	0.2853
90		0.4298	0.2979
91		0.4453	0.3108
92		0.4609	0.3238
93		0.4765	0.3371
94		0.4922	0.3506
95		0.5078	0.3643
96			0.3781
97			0.3921
98			0.4062
99			0.4204
100			0.4347
101			0.4492
102			0.4636
103			0.4782
104			0.4927

For larger n, Table A can be used with the approximate standard normal deviate $z = (t - \mu)/\sigma$ where $\mu = n(n + 1)/4$ and $\sigma = [n(n + 1)(2n + 1)/24]^{1/2}$. The table below gives μ and σ for $21 \le n \le 40$.

n	21	22	23	24	25	26	27	28	29	30
μ	115.5	126.5	138.0	150.0	162.5	175.5	189.0	203.0	217.5	232.5
σ	28.77	30.80	32.88	35.00	37.17	39.37	41.62	43.91	46.25	47.83

n	31	32	33	34	35	36	37	38	39	40
μ	248.0	264.0	280.5	297.5	315.0	333.0	351.5	370.5	390.0	410.0
σ	53.03	53.48	55.97	58.49	61.05	63.65	66.29	68.95	71.66	74.40

Source: Adapted from Table II of H. L. Harter and D. B. Owen, Eds. (1972), *Selected Tables in Mathematical Statistics*, Vol. 1, Markham Publishing Co., Chicago, with permission of The Institute of Mathematical Statistics.

Table E Cumulative Probabilities for Hypergeometric Distribution

Each table entry is the cumulative probability of a or less (left-tail) under the null distribution where $t = N/2$ if N is even and $t = (N \pm 1)/2$ if N is odd, where $N = m + n$. The entries apply to a 2 × 2 table (like that shown in Table 3.1 of Chap. 5) arranged so that $2a < m \leq t \leq n$ as a result of interchanging rows, columns, rows with columns, or lower and upper tails as necessary. For $m = 1$, use $P(A = 0) = (n - t + 1)/(n + 1)$ for any n. Other entries not included have $P = 0.0000$ or $P > 0.5$. Right-tail cumulative probabilities are found from this table using $P(A \geq a) = 1 - P(A \leq a - 1)$.

| | $n-m$ | 0 | 1 | 1 | 2 | 3 | 3 | 4 | 5 | 5 | 6 | 7 | 7 |
m	$a \backslash t-m$	0	0	1	1	1	2	2	2	3	3	3	4
2	0	0.1667	0.3000	0.1000	0.2000	0.2857	0.1429	0.2143	0.2778	0.1667	0.2222	0.2727	0.1818
	1	0.8333	0.9000	0.7000	0.8000	0.8571	0.7143	0.7857	0.8333	0.7222	0.7778	0.8182	0.7273
3	0	0.0500	0.1143	0.0286	0.0714	0.1190	0.0476	0.0833	0.1212	0.0606	0.0909	0.1224	0.0699
	1	0.5000	0.6286	0.3714	0.5000	0.5952	0.4048	0.5000	0.5758	0.4242	0.5000	0.5629	0.4371
4	0	0.0143	0.0397	0.0079	0.0238	0.0455	0.0152	0.0303	0.0490	0.0210	0.0350	0.0513	0.0256
	1	0.2429	0.3571	0.1667	0.2619	0.3485	0.1970	0.2727	0.3427	0.2168	0.2797	0.3385	0.2308
	2	0.7571	0.8333	0.6429	0.7381	0.8030	0.6515	0.7273	0.7832	0.6573	0.7203	0.7692	0.6615
5	0	0.0040	0.0130	0.0022	0.0076	0.0163	0.0047	0.0105	0.0186	0.0070	0.0128	0.0204	0.0091
	1	0.1032	0.1753	0.0671	0.1212	0.1795	0.0862	0.1329	0.1818	0.1002	0.1410	0.1833	0.1109
	2	0.5000	0.6082	0.3918	0.5000	0.5874	0.4126	0.5000	0.5734	0.4266	0.5000	0.5633	0.4367
6	0	0.0011	0.0041	0.0006	0.0023	0.0056	0.0014	0.0035	0.0068	0.0023	0.0045	0.0077	0.0031
	1	0.0400	0.0775	0.0251	0.0513	0.0839	0.0350	0.0594	0.0882	0.0430	0.0656	0.0913	0.0495
	2	0.2836	0.3835	0.2086	0.2960	0.3776	0.2308	0.3042	0.3733	0.2466	0.3100	0.3700	0.2585
	3	0.7165	0.7914	0.6166	0.7040	0.7692	0.6224	0.6958	0.7534	0.6267	0.6900	0.7415	0.6300
7	0	0.0003	0.0012	0.0002	0.0007	0.0019	0.0004	0.0011	0.0024	0.0007	0.0015	0.0028	0.0010
	1	0.0146	0.0317	0.0089	0.0203	0.0364	0.0134	0.0249	0.0399	0.0174	0.0286	0.0426	0.0209
	2	0.1431	0.2145	0.1002	0.1573	0.2178	0.1170	0.1674	0.2199	0.1299	0.1749	0.2214	0.1401
	3	0.5000	0.5952	0.4048	0.5000	0.5806	0.4194	0.5000	0.5700	0.4300	0.5000	0.5619	0.4381
8	0	0.0001	0.0004	0.0000	0.0002	0.0006	0.0001	0.0004	0.0008	0.0002	0.0005	0.0010	0.0003
	1	0.0051	0.0122	0.0030	0.0076	0.0149	0.0049	0.0099	0.0170	0.0067	0.0119	0.0188	0.0084
	2	0.0660	0.1090	0.0445	0.0767	0.1149	0.0549	0.0849	0.1192	0.0635	0.0913	0.1224	0.0706
	3	0.3096	0.3992	0.2380	0.3186	0.3950	0.2549	0.3250	0.3916	0.2678	0.3297	0.3889	0.2779
	4	0.6904	0.7620	0.6008	0.6814	0.7451	0.6050	0.6750	0.7322	0.6084	0.6703	0.7221	0.6111
9	0	0.0000	0.0001	0.0000	0.0001	0.0002	0.0000	0.0001	0.0003	0.0001	0.0002	0.0004	0.0001
	1	0.0017	0.0045	0.0010	0.0027	0.0058	0.0017	0.0038	0.0069	0.0025	0.0047	0.0079	0.0033
	2	0.0283	0.0513	0.0185	0.0349	0.0563	0.0242	0.0402	0.0602	0.0291	0.0447	0.0633	0.0335
	3	0.1735	0.2422	0.1276	0.1849	0.2449	0.1421	0.1935	0.2468	0.1535	0.2002	0.2481	0.1628
	4	0.5000	0.5859	0.4141	0.5000	0.5750	0.4250	0.5000	0.5666	0.4334	0.5000	0.5600	0.4400
10	0	0.0000	0.0000	0.0000	0.0000	0.0001	0.0000	0.0000	0.0001	0.0000	0.0001	0.0001	0.0000
	1	0.0005	0.0016	0.0003	0.0010	0.0022	0.0006	0.0014	0.0027	0.0009	0.0018	0.0032	0.0012
	2	0.0115	0.0226	0.0073	0.0150	0.0260	0.0101	0.0180	0.0287	0.0127	0.0207	0.0310	0.0151
	3	0.0894	0.1349	0.0635	0.0992	0.1402	0.0736	0.1069	0.1442	0.0820	0.1131	0.1473	0.0891
	4	0.3281	0.4100	0.2599	0.3350	0.4067	0.2735	0.3401	0.4041	0.2841	0.3441	0.4018	0.2928
	5	0.6719	0.7401	0.5900	0.6650	0.7265	0.5933	0.6599	0.7159	0.5959	0.6559	0.7072	0.5982
11	1	0.0002	0.0005	0.0001	0.0003	0.0008	0.0002	0.0005	0.0010	0.0003	0.0007	0.0013	0.0004
	2	0.0045	0.0095	0.0028	0.0061	0.0114	0.0040	0.0077	0.0130	0.0053	0.0092	0.0144	0.0065
	3	0.0431	0.0699	0.0296	0.0498	0.0749	0.0358	0.0554	0.0789	0.0412	0.0601	0.0821	0.0460
	4	0.1974	0.2632	0.1504	0.2068	0.2655	0.1628	0.2142	0.2671	0.1730	0.2200	0.2683	0.1814
	5	0.5000	0.5789	0.4211	0.5000	0.5704	0.4296	0.5000	0.5635	0.4365	0.5000	0.5579	0.4421
12	1	0.0001	0.0002	0.0000	0.0001	0.0003	0.0001	0.0002	0.0004	0.0001	0.0002	0.0005	0.0002
	2	0.0017	0.0038	0.0010	0.0024	0.0048	0.0016	0.0032	0.0056	0.0021	0.0039	0.0064	0.0027
	3	0.0196	0.0341	0.0131	0.0236	0.0377	0.0165	0.0271	0.0407	0.0197	0.0302	0.0433	0.0226
	4	0.1102	0.1566	0.0812	0.1189	0.1612	0.0906	0.1259	0.1649	0.0987	0.1318	0.1678	0.1056
	5	0.3421	0.4179	0.2772	0.3475	0.4153	0.2883	0.3518	0.4131	0.2973	0.3552	0.4112	0.3047
	6	0.6579	0.7228	0.5821	0.6525	0.7117	0.8388	0.6482	0.7027	0.5869	0.6448	0.6953	0.5888
13	1	0.0000	0.0001	0.0000	0.0000	0.0001	0.0000	0.0001	0.0001	0.0000	0.0001	0.0002	0.0001
	2	0.0006	0.0015	0.0004	0.0009	0.0019	0.0006	0.0013	0.0024	0.0008	0.0016	0.0028	0.0011
	3	0.0085	0.0157	0.0056	0.0107	0.0180	0.0073	0.0127	0.0200	0.0090	0.0145	0.0218	0.0106
	4	0.0576	0.0871	0.0412	0.0642	0.0919	0.0476	0.0697	0.0957	0.0531	0.0744	0.0990	0.0581

Table E (*continued*)

m	a	n − m 0 / t − m 0	1 / 0	1 / 1	2 / 1	3 / 1	3 / 2	4 / 2	5 / 2	5 / 3	6 / 3	7 / 3	7 / 4
	5	0.2169	0.2798	0.1697	0.2247	0.2817	0.1804	0.2311	0.2831	0.1814	0.2363	0.2842	0.1970
	6	0.5000	0.5734	0.4266	0.5000	0.5664	0.4336	0.5000	0.5607	0.4383	0.5000	0.5559	0.4441
14	1	0.0000	0.0000	0.0000	0.0000	0.0000	0.0000	0.0000	0.0000	0.0000	0.0000	0.0001	0.0000
	2	0.0002	0.0006	0.0001	0.0003	0.0008	0.0002	0.0005	0.0010	0.0003	0.0006	0.0012	0.0004
	3	0.0035	0.0070	0.0023	0.0046	0.0082	0.0031	0.0057	0.0094	0.0039	0.0067	0.0105	0.0048
	4	0.0285	0.0457	0.0199	0.0328	0.0495	0.0237	0.0366	0.0526	0.0272	0.0399	0.0554	0.0304
	5	0.1284	0.1749	0.0974	0.1362	0.1790	0.1061	0.1426	0.1823	0.1137	0.1480	0.1851	0.1202
	6	0.3532	0.4241	0.2912	0.3576	0.4219	0.3005	0.3612	0.4201	0.3082	0.3641	0.4185	0.3147
	7	0.6468	0.7088	0.5758	0.6424	0.6995	0.5781	0.6388	0.6918	0.5799	0.6359	0.6853	0.5815
15	2	0.0001	0.0002	0.0009	0.0001	0.0003	0.0001	0.0002	0.0004	0.0001	0.0002	0.0005	0.0002
	3	0.0014	0.0030	0.0092	0.0019	0.0036	0.0013	0.0024	0.0043	0.0017	0.0030	0.0049	0.0021
	4	0.0134	0.0228	0.0528	0.0160	0.0253	0.0113	0.0183	0.0276	0.0133	0.0205	0.0296	0.0153
	5	0.0716	0.1028	0.1862	0.0778	0.1072	0.0591	0.0832	0.1109	0.0646	0.0878	0.1141	0.0696
	6	0.2330	0.2933	0.4311	0.2397	0.2949	0.1956	0.2453	0.2962	0.2036	0.2499	0.2972	0.2105
	7	0.5000	0.5689	0.7067	0.5000	0.5631	0.4369	0.5000	0.5582	0.4418	0.5000	0.5541	0.4459

For larger sample sizes, use

$$P(A = 0) = \binom{n}{t} \Big/ \binom{N}{m}$$

and

$$P(A \le a) = \left[1 + \frac{tm}{n - t + 1} + \cdots + \frac{t(t + 1) \cdots (t + a - 1)}{(n - t + 1) \cdots (n - t + a)} \binom{m}{a}\right] \binom{n}{t} \Big/ \binom{N}{m}.$$

Alternatively, approximate cumulative probabilities can be found from Table A with the approximate standard normal deviate

$$z = \frac{a''d'' - b''c''}{|a'd' - b'c'|} \left[2L\left(1 - \frac{1}{6N}\right) \Big/ \left(1 + \frac{1}{6m}\right)\left(1 + \frac{1}{6n}\right)\left(1 + \frac{1}{6t}\right)\left(1 + \frac{1}{6(N - t)}\right)\right]^{1/2}$$

where a', b', c', d' are the cell entries a, b, c, d in the 2×2 table corrected by $1/2$ respectively, and

$$a'' = a' + \frac{1}{6} + \frac{0.02}{a' + 0.5} + \frac{0.01}{m + 1} + \frac{0.01}{t + 1},$$

and similarly for b'', c'', and d'', with m and t replaced by the row and column total for the entry in question, and

$$L = a' \ln \frac{a'N}{tm} + b' \ln \frac{b'N}{tn} + c' \ln \frac{c'N}{m(N - t)} + d' \ln \frac{d'N}{n(N - t)}$$

in terms of the natural logarithms ln. If common logarithms are used, replace $2L$ by $4.60517L$. This normal approximation has at least the accuracy specified below if $\min(a, b - 1, c - 1, d)$ is at least the value shown. See Ling and Pratt [1980] for the minimum guaranteed accuracy for other values of $\min(a, b - 1, c - 1, d)$ and other tail probabilities.

$\min(a, b - 1, c - 1, d)$	3	4	6	8	12	24	50
Any tail probability	$0.0^3 50$	$0.0^3 33$	$0.0^3 17$	$0.0^3 10$	$0.0^4 50$	$0.0^4 14$	$0.0^5 35$
Tail probability ≤ 0.05	$0.0^3 25$	$0.0^3 16$	$0.0^4 94$	$0.0^4 53$	$0.0^4 28$	$0.0^5 87$	$0.0^5 27$
Tail probability ≤ 0.01	$0.0^3 13$	$0.0^4 84$	$0.0^4 45$	$0.0^4 29$	$0.0^4 15$	$0.0^5 51$	$0.0^5 16$

Source: Adapted from G. J. Lieberman and D. B. Owen (1961), *Tables of the Hypergeometric Probability Distribution*, Stanford University Press, Stanford, California, with permission.

Table F Cumulative Probabilities for Wilcoxon Rank Sum Statistic

Each table entry labeled P is the left-tail cumulative probability of R_x under the null distribution for $2 \leq m \leq n \leq 10$. For $m = 1$, use $P(R_x = i) = 1/(n + 1)$ for $i = 1, \ldots, n + 1$. Right-tail cumulative probabilities are found from

$$P(R_x \geq t) = P[R_x \leq m(N + 1) - t].$$

R_x	P	R_x	P	R_x	P	R_x	P
$m = 2,$	$n = 2$	$m = 2,$	$n = 7$	8	0.182	12	0.274
				9	0.242	13	0.357
3	0.167	3	0.028	10	0.303	14	0.452
4	0.333	4	0.056	11	0.379	15	0.548
5	0.667	5	0.111	12	0.455		
		6	0.167	13	0.0545	$m = 3,$	$n = 7$
$m = 2,$	$n = 3$	7	0.250			6	0.008
		8	0.333	$m = 3,$	$n = 3$	7	0.017
3	0.100	9	0.444			8	0.033
4	0.200	10	0.556	6	0.050	9	0.058
5	0.400			7	0.100	10	0.092
6	0.600	$m = 2,$	$n = 8$	8	0.200	11	0.133
				9	0.350	12	0.192
$m = 2,$	$n = 4$	3	0.022	10	0.500	13	0.258
		4	0.044			14	0.333
3	0.067	5	0.089	$m = 3,$	$n = 4$	15	0.417
4	0.133	6	0.133			16	0.500
5	0.267	7	0.200	6	0.029		
6	0.400	8	0.267	7	0.057	$m = 3,$	$n = 8$
7	0.600	9	0.356	8	0.114	6	0.006
		10	0.444	9	0.200	7	0.012
$m = 2,$	$n = 5$	11	0.556	10	0.314	8	0.024
				11	0.429	9	0.042
3	0.048	$m = 2,$	$n = 9$	12	0.571	10	0.067
4	0.095					11	0.097
5	0.190	3	0.018	$m = 3,$	$n = 5$	12	0.139
6	0.286	4	0.036			13	0.188
7	0.429	5	0.073	6	0.018	14	0.248
8	0.571	6	0.109	7	0.036	15	0.315
		7	0.164	8	0.071	16	0.388
$m = 2,$	$n = 6$	8	0.218	9	0.125	17	0.461
		9	0.291	10	0.196	18	0.539
3	0.036	10	0.364	11	0.286		
4	0.071	11	0.455	12	0.393	$m = 3,$	$n = 9$
5	0.143	12	0.545	13	0.500	6	0.005
6	0.214					7	0.009
7	0.321	$m = 2,$	$n = 10$	$m = 3,$	$n = 6$	8	0.018
8	0.429			6	0.012	9	0.032
9	0.571	3	0.015	7	0.024	10	0.050
		4	0.030	8	0.048	11	0.073
		5	0.061	9	0.083		
		6	0.091	10	0.131		
		7	0.136	11	0.190		

Table F (continued)

R_x	P	R_x	P	R_x	P	R_x	P
12	0.105	15	0.143	14	0.024	19	0.071
13	0.141	16	0.206	15	0.036	20	0.094
14	0.186	17	0.278	16	0.055	21	0.120
15	0.241	18	0.365	17	0.077	22	0.152
16	0.300	19	0.452	18	0.107	23	0.187
17	0.364	20	0.548	19	0.141	24	0.227
18	0.432			20	0.184	25	0.270
19	0.500	$m = 4,$	$n = 6$	21	0.230	26	0.318
		10	0.005	22	0.285	27	0.367
$m = 3,$	$n = 10$	11	0.010	23	0.341	28	0.420
		12	0.019	24	0.404	29	0.473
6	0.003	13	0.033	25	0.467	30	0.527
7	0.007	14	0.057	26	0.533		
8	0.014	15	0.086			$m = 5,$	$n = 5$
9	0.024	16	0.129	$m = 4,$	$n = 9$	15	0.004
10	0.038	17	0.176	10	0.001	16	0.008
11	0.056	18	0.238	11	0.003	17	0.016
12	0.080	19	0.305	12	0.006	18	0.028
13	0.108	20	0.381	13	0.010	19	0.048
14	0.143	21	0.457	14	0.017	20	0.075
15	0.185	22	0.543	15	0.025	21	0.111
16	0.234			16	0.038	22	0.155
17	0.287	$m = 4,$	$n = 7$	17	0.053	23	0.210
18	0.346	10	0.003	18	0.074	24	0.274
19	0.406	11	0.006	19	0.099	25	0.345
20	0.469	12	0.012	20	0.130	26	0.421
21	0.531	13	0.021	21	0.165	27	0.500
		14	0.036	22	0.207		
$m = 4,$	$n = 4$	15	0.055	23	0.252	$m = 5,$	$n = 6$
10	0.014	16	0.082	24	0.302	15	0.002
11	0.029	17	0.115	25	0.355	16	0.004
12	0.057	18	0.158	26	0.413	17	0.009
13	0.100	19	0.206	27	0.470	18	0.015
14	0.171	20	0.264	28	0.530	19	0.026
15	0.243	21	0.324			20	0.041
16	0.343	22	0.394	$m = 4,$	$n = 10$	21	0.063
17	0.443	23	0.464	10	0.001	22	0.089
18	0.557	24	0.536	11	0.002	23	0.123
				12	0.004	24	0.165
$m = 4,$	$n = 5$	$m = 4,$	$n = 8$	13	0.007	25	0.214
10	0.008	10	0.002	14	0.012	26	0.268
11	0.016	11	0.004	15	0.018	27	0.331
12	0.032	12	0.008	16	0.027	28	0.396
13	0.056	13	0.014	17	0.038	29	0.465
14	0.095			18	0.053	30	0.535

Table F (*continued*)

R_x	P	R_x	P	R_x	P	R_x	P
$m = 5,$	$n = 7$	$m = 5,$	$n = 9$	33	0.220	35	0.183
				34	0.257	36	0.223
15	0.001	15	0.000	35	0.297	37	0.267
16	0.003	16	0.001	36	0.339	38	0.314
17	0.005	17	0.002	37	0.384	39	0.365
18	0.009	18	0.003	38	0.430	40	0.418
19	0.015	19	0.006	39	0.477	41	0.473
20	0.024	20	0.009	40	0.523	42	0.527
21	0.037	21	0.014				
22	0.053	22	0.021	$m = 6,$	$n = 6$	$m = 6,$	$n = 8$
23	0.074	23	0.030				
24	0.101	24	0.041	21	0.001	21	0.000
25	0.134	25	0.056	22	0.002	22	0.001
26	0.172	26	0.073	23	0.004	23	0.001
27	0.216	27	0.095	24	0.008	24	0.002
28	0.265	28	0.120	25	0.013	25	0.004
29	0.319	29	0.149	26	0.021	26	0.006
30	0.378	30	0.182	27	0.032	27	0.010
31	0.438	31	0.219	28	0.047	28	0.015
32	0.500	32	0.259	29	0.066	29	0.021
		33	0.303	30	0.090	30	0.030
$m = 5,$	$n = 8$	34	0.350	31	0.120	31	0.041
		35	0.399	32	0.155	32	0.054
15	0.001	36	0.449	33	0.197	33	0.071
16	0.002	37	0.500	34	0.242	34	0.091
17	0.003			35	0.294	35	0.114
18	0.005	$m = 5,$	$n = 10$	36	0.350	36	0.141
19	0.009			37	0.409	37	0.172
20	0.015	15	0.000	38	0.469	38	0.207
21	0.023	16	0.001	39	0.531	39	0.245
22	0.033	17	0.001			40	0.286
23	0.047	18	0.002	$m = 6,$	$n = 7$	41	0.331
24	0.064	19	0.004			42	0.377
25	0.085	20	0.006	21	0.001	43	0.426
26	0.111	21	0.010	22	0.001	44	0.475
27	0.142	22	0.014	23	0.002	45	0.525
28	0.177	23	0.020	24	0.004		
29	0.218	24	0.028	25	0.007	$m = 6,$	$n = 9$
30	0.262	25	0.038	26	0.011		
31	0.311	26	0.050	27	0.017	21	0.000
32	0.362	27	0.065	28	0.026	22	0.000
33	0.416	28	0.082	29	0.037	23	0.001
34	0.472	29	0.103	30	0.051	24	0.001
35	0.528	30	0.127	31	0.069	25	0.002
		31	0.155	32	0.090	26	0.004
		32	0.185	33	0.117	27	0.006
				34	0.147	28	0.009

Table F (*continued*)

R_x	P	R_x	P	R_x	P	R_x	P
29	0.013	44	0.246	36	0.010	50	0.176
30	0.018	45	0.281	37	0.014	51	0.204
31	0.025	46	0.318	38	0.020	52	0.235
32	0.033	47	0.356	39	0.027	53	0.268
33	0.044	48	0.396	40	0.036	54	0.303
34	0.057	49	0.437	41	0.047	55	0.340
35	0.072	50	0.479	42	0.060	56	0.379
36	0.091	51	0.521	43	0.076	57	0.419
37	0.112			44	0.095	58	0.459
38	0.136	$m = 7,$	$n = 7$	45	0.116	59	0.500
39	0.164	28	0.000	46	0.140		
40	0.194	29	0.001	47	0.168	$m = 7,$	$n = 10$
41	0.228	30	0.001	48	0.198	28	0.000
42	0.264	31	0.002	49	0.232	29	0.000
43	0.303	32	0.003	50	0.268	30	0.000
44	0.344	33	0.006	51	0.306	31	0.000
45	0.388	34	0.009	52	0.347	32	0.001
46	0.432	35	0.013	53	0.389	33	0.001
47	0.477	36	0.019	54	0.433	34	0.002
48	0.523	37	0.027	55	0.478	35	0.002
		38	0.036	56	0.522	36	0.003
$m = 6,$	$n = 10$	39	0.049			37	0.005
21	0.000	40	0.064	$m = 7,$	$n = 9$	38	0.007
22	0.000	41	0.082	28	0.000	39	0.009
23	0.000	42	0.104	29	0.000	40	0.012
24	0.001	43	0.130	30	0.000	41	0.017
25	0.001	44	0.159	31	0.001	42	0.022
26	0.002	45	0.191	32	0.001	43	0.028
27	0.004	46	0.228	33	0.002	44	0.035
28	0.005	47	0.267	34	0.003	45	0.044
29	0.008	48	0.310	35	0.004	46	0.054
30	0.011	49	0.355	36	0.006	47	0.067
31	0.016	50	0.402	37	0.008	48	0.081
32	0.021	51	0.451	38	0.011	49	0.097
33	0.028	52	0.500	39	0.016	50	0.115
34	0.036			40	0.021	51	0.135
35	0.047	$m = 7,$	$n = 8$	41	0.027	52	0.157
36	0.059	28	0.000	42	0.036	53	0.182
37	0.074	29	0.000	43	0.045	54	0.209
38	0.090	30	0.001	44	0.057	55	0.237
39	0.110	31	0.001	45	0.071	56	0.268
40	0.132	32	0.002	46	0.087	57	0.300
41	0.157	33	0.003	47	0.105	58	0.335
42	0.184	34	0.005	48	0.12	59	0.370
43	0.214	35	0.007	49	0.' ͵	60	0.406
						61	0.443
						62	0.481
						63	0.519

Table F (continued)

R_x	P	R_x	P	R_x	P	R_x	P
$m = 8, \quad n = 8$		44	0.003	50	0.010	61	0.016
		45	0.004	51	0.013	62	0.020
36	0.000	46	0.006	52	0.017	63	0.025
37	0.000	47	0.008	53	0.022	64	0.031
38	0.000	48	0.010	54	0.027	65	0.039
39	0.001	49	0.014	55	0.034	66	0.047
40	0.001	50	0.018	56	0.042	67	0.057
41	0.001	51	0.023	57	0.051	68	0.068
42	0.002	52	0.030	58	0.061	69	0.081
43	0.003	53	0.037	59	0.073	70	0.095
44	0.005	54	0.046	60	0.086	71	0.111
45	0.007	55	0.057	61	0.102	72	0.129
46	0.010	56	0.069	62	0.118	73	0.149
47	0.014	57	0.084	63	0.137	74	0.170
48	0.019	58	0.100	64	0.158	75	0.193
49	0.025	59	0.118	65	0.180	76	0.218
50	0.032	60	0.138	66	0.204	77	0.245
51	0.041	61	0.161	67	0.230	78	0.273
52	0.052	62	0.185	68	0.257	79	0.302
53	0.065	63	0.212	69	0.286	80	0.333
54	0.080	64	0.240	70	0.317	81	0.365
55	0.097	65	0.271	71	0.348	82	0.398
56	0.117	66	0.303	72	0.381	83	0.432
57	0.139	67	0.336	73	0.414	84	0.466
58	0.164	68	0.371	74	0.448	85	0.500
59	0.191	69	0.407	75	0.483	$m = 9, \quad n = 10$	
60	0.221	70	0.444	76	0.517		
61	0.253	71	0.481	$m = 9, \quad n = 9$		45	0.000
62	0.287	72	0.519			46	0.000
63	0.323	$m = 8, \quad n = 10$		45	0.000	47	0.000
64	0.360			46	0.000	48	0.000
65	0.399	36	0.000	47	0.000	49	0.000
66	0.439	37	0.000	48	0.000	50	0.000
67	0.480	38	0.000	49	0.000	51	0.000
68	0.520	39	0.000	50	0.000	52	0.000
		40	0.000	51	0.001	53	0.001
$m = 8, \quad n = 9$		41	0.000	52	0.001	54	0.001
36	0.000	42	0.001	53	0.001	55	0.001
37	0.000	43	0.001	54	0.002	56	0.002
38	0.000	44	0.002	55	0.003	57	0.003
39	0.000	45	0.002	56	0.004	58	0.004
40	0.000	46	0.003	57	0.005	59	0.005
41	0.001	47	0.004	58	0.007	60	0.007
42	0.001	48	0.006	59	0.009	61	0.009
43	0.002	49	0.008	60	0.012	62	0.011

Table F (*continued*)

R_x	P	R_x	P	R_x	P
63	0.014	$m = 10,\ n = 10$		81	0.038
64	0.017			82	0.045
65	0.022	55	0.000	83	0.053
66	0.027	56	0.000	84	0.062
67	0.033	57	0.000	85	0.072
68	0.039	58	0.000	86	0.083
69	0.047	59	0.000	87	0.095
70	0.056	60	0.000	88	0.109
71	0.067	61	0.000	89	0.124
72	0.078	62	0.000	90	0.140
73	0.091	63	0.000	91	0.157
74	0.106	64	0.001	92	0.176
75	0.121	65	0.001	93	0.197
76	0.139	66	0.001	94	0.218
77	0.158	67	0.001	95	0.241
78	0.178	68	0.002	96	0.264
79	0.200	69	0.003	97	0.289
80	0.223	70	0.003	98	0.315
81	0.248	71	0.004	99	0.342
82	0.274	72	0.006	100	0.370
83	0.302	73	0.007	101	0.398
84	0.330	74	0.009	102	0.427
85	0.360	75	0.012	103	0.456
86	0.390	76	0.014	104	0.485
87	0.421	77	0.018	105	0.515
88	0.452	78	0.022		
89	0.484	79	0.026		
90	0.516	80	0.032		

For m or n larger than 10, the probabilities are found from Table A as follows:

$$z_L = \frac{R_x + 0.5 - m(N + 1)/2}{\sqrt{mn(N + 1)/12}} \qquad z_R = \frac{R_x - 0.5 - m(N + 1)/2}{\sqrt{mn(N + 1)/12}}$$

Desired	Approximated by
Left tail probability for R_x	Right tail probability for $-z_L$
Right tail probability for R_x	Right tail probability for z_R

Source: Adapted from Table B of C. H. Kraft and C. Van Eeden (1969), *A Non-parametric Introduction to Statistics*, Macmillan Publishing Co., New York, with permission.

Table G Kolmogorov–Smirnov Two-Sample Statistic

Each table entry labeled P is the cumulative right-tail probability of mnD_{mn} under the null distribution for small P, all $2 \leq m \leq n \leq 12$ or $m + n \leq 16$, whichever comes first. For the one-sided statistics mnD_{mn}^{+} and mnD_{mn}^{-}, the corresponding probability is $P/2$ if P is very small. Each table entry in the second portion of the table, where $9 \leq m = n \leq 20$, is the smallest value of mnD_{mn} for which the cumulative right-tail probability does not exceed the selected values 0.010, 0.020, 0.050, 0.100, and 0.200 at the top. For mnD_{mn}^{+} and mnD_{mn}^{-}, each entry is approximately correct for the cumulative right-tail probability at the bottom.

m	n	mnD_{mn}	P	m	n	mnD_{mn}	P	m	n	mnD_{mn}	P
2	2	4	0.333	3	10	30	0.007	4	11	44	0.001
2	3	6	0.200			27	0.028			40	0.007
2	4	8	0.133			24	0.070			36	0.022
2	5	10	0.095			21	0.140			33	0.035
		8	0.286	3	11	33	0.005			32	0.063
2	6	12	0.071			30	0.022			29	0.098
		10	0.214			27	0.055			28	0.144
2	7	14	0.056			24	0.110	4	12	48	0.001
		12	0.167	3	12	36	0.004			44	0.005
2	8	16	0.044			33	0.018			40	0.016
		14	0.133			30	0.044			36	0.048
2	9	18	0.036			27	0.088			32	0.112
		16	0.109			24	0.189	5	5	25	0.008
2	10	20	0.030	4	4	16	0.029	5	6	30	0.004
		18	0.091			12	0.229			25	0.026
		16	0.182	4	5	20	0.016			24	0.048
2	11	22	0.026			16	0.079			20	0.108
		20	0.077			15	0.143	5	7	35	0.003
		18	0.154	4	6	24	0.010			30	0.015
2	12	24	0.022			20	0.048			28	0.030
		22	0.066			18	0.095			25	0.066
		20	0.132			16	0.181			23	0.116
3	3	9	0.100	4	7	28	0.006	5	8	40	0.002
3	4	12	0.057			24	0.030			35	0.009
		9	0.229			21	0.067			32	0.020
3	5	15	0.036			20	0.121			30	0.042
		12	0.143	4	8	32	0.004			27	0.079
3	6	18	0.024			28	0.020			25	0.126
		15	0.095			24	0.085	5	9	45	0.001
		12	0.333			20	0.222			40	0.006
3	7	21	0.017	4	9	36	0.003			36	0.014
		18	0.067			32	0.014			35	0.028
		15	0.167			28	0.042			31	0.056
3	8	24	0.012			27	0.062			30	0.086
		21	0.048			24	0.115			27	0.119
		18	0.121	4	10	40	0.002	5	10	50	0.001
3	9	27	0.009			36	0.010			45	0.004
		24	0.036			32	0.030			40	0.019
		21	0.091			30	0.046			35	0.061
		18	0.236			28	0.084			30	0.166
						26	0.126				

Table G (*continued*)

m	n	mnD_{mn}	P	m	n	mnD_{mn}	P	m	n	mnD_{mn}	P
5	11	55	0.000	6	9	54	0.000	7	8	56	0.000
		50	0.003			48	0.003			49	0.002
		45	0.010			45	0.006			48	0.005
		44	0.014			42	0.014			42	0.013
		40	0.029			39	0.028			41	0.024
		39	0.044			36	0.061			40	0.033
		35	0.074			33	0.095			35	0.056
		34	0.106			30	0.176			34	0.087
6	6	36	0.002	6	10	60	0.000			33	0.118
		30	0.026			54	0.002	7	9	63	0.000
6	7	42	0.001			50	0.004			56	0.001
		36	0.008			48	0.009			54	0.003
		35	0.015			44	0.019			49	0.008
		30	0.038			42	0.031			47	0.015
		29	0.068			40	0.042			45	0.021
		28	0.091			38	0.066			42	0.034
		24	0.147			36	0.092			40	0.055
6	8	48	0.001			34	0.125			38	0.079
		42	0.005	7	7	49	0.001			36	0.098
		40	0.009			42	0.008			35	0.127
		36	0.023			35	0.053	8	8	64	0.000
		34	0.043			28	0.212			56	0.002
		32	0.061							48	0.019
		30	0.093							40	0.087
		28	0.139							32	0.283

Right-Tail Probability for mnD_{mn} (Two-Sided Statistic)

$m = n$	0.200	0.100	0.050	0.020	0.010
9	45	54	54	63	63
10	50	60	70	70	80
11	66	66	77	88	88
12	72	72	84	96	96
13	78	91	91	104	117
14	84	98	112	112	126
15	90	105	120	135	135
16	112	112	128	144	160
17	119	136	136	153	170
18	126	144	162	180	180
19	133	152	171	190	190
20	140	160	180	200	220
	0.100	0.050	0.025	0.010	0.005

Approximate Right-Tail Probability for mnD_{mn}^{+} and mnD_{mn}^{-} (One-Sided Statistic)

For sample sizes outside the range of this table, the quantile points based on the asymptotic distribution are approximated by calculating the following for the appropriate values of m and n.

Right Tail Probability for D_{mn}

0.200	0.100	0.050	0.020	0.010
$1.07\sqrt{N/mn}$	$1.22\sqrt{N/mn}$	$1.36\sqrt{N/mn}$	$1.52\sqrt{N/mn}$	$1.63\sqrt{N/mn}$
0.100	0.050	0.025	0.010	0.005

Right Tail Probability for D_{mn}^{+} or D_{mn}^{-}

Source: Adapted from Table I of H. L. Harter and D. B. Owen, Eds. (1970), *Selected Tables in Mathematical Statistics*, Vol. 1, Markham Publishing Company, Chicago, with permission of the Institute of Mathematical Statistics.

Bibliography

Alling, David W.: Early decision in the Wilcoxon two-sample test. *J. Am. Stat. Assoc.* **58**, 713–720 (1963).

Anderson, T. W., Darling, D. A.: Asymptotic theory of certain "goodness of fit" test criteria based on stochastic processes. *Annals of Math. Stat.* **23**, 193–212 (1952).

Bahadur, R. R.: Stochastic comparison of tests. *Annals of Math. Stat.* **31**, 276–295 (1960).

Bahadur, R. R.: *Some Limit Theorems in Statistics.* (CBMS Monograph No. 4), Philadelphia: SIAM, 1971.

Barton, D. E., Mallows, C. L.: Some aspects of the random sequence. *Annals of Math. Stat.* **36**, 236–260 (1965).

Bauer, D. F.: Constructing confidence sets using rank statistics. *J. Am. Stat. Assoc.* **67**, 687–690 (1972).

Birnbaum, A.: On the foundations of statistical inference. *J. Am. Stat. Assoc.* **57**, 269–306 (1962).

Birnbaum, Z. W.: Numerical tabulation of the distribution of Kolmogorov's statistic for finite sample size. *J. Am. Stat. Assoc.* **47**, 425–441 (1952).

Birnbaum, Z. W.: On the use of the Mann–Whitney statistic. *Proc. of the Third Berkeley Symp. Math. Stat. and Probability,* Vol. I, Berkeley: Univ. Calif. 1956, pp. 13–17.

Birnbaum, Z. W., Hall, R. A.: Small sample distributions for multi-sample statistics of the Smirnov type. *Annals of Math. Stat.* **31**, 710–720 (1960).

Birnbaum, Z. W., Klose, O. M.: Bounds for the variance of the Mann–Whitney statistic. *Annals of Math. Stat.* **28**, 933–945 (1957).

Birnbaum, Z. W., Tingey, F. H.: One-sided confidence contours for probability distributions. *Annals of Math. Stat.* **22**, 592–596 (1951).

Blackman, J.: An extension of the Kolmogorov distribution. *Annals of Math. Stat.* **27**, 513–520 (1956). Correction, *ibid.* **29**, 318–324 (1958).

Blyth, C. R.: Note on relative efficiency of tests. *Annals of Math. Stat.* **29**, 898–903 (1958).

Box, G. E. P., Andersen, S. L.: Permutation theory in the derivation of robust criteria and the study of departures from assumption. *J. Royal Stat. Soc. B*, **17**, 1–34 (1955).

Bradley, James V.: *Distribution-Free Statistical Tests.* Englewood Cliffs, New Jersey: Prentice-Hall, 1968.

Brascamp, H. J., Lieb, E. H.: On extensions of the Brunn–Minkowski and Prekopa–Leindler Theorems, including inequalities for log concave functions, and with an application to the diffusion equation. *J. Func. Anal.* **22**, 366–389 (1976).

445

Bross, I. D. J.: Comment on "Does an observed sequence of numbers follow a simple rule? (Another look at Bode's law)." *J. Am. Stat. Assoc.* **66**, 562–564 (1971).

Camp, B. H.: Approximation to the point binomial. *Annals of Math. Stat.* **22**, 130–131 (1951).

Capon, J.: A note on the asymptotic normality of the Mann–Whitney–Wilcoxon statistic. *J. Am. Stat. Assoc.* **56**, 687–691 (1961a).

Capon, J.: Asymptotic efficiency of certain locally most powerful rank tests. *Annals of Math. Stat.* **32**, 88–100 (1961b).

Capon, J.: On the asymptotic efficiency of the Kolmogorov–Smirnov test. *J. Am. Stat. Assoc.* **60**, 843–853 (1965).

Carvalho, P. E. de O.: On the distribution of the Kolmogorov–Smirnov D-Statistic. *Annals of Math. Stat.* **30**, 173–176 (1959).

Chapman, Douglas G.: A comparative study of several one-sided goodness-of-fit tests. *Annals of Math. Stat.* **29**, 655–674 (1958).

Chernoff, H.: A property of some Type A regions. *Annals of Math. Stat.* **22**, 472–474 (1951).

Chernoff, H.: A measure of asymptotic efficiency for tests of a hypothesis based on the sum of observations. *Annals of Math. Stat.* **23**, 493–507 (1952).

Chernoff, H.: Large sample theory: Parametric case. *Annals of Math. Stat.* **27**, 1–22 (1956).

Chernoff, H., Savage, I. R.: Asymptotic normality and efficiency of certain nonparametric test statistics. *Annals of Math. Stat.* **29**, 972–994 (1958).

Chung, J. H., Fraser, D. A. S.: Randomization tests for a multivariate two-sample problem. *J. Am. Stat. Assoc.* **53**, 729–735 (1958).

Claypool, P. L., Holbert, D.: Accuracy of normal and Edgeworth approximations to the distribution of the Wilcoxon signed rank statistic. *J. Am. Stat. Assoc.* **69**, 255–258 (1974).

Clopper, C. J., Pearson, E. S.: The use of confidence or fiducial limits illustrated in the case of the binomial. *Biometrika*, **26**, 404–413 (1934).

Cochran, W. G.: The comparison of percentages in matched samples. *Biometrika*, **37**, 256–266 (1950).

Conover, W. J.: *Practical Nonparametric Statistics*. New York: John Wiley, 1972.

Conover, W. J.: On some methods of handling ties in the Wilcoxon signed-rank test. *J. Am. Stat. Assoc.* **68**, 985–988 (1973).

Cox, D. R.: Some problems connected with statistical inference. *Annals of Math. Stat.* **29**, 357–372 (1958a).

Cox: D. R.: The regression analysis of binary sequences. *J. Royal Stat. Soc. B*, **20**, 215–242 (1958b).

Cox, D. R.: Two further applications of a model for binary regression. *Biometrika*, **45**, 562–565 (1958c).

Cox, D. R.: *Analysis of Binary Data*. N.Y.: Halsted, 1970.

Cramér, H.: On the composition of elementary errors. *Skandinavisk Aktuarietidskrift*, **11**, 13–74, 141–180 (1928).

Cramér, H.: *Mathematical Methods of Statistics*. Princeton, New Jersey: Princeton Univ., 1946.

Crow, E. L.: Confidence intervals for a proportion. *Biometrika*, **43**, 423–435 (1956).

Daniels, H. E.: The statistical theory of the strengths of bundles of threads. *Proc. Royal Stat. Soc. A*, **183**, 405–435 (1945).

Darling, D. A.: The Kolmogorov–Smirnov, Cramér–von Mises tests. *Annals of Math. Stat.* **28**, 823–838 (1957).

Dempster, A. P.: Personal communication reported by Wilks (1962, p. 339), 1955.

Dempster, A. P.: Generalized D_n^+ statistics. *Annals of Math. Stat.* **30**, 593–597 (1959).

Depaix, M.: Distributions de déviations maximales bilatérales entre deux échantillons indépendants de même loi continue. *Comptes Rendues Acad. Sci. Paris*, **255**, 2900–2902 (1962).

Deuchler, G.: Uber die methoden der korrelationsrechung in der pädagogik und psychologie. *Zeitschrift für Pädagogische Psychologie und Experimentelle Pädagogik*, **15**, 114–131, 145–159, 229–242 (1914).

Dixon, W. J.: Power under normality of several nonparametric tests. *Annals of Math. Stat.* **25**, 610–614 (1954).

Donsker, M. D.: Justification and extension of Doob's heuristic approach to the Kolmogorov-Smirnov limit theorems. *Annals of Math. Stat.* **23**, 277–281 (1952).

Doob, J. L.: Heuristic approach to the Kolmogorov-Smirnov limit theorems. *Annals of Math. Stat.* **20**, 393–403 (1949).

Doob, J. L.: *Stochastic Processes*. New York: John Wiley, 1953.

Drion, E. F.: Some distribution free tests for the difference between empirical cumulative distribution functions. *Annals of Math. Stat.* **23**, 563–574 (1952).

Dwass, M.: Modified randomization tests for nonparametric hypotheses. *Annals of Math. Stat.* **28**, 181–187 (1957).

Dwass, M.: The distribution of a generalized D_n^+ statistic. *Annals of Math. Stat.* **30**, 1024–1028 (1959).

Edgeworth, F. Y.: On the probable errors of frequency-constants. *J. Royal Stat. Soc.* **71**, 381–397 (1908). Addendum, *ibid.*, **72**, 81–90 (1909).

Edgington, E. S.: *Statistical Inference: The Distribution-Free Approach*. New York: McGraw-Hill, 1969.

Efron, B.: Does an observed sequence of numbers follow a simple rule? (Another look at Bode's Law.) *J. Am. Stat. Assoc.* **66**, 552–559 (1971); Comments and rejoinder, *ibid.* 559–568.

Ellison, B. E.: Two theorems for inferences about the normal distribution with applications in acceptance sampling. *J. Am. Stat. Assoc.* **59**, 89–95 (1964).

Epstein, B.: Comparison of some nonparametric tests against normal alternatives with an application to life testing. *J. Am. Stat. Assoc.* **50**, 894–900 (1955).

Feller, W. E.: On the Kolomogorov–Smirnov limit theorems for empirical distributions. *Annals of Math. Stat.* **19**, 177–189 (1948).

Feller, W.: *An Introduction to Probability Theory and its Applications*. New York: John Wiley, 1969.

Fellingham, S. A., Stoker, D. J.: An approximation for the exact distribution of the Wilcoxon test for symmetry. *J. Am. Stat. Assoc.* **59**, 899–905 (1964).

Festinger, L.: The significance of difference between means without reference to the frequency distribution. *Psychometrika*, **11**, 97–105 (1946).

Fisher, R. A.: The logic of inductive inference (with discussion). *J. Royal Stat. Soc.* **98**, 39–82 (1935).

Fisher, R. A.: *Design of Experiments*. Edinburgh: Oliver & Boyd, 1966 (first edition 1935).

Fisher, R. A.: *Statistical Methods for Research Workers*. Edinburgh: Oliver and Boyd, 1970.

Fisher, R. A., Yates, F.: *Statistical Tables*. New York: Hafner, 1963.

Fisz, M.: *Probability Theory and Mathematical Statistics*. New York: John Wiley, 1963.

Fix, E., Hodges, J. L., Jr.: Significance probabilities of the Wilcoxon test. *Annals of Math. Stat.* **26**, 301–312 (1955).

Fraser, D. A. S.: Most powerful rank-type tests. *Annals of Math. Stat.* **28**, 1040–1043 (1957a).

Fraser, D. A. S.: *Nonparametric Methods in Statistics*. New York: John Wiley, 1957b.

Gastwirth, J. L.: The first-median test: A two-sided version of the control median test. *J. Am. Stat. Assoc.* **63**, 692–706 (1968).

Gastwirth, J. L., Wolff, S.: An elementary method of obtaining lower bounds on the asymptotic power of rank tests. *Annals of Math. Stat.* **39**, 2128–2130 (1968).

Gibbons, J. D.: A proposed two-sample rank test: The Psi test and its properties. *J. Royal Stat. Soc. B*, **26**, 305–312 (1964a).

Gibbons, J. D.: Effect of nonnormality on the power function of the sign test. *J. Am. Stat. Assoc.* **59**, 142–148 (1964b).

Gibbons, J. D.: On the power of two-sample rank tests on the equality of two distribution functions. *J. Royal Stat. Soc. B*, **26**, 293–304 (1964c).

Gibbons, J. D.: *Nonparametric Statistical Inference*. New York: McGraw-Hill, 1971.

Gibbons, J. D., Pratt, J. W.: *P*-values: Interpretation and methodology. *The Am. Statistician*, **29**, 20–25 (1975).

Gnedenko, B. V.: Tests of homogeneity of probability distributions in two independent samples. *Math. Nachrichten*, **12**, 26–66 (1954).

Gnedenko, B. V., Korolyuk, V. S.: On the maximum discrepancy between two empirical distributions (in Russian). *Doklady Akad. Nauk SSSR*, **80**, 525–528 (1951).

Good, I. J.: Significance tests in parallel and in series. *J. Am. Stat. Assoc.* **53**, 799–813 (1958).

Good, I. J.: A subjective evaluation of Bode's Law and an "objective" test for approximate numerical rationality. *J. Am. Stat. Assoc.* **64**, 23–49 (1969).

Goodman, L. A.: Kolmogorov–Smirnov tests for psychological research. *Psych. Bull.* **51**, 160–168 (1954).

Grizzle, J. E., Starmer, C. F., Koch, G. C.: Analysis of categorical data by linear models. *Biometrics*, **25**, 489–504 (1969).

Groeneboom, P., Oosterhoff, J.: Bahadur efficiency and probabilities of large deviations. *Statistica Neerlandica*, **31**, 1–24 (1977).

Gurland, J: An inequality satisfied by the expectation of the reciprocal of a random variable. *The Am. Statistician*, **21**, (2), 24–25 (1967).

Guttman, I.: *Statistical Tolerance Regions: Classical and Bayesian*. New York: Hafner Press, 1970.

Halperin, M., Ware, J.: Early decision in a censored Wilcoxon two-sample test for accumulating survival data. *J. Am. Stat. Assoc.* **69**, 414–422 (1974).

Hájek, J.: *Nonparametric Statistics*. San Francisco: Holden-Day, 1969.

Hájek, J., Šidák, Z.: *Theory of Rank Tests*. New York: Academic, 1967.

Harter, H. L.: Expected values of normal order statistics. *Biometrika*, **48**, 151–165 (1961).

Harter, H. L., Owen, D. B. (eds.): *Selected Tables in Mathematical Statistics*, Vol. I, Chicago: Markham Publ., 1970.

Hartigan, J. A.: Using subsample values as typical values. *J. Am. Stat. Assoc.* **64**, 1303–1317 (1969).

Harvard University Computation Laboratory: *Tables of the Cumulative Binomial Probability Distribution*. Cambridge: Harvard Univ., 1955.

Hodges, J. L., Jr.: The significance probability of the Smirnov two-sample test. *Arkiv. Mat.*, **3**, 469–486 (1957).

Hodges, J. L., Jr., Lehmann, E. L.: The efficiency of some nonparametric competitors of the t-test. *Annals of Math. Stat.* **27**, 324–335 (1956).

Hoeffding, W.: A class of statistics with asymptotically normal distribution. *Annals of Math. Stat.* **19**, 293–325 (1948).

Hoeffding, W.: On the distribution of the number of successes in independent trials. *Annals of Math. Stat.* **27**, 713–721 (1956).

Hoeffding, W.: Review of S. S. Wilks, *Mathematical Statistics. Annals of Math. Stat.* **33**, 1467–1473 (1962).

Hogg, R. V.: Adaptive robust procedures: A partial review and some suggestions for future applications and theory. *J. Am. Stat. Assoc.* **69**, 909–927 (1974).

Hollander, M.: Rank tests for randomized blocks. *Annals of Math. Stat.* **38**, 867–877 (1967).

Huber, P. J.: Robust estimation in location. *Annals of Math. Stat.* **35**, 73–101 (1964).

Iman, R. E.: Use of a t-statistic as an approximation to the exact distribution of the Wilcoxon signed ranks test statistic. *Comm. in Stat.* **3**, 795–806 (1974).

Jacobson, J. E.: The Wilcoxon two-sample statistic: Tables and bibliography. *J. Am. Stat. Assoc.* **58**, 1086–1103 (1963).

Jeffreys, H.: *Theory of Probability*, 3rd ed. Oxford: Oxford Univ., 1961.

Johnson, N. L., Kotz, S.: *Distributions in Statistics: Discrete Distributions*. New York: John Wiley, 1969.

Kac, M.: On deviations between theoretical and empirical distributions. *Proc. Nat. Academy of Sci.* **35**, 252–257 (1949).

Kadane, J. B.: For what use are tests of hypotheses and tests of significance. Introduction. *Comm. in Stat. A*, **5**, 735–736 (1976).

Karlin, S.: Decision theory for Pólya type distributions. Case of two actions, I. *Proc. Third Berkeley Symp. on Math. Stat. and Probability*, Vol. 1, Berkeley: Univ. Calif., 1956, pp. 115–129.

Karlin, S.: Pólya-type distributions, II. *Annals of Math. Stat.* **28**, 281–308 (1957a).

Karlin, S.: Pólya-type distributions, III: Admissibility for multi-action problems. *Annals of Math. Stat.* **28**, 839–860 (1957b).

Karlin, S., Rubin, H.: Distributions possessing a monotone likelihood ratio. *J. Am. Stat. Assoc.* **51**, 637–643 (1956).

Kempthorne, O.: Of what use are tests of significance and tests of hypotheses. *Comm. in Stat. A*, **5**, 763–777 (1976).

Kimball, A. W., Burnett, W. T., Jr., Doherty, D. G.: Chemical protection against ionizing radiation. I. Sampling methods for screening compounds in radiation protection studies with mice. *Radiation Research*, **7**, 1–12 (1957).

Klotz, J.: Small sample power and efficiency for the one sample Wilcoxon and normal scores tests. *Annals of Math. Stat.* **34**, 624–632 (1963).

Klotz, J.: Asymptotic efficiency of the two sample Kolmogorov–Smirnov test. *J. Am. Stat. Assoc.* **62**, 932–938 (1967).

Kolmogorov, A. N.: Sulla determinazione empirica di una legge di distribuzione. *Giorn. Inst. Ital. Attuari*, **4**, 83–91 (1933).

Korolyuk, V. S.: Asymptotic expansions for the criterion of fit of A. N. Kolmogorov and N. V. Smirnov. *Doklady Akad. Nauk SSSR*, **93**, 443–446 (1954). (*Izvestiya Akad. Nauk SSSR* Ser. Mat. **19**, 103–124 (1955).)

Korolyuk, V. S.: On the deviation of empirical distributions for the case of two independent samples. *Izvestiya Akad. Nauk SSSR* Ser. Mat., **19**, 81–96 (1955).

Kraft, C. H., Van Eeden, C.: *A Nonparametric Introduction to Statistics*. New York: Macmillan, 1968.

Kruskal, W. H.: Historical notes on the Wilcoxon unpaired two-sample test. *J. Am. Stat. Assoc.* **52**, 356–360 (1957).

Kruskal, W. H.: "Tests of Significance" in *International Encyclopedia of the Social Sciences*, **14**, 238–249 (1968). New York: The Free Press.

Lancaster, H. O.: Statistical control of counting experiments. *Biometrika*, **39**, 419–422 (1952).

Lancaster, H. O.: Significance tests in discrete distributions. *J. Am. Stat. Assoc.* **56**, 223–234 (1961).

Lehman, S. Y.: Exact and approximate distribution for the Wilcoxon statistic with ties. *J. Am. Stat. Assoc.* **56**, 293–298 (1961).

Lehmann, E. L.: The power of rank tests. *Annals of Math. Stat.* **24**, 23–43 (1953).

Lehmann, E. L.: *Testing Statistical Hypotheses*. New York: John Wiley, 1959.

Lehmann, E. L., Stein, C.: On the theory of some non-parametric hypotheses. *Annals of Math. Stat.* **20**, 28–45 (1949).

Lieberman, G. J., Owen, D. B.: *Tables of the Hypergeometric Probability Distribution.* Stanford: Stanford Univ., 1961.

Lindley, D. V.: A statistical paradox. *Biometrika,* **44**, 187–192 (1957).

Lindley, D. V.: *Bayesian Statistics, A Review.* Philadelphia: SIAM, 1971.

Ling, R. F.: A survey of the accuracy of some approximations for t, χ^2, and F tail probabilities. *J. Am. Stat. Assoc.* **73**, 274–283 (1978).

Ling, R. F., Pratt, J. W.: The accuracy of a modified Peizer approximation to the hypergeometric distribution with comparison to some other approximations. Tech. Rep. no. 348, Clemson Univ., Dept. Math. Sci., July 1980.

Loève, M.: *Probability Theory.* New York: Van Nostrand, 1955.

Madansky, A.: More on length of confidence intervals. *J. Am. Stat. Assoc.* **57**, 586–589 (1962).

Mann, H. B., Whitney, D. R.: On a test whether one of two random variables is stochastically larger than the other. *Annals of Math. Stat.* **18**, 50–60 (1947).

Mantel, N., Rahe, A. J.: Differentiated sign tests. *Internat. Stat. Rev.,* **48**, 19–28 (1980).

Massey, F. J.: A note on the estimation of a distribution function by confidence limits. *Annals of Math. Stat.* **21**, 116–119 (1950a).

Massey, F. J.: A note on the power of a nonparametric test. *Annals of Math. Stat.* **21**, 440–443 (1950b); Correction, *ibid.* **23**, 637–638 (1952).

Massey, F. J.: The distribution of the maximum deviation between two sample cumulative step functions. *Annals of Math. Stat.* **22**, 125–128 (1951a).

Massey, F. J.: The Kolmogorov-Smirnov test for goodness of fit. *J. Am. Stat. Assoc.* **46**, 68–78 (1951b).

Massey, F. J.: Distribution table for the deviation between two sample cumulatives. *Annals of Math. Stat.* **23**, 435–441 (1952).

McCornack, R. L.: Extended tables of the Wilcoxon matched pair signed rank statistic. *J. Am. Stat. Assoc.* **60**, 864–871 (1965).

McNemar, Q.: Note on the sampling error of the difference between correlated proportions or percentages. *Psychometrika,* **12**, 153–157 (1947).

Miller, L. H.: Table of percentage points of Kolmogorov statistics. *J. Am. Stat. Assoc.* **51**, 111–121 (1956).

Molenaar, W.: *Approximations to the Poisson, Binomial and Hypergeometric Distribution Functions.* Amsterdam: MC Tract 31, Mathematical Centre, 1970.

Mood, A. M.: *Introduction to the Theory of Statistics.* New York: McGraw-Hill, 1950.

Mood, A. M.: On the asymptotic efficiency of certain nonparametric two-sample tests. *Annals of Math. Stat.* **25**, 514–522, (1954).

Mood, A., Graybill, F. A.: *Introduction to the Theory of Statistics.* New York: McGraw-Hill, 1963.

Mood, A., Graybill, F. A., Boes, D. C.: *Introduction to the Theory of Statistics.* New York: McGraw-Hill, 1974.

Moses, L. E.: Nonparametric statistics for psychological research. *Psych. Bull.* **49**, 122–143 (1952).

Moses, L. E.: One sample limits of some two sample rank tests. *J. Am. Stat. Assoc.* **59**, 645–651 (1964).

Mosteller, F.: Clinical studies of analgesic drugs. *Biometrics,* **8**, 220–231 (1952).

Murphy, R. B.: Nonparametric tolerance limits. *Annals of Math. Stat.* **19**, 581–589 (1948).

National Bureau of Standards: *Tables of the Binomial Probability Distribution.* Applied Mathematics Series **6**, Wash. D. C.: U.S. Govt. Printing Office, 1949.

National Bureau of Standards: *Handbook of Mathematical Functions* Applied Mathematics Series **55**, Wash. D. C.: U.S. Govn. Printing Office, 1964.

Neyman, J.: *First Course in Probability and Statistics.* New York: Holt, 1950.

Neyman, J.: "Inductive behavior" as a basic concept of philosophy of science. *Rev. Int. Stat. Inst.* **25**, 7–22 (1957).

Neyman, J.: Tests of statistical hypotheses and their use in studies of natural phenomena. *Comm. in Stat. A*, **5**, 737–751 (1976).

Neyman, J., Pearson, E. S.: On the problem of the most efficient tests of statistical hypotheses. *Phil. Trans. of the Royal Stat. Soc. A*, **231**, 289–337 (1933).

Noether, G. E.: *Elements of Nonparametric Statistics*. New York: John Wiley, 1967.

Noether, G. E.: Some simple distribution-free confidence intervals for the center of a symmetric distribution. *J. Am. Stat. Assoc.* **68**, 716–719 (1973).

Ord, J. K.: Approximations to distribution functions which are hypergeometric series. *Biometrika*, **55**, 243–248 (1968).

Owen, D. B.: *Handbook of Statistical Tables*. Reading, Mass.: Addison–Wesley, 1962.

Paulson, E.: An approximate normalization of the analysis of variance distribution. *Annals of Math. Stat.* **13**, 233–235 (1942).

Pearson, E. S.: Some thoughts on statistical inference. *Annals of Math. Stat.* **33**, 394–403 (1962).

Pearson, E. S., Hartley, H. O. (eds.): *Biometrika Tables for Statisticians*, Vol. I. Cambridge, England: Univ. Press, 1966.

Peizer, D. B., Pratt, J. W.: A normal approximation for binomial, *F*, beta and other common, related distributions. *J. Am. Stat. Assoc.* **63**, 1416–1456 (1968).

Pitman, E. J. G.: Significance tests which may be applied to samples from any populations. *J. Royal Stat. Soc. B*, **4**, 119–130 (1937a).

Pitman, E. J. G.: Significance tests which may be applied to samples from any populations, II. The correlation coefficient test. *J. Royal Stat. Soc. B*, **4**, 225–232 (1937b).

Pitman, E. J. G.: Significance tests which may be applied to samples from any populations, III. The analysis of variance test. *Biometrika*, **29**, 322–335 (1938).

Pratt, J. W.: Remarks on zeros and ties in the Wilcoxon signed ranks procedures. *J. Am. Stat. Assoc.* **54**, 655–667 (1959).

Pratt, J. W.: On interchanging limits and integrals. *Annals of Math. Stat.* **31**, 74–77 (1960).

Pratt, J. W.: Length of confidence intervals. *J. Am. Stat. Assoc.* **56**, 549–567 (1961).

Pratt, J. W.: Robustness of some procedures for the two-sample location problem. *J. Am. Stat. Assoc.* **59**, 665–680 (1964).

Pratt, J. W.: Bayesian interpretation of standard inference situations, *J. Royal Stat. Soc. B*, **27**, 169–203 (1965).

Pratt, J. W.: A normal approximation for binomial, *F*, beta, and other common, related tail probabilities, II. *J. Am. Stat. Assoc.* **63**, 1457–1483 (1968).

Pratt, J. W.: Comment on "Post-data two sample tests of location", *J. Am. Stat. Assoc.* **68**, 104–105 (1973).

Pratt, J. W.: A discussion of the question: For what use are tests of hypotheses and tests of significance. *Comm. in Stat. A*, **5**, 779–787 (1976).

Pratt, J. W.: Concavity of the log likelihood. *J. Am. Stat. Assoc.* **76**, 103–106 (1981).

Pratt, J. W., Raiffa, H., Schlaifer, R.: The foundations of decision under uncertainty: An elementary exposition. *J. Am. Stat. Assoc.* **59**, 353–375 (1964).

Putter, J.: The treatment of ties in some nonparametric tests. *Annals of Math. Stat.* **26**, 368–386 (1955).

Pyke, R.: The supremum and infimum of the Poisson process. *Annals of Math. Stat.* **30**, 568–576 (1959).

Quade, D.: On the asymptotic power of the one-sample Kolmogorov-Smirnov tests. *Annals of Math. Stat.* **36**, 1000–1018 (1965).

Rahe, A. J.: Tables of critical values for the Pratt matched pair signed rank statistic. *J. Am. Stat. Assoc.* **69**, 368–373 (1974).

Raiffa, H., Schlaifer, R.: *Applied Statistical Decision Theory*. Boston: Div. Res., Harvard Business School, 1961.

Roberts, H. V.: For what use are tests of hypotheses and tests of significance. *Comm. in Stat. A*, **5**, 753–761 (1976).

Rosenbaum, S.: Tables for a nonparametric test of location. *Annals of Math. Stat.* **25**, 146–150 (1954).

Rustagi, J. S.: Bounds for the variance of Mann-Whitney statistics. *Annals of Math. Stat.* **13**, 119–126 (1962).

Sandiford, P. J.: A new binomial approximation for use in sampling from finite populations. *J. Am. Stat. Assoc.* **55**, 718–722 (1960).

Savage, I. R.: *Bibliography of Nonparametric Statistics.* Cambridge: Harvard Univ. Press, 1962.

Savage, L. J. *The Foundations of Statistics.* New York: John Wiley, 1954.

Scheffé, H.: A useful convergence theorem for probability distributions. *Annals of Math. Stat.* **18**, 434–438 (1947).

Scheffé, H., Tukey, J. W.: Nonparametric estimation, I. Validation of order statistics. *Annals of Math. Stat.* **16**, 187–192 (1945).

Siegel, S.: *Non-parametric Statistics for the Behavioral Sciences.* New York: McGraw-Hill, 1956.

Singer, B.: *Distribution-Free Methods for Nonparametric Methods: A Classified and Selected Bibliography.* Leicester: British Psych. Soc. 1979.

Smirnov, N. V.: Estimate of deviation between empirical distribution functions in two independent samples (in Russian). *Bull. Moscow Univ.*, **2**, 3–16 (1939).

Smirnov. N. V.: Approximation of distribution laws of random variables from empirical data (in Russian). *Uspehi Mat. Nauk*, **10**, 179–206 (1944).

Steck, G. P.: The Smirnov two sample tests as rank tests. *Annals of Math. Stat.* **40**, 1449–1466 (1969).

Stein, C.: Efficient nonparametric testing and estimation. *Proc. Third Berkeley Symp. on Math. Stat. and Probability*, Vol. 1. Berkeley: Univ. Calif., 1956, pp. 187–195.

Sterne, T. E.: Some remarks on confidence or fiducial limits. *Biometrika*, **41**, 275–278 (1954).

Stuart, A.: The comparison of frequencies in matched samples. *British J. Stat. Psych.* **10**, 29–32 (1957).

Tate, M. W., Clelland, R. C.: *Nonparametric and Shortcut Statistics*, Danville, Ill.: The Interstate Publishers & Printers, 1957.

Teichroew, D.: Tables of expected values of order statistics and products of order statistics for samples of size twenty and less from the normal distribution. *Annals of Math. Stat.* **27**, 410–426 (1956).

Terry, M. E.: Some rank order tests which are most powerful against specific parametric alternatives. *Annals of Math. Stat.* **23**, 346–366 (1952).

Tsao, C. K.: An application of Massey's distribution of the maximum deviation between two sample cumulative step functions. *Annals of Math. Stat.* **25**, 587–592 (1954).

Tukey, J. W.: Nonparametric estimation, II. Statistically equivalent blocks and tolerance regions—the continuous case. *Annals of Math. Stat.* **18**, 529–539 (1947).

Uhlmann, W.: Vergleich der hypergeometrischen mit der Binomial-Verteilung. *Metrika*, **10**, 145–158 (1966).

Uzawa, H.: Locally most powerful rank tests for two-sample problems. *Annals of Math. Stat.* **31**, 685–702 (1960).

van der Vaart, H. R.: Some extensions of the idea of bias. *Annals of Math. Stat.* **32**, 436–447 (1961).

van der Waerden, B. L.: Order tests for the two-sample problem and their power, I, II, III. *Proc. Koninklijke Nederlandse Akademie van Wetenschappen (A)*, **55** (*Indagationes Mathematicae*, **14**), 453–458 (1952); *Indagationes Mathematicae* **15**, 303–310, 311–316 (1953); correction, *Indagationes Mathematicae* **15**, 80 (1953).

van der Waerden, B. L.: Testing a distribution function. *Proc. Koninklijke Nederlandse Akademie van Wetenschappen (A)*, **56** (*Indagationes Mathematicae* **15**), 201–207 (1953).

van der Waerden, B. L.: The computation of the X-distribution. *Proc. Third Berkeley Symp. Math. Stat. and Probability*, Vol. I. Berkeley: Univ. Calif., 1956, pp. 207–208.

van der Waerden, B. L., Nievergelt, E.: *Tafeln Zum Vergleich Zweier Stichproben mittels X-test und Zeichentest*. Berlin-Gottingen-Heidelberg: Springer-Verlag, 1956.

van Eeden, C.: The relation between Pitman's asymptotic relative efficiency of two tests and the correlation coefficient between their test statistics. *Annals of Math. Stat.* **34**, 1442–1451 (1963).

von Mises, R.: *Wahrscheinlichkeitsrechnung und ihre Anwendung in der Statistik und theoretischen Physik*. Leipzig-Wien: F. Deuticke, 1931.

Walsh, J. E.: Applications of some significance tests for the median which are valid under very general conditions. *J. Am. Stat. Assoc.* **44**, 342–355 (1949a).

Walsh, J. E.: Some significance tests for the median which are valid under very general conditions. *Annals of Math. Stat.* **20**, 64–81 (1949b).

Walsh, J. E.: Nonparametric tests for median by interpolation from sign tests. *Annals of the Inst. of Stat. Math.* **11**, 183–188 (1959–60).

Walsh, J. E.: *Handbook of Nonparametric Statistics, I. Investigation of Randomness, Moments, Percentiles and Distributions*. New York: Van Nostrand, 1962a.

Walsh, J. E.: Some two-sided distribution-free tolerance intervals of a general nature. *J. Am. Stat. Assoc.* **57**, 775–784 (1962b).

Walsh, J. E.: *Handbook of Nonparametric Statistics, II: Results for Two and Several Sample Problems, Symmetry and Extremes*. New York: Van Nostrand, 1965.

Walsh, J. E.: *Handbook of Nonparametric Statistics, III: Analysis of Variance*. New York: Van Nostrand, 1968.

Wilcoxon, F.: Individual comparisons by ranking methods. *Biometrics*, **1**, 80–83 (1945).

Wilks, S. S.: *Mathematical Statistics*. New York: John Wiley, 1962.

Wilson, E. B., Hilferty, M. M.: The distribution of chi-square. *Proc. Nat. Academy of Sci.* **17**, 684–688 (1931).

Wise, M. E.: A quickly convergent expansion for cumulative hypergeometric probabilities, direct and inverse. *Biometrika*, **41**, 317–329 (1954).

Young, W. H.: On semiintegrals and oscillating successions of functions. *Proc. London Math. Soc.* (2), **9**, 286–324 (1911).

Zahl, S.: Bounds for the Central Limit Theorem error. *Ph.D. Dissertation*, Boston, Mass.: Harvard Univ. 1962.

Index

Springer Series in Statistics

Measures of Association for Cross Classifications
Leo A. Goodman and **William H. Kruskal**
1979 / 146 pp. / cloth
ISBN 0-387-**90443**-3

Statistical Decision Theory: Foundations, Concepts, and Methods
James Berger
1980 / 425 pp. / 20 illus. / cloth
ISBN 0-387-**90471**-9

Simultaneous Statistical Inference, Second Edition
Rupert G. Miller, Jr.
1981 / 299 pp. / 25 illus. / cloth
ISBN 0-387-**90548**-0

Point Processes and Queues: Martingale Dynamics
Pierre Brémaud
1981 / approx. 384 pp. / 31 illus. / cloth
ISBN 0-387-**90536**-7

Non-Negative Matrices and Markov Chains
E. Seneta
1981 / 304 pp. / cloth
ISBN 0-387-**90598**-7

Statistical Computing with APL
Francis John Anscombe
1981/426 pp./70 illus./cloth
ISBN 0-387-**90549**-9

Concepts of Nonparametric Theory
John Pratt and **Jean D. Gibbons**
1981/469 pp./23 illus./cloth
ISBN 0-387-**90582**-0